PROPERTY OF ~~████ ████~~

September 13th, 2016

Relativity
Thermodynamics
and Cosmology

Relativity Thermodynamics and Cosmology

by Richard C. Tolman

Late Professor of Physical Chemistry and Mathematical Physics at the California Institute of Technology

DOVER PUBLICATIONS, INC.
NEW YORK

Published in Canada by General Publishing Company, Ltd.
30 Lesmill Road, Don Mills, Toronto, Ontario.
Published in the United Kingdom by Constable and Company, Ltd.

This Dover edition, first published in 1987, is an unabridged and
unaltered republication of the work first published by Oxford
University Press, Oxford, England in 1934. It is reprinted by special
arrangement with Oxford University Press, Inc., 200 Madison Ave.,
New York, N.Y., 10016.

Manufactured in the United States of America
Dover Publications, Inc.
31 East 2nd Street
Mineola, N.Y. 11501

Library of Congress Cataloging-in-Publication Data

Tolman, Richard Chace, 1881–1948.
Relativity, thermodynamics, and cosmology.

Reprint. Originally published: Oxford : Clarendon Press, 1934.
Includes indexes.
1. Special relativity (Physics) 2. Thermodynamics.
3. Cosmology. 4. Electrodynamics. I. Title.
QC173.65.T65   1987   530.1′1   87-6728
ISBN 0-486-65383-8 (pbk.)

TO
GILBERT NEWTON LEWIS
PROFESSOR OF CHEMISTRY
AT THE
UNIVERSITY OF CALIFORNIA

CONTENTS

I. INTRODUCTION

§ 1. The Subject-Matter 1
§ 2. The Method of Presentation 7
§ 3. The Point of View 9

II. THE SPECIAL THEORY OF RELATIVITY

Part I. THE TWO POSTULATES AND THE LORENTZ TRANSFORMATION

§ 4. Introduction 12
§ 5. The First Postulate of Relativity 12
§ 6. The Second Postulate of Relativity 15
§ 7. Necessity for Modifying Older Ideas as to Space and Time . 17
§ 8. The Lorentz Transformation Equations . . . 18
§ 9. Transformation Equations for Spatial and Temporal Intervals.
    Lorentz Contraction and Time Dilation . . . 22
§ 10. Transformation Equations for Velocity . . . 25
§ 11. Transformation Equation for the Lorentz Contraction Factor . 27
§ 12. Transformation Equations for Acceleration . . . 27

Part II. TREATMENT OF SPECIAL RELATIVITY WITH THE HELP OF A
FOUR-DIMENSIONAL GEOMETRY

§ 13. The Space-Time Continuum 28
§ 14. The Three Plus One Dimensions of Space-Time . . 29
§ 15. The Geometry Corresponding to Space-Time . . . 30
§ 16. The Signature of the Line Element and the Three Kinds of Interval 31
§ 17. The Lorentz Rotation of Axes 32
§ 18. The Transformation to Proper Coordinates . . . 33
§ 19. Use of Tensor Analysis in the Theory of Relativity . . 34
§ 20. Simplification of Tensor Analysis in the Case of Special Relativity.
    Galilean Coordinates 37
§ 21. Correspondence of Four-Dimensional Treatment with the Pos-
    tulates of Special Relativity 39

III. SPECIAL RELATIVITY AND MECHANICS

Part I. THE DYNAMICS OF A PARTICLE

§ 22. The Principles of the Conservation of Mass and Momentum . 42
§ 23. The Mass of a Moving Particle. 43
§ 24. The Transformation Equations for Mass . . . 45
§ 25. The Definition and Transformation Equations for Force . 45
§ 26. Work and Kinetic Energy 47
§ 27. The Relations between Mass, Energy, and Momentum . 48
§ 28. Four-Dimensional Expression of the Mechanics of a Particle . 50
§ 29. Applications of the Dynamics of a Particle . . . 53
    (a) The Mass of High-Velocity Electrons . . . 53
    (b) The Relation between Force and Acceleration . . 54
    (c) Applications in Electromagnetic Theory . . . 55
    (d) Tests of the Interrelation of Mass, Energy, and Momentum 57

Part II. THE DYNAMICS OF A CONTINUOUS MECHANICAL MEDIUM
§ 30. The Principles Postulated 59
§ 31. The Conservation of Momentum and the Components of Stress t_{ij} 60
§ 32. The Equations of Motion in Terms of the Stresses t_{ij} . . 60
§ 33. The Equation of Continuity 62
§ 34. The Transformation Equations for the Stresses t_{ij} . . 62
§ 35. The Transformation Equations for the Densities of Mass and Momentum 65
§ 36. Restatement of Results in Terms of the (Absolute) Stresses p_{ij} . 69
§ 37. Four-Dimensional Expression of the Mechanics of a Continuous Medium 70
§ 38. Applications of the Mechanics of a Continuous Medium . 73
    (a) The Mass and Momentum of a Finite System . . 74
    (b) The Angular Momentum of a Finite System . . 77
    (c) The Right-Angled Lever as an Example . . 79
    (d) The Complete Static System 80

IV. SPECIAL RELATIVITY AND ELECTRODYNAMICS

Part I. ELECTRON THEORY
§ 39. The Maxwell-Lorentz Field Equations 84
§ 40. The Transformation Equations for **E, H,** and ρ . . 86
§ 41. The Force on a Moving Charge 88
§ 42. The Energy and Momentum of the Electromagnetic Field . 89
§ 43. The Electromagnetic Stresses 91
§ 44. Transformation Equations for Electromagnetic Densities and Stresses 92
§ 45. Combined Result of Mechanical and Electromagnetic Actions . 93
§ 46. Four-Dimensional Expression of the Electron Theory . . 95
    (a) The Field Equations 95
    (b) Four-Dimensional Expression of Force on Moving Charge 98
    (c) Four-Dimensional Expression of Electromagnetic Energy-Momentum Tensor 99
§ 47. Applications of the Electron Theory 99

Part II. MACROSCOPIC THEORY
§ 48. The Field Equations for Stationary Matter . . . 101
§ 49. The Constitutive Equations for Stationary Matter . . 102
§ 50. The Field Equations in Four-Dimensional Language . . 102
§ 51. The Constitutive Equations in Four-Dimensional Language . 104
§ 52. The Field Equations for Moving Matter in Ordinary Vector Language 105
§ 53. The Constitutive Equations for Moving Matter in Ordinary Vector Language 108
§ 54. Applications of the Macroscopic Theory . . . 109
    (a) The Conservation of Electric Charge . . . 109
    (b) Boundary Conditions 110
    (c) The Joule Heating Effect 112
    (d) Electromagnetic Energy and Momentum . . 113
    (e) The Energy-Momentum Tensor 115
    (f) Applications to Experimental Observations . . 116

V. SPECIAL RELATIVITY AND THERMODYNAMICS

Part I. THE THERMODYNAMICS OF STATIONARY SYSTEMS

§ 55. Introduction 118
§ 56. The First Law of Thermodynamics and the Zero Point of Energy
Content 120
§ 57. The Second Law of Thermodynamics and the Starting-Point for
Entropy Content 121
§ 58. Heat Content, Free Energy, and Thermodynamic Potential . 123
§ 59. General Conditions for Thermodynamic Change and Equilibrium 125
§ 60. Conditions for Change and Equilibrium in Homogeneous Systems 127
§ 61. Uniformity of Temperature at Thermal Equilibrium . . 130
§ 62. Irreversibility and Rate of Change 132
§ 63. Final State of an Isolated System 134
§ 64. Energy and Entropy of a Perfect Monatomic Gas . . 136
§ 65. Energy and Entropy of Black-Body Radiation . . 139
§ 66. The Equilibrium between Hydrogen and Helium . . 140
§ 67. The Equilibrium between Matter and Radiation . . 146

Part II. THE THERMODYNAMICS OF MOVING SYSTEMS

§ 68. The Two Laws of Thermodynamics for a Moving System . 152
§ 69. The Lorentz Transformation for Thermodynamic Quantities . 153
   (a) Volume and Pressure 153
   (b) Energy 154
   (c) Work 156
   (d) Heat 156
   (e) Entropy 157
   (f) Temperature 158
§ 70. Thermodynamic Applications 159
   (a) Carnot Cycle Involving Change in Velocity : . 159
   (b) The Dynamics of Thermal Radiation . . . 161
§ 71. Use of Four-Dimensional Language in Thermodynamics . 162

VI. THE GENERAL THEORY OF RELATIVITY

Part I. THE FUNDAMENTAL PRINCIPLES OF GENERAL RELATIVITY

§ 72. Introduction 165
§ 73. The Principle of Covariance 166
   (a) Justification for the Principle of Covariance . . 166
   (b) Consequences of the Principle of Covariance . . 167
   (c) Method of Obtaining Covariant Expressions . . 168
   (d) Covariant Expression for Interval . . . 169
   (e) Covariant Expression for the Trajectories of Free Particles
and Light Rays 171
§ 74. The Principle of Equivalence 174
   (a) Formulation of the Principle of Equivalence. Metric and
Gravitation 174
   (b) Principle of Equivalence and Relativity of Motion . 176
   (c) Justification for the Principle of Equivalence . . 179

(d) Use of the Principle of Equivalence in Generalizing the Principles of Special Relativity. Natural and Proper Coordinates 180

(e) Interval and Trajectory in the Presence of Gravitational Fields 181

§ 75. The Dependence of Gravitational Field and Metric on the Distribution of Matter and Energy. Principle of Mach . . 184

§ 76. The Field Corresponding to the Special Theory of Relativity. The Riemann-Christoffel Tensor 185

§ 77. The Gravitational Field in Empty Space. The Contracted Riemann-Christoffel Tensor 187

§ 78. The Gravitational Field in the Presence of Matter and Energy. 188

Part II. ELEMENTARY APPLICATIONS OF GENERAL RELATIVITY

§ 79. Simple Consequences of the Principle of Equivalence . . 192

(a) The Proportionality of Weight and Mass . . 192

(b) Effect of Gravitational Potential on the Rate of a Clock . 192

(c) The Clock Paradox 194

§ 80. Newton's Theory as a First Approximation . . . 198

(a) Motion of Free Particle in a Weak Gravitational Field . 198

(b) Poisson's Equation as an Approximation for Einstein's Field Equations 199

§ 81. Units to be Used in Relativistic Calculations . . . 201

§ 82. The Schwarzschild Line Element 202

§ 83. The Three Crucial Tests of Relativity 205

(a) The Advance of Perihelion 208

(b) The Gravitational Deflexion of Light . . . 209

(c) Gravitational Shift in Spectral Lines . . . 211

VII. RELATIVISTIC MECHANICS

Part I. SOME GENERAL MECHANICAL PRINCIPLES

§ 84. The Fundamental Equations of Relativistic Mechanics . 214

§ 85. The Nature of the Energy-Momentum Tensor. General Expression in the Case of a Perfect Fluid 215

§ 86. The Mechanical Behaviour of a Perfect Fluid . . . 218

§ 87. Re-expression of the Equations of Mechanics in the Form of an Ordinary Divergence 222

§ 88. The Energy-Momentum Principle for Finite Systems . . 225

§ 89. The Densities of Energy and Momentum Expressed as Divergences 229

§ 90. Limiting Values for Certain Quantities at a Large Distance from an Isolated System 230

§ 91. The Mass, Energy and Momentum of an Isolated System . 232

§ 92. The Energy of a Quasi-Static Isolated System Expressed by an Integral Extending Only Over the Occupied Space . . 234

Part II. SOLUTIONS OF THE FIELD EQUATIONS

§ 93. Einstein's General Solution of the Field Equations in the Case of Weak Fields 236

§ 94. Line Elements for Systems with Spherical Symmetry . 239

CONTENTS xi

§ 95. Static Line Element with Spherical Symmetry . . 241
§ 96. Schwarzschild's Exterior and Interior Solutions . . 245
§ 97. The Energy of a Sphere of Perfect Fluid . . . 247
§ 98. Non-Static Line Elements with Spherical Symmetry . . 250
§ 99. Birkhoff's Theorem 252
§ 100. A More General Line Element 253

VIII. RELATIVISTIC ELECTRODYNAMICS

Part I. THE COVARIANT GENERALIZATION OF ELECTRICAL THEORY
§ 101. Introduction 258
§ 102. The Generalized Lorentz Electron Theory. The Field Equations 258
§ 103. The Motion of a Charged Particle 259
§ 104. The Energy-Momentum Tensor 261
§ 105. The Generalized Macroscopic Theory 261

Part II. SOME APPLICATIONS OF RELATIVISTIC ELECTRODYNAMICS
§ 106. The Conservation of Electric Charge 264
§ 107. The Gravitational Field of a Charged Particle . . 265
§ 108. The Propagation of Electromagnetic Waves . . . 267
§ 109. The Energy-Momentum Tensor for Disordered Radiation . 269
§ 110. The Gravitational Mass of Disordered Radiation . . 271
§ 111. The Energy-Momentum Tensor Corresponding to a Directed Flow of Radiation 272
§ 112. The Gravitational Field Corresponding to a Directed Flow of Radiation 273
§ 113. The Gravitational Action of a Pencil of Light . . 274
(a) The Line Element in the Neighbourhood of a Limited Pencil of Light 274
(b) Velocity of a Test Ray of Light in the Neighbourhood of the Pencil 275
(c) Acceleration of a Test Particle in the Neighbourhood of the Pencil 277
§ 114. The Gravitational Action of a Pulse of Light . . . 279
(a) The Line Element in the Neighbourhood of the Limited Track of a Pulse of Light 279
(b) Velocity of a Test Ray of Light in the Neighbourhood of the Pulse 281
(c) Acceleration of a Test Particle in the Neighbourhood of the Pulse 282
§ 115. Discussion of the Gravitational Interaction of Light Rays and Particles 285
§ 116. The Generalized Doppler Effect 288

IX. RELATIVISTIC THERMODYNAMICS

Part I. THE EXTENSION OF THERMODYNAMICS TO GENERAL RELATIVITY
§ 117. Introduction 291
§ 118. The Relativistic Analogue of the First Law of Thermodynamics
§ 119. The Relativistic Analogue of the Second Law of Thermo- 292
dynamics 293

§ 120. On the Interpretation of the Relativistic Second Law of
Thermodynamics 296
§ 121. On the Interpretation of Heat in Relativistic Thermodynamics 297
§ 122. On the Use of Co-Moving Coordinates in Thermodynamic
Considerations 301

Part II. APPLICATIONS OF RELATIVISTIC THERMODYNAMICS

§ 123. Application of the First Law to Changes in the Static State of
a System 304
§ 124. Application of the Second Law to Changes in the Static State
of a System 306
§ 125. The Conditions for Static Thermodynamic Equilibrium . 307
§ 126. Static Equilibrium in the Case of a Spherical Distribution of
Fluid 308
§ 127. Chemical Equilibrium in a Gravitating Sphere of Fluid . 311
§ 128. Thermal Equilibrium in a Gravitating Sphere of Fluid . 312
§ 129. Thermal Equilibrium in a General Static Field . . 315
§ 130. On the Increased Possibility in Relativistic Thermodynamics
for Reversible Processes at a Finite Rate . . . 319
§ 131. On the Possibility for Irreversible Processes without Reaching
a Final State of Maximum Entropy 326
§ 132. Conclusion 330

X. APPLICATIONS TO COSMOLOGY

Part I. STATIC COSMOLOGICAL MODELS

§ 133. Introduction 331
§ 134. The Three Possibilities for a Homogeneous Static Universe . 333
§ 135. The Einstein Line Element 335
§ 136. The de Sitter Line Element 335
§ 137. The Special Relativity Line Element 336
§ 138. The Geometry of the Einstein Universe . . . 337
§ 139. Density and Pressure of Material in the Einstein Universe . 339
§ 140. Behaviour of Particles and Light Rays in the Einstein Universe 341
§ 141. Comparison of Einstein Model with Actual Universe . . 344
§ 142. The Geometry of the de Sitter Universe . . . 346
§ 143. Absence of Matter and Radiation from the de Sitter Universe 348
§ 144. Behaviour of Test Particles and Light Rays in the de Sitter
Universe 349
(a) The Geodesic Equations 349
(b) Orbits of Particles 351
(c) Behaviour of Light Rays in the de Sitter Universe . 353
(d) Doppler Effect in the de Sitter Universe . . 354
§ 145. Comparison of de Sitter Model with Actual Universe . . 359

Part II. THE APPLICATION OF RELATIVISTIC MECHANICS TO NON-
STATIC HOMOGENEOUS COSMOLOGICAL MODELS

§ 146. Reasons for Changing to Non-Static Models . . . 361
§ 147. Assumption Employed in Deriving Non-Static Line Element . 362

§ 148. Derivation of Line Element from Assumption of Spatial Isotropy 364
§ 149. General Properties of the Line Element . . . 370
    (a) Different Forms of Expression for the Line Element . 370
    (b) Geometry Corresponding to Line Element . . 371
    (c) Result of Transfer of Origin of Coordinates . . 372
    (d) Physical Interpretation of Line Element . . 375
§ 150. Density and Pressure in Non-Static Universe . . 376
§ 151. Change in Energy with Time 379
§ 152. Change in Matter with Time 381
§ 153. Behaviour of Particles in the Model 383
§ 154. Behaviour of Light Rays in the Model . . . 387
§ 155. The Doppler Effect in the Model 389
§ 156. Change in Doppler Effect with Distance . . . 392
§ 157. General Discussion of Dependence on Time for Closed Models 394
    (a) General Features of Time Dependence, R real, $\rho_{00} \geqslant 0$, $p_0 \geqslant 0$ 395
    (b) Curve for the Critical Function of R . . . 396
    (c) Monotonic Universes of Type M_1, for $\Lambda > \Lambda_E$ . . 399
    (d) Asymptotic Universes of Types A_1 and A_2, for $\Lambda = \Lambda_E$ 400
    (e) Monotonic Universes of Type M_2 and Oscillating Universes of Types O_1 and O_2, for $0 < \Lambda < \Lambda_E$ . . 401
    (f) Oscillating Universes of Type O_1, for $\Lambda \leqslant 0$ . . 402
§ 158. General Discussion of Dependence on Time for Open Models . 403
§ 159. On the Instability of the Einstein Static Universe . . 405
§ 160. Models in Which the Amount of Matter is Constant . . 407
§ 161. Models Which Expand from an Original Static State . 409
§ 162. Ever Expanding Models Which do not Start from a Static State 412
§ 163. Oscillating Models ($\Lambda = 0$) 412
§ 164. The Open Model of Einstein and de Sitter ($\Lambda = 0$, $R_0 = \infty$) . 415
§ 165. Discussion of Factors which were Neglected in Studying Special Models 416

Part III. THE APPLICATION OF RELATIVISTIC THERMODYNAMICS TO NON-STATIC HOMOGENEOUS COSMOLOGICAL MODELS

§ 166. Application of the Relativistic First Law . . . 420
§ 167. Application of the Relativistic Second Law . . . 421
§ 168. The Conditions for Thermodynamic Equilibrium in a Static Einstein Universe 423
§ 169. The Conditions for Reversible and Irreversible Changes in Non-Static Models 424
§ 170. Model Filled with Incoherent Matter Exerting No Pressure as an Example of Reversible Behaviour . . . 426
§ 171. Model Filled with Black-Body Radiation as an Example of Reversible Behaviour 427
§ 172. Discussion of Failure to Obtain Periodic Motions without Singular States 429
§ 173. Interpretation of Reversible Expansions by an Ordinary Observer 432

§ 174. Analytical Treatment of a Succession of Expansions and Con-
              tractions for a Closed Model with $\Lambda = 0$.   .   .   435
        (a) The Upper Boundary of Expansion   .   .   .   436
        (b) Time Necessary to Reach Maximum   .   .   .   436
        (c) Time Necessary to Complete Contraction   .   .   437
        (d) Behaviour at Lower Limit of Contraction   .   .   438
§ 175. Application of Thermodynamics to a Succession of Irreversible
              Expansions and Contractions   439

Part IV. CORRELATION OF PHENOMENA IN THE ACTUAL UNIVERSE
         WITH THE HELP OF NON-STATIC HOMOGENEOUS MODELS

§ 176. Introduction   445
§ 177. The Observational Data   446
        (a) The Absolute Magnitudes of the Nearer Nebulae   .   446
        (b) The Corrected Apparent Magnitudes for more Distant
              Nebulae   448
        (c) Nebular Distances Calculated from Apparent Magnitudes   453
        (d) Relation of Observed Red-Shift to Magnitude and Dis-
              tance   454
        (e) Relation of Apparent Diameter to Magnitude and Dis-
              tance   457
        (f) Actual Diameters and Masses of Nebulae   .   .   458
        (g) Distribution of Nebulae in Space   .   .   .   459
        (h) Density of Matter in Space   461
§ 178. The Relation between Coordinate Position and Luminosity   .   462
§ 179. The Relation between Coordinate Position and Astronomically
              Determined Distance   465
§ 180. The Relation between Coordinate Position and Apparent
              Diameter   467
§ 181. The Relation between Coordinate Position and Counts of
              Nebular Distribution   468
§ 182. The Relation between Coordinate Position and Red-shift   .   469
§ 183. The Relation of Density to Spatial Curvature and Cosmological
              Constant   473
§ 184. The Relation between Red-shift and Rate of Disappearance of
              Matter   475
§ 185. Summary of Correspondences between Model and Actual Uni-
              verse   478
§ 186. Some General Remarks Concerning Cosmological Models   .   482
        (a) Homogeneity   482
        (b) Spatial Curvature   483
        (c) Temporal Behaviour   484
§ 187. Our Neighbourhood as a Sample of the Universe as a Whole   .   486

Appendix I. SYMBOLS FOR QUANTITIES

Scalar Quantities   489
Vector Quantities   490
Tensors   490
Tensor Densities   491

Appendix II. SOME FORMULAE OF VECTOR ANALYSIS   .   . 491

Appendix III. SOME FORMULAE OF TENSOR ANALYSIS
  (a) General Notation 493
  (b) The Fundamental Metrical Tensor and its Properties . . 494
  (c) Tensor Manipulations 495
  (d) Miscellaneous Formulae 496
  (e) Formulae Involving Tensor Densities 496
  (f) Four-Dimensional Volume. Proper Spatial Volume . . 496

Appendix IV. USEFUL CONSTANTS 497

Subject Index 499

Name Index 502

Relativity
Thermodynamics
and Cosmology

for a successful quantum electrodynamics; and the second part of the chapter will be given to the development of Minkowski's macroscopic theory of moving electromagnetic media based on the extension to special relativity of Maxwell's original treatment of stationary matter.

In Chapter V, Special Relativity and Thermodynamics, we then turn to less familiar consequences of the special theory. In the first part of the chapter we consider the effect of relativity, even on the classical thermodynamics of stationary systems, in providing— through the relativistic relation between mass and energy—a natural starting-point for the energy content of thermodynamic systems, and a method for computing the energy changes accompanying physical-chemical processes from a knowledge of changes in mass. This makes it feasible to consider such problems as the thermodynamic equilibrium between hydrogen and helium, and that between matter and radiation—assuming the possibility of their interconversion—and treatments of these questions are given. In the second part of the chapter we undertake the actual extension of thermodynamics to special relativity in order to obtain a thermodynamic theory for the treatment of moving systems. Although the results which are to be derived by such an application of relativity to thermodynamics were considered by Planck and by Einstein only two years after the original presentation of the special theory, but little further attention has been paid to them. Indeed, the very essential difference between the equation

$$E = \frac{E_0}{\sqrt{(1-u^2/c^2)}} \tag{1.1}$$

giving the energy of a moving particle E in terms of its proper energy E_0 and velocity u, and the quite different equation

$$Q = Q_0\sqrt{(1-u^2/c^2)} \tag{1.2}$$

connecting a quantity of heat Q with proper heat Q_0 and velocity, has apparently not always been appreciated. The common lack of familiarity with this branch of relativity has doubtless been due to the absence of physical situations where its applications were necessary. For the later extension of thermodynamics to general relativity, nevertheless, a knowledge of the Planck-Einstein thermodynamics is essential, and at the end of this chapter we introduce a four-dimensional expression for the second law of thermodynamics in special

I

INTRODUCTION

1. The subject-matter

It is the threefold purpose of this essay, first to give a coherent and fairly inclusive account of the well-known and generally accepted portions of Einstein's theory of relativity, second to treat the extension of thermodynamics to special and then to general relativity, and third to consider the applications both of relativistic mechanics and relativistic thermodynamics in the construction and interpretation of cosmological models.

The special theory of relativity will first be developed in the next three chapters, which are devoted respectively to the kinematical, mechanical, and electromagnetic consequences of the two postulates of special relativity. In Chapter II, under the general heading 'The Special Theory of Relativity', the two postulates of the theory will be presented, together with a brief statement of the confirmatory empirical evidence in their favour; their kinematical consequences will then be developed, firstly using the ordinary language which refers kinematical occurrences to some selected set of three Cartesian axes and the set of clocks that can be pictured as moving therewith, and secondly using the more powerful quasi-geometrical language provided by the concept of a four-dimensional space-time continuum. In Chapter III, Special Relativity and Mechanics, we shall develop first the mechanics of a particle and then those of a mechanical continuum from a postulatory basis which is obtained by adding the ideas of the conservation of mass and of the equality of action and reaction to the kinematics of special relativity. No appeal to analogies with electromagnetic results will be needed to obtain the complete treatment, and the considerations will be maintained on a macroscopic level throughout. Finally, in Chapter IV, Special Relativity and Electrodynamics, we shall complete our treatment of the more familiar subject-matter of the special theory, by developing the close relationships between special relativity and electromagnetic theory. The first part of this chapter will be devoted to the incorporation of the Lorentz electron theory in the framework of special relativity, a procedure which tacitly assumes a respectable amount of validity still inherent in classical microscopic considerations in spite of the evident necessity relativity on which the extension to general relativity can later be based.

In Chapter VI, The General Theory of Relativity, we consider the fundamental principles of the general theory of relativity together with some of its more elementary applications. Part I of the chapter will treat the three corner-stones—the principle of covariance, the principle of equivalence, and the hypothesis of Mach—on which the theory rests. In agreement with the point of view first stated by Kretchsmann, the principle of covariance will be regarded as having a logically formal character which can imply no necessary physical consequences, but at the same time in agreement with Einstein we shall emphasize the importance of using covariant language in searching for the axioms of physics, in order to eliminate the insinuation of unrecognized assumptions which might otherwise result from using the language of particular coordinates. The discussion of the principle of equivalence will emphasize not only its empirical justification as an immediate and natural generalization of Galileo's discovery that all bodies fall at the same rate, but will also lay stress on the philosophical desirability of the principle in making it possible to maintain the general idea of the relativity of all kinds of motion including accelerations and rotations as well as uniform velocities. The designation 'Mach hypothesis' will be used to denote the general idea that the geometry of space-time is determined by the distribution of matter and energy, so that some kind of field equations connecting the components of the metrical tensor $g_{\mu\nu}$ with those of the energy-momentum tensor $T_{\mu\nu}$ are in any case implied. In presenting the field equations actually chosen by Einstein, the cosmological or Λ-term will be introduced and retained in many parts of the later treatment, not because of direct empirical or theoretical evidence for the existence of this term, but rather on account of the logical possibility of its existence and the necessity for its presence in the case of certain cosmological models which at least deserve discussion. Part II of Chapter VI will be given to elementary applications of general relativity. These will include a discussion of the clock paradox which proved so puzzling during the interval between the developments of the special and general theories of relativity. Treatment will also be given to Newton's theory of gravitation as a first and very close approximation to Einstein's theory, and the three crucial tests of general relativity will be considered.

Chapter VII, Relativistic Mechanics, will be divided into two parts on general mechanical principles and on solutions of the field equations. In Part I, after illustrating the nature of the energy-momentum tensor and of the fundamental equations of mechanics by application to the behaviour of a perfect fluid, the equations of mechanics will be re-expressed in the form containing the pseudotensor density of potential gravitational energy and momentum t^ν_μ permitting us then to obtain conservation laws for Einstein's generalized expressions for energy and momentum, to exhibit the relation between energy and gravitational mass, and to show the reduction of the energy of a system in the case of weak fields to the usual Newtonian form including potential gravitational energy. In Part II of Chapter VII, Einstein's general solution for the field equations in the case of weak fields will first be presented. This will then be followed by a discussion of the properties of the solutions that can be obtained in special cases of spherical symmetry and the like, including useful explicit expressions for the Christoffel symbols and components of the energy-momentum tensor which then apply.

Chapter VIII, Relativistic Electrodynamics, will present the further extensions to general relativity both for the Lorentz electron theory and for the Minkowski macroscopic theory. This will be followed by a number of applications including the derivation of an expression for the relativistic energy-momentum tensor for blackbody radiation, together with discussions of the gravitational interaction of light rays and particles, and of the generalized Doppler effect, these latter being matters of special importance for the interpretation of astronomical findings.

Chapter IX, Relativistic Thermodynamics, considers the extension of thermodynamics from special to general relativity together with its applications. The principles of relativistic mechanics themselves are taken as furnishing the analogue of the ordinary first law of classical thermodynamics; and the analogue of the second law is provided by the covariant generalization of the four-dimensional form in which the second law can be expressed in the case of special relativity. Since the above choice for the analogue of the first law introduces only generally accepted results of relativity, the whole character of relativistic thermodynamics is determined by the relativistic second law. The axiom chosen for this law is hence carefully examined as to meaning; its present status is discussed as being the direct

covariant re-expression and therefore the most probable generalization of the ordinary second law; and its future status as a postulate
to be verified or rejected on empirical grounds is emphasized. Following this discussion, applications are made to illustrate the characteristic differences between the results of relativistic thermodynamics,
and those which might at first sight seem probable on the basis of
a superficial extrapolation of conclusions familiar in the classical
thermodynamics. Thus in the case of static systems, although we
shall find the physical-chemical equilibrium between reacting substances—as measured by a local observer—unaltered from that which
would be predicted classically, we shall find on the other hand as a
new phenomenon the necessity for a temperature gradient at thermal
equilibrium to prevent the flow of heat from regions of higher to those
of lower gravitational potential, in agreement with the qualitative
idea that all forms of energy have weight as well as mass. Turning to
non-static systems we shall then show the possibility for a limited
class of thermodynamic processes which can occur both reversibly
and at a finite rate—in contrast to the classical requirement of an
infinitely slow rate to secure that maximum efficiency which would
permit a return both of the system and its surroundings to their
initial state. We shall later find that the principles of relativistic
mechanics themselves provide a justification for this new thermodynamic conclusion, since they permit the construction of cosmological
models which would expand to an upper limit and then return with
precisely reversed velocities to earlier states. Finally, in the case of
irreversible processes taking place at a finite rate, we shall discover
possibilities for a continuous increase in entropy without ever reaching an unsurpassable value of that quantity—in contrast to the
classical conclusion of a final quiescent state of maximum entropy.
This new kind of thermodynamic behaviour, which may be regarded
as mainly resulting from the known modification of the principle of
energy conservation by general relativity, will also find later illustration among the cosmological models predicted as possible by the
principles of relativistic mechanics.

In Chapter X, Application to Cosmological Models, we complete
the text except for some appendices containing useful formulae and
constants. In the first part of this chapter we shall show that the
only possible static homogeneous models for the universe are the
original ones of Einstein and de Sitter, and shall discuss some of their

properties which are important without reference to the adequacy of the models as pictures of the actual universe. We shall then turn to the consideration of non-static homogeneous models which can be constructed so as to exhibit a number of the properties of the actual universe, including, of course, the red shift in the light from the extra-galactic nebulae. Special attention will be given to the method of correlating the properties of such models with the results of astronomical observations although the details for obtaining the latter will not be considered. Attention will also be paid to the theoretically possible properties of such models, without primary reference to their immediate applicability in the correlation of already observed phenomena, since no models at the present stage of empirical observation can supply more than very provisional pictures of the actual universe.

The most important omission in this text, from the subjects usually included in applications of the special theory of relativity, is the relativistic treatment of the statistical mechanics of a gas, as developed by Jüttner and to some extent by the present writer.† The omission is perhaps justified by our desire in the present work to avoid microscopic considerations as far as possible, and by the existing absence of many physical situations where the use of this logically inevitable extension of relativity theory has as yet become needed.

In the case of the general theory of relativity, the most important omission lies in neglecting the attempts which have been made to construct a unified field theory, in which the phenomena of electricity as well as gravitation would both be treated from a combined 'geometrical' point of view. Up to the present, nevertheless, these attempts appear either to be equivalent to the usual relativistic extension of electromagnetic theory as given in the present text, or to be—although mathematically interesting—of undemonstrated physical importance. Furthermore, it is hard to escape the feeling that a successful unified field theory would involve microscopic considerations which are not the primary concern of this book.

The most important inclusions, as compared with older texts on relativity, consist in the extension of thermodynamics to general relativity, and the material on non-static models of the universe. Other additions are provided by the calculations of thermodynamic

† Jüttner, *Ann. d. Physik*, **34**, 856 (1911); Tolman, *Phil. Mag.*, **28**, 583 (1914).

equilibria with the help of the mass-energy relation of special relativity, by the demonstration of the reduction of the relativistic
expression for energy in the case of weak fields to the Newtonian
expression *including* potential gravitational energy, by explicit expressions given for the components of the energy-momentum tensor
in the case of special fields, and by the treatments given to the
energy-momentum tensor for radiation and to the gravitational interaction of light rays and particles.

2. The method of presentation

In the presentation of material, the endeavour will be made to
emphasize the physical nature of assumptions and conclusions and the
physical significance of their interconnexion, rather than to lay stress
on mathematical generality or even, indeed, on mathematical rigour.
The exposition will of course make use of the language and methods
of tensor analysis, a table of tensor formulae being given in Appendix
III to assist the reader in this connexion. No brief will be held,
however, for the fallacious position that the possibilities of covariant
expression are exhausted by the use of tensor language; and no
hesitation will be felt in introducing Einstein's pseudo-tensor density
t^ν_μ of potential gravitational energy and momentum in order to secure
quantities obeying conservation laws, which can be taken as the
relativistic analogues of energy and momentum.

To make sure that reader and writer are not substituting a satisfaction in mathematical complications or in geometrical analogies for
the main physical business at hand, the frequent translation of mathematical expressions into physical language will be undertaken. Stress
will be laid on the immediate physical significance of proper quantities
such as proper lengths, times, temperatures, macroscopic densities,
etc., whose values can be determined by a local observer using familiar
methods of measurement. Special attention will be given to the procedure for relating the coordinate position of nebulae with actual
astronomical estimates of distance.

In presenting the special theory of relativity no particular relation
will be assumed between the units of length and time, and the
formulae obtained will explicitly contain the velocity of light c. In
going over to the general theory of relativity, however, units will be
assumed which give both the velocity of light and the constant of
gravitation the values unity. This introduces a gain in simplicity of

mathematical form which is partially offset by the loss in immediate physical significance and applicability. The translation of results into ordinary physical units will be facilitated, however, by the table in Appendix IV.

The method of presenting the mechanics of a particle will be similar to that first developed by Professor Lewis and the present writer which obtains a basis for the treatment by combining the kinematics of special relativity with the conservation laws for mass and momentum. The Laue mechanics for a continuous medium will then be obtained by the further development of these same ideas, using the transformation equations for force provided by particle dynamics. This method seems to afford a more direct mechanical insight than methods based on analogies with electromagnetic relations, or on those starting from some variational principle as was used for example by the present writer in an earlier book.†

To turn to more general features of the method of presentation, the ideal treatment for such a highly developed subject as the theory of relativity would perhaps be a strictly deductive one. In such a method we should start with a set of indefinables, definitions, and postulates and then construct a logical universe of discourse. The indefinables and definitions would provide the subject-matter in this universe of discourse, and the postulates together with the theorems, derived from them, with the help of logic or other discipline more fundamental than that of the field of interest, would provide the significant assertions that could be made concerning the subject-matter. The usefulness of this logical construct in explaining the phenomena of the actual world would then depend on the success with which we could set up a one-to-one correspondence between the subject-matter and assertions in our universe of discourse and the elements and regularities observed in actual experience, in other words, on the success with which we could use the construct as a representative map for finding our way around in the external world. Although the attempt will not be made in this book to construct such a logical universe of discourse, and no attention will be paid to matters so pleasant to the logician as the search for the smallest number of mutually independent and compatible postulates, it is nevertheless hoped that the method of exposition will benefit from a recognition of this ideal.

† *The Theory of the Relativity of Motion*, University of California Press, 1917.

3. The point of view

Throughout the essay a macroscopic and phenomenological point of view will be adopted as far as feasible. This is made possible in the case of relativistic mechanics by a treatment of mechanical media which defines the energy-momentum tensor in terms of such quantities as the proper *macroscopic* density of matter ρ_{00} and the proper pressure p_0 which could be directly measured by a local observer. The use of the proper *microscopic* density of matter ρ_0 will be avoided. In the case of relativistic thermodynamics the treatment is, of course, naturally macroscopic on account of the essential nature of that science when we do not undertake any statistical mechanical interpretations. Thus the quantity ϕ_0 will be taken as the entropy density of the thermodynamic fluid or working substance as determined by a local observer at the point and time of interest, using ordinary thermodynamic methods and introducing no conceptual division of the fluid into such elements as atoms and light quanta. In the case of electrodynamics, however, the macroscopic point of view cannot be entirely maintained, since, in spite of the use that can be made of the Minkowski phenomenological electrodynamics of moving media, we have to be interested in the propagation of electromagnetic waves of such high frequency that some form of quantum electrodynamics will ultimately be necessary for their satisfactory treatment.

We of course accept Einstein's theory of relativity as a valid basis on which to build. In the case of the special theory of relativity the observational verification of the foundations provided by the Michelson-Morley experiment, by Kennedy's time transformation experiment, and by de Sitter's analysis of the orbits of double stars, and in the case of the general theory the observational verification of predictions provided by the motion of the perihelion of Mercury, by the bending of light in passing the sun, and by the effect of differences in gravitational potential on the wave-length of light are sufficient to justify such an acceptance. Future changes in the structure of theoretical physics are of course inevitable. Nevertheless, the variety of the tests to which the theory of relativity has been subjected, combined with its inner logicality, are sufficient to make us believe that further advances must incorporate enough of the present theory of relativity to make it a safe provisional foundation for macroscopic considerations.

In the present stage of physics it appears probable that the most serious future modifications in the theory of relativity will occur in the treatment of microscopic phenomena involving the electric and gravitational fields in the neighbourhood of individual elementary particles. Here some fusion of the points of view of the present theory of relativity and of the quantum mechanics will be necessary, which might be brought about, as Einstein is inclined to believe, by an explanation of quantum phenomena as the statistical result to be expected on the basis of a successful unified field theory, or, as the proponents of the quantum mechanics are more inclined to believe, by some unified extension of quantum mechanics and quantum electrodynamics. In any case it seems certain that the present form of the theory of relativity is not suitable for the treatment of microscopic phenomena. Fortunately for the consideration of problems in celestial mechanics and cosmology, we do not need to consider the difficulties that might thus arise since the scale of our interest is so large that the phenomena are in any case most naturally treated from a macroscopic point of view.

As a further remark concerning the point of view adopted it may be well to emphasize at this point the highly abstract and idealized character of the conceptual models of the universe which we shall study in the last chapter. The models will always be much simpler than our actual surroundings, neglecting for example local details in the known structure of the universe and replacing the actual disposition of the material therein by a continuous distribution of fluid. The reason for such idealization lies, of course, in the simplification which it introduces into the mathematical treatment. The procedure is analogous to the introduction, for example, of rigid weightless levers into the considerations of the older mechanics, or perfectly elastic spherical molecules into the simple kinetic theory, and is justified in so far as our physical intuition is successful in retaining in the simplified picture the essential elements of the actual situation.

In addition to the introduction of fairly obvious simplifications in constructing cosmological models, it will also be necessary to introduce assumptions concerning features which are as yet unknown in the actual universe. Thus, since the distribution of the extra-galactic nebulae has been found to be roughly uniform out to some 10^8 light-years, we shall usually assume a homogeneous distribution of material throughout the whole of our models, even though we shall emphasize that this may not be true for the actual universe. Furthermore, we would assume a specially simple form when described with the help of a system of coordinates at rest in the ether.

In the light of the ether theory the necessity thus arises for an experimental verification of the postulate or the conclusions that can be drawn from it. The usefulness of a deductive branch of physics depends on the success with which it can be used as a representative map for correlating the phenomena of the external world. Hence a direct experimental test of the postulates is not necessary provided the conclusions can be verified. Nevertheless, a feeling of greater intellectual satisfaction is obtained when the postulates themselves are chosen in such a way as to permit reasonably direct experimental tests.

Fortunately the direct experimental verification of this postulate of the special theory of relativity may now be regarded as extremely satisfactory. In the first place we must put the well known Michelson-Morley experiment, which on the basis of the theory of a fixed ether should have led to a detection of the velocity of the earth's motion through that medium. The null effect obtained in the original performance of this experiment and in all the fairly numerous repetitions, except those of Miller,† leave little doubt that no velocity through the ether can thus be detected, even of the magnitude of the 30 km. per sec. which should certainly arise from the earth's known rotation in its orbit. Among recent repetitions, that of Kennedy‡ appears extremely satisfactory and has reduced the observational error of the null effect to the order of ± 2 km. per sec. or less.

As is well known, the result thus obtained in the Michelson-Morley experiment could be explained by itself alone, without giving up the notion of a fixed ether, by assuming that bodies moving through this medium suffer the so-called Lorentz-Fitzgerald contraction in the direction of their motion, which would produce just the necessary distortion in a moving Michelson interferometer to lead to a null effect. For this reason it is specially satisfactory that we can now put in the second place as a part of the direct verification of our postulate

---

† For a summary up to 1926, see Miller, *Science*, **63**, 433 (1926). This work also shows no effect of ether drift as great as would be expected to accompany the full velocity of the earth's motion in its orbit. An effect corresponding to a velocity of about 10 km. per sec. along an axis with its apex in the constellation Draco is reported, however, and interpreted as possibly due to a velocity through the ether of 200 km. per sec. or more, whose effects are partially compensated by the Lorentz-Fitzgerald contraction. A still later account of Miller's work has just been published in *Reviews of Modern Physics*, **5**, 203 (1933).

‡ Kennedy, *Proc. Nat. Acad.* **12**, 621 (1926). Illingworth, *Phys. Rev.* **30**, 692 (1927).

an experiment devised by Kennedy, which on the basis of a fixed ether and a real Lorentz-Fitzgerald contraction should still lead to a detection of the motion of the earth through that ether. The apparatus for this experiment consists of a Michelson interferometer with the two arms as unequal in length as feasible, so that the two beams which recombine to give interference fringes have a considerable difference in the time required to pass from the source to the point of recombination. Assuming a fixed ether, but allowing for the Lorentz-Fitzgerald contraction associated with motion through this medium, analysis then shows that the difference in time of travel for the two beams would depend in a very simple way on the difference in length of the two arms and on the velocity of the apparatus through the ether. Hence, provided the period of the light source does not itself depend on this velocity, we should expect a shift in the fringe pattern to accompany the diurnal changes in the velocity of the apparatus through the ether produced by the earth's revolution on its axis, and the annual changes produced by its rotation in its orbit. The experiment was of course a very difficult one to perform, but the final results of Kennedy and Thorndike† have satisfactorily demonstrated a null effect to the order of the experimental error, which corresponds to a velocity of only about ± 10 km. per sec.

In addition to the Michelson-Morley and Kennedy-Thorndike experiments there have been a considerable number of other types of experiment devised to detect the motion of the earth through the ether, all of which have led to negative results.‡ Some of these are of considerable interest, but the two tests devised by Michelson and Kennedy are the most important and the most simply related to the ideas as to space and time which have been embodied in the special theory of relativity. To account for these two experiments on the basis of a fixed ether it would be necessary to introduce ingenious assumptions as to a change in length or Lorentz-Fitzgerald contraction just sufficient to give a null effect in the Michelson experiment, and as to a change in period or time dilation just sufficient to give a null effect in the Kennedy experiment—all to the end of retaining a fixed ether so devilishly constructed that its existence could never

† Kennedy and Thorndike, *Phys. Rev.* **42**, 400 (1932).
‡ See J. Laub, 'Über die experimentellen Grundlagen des Relativitätsprinzips': *Jahrb. der Radioaktivität u. Elektronik*, **7**, 405 (1910).

be detected. In the theory of relativity, however, we proceed at the start from the basis that absolute velocity can have no significance and hence find nothing to trouble us in the result of these experiments. In the course of the development of the theory we shall obtain, moreover, the simple and unforced counterparts of the assumptions as to change in length and period, which would have to be introduced in an artificial and arbitrary manner in order to retain the notion of a fixed ether.

6. The second postulate of relativity

In addition to the first postulate the special theory of relativity depends on a second postulate, which states that *the velocity of light in free space is the same for all observers, independent of the relative velocity of the source of light and the observer.* This postulate can be looked at as the result of combining the principle familiar to the ether theory of light, that the velocity of light is independent of the velocity of its source, with the idea resident in the first postulate which makes it impossible to assign any significance to the absolute velocity of the source but does permit us to speak of the relative velocity of the source and observer.

It is important to note that the essential quality of the second postulate may thus be regarded as having been provided by a theory of light which assumed space to be filled with some form of ether, while the first postulate of relativity may be regarded as the natural consequence of the Newtonian point of view of the emptiness of free space. It is not surprising that the combination of principles of such different character should have led to a modification in our ideas as to the nature of time and space.

At the time of Einstein's development of the special theory of relativity no experimental evidence had been assembled to show that the velocity of light *is* independent of the velocity of its source, and the adoption of the principle was due to its familiarity in the wave theory of light. At the present time, however, the experimental evidence is sufficient to exclude very definitely the most natural alternative proposal namely, that the velocity of light and the velocity of its source are additive, as assumed in the so-called emission theories of light.

As the most important evidence against the hypothesis that the velocity of light and the velocity of its source are additive we must put the considerations of Comstock† and de Sitter‡ concerning the orbits of double stars. If the velocity of light did depend additively on the velocity of its source, it is evident in the case of distant doublets that the time taken for light to reach the earth from a given member of the pair would be greatly shortened when this member is approaching the earth at the time of emission and greatly lengthened when the member is receding. The analysis of de Sitter shows that the distortion of the observations thus produced would have the effect of introducing a spurious eccentricity into the calculated orbit; and from the existence of doublets of small observed eccentricity the conclusion is drawn that the velocity of light could at the most be changed by only a small fraction of the velocity of the source.

In addition to this very satisfactory evidence in favour of the principle of the constancy of the velocity of light, there are a number of optical experiments which show the untenability of different forms which have been proposed for the emission theory of light. These proposed forms of emission theory agree in assuming that the velocity of light from a moving source is to be taken as the vector sum of the ordinary velocity of light c and the velocity of the source v but vary in their assumptions as to the velocity of light after reflection from a mirror. The three assumptions which have been particularly considered are (1) that the excited portion of the mirror acts as a new source and that reflected light has the same velocity c relative to the mirror as that of light relative to its original source,§ (2) that reflected light has the velocity c relative to the mirror image of its source,‖ and (3) that light retains throughout its whole path the velocity c with respect to its original source.††

Optical experiments contradicting the first two of the above possible assumptions have not been difficult to find.‡‡ The third of the above assumptions formed, however, the basis of a fairly complete emission theory which was developed by Ritz, and optical experiments to test it are difficult to perform since they are dependent on effects of the second order in v/c. It has been pointed out, however, by La Rosa

† Comstock, *Phys. Rev.* **30**, 267 (1910).
‡ de Sitter, *Proc. Amsterdam Acad.* **15**, 1297 (1913); ibid. **16**, 395 (1913).
§ Tolman, *Phys. Rev.* **31**, 26 (1910).
‖ J. J. Thomson, *Phil. Mag.* **19**, 301 (1910). Stewart, *Phys. Rev.* **32**, 418 (1911).
†† Ritz, *Ann. de Chim. et Phys.* **13**, 145 (1908).
‡‡ Tolman, *Phys. Rev.* **31**, 26 (1910); ibid. **35**, 136 (1913). Marjorana, *Phil. Mag.* **35**, 163 (1918); ibid. **37**, 145 (1919).

and by the present writer† that a repetition of the Michelson-Morley experiment, using light coming originally from the sun rather than from a terrestrial source, should lead to a fringe shift corresponding to the earth's velocity in its orbit around the sun, if the Ritz theory were correct. In the repetitions of the Michelson-Morley experiment which have been made by Miller tests were made using light from the sun and no effect of the kind predicted was observed.‡ In any case, of course, we have the astronomical evidence of de Sitter against all forms of the emission theory.

7. Necessity for modifying older ideas as to space and time

Let us now accept the two postulates of special relativity as experimentally justified and inquire into the effect they have on our ideas

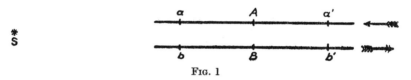

Fig. 1

as to the nature of space and time. Since the first postulate is a natural consequence of the Newtonian point of view that free space is empty, and the second postulate is a natural outcome of the opposing idea that space is everywhere filled with a fixed ether, we can expect the combination of the two postulates to lead to consequences which do not agree with our uninformed intuitions as to the nature of space and time. We shall illustrate this in the present section by a simple example.

Consider a source of light S (Fig. 1), and two systems, A moving *towards* the source S, and B moving *away* from it. Observers on the two systems mark off some given distance aa' and bb', say one kilometre, on each of the systems in the direction of the source in order to measure the velocity of light by determining the time taken for it to travel from a to a' and from b to b'.

In accordance with the first postulate of relativity we cannot

† La Rosa, *Phys. Zeits.* **13**, 1129 (1912). Tolman, *Phys. Rev.* **35**, 136 (1912).
‡ Miller, *Proc. Nat. Acad.* **11**, 306 (1925). Professor Miller informs the writer that five sets of observations were made using sunlight at Mount Wilson in 1924: July 1 at 4.45 p.m.; July 8 at 2.45 p.m.; and at 5.55 p.m.; July 9 at 9.00 a.m.; July 26 at 9.30 a.m. By comparing these observations with those made using an acetylene lamp just before or after the sunlight experiments, the results obtained were found to be substantially the same.

assign any significance to the absolute velocities of the two systems, but can speak of their velocities relative to the source. And in accordance with the second postulate of relativity the measured velocity of light must be independent of this relative velocity between source and observer.

Hence we are led to the conclusion that the time taken for the light to travel from a to a' shall measure the same as that for the light to travel from b to b', in spite of the fact that A is moving towards the source and B away from it. This result seems to contradict the simple conclusions of common sense. Hence if the two postulates of relativity are true it is evident that our natural intuitions as to the nature of space and time are not completely correct, presumably because they are based on a too limited ancestral experience—human and animal—with spatial and temporal phenomena.

In view of the experimental verification of the two postulates of special relativity, the example makes evident the necessity for a detailed study of the relations connecting the spatial and temporal measurements made by observers in relative motion. This we shall undertake in the next section. We shall gain thereby not only correct methods for the treatment of such measurements, but ultimately improved spatial and temporal intuitions as well.

8. The Lorentz transformation equations

To study the fundamental problem of the relations connecting the spatial and temporal measurements made by observers in relative motion, let us consider two systems of space-time coordinates S and S' (Fig. 2) in relative motion with the velocity V,† which for convenience may be taken as in the x-direction. Each system is provided with a set of right-angled Cartesian axes, as indicated in the figure, and with a set of clocks distributed at convenient intervals throughout the system and moving with it.

The *position* of any given point in space at which some event occurs can be specified by giving its spatial coordinates x, y, and z with respect to the axes of system S, or its coordinates x', y', and z' with respect to system S'. And the *time* at which the event occurs can be specified by giving the clock readings t or t' in the two systems.

For convenience the two systems are chosen so that the Cartesian

† We shall use the capital letter V to designate the relative velocity of the two systems of axes, and the small letters u, u', etc., to designate the velocity of a point relative to the coordinate systems.

axes OX and $O'X'$ lie along the same line, and for further simplification
the starting-point for time measurements in the two systems is taken
so that t and t' are equal to zero when the two origins O and O' are
in coincidence.

The specific problem that now arises is to obtain a set of transformation equations connecting the variables of the two systems
which will make it possible to transform the description of any given

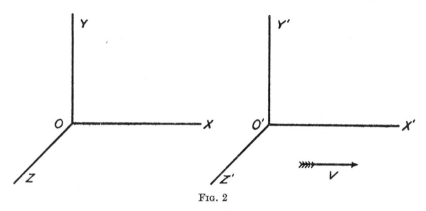

Fɪɢ. 2

kinematical occurrence from the variables of the one system to those
of the other. In other words, if some *given* kinematical occurrence
has been measured by an observer moving with system S' and
described in terms of the quantities x', y', z', and t', we desire a set
of expressions for these quantities which on substitution will give a
correct description of the *same* occurrence in terms of the variables
x, y, z, and t, used by an observer moving with system S.

The correct expressions for this purpose were first obtained by
Lorentz, and hence are usually called the Lorentz transformation
equations, although their full significance from the point of view of
the relativity of motion was first appreciated by Einstein. They may
be written in the form

$$x' = \frac{x - Vt}{\sqrt{(1 - V^2/c^2)}},$$
$$y' = y,$$
$$z' = z, \tag{8.1}$$
$$t' = \frac{t - xV/c^2}{\sqrt{(1 - V^2/c^2)}},$$

where V is the relative velocity of the two systems and c the velocity of light. Or, by solving for the unprimed quantities in terms of the primed quantities, they may be written in the form

$$x = \frac{x'+Vt'}{\sqrt{(1-V^2/c^2)}},$$

$$y = y',$$

$$z = z',$$ \hfill (8.2)

$$t = \frac{t'+x'V/c^2}{\sqrt{(1-V^2/c^2)}}.$$

A unique derivation of these equations from the first and second postulate's of relativity, making use of obvious assumptions as to the validity of Euclidean geometry, the homogeneity of space and time, etc., can be obtained. We may content ourselves now, however, with pointing out that the equations do agree with the two postulates of relativity.

In accordance with the first postulate of relativity, since absolute velocity has no significance, the two systems S and S' must be entirely equivalent for the description of physical occurrences. Hence the transformation equations for changing from the variables of system S to those of system S' must have exactly the same form as those for the reverse transformation, except of course for the sign of the relative velocity V. This condition, however, is evidently met since the set of equations (8.2) which are obtained by solving the set (8.1) are seen to be of unchanged form except for the substitution of $-V$ in place of $+V$.

In accordance with the second postulate of relativity, the velocity of light must measure the same in both systems of coordinates. To show that this is the case, we first call attention to the important fact that the Lorentz equations have been so constructed as to make the quantity
$$dx^2+dy^2+dz^2-c^2\,dt^2 \hfill (8.3)$$
an invariant for the transformation. This is evident since on substituting equations (8.2) in (8.3) we obtain

$$dx^2+dy^2+dz^2-c^2\,dt^2$$

$$= \left(\frac{dx'+V\,dt'}{\sqrt{(1-V^2/c^2)}}\right)^2+dy'^2+dz'^2-c^2\left(\frac{dt'+(V\,dx'/c^2)}{\sqrt{(1-V^2/c^2)}}\right)^2$$

$$= dx'^2+dy'^2+dz'^2-c^2\,dt'^2. \hfill (8.4)$$

The invariance of this expression, however, immediately shows that the velocity of light will measure the same in both systems, since, if we have any impulse travelling with the velocity c with respect to system S in accordance with the equation

$$dx^2+dy^2+dz^2-c^2\,dt^2 = 0, \qquad (8.5)$$

we shall also have it travelling with the velocity c with respect to system S' in accordance with the equation

$$dx'^2+dy'^2+dz'^2-c^2\,dt'^2 = 0. \qquad (8.6)$$

We thus see that the Lorentz transformation equations are in accord with the two postulates of relativity. It should also be noted that they are in accord with our ideas as to the homogeneity of space and time. Furthermore, when the relative velocity between the systems V is small compared with that of light they reduce to the so-called Galilean transformation equations,

$$x' = x-Vt,$$
$$y' = y,$$
$$z' = z, \qquad (8.7)$$
$$t' = t,$$

which we might expect to hold on the basis of an intuition founded on a past experience limited to low velocities, and which were implicit in the ideas of Galileo and Newton as to the nature of space and time.

It should also be remarked that the set of Lorentz transformations between all systems in unaccelerated uniform motion form a group, such that the combined result of successive transformations is equivalent to a single transformation from the original to the final system of coordinates. It may also be pointed out that the transformation becomes imaginary for relative velocity between the systems V greater than the velocity of light c, a result which is consistent with that of a following section showing that c is to be regarded as an upper limit for the possible relative velocities between material systems.

With the help of simple manipulations we may now obtain from the Lorentz transformation equations a number of further equations which will prove useful for transforming the measurements of geometrical or kinematical quantities which depend on the coordinates, and which will permit some simple physical interpretations.

9. Transformation equations for spatial and temporal intervals. Lorentz contraction and time dilation

By the simple differentiation of equations (8.1) we obtain

$$dx' = \frac{dx - V\,dt,}{\sqrt{(1 - V^2/c^2)}},$$
$$dy' = dy, \qquad\qquad (9.1)$$
$$dz' = dz,$$
$$dt' = \frac{dt - V\,dx/c^2}{\sqrt{(1 - V^2/c^2)}},$$

where the differential quantities dx, dy, dz, dt, and dx', dy', dz', dt' are to be interpreted as giving the measurements in the two systems of the spatial and temporal intervals which correspond to the difference in position and time of some *given* pair of neighbouring events. With the help of these equations we can now easily draw conclusions as to the intercomparison of measuring sticks and clocks in the two systems.

Consider two measuring sticks held parallel to the x-axis, one in each of the two systems, in such a way that their scale divisions can be compared as the two sticks slide past each other; and consider as the events to be observed the coming into coincidence of division marks on one of the measuring sticks with division marks on the other.

Let us first determine how a length dx' laid off on the measuring stick in system S' will appear when measured in system S. To do this we must consider coincidences, *which appear simultaneous in system S*, between the end points of dx' and division marks on the measuring stick in system S. Since the coincidences are simultaneous in system S, we shall have

$$dt = 0, \qquad\qquad (9.2)$$

and by substitution in (9.1) obtain

$$dx' = \frac{dx}{\sqrt{(1 - V^2/c^2)}} \quad \text{or} \quad dx = dx'\sqrt{(1 - V^2/c^2)}. \qquad (9.3)$$

We conclude that a measuring stick travelling with system S' and measuring dx' in the units of that system will measure the shorter length $dx'\sqrt{(1 - V^2/c^2)}$ in the units of system S when the simultaneous positions of its ends are observed in that system.

Let us next determine how a length dx laid off on the measuring stick in system S will appear when measured in system S'. To do this we must now consider coincidences, *which appear simultaneous in system S'*, between the end points of dx and division marks on the measuring stick in system S'. Since the coincidences are simultaneous in system S', we shall have in accordance with the last equation (9.1)

$$dt' = \frac{dt - V\,dx/c^2}{\sqrt{(1 - V^2/c^2)}} = 0, \qquad (9.4)$$

and substituting into the first equation (9.1) shall this time obtain

$$dx' = dx\,\sqrt{(1 - V^2/c^2)}. \qquad (9.5)$$

We conclude that a metre stick travelling with system S measures shorter in the same ratio as before when the simultaneous positions of its ends are observed in the other system S'.

The two situations are symmetrical and in entire agreement. In both cases we find that a metre stick measures shorter in the ratio $\sqrt{(1 - V^2/c^2)} : 1$, when moving with the velocity V past the system in which the observation of length is being made, than when measured in a system in which it is at rest.

Accepting the two postulates of relativity, this result, which may be called the Lorentz contraction, is to be regarded as an entirely *real* one which except for experimental difficulties could be verified by direct observation of the kind just described. The result differs from the contraction originally postulated by Lorentz and Fitzgerald to explain the Michelson-Morley experiment, since the present result gives a symmetrical relation between two measuring sticks in relative motion, while the hypothesis of Lorentz and Fitzgerald required a change in length for a single metre stick depending on its actual velocity through a real fixed ether.

Turning now to the second and third equations in the set (9.1), we note at once that there will be no disagreement as to measurements made in the two systems of coordinates of distances at right angles to the line of motion. There is thus no change in length for a metre stick which is moving perpendicular to its length past the system of coordinates in which it is to be measured. This is in immediate agreement with the possibility for a direct comparison of the lengths of two metre sticks in relative motion at right angles to their extension, since in this case the judgement that the two ends of the one metre stick had passed through coincidence with the two ends of the other could not depend on the motion of the observer.

Equations (9.1) can also be used to provide conclusions as to the intercomparison of clocks in relative motion. Let us first determine how a time interval dt' measured on a *single* clock in system S' between two events, which occur at the same point in S', will measure with the clocks of system S. Since the two events occur at the same point in S', we have from the first of equations (9.1)

$$dx' = \frac{dx - V\,dt}{\sqrt{(1 - V^2/c^2)}} = 0, \tag{9.6}$$

and substituting this into the fourth of the equations we easily obtain

$$dt' = dt\sqrt{(1 - V^2/c^2)} \quad \text{or} \quad dt = \frac{dt'}{\sqrt{(1 - V^2/c^2)}}. \tag{9.7}$$

We conclude that the time interval between two events which has the duration dt' when measured with a given clock in system S', will have the longer duration $dt'/\sqrt{(1 - V^2/c^2)}$ when measured by the clocks in system S.

Similarly we may determine how a time interval dt which can be measured on a single clock in system S between two events, which occur at the same point in system S, will measure with the clocks of system S'. In this case since the two events occur at the same point in system S we have

$$dx = 0 \tag{9.8}$$

and substituting in the fourth of equations (9.1) immediately obtain

$$dt' = \frac{dt}{\sqrt{(1 - V^2/c^2)}}. \tag{9.9}$$

Again we conclude that the time interval between two events which has the duration dt when measured with a given clock has a longer duration when measured by clocks relative to which the first clock is moving.

The two situations, in the case of the clocks as in the case of the measuring sticks, are symmetrical and in entire agreement. In both cases the seconds of the single clock appear lengthened in the ratio $1 : \sqrt{(1 - V^2/c^2)}$ when it is moving with the velocity V past the clocks with which it is being compared.

This *time dilation* and the conclusions as to the setting of clocks which can be shown to go with it are to be regarded except for experimental difficulties as an entirely verifiable mutual property of systems of clocks in relative motion, even as the Lorentz contraction could be regarded as a verifiable mutual property of metre sticks in relative motion. Furthermore just as the Michelson-Morley experiment can be regarded as a direct test of the Lorentz contraction, the Kennedy-Thorndike experiment can be regarded as a direct test of time dilation.

Before leaving this section it will be well to put the fourth of equations (9.1) in another form which is often useful. Dividing through by dt we can write

$$\frac{dt'}{dt} = \frac{1 - \dfrac{V}{c^2}\dfrac{dx}{dt}}{\sqrt{(1 - V^2/c^2)}} = \frac{1 - \dfrac{V\dot{x}}{c^2}}{\sqrt{(1 - V^2/c^2)}}, \tag{9.10}$$

which connects the measurements dt' and dt of the time interval in the two systems S' and S between neighbouring events which occur at neighbouring points in space. The spatial interval between the two events, when measured in system S, has as its x-component the distance which would be travelled with the component velocity $\dot{x}$ in the time dt.

10. Transformation equations for velocity

With the help of equations (8.1) and (9.10) we can now easily obtain expressions for transforming measurements of velocity from the one system of coordinates to the other. Differentiating the first three of equations (8.1) with respect to t' and substituting the value for dt'/dt given by (9.10) we easily arrive at the results

$$\dot{x}' = \frac{\dot{x} - V}{1 - \dot{x}V/c^2} \quad \text{or} \quad u_x' = \frac{u_x - V}{1 - u_x V/c^2},$$

$$\dot{y}' = \frac{\dot{y}\sqrt{(1 - V^2/c^2)}}{1 - \dot{x}V/c^2} \qquad u_y' = \frac{u_y\sqrt{(1 - V^2/c^2)}}{1 - u_x V/c^2}, \tag{10.1}$$

$$\dot{z}' = \frac{\dot{z}\sqrt{(1 - V^2/c^2)}}{1 - \dot{x}V/c^2} \qquad u_z' = \frac{u_z\sqrt{(1 - V^2/c^2)}}{1 - u_x V/c^2},$$

where the placing of a dot over a quantity has the significance of differentiation with respect to the time in the particular system of coordinates involved, so that we have for example for the component velocities in the x-direction in the two systems the different forms of expression

$$u_x = \dot{x} = dx/dt \quad \text{and} \quad u_x' = \dot{x}' = dx'/dt'.$$

The significance of these transformation equations is as follows: If for an observer in system S a point is found to be moving with the uniform velocity ($\dot{x}$, $\dot{y}$, $\dot{z}$) its velocity ($\dot{x}'$, $\dot{y}'$, $\dot{z}'$) as measured by an observer in system S' can be calculated from the equations (10.1).

Reciprocal equations for transformation in the opposite direction can of course be obtained by solving for the unprimed quantities in terms of the primed, and in accordance with the first postulate of relativity agree with that which results from interchanging primed and unprimed letters and changing the sign of V. It is often most convenient to have the transformation equations in the form in which they are solved for the unprimed quantities since this leads more readily to final expressions without the primes. For this reason it will be best to write down the reciprocal equations to (10.1), and from now on to give our remaining transformation equations in the form in which they are solved for the unprimed quantities. We obtain from (10.1)

$$u_x = \frac{u_x' + V}{1 + u_x' V/c^2},$$

$$u_y = \frac{u_y' \sqrt{(1 - V^2/c^2)}}{1 + u_x' V/c^2}, \qquad (10.2)$$

$$u_z = \frac{u_z' \sqrt{(1 - V^2/c^2)}}{1 + u_x' V/c^2}.$$

The foregoing transformation equations immediately indicate that the velocity of light c may be regarded as an upper limit of possible velocities. The result is most readily seen if we use the equations in their second form (10.2) in which the relative velocity of the two systems occurs with the positive sign. In accordance with the first of these equations, even if we give the velocity of system S' past S the limiting value c and take a particle which itself has the limiting velocity $u_x' = c$ in the same direction with respect to system S', the measured velocity with respect to system S will still be only

$$u_x = \frac{c + c}{1 + c^2/c^2} = c, \qquad (10.3)$$

the velocity of light.

In addition to this indication that the velocity of light is to be regarded as an upper limit, we shall find later that it would take an infinite amount of energy to give a material particle the velocity of light with respect to a system in which it was originally at rest. Furthermore, retaining our ideas as to cause and effect as being essentially valid for macroscopic considerations, it can be shown that causal impulses cannot be transmitted with a velocity greater than light, since it would then be possible to find systems of coordinates in which the effect would precede the cause.[†]

11. Transformation equations for the Lorentz contraction factor

The quantity $\sqrt{(1-u^2/c^2)}$, which is the Lorentz contraction factor for an object moving with the velocity u with respect to a given system of coordinates, is sufficiently important to justify writing down the transformation equation for it which can be obtained from (10.2), namely,

$$\sqrt{(1-u^2/c^2)} = \frac{\sqrt{(1-u'^2/c^2)}\sqrt{(1-V^2/c^2)}}{1+u'_x V/c^2}, \tag{11.1}$$

where

$$u^2 = u_x^2 + u_y^2 + u_z^2. \tag{11.2}$$

12. Transformation equations for acceleration

By the further differentiation of equations (10.2) transformation equations for acceleration are obtained which can be written in the form

$$\dot{u}_x = \left(1+\frac{u'_x V}{c^2}\right)^{-3}\left(1-\frac{V^2}{c^2}\right)^{\frac{3}{2}} \dot{u}'_x,$$

$$\dot{u}_y = \left(1+\frac{u'_x V}{c^2}\right)^{-2}\left(1-\frac{V^2}{c^2}\right)\dot{u}'_y - u'_y \frac{V}{c^2}\left(1+\frac{u'_x V}{c^2}\right)^{-3}\left(1-\frac{V^2}{c^2}\right)\dot{u}'_x, \tag{12.1}$$

$$\dot{u}_z = \left(1+\frac{u'_x V}{c^2}\right)^{-2}\left(1-\frac{V^2}{c^2}\right)\dot{u}'_z - u'_z \frac{V}{c^2}\left(1+\frac{u'_x V}{c^2}\right)^{-3}\left(1-\frac{V^2}{c^2}\right)\dot{u}'_x.$$

Whereas it can be seen from equations (10.2) that a constant velocity in system S' implies a constant velocity in system S, it is interesting to note from equations (12.1) that a constant acceleration with respect to system S' would not in general imply a constant acceleration in system S, since the component accelerations in S depend not only on the accelerations in S' but also on the component velocities in that system which would be changing with the time.

It will be appreciated of course that both the transformation equations for velocity (10.2) and for acceleration (12.1) must be applied in general to the motion of a particle at some specific identifiable point on its path.

[†] See for example, Tolman, *The Theory of the Relativity of Motion*, § 52, The University of California Press, 1917.

II

THE SPECIAL THEORY OF REIATIVITY (contd.)

13. The space-time continuum

It is evident from the foregoing discussion of the consequences of
the two postulates of relativity that spatial and temporal measurements are linked together in a very intimate manner. This appears
clearly when we contrast the simple Galilean transformation equations
(8.7) with the Lorentz transformation equations (8.1). For example,
the Galilean time transformation equation

$$t' = t \tag{13.1}$$

would indicate a universal time equally suitable for use by all
observers, while the corresponding Lorentz equation

$$t' = \frac{t - xV/c^2}{\sqrt{(1 - V^2/c^2)}} \tag{13.2}$$

indicates that there is no single universal time equally suitable for all
observers, but rather that the process of changing from one set of
Cartesian axes to another for making spatial measurements should be
accompanied by a change in the apparatus for time measurement, if
the laws of physics are to have the same expression in the two systems
of coordinates.

An acceptance of the two postulates of relativity thus shows that
the older notion of space and time as two independently existing
continua—a three-dimensional continuum for the spatial location
of events and an independent one-dimensional continuum for the
temporal location of events—is a conceptual idea which we cannot
now expect will be entirely successful for the correlation of spatial
and temporal experiences. The possible alternative concept of space-
time as a combined four-dimensional continuum, first introduced by
Minkowski, has, however, proved very valuable.

We must now turn to the method of expressing the facts of special
relativity which can be obtained from this new conceptual apparatus.
The importance of the method, which can hardly be overestimated,
lies in several directions. The method is of great assistance in building
up a set of appropriate space-time intuitions. The method avoids the

singling out of a particular axis as the direction for the relative motion of coordinate systems as has been done in the previous parts of this chapter. The quasi-geometrical language used in treating the mathematics of the four-dimensional continuum is seldom misleading and often very suggestive and helpful. Finally, without this language Einstein's development of the general theory of relativity would have been seriously hampered.

Although in the remainder of this chapter we shall mainly consider the mathematics of the so-called 'flat' space-time continuum appropriate for the facts of special relativity, the results which can be obtained therefrom are fundamental for the later treatment of the 'curved' space-time of general relativity.

14. The three plus one dimensions of space-time

To appreciate the nature of the space-time continuum it is advisable to introduce at once the language of a conceptual four-dimensional geometry. With the help of this language we can regard space-time as itself corresponding to a hyper-space of four dimensions, which could be provided with mutually perpendicular axes for plotting the values of the four quantities x, y, z, and t that can be used in describing spatial-temporal occurrences. In accordance with this language the position *where* an event occurs and the instant *when* it occurs would both be represented by the location of a single point in the four-dimensional continuum.

In using this language it is important to guard against the fallacy of assuming that all directions in the hyper-space are equivalent, and of assuming that extension in time is of the same nature as extension in space merely because it may be convenient to think of them as plotted along perpendicular axes. A similar fallacy would be to assume that pressure and volume are the same kind of quantity because they are plotted at right angles in the diagram on a pv indicator card. That there must be a difference between the spatial and temporal axes in our hyper-space is made evident, by contrasting the physical possibility of rotating a metre stick from an orientation where it measures distances in the x-direction to one where it measures distances in the y-direction, with the impossibility of rotating it into a direction where it would measure time intervals—in other words the impossibility of rotating a metre stick into a clock.

In accordance with this difference, time should in no sense be considered as the fourth dimension of space, but rather as one, and at that a *unique* one, of the four dimensions of space-time. This distinction is often emphasized by speaking of the space-time continuum as (3+1)-dimensional rather than merely as four-dimensional. The (3+1)-dimensional character of the space-time continuum finds expression at the start in the kind of geometry used, as will be seen in a later section (§ 16).

15. The geometry corresponding to space-time

The geometry chosen as corresponding to the space-time continuum, i.e. the kind of mathematics used, must be appropriate to serve as a means for expressing the conclusions drawn from the two postulates of relativity. As an essential and fundamental element in these conclusions we shall take the invariance with respect to the Lorentz transformation of the expression

$$dx^2 + dy^2 + dz^2 - c^2\,dt^2 \tag{15.1}$$

which was proved in (8.4), and shall choose a geometry which is conceptually constructed to correspond to this invariance.

To do this we shall characterize our geometry by taking†

$$ds^2 = -dx^2 - dy^2 - dz^2 + c^2\,dt^2 \tag{15.2}$$

as the expression for an element of interval in our four-dimensional hyper-space in terms of x, y, z, and t. Since a given element of interval in a space is a conceptual entity which exists independent of any particular choice of axes it is invariant for all transformations of coordinates. Hence the choice of equation (15.2) as our starting-point preserves the desired invariance not only for the group of Lorentz transformations which will leave the right-hand side unchanged in form, but for all possible transformations of coordinates as well. This additional property will be of significance when we come to the consideration of the general theory of relativity.

Since the entire nature of a geometry‡ is known to be determined by the form of its line element, the choice of (15.2) has completely fixed the character of the geometry we are to use; and we may now examine some of its simpler properties and inquire into its actual usefulness for expressing the conclusions of special relativity.

---

† It is of course a mere matter of convention whether we assign the negative sign to the spatial components and the positive sign to the temporal components. We have followed here the more usual practice.

‡ Except for further possible assumptions as to connectivity and the identification of points.

16. The signature of the line element and the three kinds of interval

Examining the expression for the line element (15.2) we note that the quadratic form chosen is characterized by the negative signs of the spatial components dx^2, dy^2, and dz^2 and the positive sign of the temporal component $c^2\,dt^2$. This difference in sign may be regarded as reflecting the difference in the nature of spatial and temporal extension already emphasized above.

Since the signature of the quadratic form—minus two—corresponding to the three negative signs and one positive sign, cannot be changed by any *real* transformation of coordinates, the distinction between spatial and temporal coordinates will always be preserved, and we shall encounter no difficulties in differentiating the time-like coordinate from the others by examining the signs. If we allow an *imaginary* transformation of coordinates the signature of the quadratic form will be changed but the distinction between coordinates can then be determined if we know their real or imaginary correlation with the physical process of counting off division points along the actual axes.

Introducing into (15.2) the imaginary transformation

$$x = i\bar{x} \qquad y = i\bar{y} \qquad z = i\bar{z} \qquad ct = \bar{u}, \tag{16.1}$$

we obtain

$$ds^2 = d\bar{x}^2 + d\bar{y}^2 + d\bar{z}^2 + d\bar{u}^2 \tag{16.2}$$

with a change in signature to plus four. In accordance with this simple form, the geometry used in special relativity is often spoken of as that of a four-dimensional Euclidean (flat) space. The form (16.2) has also been used with the idea of simplifying the mathematical treatment. This procedure, however, introduces really but little simplification together with some chance for confusion, and often necessitates a transformation back to the original coordinates before making physical applications. We shall not find occasion to use it in this book.

Returning now to the original form of the line element

$$ds^2 = -dx^2 - dy^2 - dz^2 + c^2\,dt^2 \tag{16.3}$$

we note, in contrast to geometries where the signature is equal to the number of coordinates, the possibility for more than one kind of interval, depending on the relative magnitude of the spatial and temporal components. In the present case we shall call the interval *space-like, time-like,* or *singular* according as $dx^2 + dy^2 + dz^2$ is respectively *greater than, less than,* or *equal to $c^2 dt^2$.*

In the case of a space-like interval, a Lorentz transformation to so-called proper coordinates can always be found (see § 18) which will reduce the temporal component to zero, so that we can regard the magnitude of a space-like interval as physically determinable by comparison with a suitably moving and oriented metre stick. Similarly, in the case of a time-like interval we have the possibility of determining the magnitude by comparison with a clock. The magnitude of singular intervals is in any case zero.

This possibility for a direct and unique determination of the magnitude of intervals by an appropriate physical measurement is in agreement with their invariance to coordinate transformations. In addition it provides means for the physical interpretation of the geometric results.

17. The Lorentz rotation of axes

In using the geometry corresponding to the space-time continuum, we are of course not limited to any particular system of coordinates x, y, z, and t; but can transform at will to any other set of four coordinates whose functional dependence on the original coordinates is known. Of the various possible transformations, we shall wish to consider for the purposes of special relativity only those which leave the expression for the element of interval in terms of the coordinates

$$ds^2 = -dx^2 - dy^2 - dz^2 + c^2 dt^2 \qquad (17.1)$$

in the same simple form as a sum of squares without cross products, and shall leave the consideration of more general kinds of transformation until it becomes necessary for the purposes of general relativity. Or in more geometrical language, since the flat space-time considered in special relativity makes it possible to use rectangular coordinates in which the expression for the line element preserves the simple form (17.1), there will be no advantage in introducing curvilinear coordinates until we come to the curved space-time of general relativity.

The changes of coordinates which leave the form (17.1) unchanged include: the transformations which can be regarded geometrically as a transfer of origin, such, for example, as would be given by

$$x' = x + x_0 \qquad y' = y \qquad z' = z \qquad t' = t, \qquad (17.2)$$

where x_0 is a constant; the transformations which can be regarded as a spatial rotation of axes, such, for example, as would be given by

$$x' = x\cos\theta + y\sin\theta \qquad y' = y\cos\theta - x\sin\theta \qquad z' = z \qquad t' = t,$$
$$(17.3)$$

where θ is the angle of rotation in the xy-plane; and the Lorentz transformations, which can be regarded as a change in the velocity of the spatial axes, of which we have already had the example,

$$x' = \frac{x - Vt}{\sqrt{(1 - V^2/c^2)}} \qquad y' = y \qquad z' = z \qquad t' = \frac{t - xV/c^2}{\sqrt{(1 - V^2/c^2)}}.$$
$$(17.4)$$

That (17.1) and (17.2) will leave the right-hand side unchanged in form is evident on inspection, and that the transformation (17.4) does not change the form has already been shown by (8.4).

The transformation (17.4) can be expressed in the form

$$x' = x\cosh\phi - ct\sinh\phi \qquad y' = y \qquad z' = z$$
$$ct' = ct\cosh\phi - x\sinh\phi, \qquad (17.5)$$

where $\qquad \phi = \cosh^{-1}\frac{1}{\sqrt{(1 - V^2/c^2)}}.$ $\qquad (17.6)$

On account of the similarity between (17.3) and (17.5) we could speak of the latter as an imaginary rotation in the xt-plane, and use the term Lorentz rotation of axes as descriptive of the Lorentz transformation.

18. The transformation to proper coordinates

Among the different possible Lorentz transformations we shall often be interested in those which will give a change to so-called *proper coordinates* for the particular interval ds in which we may be interested, If the interval is *space-like* in character, the time component will then be zero in proper coordinates, and if it is *time-like* in character the spatial components will be zero in proper coordinates.

This transformation to proper coordinates can always be made. Consider an interval the square of whose magnitude is given in the original coordinates by the expression

$$ds^2 = -dx^2 + c^2\,dt^2, \qquad (18.1)$$

where merely for simplicity a spatial rotation of axes has previously been made, if necessary, to eliminate the y and z components. And consider the transformation equations (9.1) which give us

$$dx' = \frac{dx - (V/c)\, c\, dt}{\sqrt{(1 - V^2/c^2)}}, \qquad (18.2)$$

$$c\, dt' = \frac{c\, dt - (V/c)\, dx}{\sqrt{(1 - V^2/c^2)}}. \qquad (18.3)$$

If the interval (18.1) is space-like in character, the absolute magnitude of dx will be greater than that of $c\, dt$ and we can evidently choose a value of (V/c), less than the possible upper limit of ± 1, which will make (18.3) equal to zero, so that the expression for the interval will reduce to

$$ds^2 = -dx'^2 \qquad (18.4)$$

when we transform to the primed coordinates. On the other hand, if the interval is time-like in character, the absolute magnitude of $c\, dt$ will be greater than that of dx and we can choose a value of (V/c) which will make (18.2) equal to zero, so that the expression for the interval will reduce to

$$ds^2 = dt'^2. \qquad (18.5)$$

In accordance with (18.4) and (18.5) by transforming to proper coordinates, i.e. changing to axes moving with the appropriate velocity, we can determine the value of any space-like interval by direct measurement with a suitably oriented and moving metre stick, and determine the value of any time-like interval by direct measurement with a suitably moving clock. As remarked above this provides a means for the physical interpretation of the mathematical results obtained from the geometry.

19. Use of tensor analysis in the theory of relativity

One of the great advantages of our present quasi-geometrical methods lies in the readiness with which we may now use the language of tensor analysis for the treatment of physical problems. A collection of the formulae of tensor analysis will be found in Appendix III, and in the present section it will be sufficient to consider the definitions from which all the properties of tensors can be derived, and then point out in the next section certain simplifications which can be introduced in the case of the flat space-time of special relativity.

In a space or continuum of four dimensions, corresponding to the four generalized coordinates (x^1, x^2, x^3, x^4), a tensor of rank r can be defined as a collection of 4^r quantities associated with a given point in the continuum, whose values are transformed in accordance with certain definite rules when any new set of coordinates (x'^1, x'^2, x'^3, x'^4) are introduced as functions of the original coordinates by the equations

$$x'^1 = x'^1(x^1, x^2, x^3, x^4)$$
$$x'^2 = x'^2(x^1, x^2, x^3, x^4)$$
$$x'^3 = x'^3(x^1, x^2, x^3, x^4)$$
$$x'^4 = x'^4(x^1, x^2, x^3, x^4).$$

(19.1)

A tensor of *rank zero*, or scalar, S will be defined as a single quantity whose value is unaltered by the transformation of coordinates in accordance with the equation

$$S' = S.$$

(19.2)

A *contravariant* tensor of *rank one*, or vector, A^α will be defined as a collection of four quantities

$$A^\alpha = (A^1, A^2, A^3, A^4),$$

(19.3)

whose values are changed by the transformation of coordinates in accordance with the equation

$$A'^\mu = \frac{\partial x'^\mu}{\partial x^\alpha} A^\alpha,$$

(19.4)

where $(\partial x'^\mu / \partial x^\alpha)$ is the value obtained from (19.1) corresponding to the given point in the continuum, and the double occurrence of the 'dummy' suffix α will be taken to denote a summation over the values $\alpha = 1, 2, 3, 4$. And a *covariant* tensor of *rank one* B_α will be defined as a collection of four quantities

$$B_\alpha = (B_1, B_2, B_3, B_4),$$

(19.5)

whose values are transformed in accordance with

$$B'_\mu = \frac{\partial x^\alpha}{\partial x'^\mu} B_\alpha.$$

(19.6)

A *contravariant* tensor of *rank two* $T^{\alpha\beta}$ will be defined as a collection of sixteen quantities

$$T^{\alpha\beta} = \begin{matrix} T^{11} & T^{12} & T^{13} & T^{14} \\ T^{21} & T^{22} & T^{23} & T^{24} \\ T^{31} & T^{32} & T^{33} & T^{34} \\ T^{41} & T^{42} & T^{43} & T^{44}, \end{matrix}$$

(19.7)

whose values are transformed in accordance with

$$T'^{\mu\nu} = \frac{\partial x'^\mu}{\partial x^\alpha} \frac{\partial x'^\nu}{\partial x^\beta} T^{\alpha\beta}.$$

(19.8)

And a *covariant* tensor of the same rank $S_{\alpha\beta}$ will be defined as a collection of sixteen quantities which are transformed in accordance with

$$S'_{\mu\nu} = \frac{\partial x^\alpha}{\partial x'^\mu} \frac{\partial x^\beta}{\partial x'^\nu} S_{\alpha\beta}. \tag{19.9}$$

Tensors of mixed contravariant and covariant nature or of higher rank can be similarly defined in accordance with the general expression

$$T'^{\mu\nu\ldots}_{\rho\sigma\ldots} = \frac{\partial x'^\mu}{\partial x^\alpha} \frac{\partial x'^\nu}{\partial x^\beta} \frac{\partial x^\delta}{\partial x'^\rho} \frac{\partial x^\epsilon}{\partial x'^\sigma} \ldots T^{\alpha\beta\ldots}_{\delta\epsilon\ldots} \tag{19.10}$$

The double occurrence of dummy suffixes in a given term of a tensor expression will always be taken to denote summation over the four values $1, 2, 3, 4$. Scalars are not necessarily to be regarded as located at any given point in the continuum, but tensors of higher rank must in general be thought of as associated with some given point, since the transformation factors $(\partial x'^\mu/\partial x^\alpha)$ etc. will in general be different at different points in the continuum. Tensor fields may of course be constructed, in which a value of the field tensor is associated with each point in the continuum.

In case the continuum has the *metrical properties* afforded by an expression

$$ds^2 = g_{\mu\nu} dx^\mu dx^\nu \qquad (g_{\mu\nu} = g_{\nu\mu}) \tag{19.11}$$

for the scalar measure of the element of interval ds corresponding to the infinitesimal vector dx^μ, the fundamental metrical tensor $g_{\mu\nu}$ will be of special importance in the analysis. With it are associated the quantity g (not a scalar) which is defined as the determinant

$$g = |g_{\mu\nu}| \tag{19.12}$$

and the contravariant tensor $g^{\mu\nu}$ which is defined as the normalized minor of $g_{\mu\nu}$

$$g^{\mu\nu} = \frac{|g_{\mu\nu}|\,\text{minor}}{g}. \tag{19.13}$$

With the help of these two fundamental tensors we may now *define* the method of raising and lowering indices, so as to obtain *associated* tensors of different degrees of covariance or contravariance, as given by the equations

$$T^{\ldots\alpha\ldots}_{\ldots\ldots\ldots} = g^{\alpha\beta} T^{\ldots\ldots\ldots}_{\ldots\beta\ldots}$$

and

$$T^{\ldots\ldots\ldots}_{\ldots\alpha\ldots} = g_{\alpha\beta} T^{\ldots\beta\ldots}_{\ldots\ldots\ldots} \tag{19.14}$$

This completes the definitions necessary as a basis for tensor analysis and all further properties of tensors and rules of analysis can be obtained therefrom. Thus all the methods given in Appendix III for operating on tensors to obtain new tensors by addition, multiplication, contraction, covariant differentiation, etc., can all of them be verified by showing that the result obtained has components which transform on change of coordinates in accordance with the rules of transformation by which tensors were defined above.

The great advantages of tensor analysis as a tool for mathematical physics arise in two ways. In the first place it gives a very condensed and convenient language for the expression of physical laws. Thus the single tensor equation

$$R^\tau_{\mu\nu\sigma} = 0 \qquad (19.15)$$

is itself a representation of the 256 different equations that are obtained by assigning the different values $1, 2, 3, 4$ to μ, ν, σ, and τ, and results may be obtained with the help of tensor analysis which would be extremely hard to calculate by 'long-hand' methods. In the second place the expression of a physical law by a tensor equation has exactly the same form in *all* coordinate systems, since it is readily seen from the general transformation rule (19.10) that any tensor equation

$$T^{\mu\nu\cdots}_{\rho\sigma\cdots} = 0 \qquad (19.16)$$

will be changed into an expression of just the same form

$$T'^{\mu\nu\cdots}_{\rho\sigma\cdots} = 0 \qquad (19.17)$$

when the coordinates are transformed from (x^1, x^2, x^3, x^4) to (x'^1, x'^2, x'^3, x'^4). The relations of this very convenient property to the postulates of the special and general theories of relativity will be more closely considered in § 21 and in § 73.

20. Simplification of tensor analysis in the case of special relativity. Galilean coordinates

In the case of the flat space-time continuum of the special theory of relativity, certain simplifications in the use of tensor analysis are possible since in accordance with (15.2) we can then reduce the general expression for the element of interval (19.11) to the specially simple form

$$ds^2 = -(dx^1)^2 - (dx^2)^2 - (dx^3)^2 + (dx^4)^2, \qquad (20.1)$$

provided we introduce so-called *Galilean coordinates* defined in terms of our previous spatial and temporal variables (x, y, z, t) by the equations

$$x^1 = x \qquad x^2 = y \qquad x^3 = z \qquad x^4 = ct. \qquad (20.2)$$

In terms of these new coordinates the Lorentz transformation (17.4) corresponding to the change to a new set of spatial axes moving relative to the original ones in the x-direction with the velocity V can be written in the form

$$x'^1 = \frac{x^1 - x^4 V/c}{\sqrt{(1 - V^2/c^2)}} \qquad x'^2 = x^2 \qquad x'^3 = x^3 \qquad x'^4 = \frac{x^4 - x^1 V/c}{\sqrt{(1 - V^2/c^2)}}$$

$$(20.3)$$

and the values for the factors $(\partial x'^\mu/\partial x^\alpha)$ etc. used in accordance with (19.10) for the transformation of tensors from the one system of coordinates to the other reduce for this simple case to

$$\frac{\partial x'^1}{\partial x^1} = \frac{\partial x'^4}{\partial x^4} = \frac{1}{\sqrt{(1 - V^2/c^2)}}$$

$$\frac{\partial x'^1}{\partial x^4} = \frac{\partial x'^4}{\partial x^1} = -\frac{V/c}{\sqrt{(1 - V^2/c^2)}} \qquad (20.4)$$

$$\frac{\partial x'^2}{\partial x^2} = \frac{\partial x'^3}{\partial x^3} = 1,$$

with all others zero.

Furthermore, when using the Galilean coordinates (20.2) appropriate to special relativity, it should be noted that the Lorentz contraction factor $\sqrt{(1 - u^2/c^2)}$, corresponding to a point moving with the velocity u, is given in accordance with (20.1) by the very simple expression

$$\sqrt{\left(1 - \frac{u^2}{c^2}\right)} = \frac{ds}{dx^4}, \qquad (20.5)$$

where the time-like interval ds is an element of the four-dimensional trajectory of the moving point.

In addition, in the case of special relativity, since the metrical tensor corresponding to the formula for the interval (20.1) has the simple Galilean values

$$g_{11} = g_{22} = g_{33} = -1 \qquad g_{44} = 1 \qquad (20.6)$$

$$g_{\mu\nu} = 0 \quad (\mu \neq \nu)$$

the raising and lowering of suffixes in accordance with (19.14) will, in the case of the coordinates (20.2), result only in a change of sign for certain of the components. Thus it will be found on applying the rules that the associated vectors A_μ and A^μ are connected by the simple relations

$$A_i = -A^i \quad (i = 1, 2, 3) \qquad A_4 = A^4 \qquad (20.7)$$

and the associated tensors $T_{\mu\nu}$ and $T^{\mu\nu}$ are connected by the relations

$$T_{\mu\nu} = T^{\mu\nu} \text{ (except for } T_{i4} = -T^{i4} \text{ and } T_{4i} = -T^{4i}; i = 1, 2, 3)$$

$$(20.8)$$

Finally, in the case of special relativity, it should be noted that several tensor operations are much simplified when the coordinates (20.2) are used. Thus the process of constructing a new tensor by covariant differentiation as given by equation (33) in Appendix III takes a very simple form in these coordinates, and we can write for example for the covariant derivative of $T^{\mu\nu}$

$$(T^{\mu\nu})_\alpha = \frac{\partial T^{\mu\nu}}{\partial x^\alpha}. \tag{20.9}$$

Similarly for the divergence or contracted covariant derivative we can write

$$(T^{\mu\nu})_\nu = \frac{\partial T^{\mu\nu}}{\partial x^\nu} \tag{20.10}$$

instead of the complicated expressions that would be necessary in more general coordinates.

These simplifications in tensor analysis for special relativity are of considerable convenience.

21. Correspondence of four-dimensional treatment with the postulates of special relativity

To complete our consideration of the geometrical four-dimensional method of treating the special theory of relativity, we must now point out its correspondence with the two postulates of the special theory. This is an extremely simple matter.

In accordance with the discussion of § 5 the *first postulate* of special relativity will be satisfied if the laws of physics, in the absence of gravitational action, are the same for all observers in uniform relative motion. This, however, can be achieved with our present methods if we can state these laws in the form of tensor equations, using therein tensors whose components have the same physical significance for all systems of coordinates that correspond to different sets of Cartesian axes in uniform relative motion. Since tensor equations if true in one system of coordinates are true in all systems of coordinates (see 19.16, 17), we shall then obtain the desired correspondence with the first postulate, provided of course that our tensors have the character stated.

The actual problem of constructing tensors whose components have the same physical significance in different systems of coordinates, corresponding to sets of axes in uniform relative motion, can be met in three different ways. In the first method of proceeding we *define* the tensor by stating the physical quantities which are to be taken as the components of the tensor in question referred to an *arbitrary* set of coordinates as given by (20.2), and then show by actually performing the Lorentz transformation (20.3) that the components are transformed to the corresponding physical quantities referred to other systems of coordinates. In the second method of proceeding we *define* the tensor by stating the physical quantities which are to be taken as the components of the tensor in question referred to *proper* coordinates with respect to which the material to which the tensor applies is at rest; on account of the unique position of proper coordinates this will of course assure the same physical significance for the components in coordinates corresponding to different states of motion. In the third method of proceeding we construct the tensor of interest by the rules of tensor manipulation from simpler ones whose physical significance in different sets of coordinates is already known. As simple examples of such tensors, which may be used for constructing further tensors, we have the scalar element of interval ds, the contravariant vector corresponding to a small coordinate displacement dx^μ, and the contravariant vectors of generalized 'velocity' and 'acceleration' dx^μ/ds and d^2x^μ/ds^2, where ds is the time-like interval which is an element of the four-dimensional trajectory of a moving point.

The correspondence of our four-dimensional method with the *second postulate* of special relativity is even simpler. In accordance with this postulate the velocity of light in free space must measure the same for different observers in uniform relative motion, and this result is secured by the way in which we originally defined the character of the space-time continuum for special relativity in § 15.

In accordance with (15.2) the element of interval in this continuum, using a given system of ordinary spatial and temporal coordinates (x, y, z, t), is given by

$$ds^2 = -dx^2 - dy^2 - dz^2 + c^2\, dt^2 \qquad (21.1)$$

and the four-dimensional trajectory of a light impulse, travelling with the velocity c, will hence be characterized by taking the value

$$ds = 0 \qquad (21.2)$$

for any element of the trajectory, since the substitution of (21.2) in (21.1) at once leads to the relation

$$\left(\frac{dx}{dt}\right)^2 + \left(\frac{dy}{dt}\right)^2 + \left(\frac{dz}{dt}\right)^2 = c^2. \qquad (21.3)$$

If, however, we now transform to any other system of coordinates (x', y', z', t'), corresponding to a new set of axes in uniform motion relative to original ones, we know that the *form of expression* for the interval will still be the same on account of the nature of the Lorentz transformation, and that the *value* of the interval will still be the same on account of the scalar character of ds. Hence also in these new coordinates the velocity of light will be given by

$$\left(\frac{dx'}{dt'}\right)^2 + \left(\frac{dy'}{dt'}\right)^2 + \left(\frac{dz'}{dt'}\right)^2 = c^2 \tag{21.4}$$

as is required by the second postulate.

Our four-dimensional geometry has thus provided us with a very useful language for treating the facts of special relativity, which we shall not hesitate to use whenever it proves more convenient than the older language. In addition it is a language which is almost indispensable for the treatment of general relativity.

III

SPECIAL RELATIVITY AND MECHANICS

Part I. THE DYNAMICS OF A PARTICLE

22. The principles of the conservation of mass and momentum

We must now consider the effect of the special theory of relativity in modifying the older Newtonian mechanics. We shall first treat the mechanics of particles, sufficiently for our later needs, and then consider in Part II the dynamics of a continuous mechanical medium.

As a postulatory basis for the mechanics of interacting particles we may take the two principles of the conservation of mass and momentum, in conjunction with the foregoing kinematical results of special relativity.

In accordance with these two conservation laws, the total mass of a system of particles must remain constant as the particles act on each other in agreement with the equation

$$\sum m = \text{const.}, \qquad (22.1)$$

where the summation $\sum$ is to be taken over the masses m of all the particles in the system, and the components of the total momentum of the system in the x, y, and z directions must also remain constant in agreement with the equations

$$\sum mu_x = \text{const.},$$
$$\sum mu_y = \text{const.}, \qquad (22.2)$$
$$\sum mu_z = \text{const.},$$

where the summations are to be taken over the components of momenta of all the individual particles.† And in accordance with the principles of relativity these equations must hold true in all sets of coordinates in uniform relative motion.

Since the Newtonian system of mechanics also included the ideas of the relativity of motion and of the conservation of mass and momentum, equations (22.1) and (22.2) would also hold in Newtonian theory in all sets of coordinates in uniform motion. There is nevertheless an important difference between Newtonian and relativistic

---

† Our present considerations apply to systems of particles which could interact only by collision. We are not yet concerned with more complicated systems where a continuous distribution of mass and momentum might have to be assigned to the field.

mechanics owing to the difference in the transformation equations which would be applicable in changing from one set of moving coordinates to another. In Newtonian mechanics we should use the simple Galilean transformation equations (8.7) and should find it possible to satisfy equations(22.1) and (22.2) in all systems of coordinates on the assumption that the mass of a particle is a constant independent of its velocity. In relativistic mechanics, however, we must use the more complicated Lorentz transformation equations (8.1), and shall then find it possible to satisfy equations (22.1) and (22.2) only on the assumption that the mass of a particle depends on its velocity, as will be shown in the next section.

23. The mass of a moving particle

In order to show that the mass of a particle must depend on its velocity, if the conservation laws are to hold in all systems of coordinates, we shall first consider the conservation of mass and momentum, in two different systems of coordinates S' and S, for the case of a very simple head-on collision between two similar elastic particles.

In the first system of coordinates, for convenience the primed system S', let the two particles be moving before collision with the velocities $+u'$ and $-u'$ parallel to the x-axis in such a way that a head-on encounter can occur. Since by hypothesis the two particles are perfectly similar and elastic, it is evident that they will first be brought to rest on collision and then rebound under the action of the elastic forces developed, moving back over their original paths with the respective velocities $-u'$ and $+u'$ of the same magnitude as before but reversed in direction. In this system of coordinates the collision is obviously such as to satisfy the conservation laws of mass and momentum.

Let us now change to a second system of coordinates S moving relative to the first in the x-direction with the velocity $-V$. Using this new system of coordinates, let us denote by u_1 and u_2 the velocities of the two particles before collision, and allowing for the possibility that mass may depend on velocity let us denote by m_1 and m_2 the masses of the two particles before collision. Furthermore, let us denote by M the sum of the masses of the two particles at the instant in the course of the collision when they have come to relative rest, and are hence both moving with the velocity $+V$ with respect to our present system of coordinates, S.

In accordance with the conservation laws, which must also hold in this new system of coordinates, the total mass and total momentum of the two particles must be the same before collision and at the instant of relative rest, so that we can evidently write

$$m_1 + m_2 = M \qquad (23.1)$$

and

$$m_1 u_1 + m_2 u_2 = MV. \qquad (23.2)$$

In addition, however, using the transformation equation for velocity given by (10.2) we can write for the velocities u_1 and u_2, in terms of their values $+u'$ and $-u'$ with respect to the original coordinates S, the expressions

$$u_1 = \frac{u'+V}{1+u'V/c^2} \quad \text{and} \quad u_2 = \frac{-u'+V}{1-u'V/c^2}. \qquad (23.3)$$

And by combining these three equations and solving for the ratio of the two masses, we easily obtain

$$\frac{m_1}{m_2} = \frac{1+u'V/c^2}{1-u'V/c^2}, \qquad (23.4)$$

which with the help of the transformation equation (11.1) gives us

$$\frac{m_1}{m_2} = \frac{\sqrt{(1-u_2^2/c^2)}}{\sqrt{(1-u_1^2/c^2)}}. \qquad (23.5)$$

In accordance with this result the masses of the two particles, which by hypothesis have the same value, say m_0, when at rest, become inversely proportional to $\sqrt{(1-u^2/c^2)}$ when moving with the velocity u, so that we may now write

$$m = \frac{m_0}{\sqrt{(1-u^2/c^2)}} \qquad (23.6)$$

as the desired expression for the mass m of a moving particle in terms of its velocity u and mass at rest m_0.

Although this derivation[†] of the expression for the mass of a moving particle depends on the consideration of a simple type of head-on collision for the two particles, it can also be shown quite easily, nevertheless, that the same expression is also directly obtained from the consideration of a glancing transverse collision,[‡] and in addition that the expression with u taken as the total velocity is sufficient to secure the conservation of mass and momentum in all systems of coordinates for any kind of collision between two particles.[§]

† Tolman, *Phil. Mag.* **23**, 375 (1912).
‡ Lewis and Tolman, *Phil. Mag.* **18**, 510 (1909).   § Tolman, loc. cit.

We have, moreover, of course the experimental verification of the expression in the case of the mass of moving electrons to which we shall call attention in § 29. We shall hence have no hesitation in accepting the expression as correct in general for the mass of a moving particle.

It is of interest to note in accordance with (23.6) that the mass of a particle would become infinite at the velocity of light. This is an agreement with our previous findings in § 10 that the velocity of light is to be regarded as an upper limit of possible velocities.

It may also be remarked in concluding the present section, that our discussion already indicates that we shall have to ascribe mass to the potential energy of elastic deformation, in order to retain the conservation laws of mass and momentum. This is evident from the fact that the foregoing equations for the head-on collision lead to the result

$$M > \frac{2m_0}{\sqrt{(1 - V^2/c^2)}}, \tag{23.7}$$

which shows that the total mass of the two particles at the instant during the course of the collision when they have come to relative rest is greater than would be calculated from their velocity V and total undeformed rest-mass $2m_0$.

24. The transformation equations for mass

In accordance with equation (23.6) the mass of a given particle will measure differently in different sets of coordinates since the velocity will be different. From the transformation equation for the factor $\sqrt{(1 - u^2/c^2)}$ given by (11.1) we easily obtain for the transformation of masses the result

$$m = m' \frac{(1 + u'_x V/c^2)}{\sqrt{(1 - V^2/c^2)}}. \tag{24.1}$$

And by differentiating with respect to the time and simplifying we obtain

$$\frac{dm}{dt} = \frac{dm'}{dt'} + \frac{m'V}{c^2}(1 + u'_x V/c^2)^{-1}\frac{du'_x}{dt'} \tag{24.2}$$

as a transformation equation for the rate at which the mass of a particle is changing owing to change in velocity.

25. The definition and transformation equations for force

Since the mass of a moving particle will change with its velocity, it is no longer possible as in Newtonian mechanics to define force *both* as mass times acceleration *and* as rate of change of momentum. It proves to be most convenient to take the latter definition, since the principle of the equality of action and reaction for forces then becomes identical with the principle of the conservation of momentum which we took as an axiom.

We shall hence write as the equation of definition for the force **F** acting on a particle of mass m and velocity **u** the vector expression†

$$\mathbf{F} = \frac{d}{dt}(m\mathbf{u}) = \frac{d}{dt}\left(\frac{m_0\mathbf{u}}{\sqrt{(1-u^2/c^2)}}\right), \qquad (25.1)$$

or in scalar form

$$F_x = \frac{d}{dt}(m\dot{x}) = \frac{d}{dt}\left(\frac{m_0 u_x}{\sqrt{(1-u^2/c^2)}}\right),$$

$$F_y = \frac{d}{dt}(m\dot{y}) = \frac{d}{dt}\left(\frac{m_0 u_y}{\sqrt{(1-u^2/c^2)}}\right), \qquad (25.2)$$

$$F_z = \frac{d}{dt}(m\dot{z}) = \frac{d}{dt}\left(\frac{m_0 u_z}{\sqrt{(1-u^2/c^2)}}\right).$$

It will be noted in accordance with this definition that in general force and acceleration will not be in the same direction as was the case in Newtonian mechanics. The advantages of the definition are, however, very great, not only because it preserves the principle of the equality of action and reaction but because it also can be shown to simplify the interpretation of electromagnetic phenomena (see for example § 29).

Since we have already obtained transformation equations (10.2) (12.1) (24.1) (24.2) for all the quantities occurring on the right-hand side of (25.2) we can now also readily obtain transformation equations for the components of force which can be written in the form

$$F_x = F'_x + \frac{u'_y V}{c^2 + u'_x V}F'_y + \frac{u'_z V}{c^2 + u'_x V}F_z,$$

$$F_y = \frac{c^2\sqrt{(1-V^2/c^2)}}{c^2 + u'_x V}F'_y, \qquad (25.3)$$

$$F_z = \frac{c^2\sqrt{(1-V^2/c^2)}}{c^2 + u'_x V}F'_z.$$

These transformation equations have been derived for the particular case of the forces acting on a particle to change its state of motion.

---

† Note the inclusion of m_0 inside the bracket which is to be differentiated. This makes the expression applicable also in cases where the proper mass of the particle varies, as it might, for example, from an inflow of heat.

Nevertheless, it is to be noted that particles can have their state of motion changed not only by interaction with each other, but by interaction with other larger mechanical systems or with electromagnetic systems as well. Hence, since we shall wish to retain the equality of action and reaction and thus the conservation of momentum in all branches of physics, it is evident that these same transformation equations must hold for all kinds of forces and all kinds of systems on which they may act. This conclusion will be of great importance in extending our system of dynamics to include the mechanics of a continuous medium.

26. Work and kinetic energy

As in the older mechanics we shall find it convenient to define the work done on a particle as equal to the force acting multiplied by the distance through which the particle is displaced in the direction of the action, as given by the equation

$$dW = \mathbf{F} \cdot d\mathbf{r} \tag{26.1}$$

where $\mathbf{r}$ is the radius vector determining the position of the particle. We shall also define the energy given to a particle by the action of a force as equal to the work done on it.

In case we do work on a free particle we can easily evaluate its increase in kinetic energy in terms of change in velocity. Introducing into (26.1) the expression for force given by (25.1), we can write for the increase in kinetic energy

$$dE = m\frac{d\mathbf{u}}{dt} \cdot d\mathbf{r} + \frac{dm}{dt}\mathbf{u} \cdot d\mathbf{r}$$
$$= m\mathbf{u} \cdot d\mathbf{u} + \mathbf{u} \cdot \mathbf{u}\, dm$$
$$= mu\, du + u^2\, dm. \tag{26.2}$$

And substituting the expression for mass as a function of velocity given by (23.6) this becomes

$$dE = \frac{m_0 u}{(1-u^2/c^2)^{\frac{1}{2}}}\, du + \frac{m_0 u^3/c^2}{(1-u^2/c^2)^{\frac{3}{2}}}\, du$$
$$= \frac{m_0 u\, du}{(1-u^2/c^2)^{\frac{3}{2}}}. \tag{26.3}$$

We thus see, just as in Newtonian mechanics, that the kinetic energy given to a particle is solely a function of its change in velocity independent of the particular way in which this change is brought about. Furthermore, in accordance with equation (26.1) and the principle of the equality of action and reaction, it is evident when two particles interact by elastic collision that the increase in kinetic energy of the one will be equal to the decrease in kinetic energy of the other, so that we shall also have in relativistic mechanics an analogue of the older principle of the conservation of *vis-viva* for elastic encounters.

Integrating expression (27.3) from zero to u, we obtain for the total kinetic energy of a particle of rest-mass m_0 moving with the velocity u

$$E = \frac{m_0 c^2}{\sqrt{(1-u^2/c^2)}} - m_0 c^2, \tag{26.4}$$

which reduces at velocities small compared with that of light, as would be expected, to the familiar Newtonian expression

$$E = \tfrac{1}{2} m_0 u^2. \tag{26.5}$$

27. The relations between mass, energy, and momentum

We must now consider a very important relation between mass and energy which was quite unknown to the Newtonian mechanics. In accordance with §§ 23 and 26, the mass and energy of a particle are both dependent on the velocity and increase with it. And if we substitute the expression for mass as a function of velocity given by (23.6) into the expressions (26.3) and (26.4) for increase in kinetic energy and total kinetic energy, we easily obtain the remarkably simple relation

$$dE = c^2 \, dm \tag{27.1}$$

for the increase in the kinetic energy of a particle in terms of its increase in mass, together with

$$E = c^2(m - m_0) \tag{27.2}$$

for its total kinetic energy in terms of the increase in mass of the particle over that which it has at rest. In accordance with these equations the change in kinetic energy in ergs is equal to the change in mass in grammes multiplied by the square of the velocity of light in centimetres per second.

We must now investigate the implied and suggested consequences of this remarkable proportionality between increased mass and kinetic energy. Since we shall take the principle of the conservation of mass not only as a fundamental postulate for a system of particles but for systems in general as well, this proportionality between increased mass and kinetic energy immediately implies in general that any isolated system will always retain the same possibility of furnishing kinetic energy, without any alteration as to the theoretical amount available, although perhaps with some change as to the readiness of availability. *Hence we can regard the principle of the conservation of mass as itself guaranteeing the principle of the conservation of energy.*

Furthermore, the proportionality between kinetic energy and increase in mass, together with the principles of the conservation of both mass and energy, immediately suggests that energy in any form always has the corresponding amount of mass immediately associated with it. Thus, for example, when a moving particle is brought to rest and hence loses both its increased mass $(m-m_0)$ and kinetic energy $c^2(m-m_0)$, it seems reasonable to assume that this mass and energy, which are associated together in the moving particle and which leave it in association when the particle loses its motion, will still remain always associated together. Indeed if the particle is brought to rest by elastic transfer of energy to other particles, as in the case of viscous forces arising from collisions with hypothetical elastic molecules, the considerations of § 26 are sufficient to show that the mass and energy do pass on in association to other particles. And in addition we have already seen in § 23 that we must ascribe mass to the potential energy generated during the course of an elastic collision (see 23.7). *Hence in what follows we shall postulate in general that a quantity of energy E always has immediately associated with it a mass m of the amount*

$$m = \frac{E}{c^2}. \tag{27.3}$$

In addition as a further consequence of the association of mass with any given quantity of energy, as given by equation (27.3), it would also appear natural to assume the reciprocal relation of an association of energy with any given quantity of mass. *This we shall do in what follows by postulating the relation*

$$E = mc^2 \tag{27.4}$$

for the energy E associated with a mass of any kind m. This relation, which would imply an enormous store of energy $m_0 c^2$ still resident in a particle even when it is brought to rest, appears somewhat more strained than our previous considerations, but nevertheless logically plausible.

Finally, as an important consequence of this association of mass and energy, it is evident that the transfer of energy will necessarily involve the presence of momentum. For example, if we have a quantity of energy E which is being bodily transferred with the velocity u we can write for the associated momentum

$$\mathbf{G} = m\mathbf{u} = \frac{E}{c^2}\,\mathbf{u}. \qquad (27.5)$$

In addition to the transfer of energy by the bodily motion of the system containing it, we shall also wish to allow, however, for the transfer of energy when forces do work on a moving system. Thus if we consider a rod moving parallel to its length with forces acting on the two ends in such a way that work is done on the rear end and delivered at the forward end, it is evident that in addition to the transfer of the energy content of the rod by its forward motion, there is a further flow of energy down the rod because of the action of these forces. *In order to allow for the momentum associated with all forms of energy transfer we shall then write*

$$\mathbf{g} = \frac{\mathbf{s}}{c^2} \qquad (27.6)$$

as a general relation between density of momentum **g** *and density of energy flow* s. This expression contains no restrictions as to the mechanism of the energy transfer and will be fundamental for our later work.

28. Four-dimensional expression of the mechanics of a particle

The foregoing discussion contains all the underlying principles that are necessary for treating the mechanics of a particle, and we may now show the simplicity with which they can be expressed with the help of the four-dimensional language developed at the end of the preceding chapter.

Returning to our fundamental idea of a four-dimensional space-time continuum, we can write, in accordance with (20.1),

$$ds^2 = -(dx^1)^2 - (dx^2)^2 - (dx^3)^2 + (dx^4)^2 \qquad (28.1)$$

as an expression for an infinitesimal line element ds in this continuum in terms of the rectangular so-called Galilean coordinates

$$x^1 = x \quad x^2 = y \quad x^3 = z \quad x^4 = ct, \qquad (28.2)$$

and may then define the four-dimensional 'momentum' of a particle as the product of its rest-mass m_0 and its four-dimensional 'velocity' dx^μ/ds

$$m_0 \frac{dx^\mu}{ds} = \left(m_0 \frac{dx^1}{ds}, \ m_0 \frac{dx^2}{ds}, \ m_0 \frac{dx^3}{ds}, \ m_0 \frac{dx^4}{ds} \right). \tag{28.3}$$

Working out expressions for the four components of this vector, however, in terms of our usual coordinates x, y, z, and t as given by (28.2), we easily obtain

$$m_0 \frac{dx^1}{ds} = \frac{m_0}{c\sqrt{(1-u^2/c^2)}} \frac{dx}{dt},$$

$$m_0 \frac{dx^2}{ds} = \frac{m_0}{c\sqrt{(1-u^2/c^2)}} \frac{dy}{dt},$$

$$m_0 \frac{dx^3}{ds} = \frac{m_0}{c\sqrt{(1-u^2/c^2)}} \frac{dz}{dt}, \tag{28.4}$$

$$m_0 \frac{dx^4}{ds} = \frac{m_0}{\sqrt{(1-u^2/c^2)}},$$

where
$$u^2 = \left(\frac{dx}{dt}\right)^2 + \left(\frac{dy}{dt}\right)^2 + \left(\frac{dz}{dt}\right)^2. \tag{28.5}$$

Hence we see at once that our fundamental principles of the conservation of the components of momentum $m_0 u_x/\sqrt{(1-u^2/c^2)}$, etc., of mass $m_0/\sqrt{(1-u^2/c^2)}$, and of energy $m_0 c^2/\sqrt{(1-u^2/c^2)}$ can all of them be obtained for interacting particles by the simple requirement

$$\sum m_0 \frac{dx^\mu}{ds} = \text{const.}, \tag{28.6}$$

where the summation $\sum$ is to be taken over all the particles of the system. This expression is *not* a tensor equation, since the left-hand side is a sum of vectors taken at different points in space-time (see § 19), and the right-hand side is not even a tensor in form. The equation is valid, however, for the particular kind of coordinates (28.2), and illustrates, moreover, the condensation which can be achieved with the help of four-dimensional language.

If, nevertheless, we consider a *single particle in free space* unacted on by other bodies we can obtain a very simple and important tensor equation to describe its motion. For such a particle, it is evident from (28.6) that the motion will be given by

$$\frac{dx^\mu}{ds} = \text{const.} \tag{28.7}$$

In the rectangular coordinates (28.2) being used this is, however, the equation for a straight line or geodesic. And this result can now be expressed in the general form

$$\delta \int ds = 0, \tag{28.8}$$

which is a tensor (scalar) equation, valid in all systems of coordinates.

This result that the four-dimensional track of a *free particle* is a geodesic will be very important when we come to the general theory of relativity. In the case of a *ray of light*, we can take the track as being not only a geodesic, but with the additional restriction

$$ds = 0 \tag{28.9}$$

already discussed in § 21.

Also in the case of a *particle acted on by a force* we can make use of tensor language by considering a contravariant vector F^μ, the so-called Minkowski force, which can be defined by the equation

$$F^\mu = c^2 \frac{d}{ds}\left(m_0 \frac{dx^\mu}{ds}\right), \tag{28.10}$$

where m_0 is the proper mass of the particle as measured by a local observer, dx^μ/ds is its generalized velocity, and the differentiation $d(\ )/ds$ with respect to its four-dimensional trajectory is purposely taken so as to include possible changes in the proper mass m_0 of the particle due, for example, to the generation of heat within it.

The above expression is to be regarded as a tensor equation defining F^μ in all systems of coordinates. In the particular kind of coordinates given by (28.2) it is easy to calculate for the individual components the values

$$F^1 = \frac{1}{\sqrt{(1-u^2/c^2)}} \frac{d}{dt}\left(\frac{m_0 u_x}{\sqrt{(1-u^2/c^2)}}\right),$$

$$F^2 = \frac{1}{\sqrt{(1-u^2/c^2)}} \frac{d}{dt}\left(\frac{m_0 u_y}{\sqrt{(1-u^2/c^2)}}\right),$$

$$F^3 = \frac{1}{\sqrt{(1-u^2/c^2)}} \frac{d}{dt}\left(\frac{m_0 u_z}{\sqrt{(1-u^2/c^2)}}\right),$$

$$F^4 = \frac{1}{\sqrt{(1-u^2/c^2)}} \frac{d}{dt}\left(\frac{m_0 c^2}{c\sqrt{(1-u^2/c^2)}}\right),$$

$$\tag{28.11}$$

where u is the ordinary velocity of the particle. Hence, remembering the expressions for the ordinary components of force given by (25.2) and the relation between mass and energy given by (27.4) we can now write for the components of F^μ in our present coordinates the simple expressions

$$F^\mu = \left(\frac{F_x}{\sqrt{(1-u^2/c^2)}}, \frac{F_y}{\sqrt{(1-u^2/c^2)}}, \frac{F_z}{\sqrt{(1-u^2/c^2)}}, \frac{1}{\sqrt{(1-u^2/c^2)}} \frac{dE}{c\,dt} \right).$$

$$(28.12)$$

The possibility thus demonstrated of using the components of ordinary force to construct a four-dimensional vector proves to be quite useful. In accordance with the discussion of § 25, forces of any origin whatever must all obey the same transformation laws, and will hence all share in this demonstrated property. The knowledge thus provided as to the nature of forces can be very helpful, especially when further information may be lacking. [See § 54 (c).]

29. Applications of the dynamics of a particle

This completes the development of the dynamics of a particle as far as will be needed for our later considerations. The results are to be accepted not only on the basis of the experimental verification which they have received in those cases where it has been possible to test differences between the predictions of relativistic and New-tonian mechanics, but also on the basis of the inner logicality of the theory which has led to them and the harmony of this theory with the rest of physics. The achievement of this logicality and harmony depends on the reconciliation of so many factors that we can feel con-siderable confidence in accepting results of the theory when necessary prior to their experimental verification.

To conclude the treatment we may now briefly consider a few applications of the dynamics of a particle which will illustrate both the contact of the theory with experiment, and the logicality and harmony mentioned above.

(a) **The mass of high-velocity electrons.** The increase in the mass of a particle with increasing velocity, which was obtained in § 23, is fundamental for relativistic mechanics and forms the basis which implies or suggests the further development. For this reason it is specially satisfactory that the expression $m_0/\sqrt{(1-u^2/c^2)}$ given by (23.6) for the mass of a moving particle has now received good experimental verification in the considerable number of measure-ments which have been made on high-velocity β and cathode particles, since the original more or less qualitative discovery by Kaufmann of a dependence of mass on velocity. A fairly recent description and critique of these measurements will be found in the *Handbuch der Physik*.[†]

(b) **The relation between force and acceleration.** As already noted in § 25, if we define the force acting on a particle as its rate of change of momentum as given by the equation

$$\mathbf{F} = \frac{d}{dt}(m\mathbf{u}) = m\frac{d\mathbf{u}}{dt} + \mathbf{u}\frac{dm}{dt} = \frac{m_0}{\sqrt{(1-u^2/c^2)}}\frac{d\mathbf{u}}{dt} + \frac{d}{dt}\left\{\frac{m_0}{\sqrt{(1-u^2/c^2)}}\right\}\mathbf{u}$$

(29.1)

it is evident in relativistic mechanics that the force $\mathbf{F}$ and acceleration $d\mathbf{u}/dt$ will in general not be in the same direction, as was the case in Newtonian mechanics.

The resolution of the force into components parallel to the acceleration and parallel to the velocity, as given by (29.1), makes the reason for this changed state of affairs immediately apparent. Since the acceleration itself will in general lead to a change in the mass of the particle, we must expect a change in momentum in the direction of the already existing velocity $\mathbf{u}$ as well as in the direction of the acceleration $d\mathbf{u}/dt$. Hence components of force will be needed in general both in the direction of the acceleration and of the existing velocity.

The force may also be resolved into components parallel and perpendicular to the acceleration. If, for example, we have a particle moving in the xy-plane with the velocity

$$\mathbf{u} = u_x\mathbf{i} + u_y\mathbf{j}$$

(29.2)

and desire to accelerate it in the y-direction, it can easily be shown[‡] that we must apply, in addition to a component of force F_y in the desired direction, an additional component F_x at right angles which will be given by the relation

$$F_x = \frac{u_x u_y}{c^2 - u_x^2}F_y.$$

(29.3)

This method of resolving the force is also sometimes useful in giving an insight into the relations between force and acceleration. The extra component in the x-direction is necessary, when the particle already has a component of velocity u_x in that direction, in order to take care of changes in momentum in that direction, arising from

† See report by Gerlach, *Handbuch der Physik*, xxii, Berlin, 1926.
‡ Tolman, *Phil. Mag.* **22**, 458 (1911).

changes in mass even when the velocity in that direction remains constant.

In accordance with (29.1) or (29.3) it will be seen that force and acceleration will be in the same direction for the two special cases of a *transverse acceleration* in which the force is applied at right angles to the existing velocity, and of a *longitudinal acceleration* in which the force is applied in the same direction as the existing velocity. For a transverse acceleration equation (29.1) reduces to

$$\mathbf{F} = \frac{m_0}{\sqrt{(1-u^2/c^2)}} \frac{d\mathbf{u}}{dt}, \tag{29.4}$$

and for a longitudinal acceleration it reduces to

$$\mathbf{F} = \frac{m_0}{(1-u^2/c^2)^{\frac{3}{2}}} \frac{d\mathbf{u}}{dt}. \tag{29.5}$$

An examination of these equations shows the reason why

$$m_0/\sqrt{(1-u^2/c^2)}$$

has sometimes been called the transverse mass of a particle and $m_0/(1-u^2/c^2)^{\frac{3}{2}}$ the longitudinal mass. It should be emphasized, however, that it is only the first of these quantities $m_0/\sqrt{(1-u^2/c^2)}$ which can be regarded as a fundamental expression for the mass of a particle, since this is the quantity which will give the momentum when multiplied by the velocity of the particle, and is the quantity which is conserved when particles interact by collision.

(c) **Applications in electromagnetic theory.** Although a fundamental discussion is necessary for a complete development of the principles of electromagnetic theory, it is interesting to point out in passing that certain special electrical problems can be advantageously treated with the help of the dynamics of a particle.

As a typical problem of this kind we may consider the calculation of the force with which a charge e moving with the uniform velocity V, for simplicity taken in the x-direction, would act on a second charge e_1 in its neighbourhood. To treat this problem with the help of the dynamics of a particle, we may first take a system of coordinates in which the charge e is at rest so that it may then be regarded as surrounded by a simple electrostatic field. In this original system of coordinates the force on e_1 can be calculated very simply with the help of Coulomb's inverse square law of electrostatic repulsion, and by making use of the transformation equations for force (25.3) we can then change to the desired system of coordinates in which the charge e is in motion.

For the case in hand if the charge e is taken as at the origin of coordinates at the instant of interest and as moving along the x-axis with the uniform velocity V, the force on e_1 can readily be shown† by this method to be given by

$$F_x = \frac{ee_1}{s^3}\left(1 - \frac{V^2}{c^2}\right)\left\{x + \frac{V}{c^2}(yu_y + zu_z)\right\},$$

$$F_y = \frac{ee_1}{s^3}\left(1 - \frac{V^2}{c^2}\right)\left(1 - \frac{u_x V}{c^2}\right)y, \qquad (29.6)$$

$$F_z = \frac{ee_1}{s^3}\left(1 - \frac{V^2}{c^2}\right)\left(1 - \frac{u_x V}{c^2}\right)z,$$

where x, y, z, and u_x, u_y, u_z denote the coordinates and components of velocity of e_1 with respect to this system of axes, and s is an abbreviation for

$$s = \sqrt{\{x^2 + (1 - V^2/c^2)(y^2 + z^2)\}}. \qquad (29.7)$$

The result is the same as can be obtained by the more usual method of first computing the electric and magnetic fields produced by the moving charge e and then determining the force which they exert on the charge e_1 which is itself moving through them. The present treatment shows that the somewhat complicated action of these electric and magnetic fields on the charge e_1 can be regarded as a simple electrostatic action by a suitable choice of coordinates. The general relations between electric and magnetic field strengths in different systems of coordinates will be treated in the following chapter on electromagnetic theory.

A further illustration of the methods of applying the dynamics of a particle to electromagnetic problems can be obtained if we again consider the charge e as constrained to move along the x-axis with the uniform velocity V, and take the charge e_1 as located at the instant of interest on the y-axis at $y = y$ and moving in the x-direction with the same velocity $u_x = V$ as the charge e itself, and having any desired component of velocity u_y in the y-direction. Under these circumstances it is evident from the simple qualitative considerations placed at our disposal by the theory of relativity, that the charge e_1 should merely receive an acceleration in the y-direction and retain unchanged its component of velocity in the x-direction, since from

† Tolman, *Phil. Mag.* **25**, 150 (1913).

the point of view of an observer moving along with e the phenomenon is merely one of ordinary electrostatic repulsion. It is interesting to see in detail, however, how this comes about.

Substituting the values given above for the coordinates and components of velocity of e_1 into (29.6), we obtain for the components of force acting on e_1

$$F_x = \frac{ee_1}{s^3}\left(1 - \frac{V^2}{c^2}\right)\frac{Vu_y}{c^2}y$$

and

$$F_y = \frac{ee_1}{s^3}\left(1 - \frac{V^2}{c^2}\right)^2 y,$$

(29.8)

and at first sight are surprised to find any component of force in the x-direction, since we expect the acceleration to be solely in the y-direction. In accordance with the preceding section we remember, however, that in general force and acceleration are not in the same direction, and by combining the two equations above we easily obtain

$$F_x = \frac{Vu_y}{c^2 - V^2}F_y,$$

(29.9)

which, with $u_x = V$, is the relation (29.3) between the components of force that we have already obtained as the necessary condition for acceleration solely in the y-direction.

Other applications of particle dynamics to electromagnetic problems will suggest themselves to the reader.

(d) Tests of the interrelation of mass, energy, and momentum. The relations between mass, energy, and momentum obtained in § 27 are among the most important conclusions that have been drawn from the Einstein theory of relativity. There are several points of contact between these relations and experiment which we may now consider.

The first of these relations was that connecting increase in kinetic energy with increase in mass as given by equation (27.1). From a qualitative point of view since increase in velocity will certainly lead to increase in kinetic energy, it is evident that all of the experiments on the increase of mass with increase in velocity are in agreement with the general idea that increase in energy and increase in mass go hand in hand. Among these experiments on the relation between mass and velocity, however, were those of Hupka† in which the particles received their velocity by acceleration through a measured potential drop, the velocity then being calculated by equating the relativistic expression for kinetic energy (26.4) to the work done by

† Hupka, *Ann. der Physik*, **31**, 169 (1910).

the electric forces that produced the acceleration. Hence these particular experiments can also be regarded as a quantitative verification of the relation between increased mass and kinetic energy, provided we accept the simplest principles of electrical theory.

Turning next to the more general ideas, embodied in equations (27.3) and (27.4), that all kinds of energy have the associated mass E/c^2 and all kinds of mass the associated energy mc^2, it is evident that obvious macroscopic tests, such for example as would be given by measurements of heat content and inertia, hold little promise owing to the great size of the conversion factor c^2.

In the field of atomic physics, however, the range of validity for such ideas has recently been strikingly extended. Thus, the qualitative suggestion,[‡] that the energy of the incoming cosmic rays might provide for the internal rest-masses of the pairs of positive and negative electrons observed by Anderson[†] and by Blackett and Occhialini,[‡] has now been supplemented by the results obtained with γ-rays by Anderson and Neddermeyer[§] which give quantitative indication that the known energy of the γ-rays is sufficient to account for the rest-masses plus the kinetic energies of the pairs of positive and negative electrons that appear. Furthermore, the long entertained possibility for intranuclear processes accompanied by a transformation of rest-mass into familiar forms of energy, has now received excellent quantitative confirmation by the measurements of Oliphant, Kinsey and Rutherford,[||] which show that the decreases in mass, when the two isotopes of lithium Li^7 and Li^6 combine respectively with the isotopes of hydrogen H^1 and H^2 to form helium He^4, are just sufficient to account for the kinetic energy of the pairs of α-particles formed.

Turning finally, moreover, to the relation of momentum with transfer of energy (27.5), which was itself based on the assumption of an equivalence between mass and energy, we have the quantitative and beautiful experimental verification provided by measurements of light pressure. These show with considerable exactness that we have in the case of radiation the theoretically expected relation (27.6) between density of momentum and density of energy flow.

† Anderson, *Science*, **76**, 238 (1932).

‡ Blackett and Occhialini, *Proc. Roy. Soc.*, A **139**, 69 (1933).

§ Anderson and Neddermeyer, *Phys. Rev.*, **43**, 1034 (1933). For more complete theory see Oppenheimer and Plesset, *Phys. Rev.*, **44**, 53 (1933).

|| Oliphant, Kinsey and Rutherford, *Proc. Roy. Soc.*, A **141**, 722 (1933). See also Bainbridge, *Phys. Rev.*, **44**, 123 (1933).

III

SPECIAL RELATIVITY AND MECHANICS (*contd.*)

Part II. THE DYNAMICS OF A CONTINUOUS MECHANICAL MEDIUM

30. The principles postulated

In the classical Newtonian mechanics after treating the dynamics of particles it was customary to proceed to a development of the dynamics of rigid bodies whose state could be specified by the six coordinates which would give the position and orientation of the body and the six corresponding momenta. In relativistic mechanics, however, it is evident as soon as we consider bodies of a finite size that in general an infinite number of variables will be necessary to determine their state, since disturbances set up in one part of the body can only be transmitted to other parts with a velocity less than that of light. In relativistic mechanics the most nearly rigid body we can think of would be one in which disturbances are propagated with the limiting velocity of light, and the older idea of a completely rigid body whose parts would act together as a whole is no longer a legitimate abstraction. We may hence proceed at once to a development of the mechanics of a continuous medium, the resulting theory being due originally to the work of Laue.

As the postulatory basis for this development we shall take the principles of the special theory of relativity and the two principles of the conservation of mass and momentum in all systems of coordinates used in developing the mechanics of a particle, and in addition shall combine these with the conclusions as to the transformation equations for forces and the relations between mass and energy which were provided by the mechanics of a particle.

In accordance with this basis the theory of the mechanics of an elastic continuum can be regarded as a natural extension of the mechanics of a particle. The theory cannot, however, be regarded in any sense as deduced from the mechanics of particles, since we shall make no attempt to derive the properties of a continuum from the relativistic behaviour of the particles or molecules out of which the continuum might be thought of as composed. Even in the older Newtonian mechanics the attempt to obtain a rigorous derivation of the mechanics of an elastic continuum from that for particles was perhaps not entirely satisfactory, and at the present time such an attempt would be complicated not only by the facts of relativity, but also by the necessity of applying quantum mechanics to the behaviour of the ultimate particles. For these reasons it has seemed best to obtain the mechanics of a continuum from its own postulatory basis—as given above—with the help of a macroscopic treatment that avoids the necessity for quantum mechanics. This we proceed to do.

31. The conservation of momentum and the components of stress t_{ij}

The first item in our postulatory basis to which we shall wish to pay attention is the principle of the conservation of momentum. To secure the validity of this principle we shall again regard force as equal to rate of change of momentum and require an equality between action and reaction in the interior of our elastic medium. Let us now see in detail how this is to be done.

Considering a given set of Cartesian axes x, y, z, let us first define the components of stress t_{ij} at any point in our medium as the nine quantities

$$t_{ij} = \begin{matrix} t_{xx} & t_{xy} & t_{xz} \\ t_{yx} & t_{yy} & t_{yz} \\ t_{zx} & t_{zy} & t_{zz} \end{matrix} \qquad (31.1)$$

which give the normal and tangential components of force exerted by the medium on unit surfaces at the point in question, in accordance with the usual understanding, that the symbol t_{ij} denotes the component of force parallel to the i-axis exerted on unit surface normal to the j-axis by the material lying on the side of this surface corresponding to *smaller* values of the coordinate x_j.

With this definition of the components of stress t_{ij}, the principle of the equality of action and reaction can then be maintained by taking $-t_{ij}$ as the component of force parallel to the i-axis exerted on unit surface normal to the j-axis, by the material on the side of the surface corresponding to *larger* values of the coordinate x_j. And this will be done in what follows as will be seen in the next section.

32. The equations of motion in terms of the stresses t_{ij}

With the help of the foregoing we may now obtain an expression for the equations of motion of the medium in terms of the stress t_{ij}.

On the one hand, we may calculate the net force acting on a unit cube of the medium by considering the difference in the stress acting on the parallel surfaces by which the cube is bounded. For example, if we are considering the component of force in the x-direction and fix our attention on the pair of faces perpendicular to the y-axis, we can take t_{xy} as the force exerted on the lower of these two surfaces and in accordance with the postulated equality of action and reaction can take $-(t_{xy}+\partial t_{xy}/\partial y)$ as the force exerted on the upper surface. Hence for the net contribution of this pair of surfaces to the component of force in the x-direction we shall have $-\partial t_{xy}/\partial y$, and summing for all three pairs of parallel surfaces can write

$$f_x = -\frac{\partial t_{xx}}{\partial x} - \frac{\partial t_{xy}}{\partial y} - \frac{\partial t_{xz}}{\partial z} \qquad (32.1)$$

for the total force in the x-direction acting on a unit cube of the material. Or generalizing, we can write

$$f_i = -\frac{\partial t_{ij}}{\partial x_j} \qquad (32.2)$$

for the component of force acting in the ith direction on unit volume, where the double occurrence of the dummy suffix j indicates summation for the three coordinates x, y, z.

On the other hand, since f_i is the component of force on unit volume we can take $f_i \, \delta v$ as the force on a small element of the material of volume δv, and equate this to the rate of change of the momentum of the element in accordance with the expression

$$f_i \, \delta v = \frac{d}{dt}(g_i \, \delta v), \qquad (32.3)$$

where g_i is the density of momentum at the point in question parallel to the i-axis.

Combining (32.2) and (32.3), we can then write the equations of motion for the element δv in the form

$$-\frac{\partial t_{ij}}{\partial x_j} \delta v = \frac{d}{dt}(g_i \, \delta v)$$

$$= \frac{dg_i}{dt} \delta v + g_i \frac{d}{dt}(\delta v). \qquad (32.4)$$

This expression can be simplified, however, since we can evidently write for the rate of change in the momentum density of the element

$$\frac{dg_i}{dt} = \frac{\partial g_i}{\partial t} + u_x \frac{\partial g_i}{\partial x} + u_y \frac{\partial g_i}{\partial y} + u_z \frac{\partial g_i}{\partial z}$$

$$= \frac{\partial g_i}{\partial t} + u_j \frac{\partial g_i}{\partial x_j}, \qquad (32.5)$$

where the first term arises from the rate of change at the point in question and the second term from the motion of the element with the components of velocity u_j. And for the rate of change of volume we can write

$$\frac{d}{dt}(\delta v) = \left(\frac{\partial u_x}{\partial x}+\frac{\partial u_y}{\partial y}+\frac{\partial u_z}{\partial z}\right)\delta v$$

$$= \frac{\partial u_j}{\partial x_j}\delta v. \tag{32.6}$$

Substituting (32.5) and (32.6) in (32.4) and simplifying, we then obtain the equations of motion for our medium in the simple form desired

$$-\frac{\partial t_{ij}}{\partial x_j} = \frac{\partial g_i}{\partial t}+u_j\frac{\partial g_i}{\partial x_j}+g_i\frac{\partial u_j}{\partial x_j}$$

$$= \frac{\partial g_i}{\partial t}+\frac{\partial}{\partial x_j}(g_i u_j). \tag{32.7}$$

The result is a general representation for the three separate equations that correspond to taking the subscript i as $x, y,$ or z, and summing for the three axes in the case of the dummy subscript j.

33. The equation of continuity

The foregoing three equations were obtained as the outcome of our postulate as to the conservation of momentum, and we may now supplement them with the help of the principle of the conservation of mass by the equation of continuity

$$\frac{\partial g_x}{\partial x}+\frac{\partial g_y}{\partial y}+\frac{\partial g_z}{\partial z} = -\frac{\partial \rho}{\partial t} \tag{33.1}$$

or

$$\frac{\partial g_j}{\partial x_j} = -\frac{\partial \rho}{\partial t}, \tag{33.2}$$

where ρ is the density of mass at the point in question. Since the density of momentum g is by definition equal to the density of mass flow, this equation is an evident expression of our postulate as to the conservation of mass.

34. The transformation equations for the stresses t_{ij}

With the help of the two conservation laws of mass and momentum, we have thus obtained the equations of motion (32.7) and the equation of continuity (33.2) for a continuous medium, and in accordance with the first postulate of relativity, equations of this same form will apply to the behaviour of the medium in all systems of coordinates in uniform relative motion. In order, however, to make any use of these relations connecting the quantities—stress t_{ij}, density of momentum g, and density of mass ρ—we must now show how the values of these quantities referred to any given system of coordinates are to be determined. To accomplish this, we shall obtain in the present section transformation equations which will permit a calculation of the components of stress t_{ij} in terms of the components t_{ij}^0 as they would be directly measured by an observer moving along with the medium at the point of interest; and in the next section we shall obtain transformation equations which will similarly permit the calculation of the other quantities g and ρ in terms of quantities which could be directly measured by ordinary methods.

These transformation equations for t_{ij}, g, and ρ will themselves be based of course on our previous study of the Lorentz transformation and the conclusions drawn therefrom. And it should perhaps be emphasized that it is this introduction of the Lorentz transformation which determines the essential character of the relativistic mechanics of a continuum, since the equations of motion (32.7) and continuity (33.2) would also be true in Newtonian mechanics in all systems of coordinates if we should use the Galilean transformation instead of the Lorentz transformation.

Let us now consider the transformation for the components of the stress t_{ij} from one system of coordinates to another. Since these components of stress have themselves been defined in terms of forces and the areas on which they act, we are already well prepared to calculate the transformation equations for these quantities. In transforming the expressions for the areas we shall merely have to allow for the Lorentz contraction (§ 9), which was an immediate result of the fundamental transformation equations for spatial and temporal measurements. And in transforming the expressions for components of force we can use the results of § 25, since as already pointed out in that section the transformation equations for forces of any origin must be the same if we are to retain the conservation of momentum in general and in all systems of coordinates.

For simplicity let us assume that our original system of coordinates S has been oriented so that the material, at the point of interest in the medium, will be moving with respect to this system with the velocity u, parallel to the x-axis without components of velocity in the y- and z-directions. And let us take as our second system, so-called proper coordinates S^0, also moving in the x-direction with respect to S with the velocity

$$u = V, \qquad (34.1)$$

so that the material at the point and time of interest will be at rest in system S^0 in accordance with the equations

$$u_x^0 = u_y^0 = u_z^0 = 0. \qquad (34.2)$$

We may now easily secure expressions for the components of stress t_{ij} with respect to S in terms of the components t_{ij}^0 with respect to S^0. Substituting the expressions for velocity (34.1) and (34.2) into the transformation equations for force (25.3), we at once obtain

$$F_x = F_x^0 \qquad F_y = F_y^0\sqrt{(1-u^2/c^2)} \qquad F_z = F_z^0\sqrt{(1-u^2/c^2)} \qquad (34.3)$$

as transformation equations connecting measurements of force in the two systems; and noting that the Lorentz contraction (9.3) will affect the transformation of areas normal to the y- and z-axes but not those normal to the x-axis, we can write

$$A_x = A_x^0 \qquad A_y = A_y^0\sqrt{(1-u^2/c^2)} \qquad A_z = A_z^0\sqrt{(1-u^2/c^2)} \qquad (34.4)$$

for the transformation of areas normal to the directions indicated by the subscripts. Returning then to our original definition of the components of stress (31.1) in terms of force per unit area, we easily see that the transformation equations will be†

$$t_{xx} = t_{xx}^0 \qquad t_{xy} = \frac{t_{xy}^0}{\sqrt{(1-u^2/c^2)}} \qquad t_{xz} = \frac{t_{xz}^0}{\sqrt{(1-u^2/c^2)}}$$

$$t_{yx} = t_{yx}^0\sqrt{(1-u^2/c^2)} \qquad t_{yy} = t_{yy}^0 \qquad t_{yz} = t_{yz}^0$$

$$t_{zx} = t_{zx}^0\sqrt{(1-u^2/c^2)} \qquad t_{zy} = t_{zy}^0 \qquad t_{zz} = t_{zz}^0 \qquad (34.5)$$

specialized, of course, by the simplification that the direction of axes in system S has been chosen so as to make the velocity u of the material at the point of interest parallel to the x-axis.

Owing to the circumstance that the velocity of the material is zero at the point of interest with respect to the proper coordinates S^0, the ordinary principles of Newtonian mechanics can be applied in that system, which lead, as is well known, to the symmetry of the stress tensor t_{ij}^0 so that we have

$$t_{xy}^0 = t_{yx}^0 \qquad t_{yz}^0 = t_{zy}^0 \qquad t_{zx}^0 = t_{xz}^0 \qquad (34.6)$$

† These equations for the transformation of stresses differ from those given by Tolman, *The Theory of the Relativity of Motion*, § 122, since the stresses were there defined with reference to unit *proper* volume of the material. The present definition in terms of force per unit area as measured in either system of coordinates is chosen to agree with the usage of Laue, *Das Relativitätsprinzip*, second edition, Braunschweig, 1913.

in system S^0. We note then in accordance with (34.5) the important conclusion that the components of the stress in system S will not give a symmetrical array. So that in general when the point of interest is moving with respect to the coordinate system we can expect to find

$$t_{ij} \neq t_{ji}. \tag{34.7}$$

The great importance of the transformation equations for the components of stress (34.5) lies in the possibility which they provide for correlating the stress in rapidly moving material with the known behaviour of stress in stationary material.

35. The transformation equations for the densities of mass and momentum

In addition to the above equations (34.5) which permit us to calculate the stress at any point in our medium in terms of the stress as measured by an observer moving with the material at that point, we shall also desire—as pointed out at the beginning of the last section—equations which will permit us to calculate the densities of mass ρ and momentum g_i in terms of quantities which could be measured by an observer moving with the material. To obtain these relations will be a somewhat long and complicated task, and in carrying out the deduction we shall have to make use of the relativistic relations between mass, energy, and momentum which is the remaining part of the postulatory basis stated in § 30, which has not yet been employed.

With the help of these relations between mass, energy, and momentum we shall first obtain an expression for the momentum of a moving portion of our medium in terms of its mass (or energy), velocity, and state of stress. This expression for momentum will then permit us to calculate the force acting on a stressed portion of the medium when its momentum and velocity are changed, and hence to calculate the work done and increase in energy when the material is brought from zero velocity up to the actual velocity of interest. We shall then be in a position to compute the mass, energy, and momentum of the moving material in terms of its velocity and its mass, energy, and state of stress as they appear to an observer moving with it. We now turn to the derivation which can be obtained along these lines.

In accordance with our ideas as to the connexion between density of momentum and density of energy flow as given by (27.6), it is evident that the momentum of a moving portion of material when subjected to stress will be due not only to the bodily motion of the mass which it contains but also to the density of energy flow arising from the work done by the forces of stress that act on its moving faces. Thus, if we have material of density ρ moving with the velocity u which we take for simplicity as parallel to the x-axis, we can write for its density of momentum in the x-, y-, and z-directions

$$g_x = \rho u + \frac{t_{xx} u}{c^2},$$

$$g_y = \frac{t_{xy} u}{c^2}, \tag{35.1}$$

$$g_z = \frac{t_{xz} u}{c^2},$$

since $t_{xx} u$, $t_{xy} u$, and $t_{xz} u$ are evidently the densities of energy flow in the directions indicated due to the action of the forces of stress, and division by c^2 will be necessary owing to the difference in units for the measurement of mass and energy. It is an important and interesting result of relativistic mechanics that there will be in general components of momentum in a stressed body at right angles to the direction of motion.

For the total momentum of a small portion of the medium of volume v we can then write in accordance with (35.1) the expressions

$$G_x = \frac{E + t_{xx} v}{c^2} u,$$

$$G_y = \frac{t_{xy} v}{c^2} u, \tag{35.2}$$

$$G_z = \frac{t_{xz} v}{c^2} u,$$

where for later convenience we have expressed the total mass in terms of the energy E divided by c^2. And from the definition of force as equal to the rate of change of momentum we can write

$$F_x = \frac{d}{dt}\left(\frac{E + t_{xx} v}{c^2} u\right),$$

$$F_y = \frac{d}{dt}\left(\frac{t_{xy} v}{c^2} u\right), \tag{35.3}$$

$$F_z = \frac{d}{dt}\left(\frac{t_{xz} v}{c^2} u\right)$$

for the components of force which would have to be applied to the material in the volume v in order to change its velocity u parallel to the x-axis.

We are now ready to calculate the work done and energy input necessary to bring a given portion of our stressed material from zero velocity up to the velocity of interest. Let us start with material having the volume v^0, energy content E^0, and stress t_{ij}^0 and bring it from zero velocity to that of interest by an adiabatic acceleration parallel to the x-axis which leaves the condition of the material (i.e. v^0, E^0, and t_{ij}^0) unchanged when measured by an observer moving with the material. In accordance with the Lorentz contraction (9.3) we can write for the volume at the velocity u

$$v = v^0 \sqrt{(1-u^2/c^2)}, \tag{35.4}$$

and in accordance with the transformation equations for stress (34.5) shall have

$$t_{xx} = t_{xx}^0 \tag{35.5}$$

throughout the course of the acceleration. For the rate of energy increase we can then evidently write

$$\frac{dE}{dt} = F_x \frac{dx}{dt} - t_{xx} \frac{dv}{dt}, \tag{35.6}$$

where the first term is the rate at which work is done by the action of the force which produces the acceleration, and the second term is the rate at which work is done by the forces of stress which act on a volume which is decreasing in its length parallel to the x-axis owing to the Lorentz contraction.

Writing u in place of dx/dt, and substituting the expression for F_x given by (35.3) we can then re-express (35.6) in the form

$$\frac{dE}{dt} = \frac{dE}{dt}\frac{u^2}{c^2} + E\frac{u}{c^2}\frac{du}{dt} + t_{xx}\frac{u^2}{c^2}\frac{dv}{dt} + t_{xx}v\frac{u}{c^2}\frac{du}{dt} - t_{xx}\frac{dv}{dt},$$

where t_{xx} has been treated as a constant in accordance with (35.5). This can easily be rewritten in the form

$$\left(1-\frac{u^2}{c^2}\right)\frac{d}{dt}(E+t_{xx}v) = (E+t_{xx}v)\frac{u}{c^2}\frac{du}{dt},$$

which can readily be integrated between zero velocity and u to give us the final result

$$E+t_{xx}v = \frac{E^0+t_{xx}^0 v^0}{\sqrt{(1-u^2/c^2)}}, \tag{35.7}$$

where the superscript ^{0} indicates the values of the quantities involved as measured by an observer at rest with respect to the medium.

This last equation, however, now permits us to write the desired expressions for the densities of mass and momentum. Dividing (35.7) by the volume v, noting the relation between v and v^0 given by (35.4) introducing the equality of t_{xx} and t^0_{xx} given by (35.5), and changing from the density of energy to that of mass with the help of the factor c^2, we can easily obtain for the density of mass

$$\rho = \frac{\rho_{00} + t^0_{xx} u^2/c^4}{1 - u^2/c^2}, \qquad (35.8)$$

where ρ_{00} is the proper density of the material as measured by an observer moving with it. And combining this result with (35.1) and (34.5) we obtain for the densities of momentum parallel to the three axes

$$g_x = \frac{c^2\rho_{00} + t^0_{xx}}{1 - u^2/c^2} \frac{u}{c^2},$$

$$g_y = \frac{t^0_{xy}}{\sqrt{(1 - u^2/c^2)}} \frac{u}{c^2}, \qquad (35.9)$$

$$g_z = \frac{t^0_{xz}}{\sqrt{(1 - u^2/c^2)}} \frac{u}{c^2}.$$

These are the desired expressions which will permit us to calculate the densities of mass ρ and momentum g_i, at a point in a medium moving with the velocity u, in terms of this velocity and the density ρ_{00} and stress t^0_{ij} as measured by an observer moving with the material. The equations are specialized for simplicity by a choice of coordinates such that the direction of motion is parallel to the x-axis, but are otherwise general.

It should be specially noted that these equations have been derived without any reference to the microscopic behaviour of the ultimate particles of which the material might be thought of as composed, and the quantities occurring therein, such as density, velocity, and stress, are to be regarded as macroscopically measured. To emphasize this we have used the symbol ρ_{00} to designate the proper *macroscopic* density of the material as measured by a local observer, since the symbol ρ_0 with a single subscript is usually used to designate a hypothetical *microscopic* density. As mentioned in § 30, by adopting a macroscopic treatment, we have avoided the necessity for a quantum-mechanical treatment of the behaviour of the ultimate particles.

36. Restatement of results in terms of the (absolute) stress p_{ij}

The foregoing transformation equations for the components of stress and the densities of mass and momentum, together with the equations of motion and continuity, evidently provide a complete apparatus for treating the mechanics of a continuous medium. Nevertheless this apparatus may be put in a specially simple form, as will be shown below, if we now define a new array of quantities p_{ij} by the equations

$$p_{ij} = t_{ij} + g_i u_j, \tag{36.1}$$

where t_{ij} are the components of stress at the point in question as previously defined, and g_i and u_j are the indicated components of momentum density and velocity at that point.

In accordance with this definition, together with the relation (34.6), we have in the special case of proper coordinates, which are moving with the point of interest, the simple relations

$$p_{ij}^0 = p_{ji}^0 = t_{ij}^0 = t_{ji}^0, \tag{36.2}$$

and making use of this result, together with the transformation equations for stress (34.5) and momentum density (35.9), we easily calculate for more general coordinates, in which the material at the point of interest is moving with the velocity u in the x-direction, the transformation equations

$$p_{xx} = \frac{p_{xx}^0 + \rho_{00} u^2}{1 - u^2/c^2} \qquad p_{xy} = \frac{p_{xy}^0}{\sqrt{(1 - u^2/c^2)}} \qquad p_{xz} = \frac{p_{xz}^0}{\sqrt{(1 - u^2/c^2)}}$$

$$p_{yx} = \frac{p_{yx}^0}{\sqrt{(1 - u^2/c^2)}} \qquad p_{yy} = p_{yy}^0 \qquad p_{yz} = p_{yz}^0 \qquad (36.3)$$

$$p_{zx} = \frac{p_{zx}^0}{\sqrt{(1 - u^2/c^2)}} \qquad p_{zy} = p_{zy}^0 \qquad p_{zz} = p_{zz}^0.$$

Furthermore, the transformation equations for density of mass and momentum (35.8, 9) may now be re-expressed in the form

$$\rho = \frac{\rho_{00} + p_{xx}^0 u^2/c^4}{1 - u^2/c^2} \tag{36.4}$$

and

$$g_x = \frac{c^2 \rho_{00} + p_{xx}^0}{1 - u^2/c^2} \frac{u}{c^2} \qquad g_y = \frac{p_{xy}^0}{\sqrt{(1 - u^2/c^2)}} \frac{u}{c^2} \qquad g_z = \frac{p_{xz}^0}{\sqrt{(1 - u^2/c^2)}} \frac{u}{c^2}. \tag{36.5}$$

Finally, with the help of the definition (36.1), the equations of motion (32.7) can be expressed in the new language in the extremely simple form

$$\frac{\partial p_{ij}}{\partial x_j}+\frac{\partial g_i}{\partial t} = 0 \qquad (36.6)$$

and the equation of continuity (33.2) may again be written

$$\frac{\partial g_j}{\partial x_j}+\frac{\partial \rho}{\partial t} = 0. \qquad (36.7)$$

Since the equations (36.2, 3, 4, 5) permit us to compute all the quantities occurring in the equations of motion and continuity (36.6, 7), in terms of quantities measurable by ordinary methods by a local observer moving with the material, we now have in a compact and convenient form all that is needed for treating the mechanics of a continuous medium. The transformation equations (36.3, 4, 5) are specialized for simplicity to the slight extent that we have chosen our axes in such a way that the velocity u of the medium at the point in question is parallel to the x-axis, but are otherwise general.

It is of interest to note that although the stress t_{ij} as originally defined in terms of the forces exerted by the medium on unit area did not give a symmetrical array of quantities (34.5) except in the case of proper coordinates, nevertheless the new quantities p_{ij} do give a symmetrical array in all coordinates as shown by (36.2, 3).

Since the forces corresponding to the t_{ij} are those which one portion of the medium exerts on another, the surfaces on which the t_{ij} act are at rest relative to the medium. For this reason the quantities t_{ij} are sometimes called the components of *relative* stress. On the other hand, the new quantities p_{ij} determine in accordance with (36.6) the rate of change of momentum density at a given point fixed in space as referred to the coordinate system. For this reason the quantities p_{ij} are sometimes called the components of *absolute* stress, as was done in the heading of this section.

The introduction of the new quantities p_{ij} is of great advantage in now permitting a further re-expression of the apparatus for treating the mechanics of a continuous medium in very simple four-dimensional language with the help of a generalized symmetrical four-dimensional tensor, a matter to which we now turn in the following section.

37. Four-dimensional expression of the mechanics of a continuous medium

To obtain an apparatus for treating the mechanics of a continuous medium in four-dimensional language we shall now return to our fundamental idea of a four-dimensional space-time continuum with the system of Galilean coordinates (x^1, x^2, x^3, x^4) which are related to our previous spatial and temporal coordinates by the expressions (19.1)

$$x^1 = x \quad x^2 = y \quad x^3 = z \quad x^4 = ct \qquad (37.1)$$

and introduce a symmetrical four-dimensional tensor $T^{\mu\nu}$—the so-called energy-momentum tensor—for describing the condition of the mechanical medium at any given point in space and time. The ten independent components of this symmetrical tensor $T^{\mu\nu}$ will be taken in such a way as to be very simply related to the ten quantities p_{ij}, g_i, and ρ, used above in treating the mechanics of a continuum, and so as to lead to a single very simple tensor equation which will be equivalent to the three equations of motion and the equation of continuity necessary for the previous treatment.

To define the energy-momentum tensor $T^{\mu\nu}$ in terms of our previous quantities, we shall first consider proper coordinates $(x_0^1, x_0^2, x_0^3, x_0^4)$ such that the medium at the point and time of interest has zero spatial velocities with respect to these particular coordinates

$$\frac{dx_0^1}{ds} = \frac{dx_0^2}{ds} = \frac{dx_0^3}{ds} = 0, \qquad (37.2)$$

and then state that in these coordinates the tensor reduces so that its components have the simple values

$$
\begin{aligned}
T_0^{\alpha\beta} = \quad & p_{xx}^0 \quad p_{xy}^0 \quad p_{xz}^0 \quad 0 \\
& p_{yx}^0 \quad p_{yy}^0 \quad p_{yz}^0 \quad 0 \\
& p_{zx}^0 \quad p_{zy}^0 \quad p_{zz}^0 \quad 0 \\
& \ \ 0 \qquad 0 \qquad 0 \qquad c^2\rho_{00}.
\end{aligned}
\qquad (37.3)
$$

It is evident that this statement completely defines the components of the tensor at the point in question in all systems of coordinates, since we can write in accordance with our general equation (19.8)

$$T^{\mu\nu} = \frac{\partial x^\mu}{\partial x_0^\alpha} \frac{\partial x^\nu}{\partial x_0^\beta} T_0^{\alpha\beta} \qquad (37.4)$$

as an expression for obtaining the components of this tensor in any desired new system of coordinates (x^1, x^2, x^3, x^4), from their values in proper coordinates as given by (37.3).

We are now ready to investigate the usefulness of this newly defined four-dimensional tensor. If we transform from the original coordinates $(x_0^1, x_0^2, x_0^3, x_0^4)$ in which the material was at rest at the point of interest to a new set (x^1, x^2, x^3, x^4) in which the material is moving parallel to the x-axis with the velocity u, we must set

$$x^1 = \frac{x_0^1 + ux_0^4/c}{\sqrt{(1-u^2/c^2)}} \quad x^2 = x_0^2 \quad x^3 = x_0^3 \quad x^4 = \frac{x_0^4 + ux_0^1/c}{\sqrt{(1-u^2/c^2)}} \quad (37.5)$$

in accordance with the Lorentz transformation equations (20.3) with V put equal to $-u$. We then obtain in agreement with (20.4) for the differential coefficients that occur in the transformation equation (37.4) as the only non-vanishing values

$$\frac{\partial x^1}{\partial x_0^1} = \frac{\partial x^4}{\partial x_0^4} = \frac{1}{\sqrt{(1-u^2/c^2)}},$$

$$\frac{\partial x^1}{\partial x_0^4} = \frac{\partial x^4}{\partial x_0^1} = \frac{u/c}{\sqrt{(1-u^2/c^2)}}, \quad (37.6)$$

$$\frac{\partial x^2}{\partial x_0^2} = \frac{\partial x^3}{\partial x_0^3} = 1.$$

Substituting these into (37.4) with the $T_0^{\alpha\beta}$ as given by (37.3) we can then readily obtain for the components of the tensor in the new coordinates the values

$$T^{\mu\nu} = \begin{array}{cccc} \dfrac{p_{xx}^0 + \rho_{00} u^2}{1 - u^2/c^2} & \dfrac{p_{xy}^0}{\sqrt{(1-u^2/c^2)}} & \dfrac{p_{xz}^0}{\sqrt{(1-u^2/c^2)}} & \dfrac{c^2\rho_{00} + p_{xx}^0}{1 - u^2/c^2}\dfrac{u}{c} \\[3ex] \dfrac{p_{yx}^0}{\sqrt{(1-u^2/c^2)}} & p_{yy}^0 & p_{yz}^0 & \dfrac{p_{xy}^0}{\sqrt{(1-u^2/c^2)}}\dfrac{u}{c} \\[3ex] \dfrac{p_{zx}^0}{\sqrt{(1-u^2/c^2)}} & p_{zy}^0 & p_{zz}^0 & \dfrac{p_{xz}^0}{\sqrt{(1-u^2/c^2)}}\dfrac{u}{c} \\[3ex] \dfrac{c^2\rho_{00} + p_{xx}^0}{1 - u^2/c^2}\dfrac{u}{c} & \dfrac{p_{xy}^0}{\sqrt{(1-u^2/c^2)}}\dfrac{u}{c} & \dfrac{p_{xz}^0}{\sqrt{(1-u^2/c^2)}}\dfrac{u}{c} & \dfrac{c^2\rho_{00} + p_{xx}^0 u^2/c^2}{1 - u^2/c^2} \end{array},$$
$$(37.7)$$

and comparing these values with those given by the transformation equations (36.3, 4, 5) in the previous section, we see that this reduces to the simple symmetrical array

$$T^{\mu\nu} = \begin{array}{cccc} p_{xx} & p_{xy} & p_{xz} & cg_x \\ p_{yx} & p_{yy} & p_{yz} & cg_y \\ p_{zx} & p_{zy} & p_{zz} & cg_z \\ cg_x & cg_y & cg_z & c^2\rho, \end{array} \quad (37.8)$$

the components of the tensor thus having the same physical significance in all systems of Galilean coordinates.

The usefulness of this energy-momentum tensor becomes at once apparent since our earlier equations of motion and continuity can now be written in the extremely simple combined form

$$\partial T^{\mu\nu}/\partial x^\nu = 0. \tag{37.9}$$

Noting the values of the coordinates given by (37.1), it is easily seen that the four equations, obtainable from (37.9) by assigning to μ the different values 1, 2, 3, 4 and summing over the dummy suffix ν, are as a matter of fact identical with our previous equations of motion and continuity (36.6, 7).

Since $\partial T^{\mu\nu}/\partial x^\nu$ is the expression in our present simple coordinates for the contracted covariant derivative of $T^{\mu\nu}$, this final equation can be regarded as expressing a tensor relation, and in accordance with the discussion of § 21 this property is sufficient to secure agreement with the postulates of relativity. Hence this equation, together with the definition of $T^{\mu\nu}$ given by (37.3), may now itself be taken as a very satisfactory postulatory basis for the mechanics of a continuous medium. This starting-point will be of special value when we come to the general theory of relativity.

38. Applications of the mechanics of a continuous medium

It is evident that the system of mechanics, whose development we have now completed, differs in important respects from Newtonian mechanics. These differences can be most clearly seen with the help of equations (36.4, 5) which show not only that the mass of a moving body would depend on its velocity as already found in the case of particles, but in addition that the mass and momentum would depend on the state of stress, and that there would in general be components of momentum in a stressed body at right angles to the direction of motion.

A direct experimental test of these additional differences between Newtonian and relativistic mechanics would be very important, especially as much of the mechanics of general relativity will have to be founded on our present results. It will be noted, moreover, in accordance with the equations mentioned that these new differences from Newtonian mechanics would exist even at low velocities provided the stresses $p_{ij}^0 = t_{ij}^0$ were great enough. Nevertheless, there are at present no simple mechanical examples known in which these stresses are large enough to produce appreciable deviations from Newtonian mechanics. This is unfortunate from the point of view of obtaining a direct experimental verification of the new mechanics, but of course, on the other hand, signifies that there are no known mechanical phenomena which disagree with the new mechanics.

Since simple direct tests of the mechanics under discussion are not feasible, our trust in its conclusions must be dependent on the coherence of the theory with the rest of physics and on its own internal coherence. The presence of such coherence has—it is believed—been made evident in the method of development which we have chosen to obtain the theory. The feeling that such coherence exists will, however, be further strengthened if we now develop some of the consequences of the theory, and show as far as possible their rational nature and relation to other fields of physics.

(a) **The mass and momentum of a finite system.** Starting with the equations of motion and continuity in the differential form in which they have been derived, let us first consider the results that can then be obtained by integrating over a finite volume.

Considering first the equation of continuity (33.1) and carrying out an integration over a definite *fixed volume in space*, we can write for the rate of change in the mass inside that volume

$$\frac{dm}{dt} = \int \frac{\partial \rho}{\partial t}\, dv = -\iiint \left(\frac{\partial g_x}{\partial x} + \frac{\partial g_y}{\partial y} + \frac{\partial g_z}{\partial z} \right) dx\,dy\,dz, \qquad (38.1)$$

or by performing a part of the integration we can write this in the form

$$\frac{dm}{dt} = \int \frac{\partial \rho}{\partial t}\, dv = -\iint |g_x|_x^{x'}\, dy\,dz - \iint |g_y|_y^{y'}\, dx\,dz - \iint |g_z|_z^{z'}\, dx\,dy,$$
$$(38.2)$$

where the limits of integration at the boundary of the volume considered have been denoted by x, x', etc. We have thus related the rate of change of the mass inside the spatial volume considered to the density of flow across the boundary. For an *isolated system* this will give us the principle of the conservation of mass and also the principle of the conservation of energy on account of the interrelation of those two quantities.

In a similar way starting with the equations of motion in their original form we can obtain information by integration as to the rate of change in the momentum of a finite system. Here, however, we have two possibilities of procedure corresponding to the two different forms in which the equations of motion have been expressed—originally in terms of the stresses t_{ij} and later in terms of the p_{ij}.

Let us first consider the equations of motion in their later rather simpler form (36.6). We can then again integrate over a definite *fixed volume in space*, and obtain for the rate of change of the *i*th component of momentum inside this volume

$$\frac{dG_i}{dt} = \int \frac{\partial g_i}{\partial t}\, dv = -\int\int\int \left(\frac{\partial p_{ix}}{\partial x}+\frac{\partial p_{iy}}{\partial y}+\frac{\partial p_{iz}}{\partial z}\right) dxdydz, \quad (38.3)$$

or by performing a part of the integration we can write this in the form

$$\frac{dG_i}{\partial t}$$

$$= \int \frac{\partial g_i}{\partial t}\, dv = -\int\int |p_{ix}|_x^{x'}\, dydz - \int\int |p_{iy}|_y^{y'}\, dxdz - \int\int |p_{iz}|_z^{z'}\, dxdy, \quad (38.4)$$

where the limits of integration at the boundary of the volume considered have been denoted by x, x', etc. We have thus related the rate of change of the momentum inside the spatial volume considered to the values of p_{ij} at the boundary. For an *isolated system* this will of course give us the principle of the conservation of momentum.

We may also consider the equations of motion in their earlier form (32.4)

$$\frac{d}{dt}(g_i\, \delta v) = -\frac{\partial t_{ij}}{\partial x_j}\, \delta v, \quad (38.5)$$

which gives the rate of change in the momentum of the element of material lying at the time in question in the volume δv, instead of the rate of change in the density of momentum at the point of location. We can then integrate, this time over the *material located at the instant under consideration inside a given boundary* instead of over a fixed volume in space, and obtain for the rate of change in the momentum of this material

$$\frac{d|G_i|}{dt} = \frac{d}{dt}\int g_i\, dv = -\int\int\int \left(\frac{\partial t_{ix}}{\partial x}+\frac{\partial t_{iy}}{\partial y}+\frac{\partial t_{iz}}{\partial z}\right) dxdydz, \quad (38.6)$$

or by performing a part of the integration we can write this in the form

$$\frac{d|G_i|}{dt}$$

$$= \frac{d}{dt}\int g_i\, dv = -\int\int |t_{ix}|_x^{x'}\, dydz - \int\int |t_{iy}|_y^{y'}\, dxdz - \int\int |t_{iz}|_z^{z'}\, dxdy, \quad (38.7)$$

where the limits of integration at the boundary have been denoted
by x, x', etc., and we used the symbol $|G_i|$ to denote the momentum
of a given amount of the material as distinct from the momentum G_i
inside a given spatial volume. Equation (38.7) relates the rate of
change in the momentum of a given amount of the material to the
forces acting on its surface, and reduces again to the principle of the
conservation of momentum for an isolated system.

To conclude the present section it may also be pointed out that the
results given by the equations (38.2) and (38.4) for the rate of change
in the mass and the three components of momentum inside a given
volume of space can be expressed with the help of our four-dimensional tensor language by a single generalized equation. For this
purpose we start with the equations of motion and continuity in the
very simple form given by (37.9)

$$\partial T^{\mu\nu}/\partial x^\nu = 0, \tag{38.8}$$

where $T^{\mu\nu}$ is the energy-momentum tensor as defined in § 37. By
integrating over the spatial coordinates x^1, x^2, x^3 for the spatial
volume of interest we obtain

$$\iiint \left(\frac{\partial T^{\mu 1}}{\partial x^1} + \frac{\partial T^{\mu 2}}{\partial x^2} + \frac{\partial T^{\mu 3}}{\partial x^3} + \frac{\partial T^{\mu 4}}{\partial x^4} \right) dx^1 dx^2 dx^3 = 0, \tag{38.9}$$

and by performing a part of the integration and rearranging we can
then write

$$\iiint \frac{\partial T^{\mu 4}}{\partial x^4} dx^1 dx^2 dx^3$$

$$= - \iint |T^{\mu 1}|_{x^1}^{x'^1} dx^2 dx^3 - \iint |T^{\mu 2}|_{x^2}^{x'^2} dx^1 dx^3 - \iint |T^{\mu 3}|_{x^3}^{x'^3} dx^1 dx^2, \tag{38.10}$$

where the limits of integration at the boundary have been denoted
by x^1, x'^1, etc.

This result may be called the energy-momentum principle as
applied to a finite region. Noting the values of $T^{\mu\nu}$ as given by
(37.8), we see that with μ taken as 1, 2, or 3 it relates the rate of
change with the time x^4 of a component of momentum within the
region with conditions at the boundary, and is equivalent to the three
equations given by (38.4). And with μ taken as 4 it relates the rate
of change of mass or energy with conditions at the boundary and is
equivalent to (38.2). The equation thus agrees with the idea that
changes in the mass, energy, and momentum within a given region
are due to flow across the boundary, and in the case of an isolated

system is seen to reduce to the principles of the conservation of mass, energy, and momentum.

(b) **The angular momentum of a finite system.** In the case of angular momentum we shall be primarily interested in the amount of momentum associated with a finite system rather than that located in a fixed region in space. For such a system we may then define the component of angular momentum, taken for specificity around the z-axis, in the usual way as given by the integral

$$M_z = \int (xg_y - yg_x)\, dv \qquad (38.11)$$

taken over the *material composing the system* in which we are interested; where x and y are coordinates of the element of the material dv and g_x and g_y are the indicated components of density of momentum at that point. And differentiating this with respect to the time, using (38.5) for the rate of change of the momentum of the element dv, we easily obtain

$$\frac{dM_z}{dt} = \frac{d}{dt} \int (xg_y - yg_x)\, dv = \int \left(-x\frac{\partial t_{yj}}{\partial x_j} + y\frac{\partial t_{xj}}{\partial x_j} + u_x g_y - u_y g_x \right) dv$$
$$(38.12)$$

as an expression for the rate of change with time of this component of the total angular momentum of the system.

This relation can be re-expressed to advantage, however, to show the effect of forces external to the system in changing its angular momentum with the help of a somewhat complicated transformation. Considering the two first terms, we can obtain with the help of a partial integration

$$\int \left(-x\frac{\partial t_{yj}}{\partial x_j} + y\frac{\partial t_{xj}}{\partial x_j} \right) dv$$
$$= \iiint \left(-x\frac{\partial t_{yx}}{\partial x} - x\frac{\partial t_{yy}}{\partial y} - x\frac{\partial t_{yz}}{\partial z} + y\frac{\partial t_{xx}}{\partial x} + y\frac{\partial t_{xy}}{\partial y} + y\frac{\partial t_{xz}}{\partial z} \right) dx\,dy\,dz$$
$$= \iint |-xt_{yx} + yt_{xx}|_x^{x'}\, dy\,dz + \iint |-xt_{yy} + yt_{xy}|_y^{y'}\, dx\,dz +$$
$$+ \iint |-xt_{yz} + yt_{xz}|_z^{z'}\, dx\,dy + \iiint (t_{yx} - t_{xy})\, dx\,dy\,dz, \quad (38.13)$$

where the limits of integration at the boundary of the system have been denoted by x, x', etc. In addition we can write with the help of (36.1)

$$t_{yx} - t_{xy} = p_{yx} - g_y u_x - p_{xy} + g_x u_y = -(u_x g_y - u_y g_x) \quad (38.14)$$

on account of the symmetry of p_{ij}. Substituting (38.13) and (38.14) into (38.12) we then obtain the desired result

$$dM_z/dt = -\iint |xt_{yx}-yt_{xx}|_x^{x'} \, dydz -$$
$$-\iint |xt_{yy}-yt_{xy}|_y^{y'} \, dxdz - \iint |xt_{yz}-yt_{xz}|_z^{z'} \, dxdy. \quad (38.15)$$

In accordance with this expression, we see that the rate of change of the angular momentum of the system is equal to the turning moment of the exterior forces which act on it from the outside. Furthermore, for an *isolated system* in which these forces vanish, the equation evidently reduces to the principle of the conservation of angular momentum $dM_z/dt = 0.$ (38.16)

To complete our consideration of angular momentum we must now point out an important difference between relativistic and Newtonian mechanics. Returning to our original expression for angular momentum (38.11), let us consider a system in a steady state of uniform motion in a straight line, the momentum **g** dv of each element of volume being a constant independent of the time. Under these circumstances it might be expected that the angular momentum of the system would also be a constant independent of the time. Nevertheless, differentiating (38.11) with respect to the time, allowing for the constancy of **g** dv for each element, we obtain for the rate of change of the angular momentum with the time the actual result

$$dM_z/dt = \int (u_x g_y - u_y g_x) \, dv. \quad (38.17)$$

In Newtonian mechanics, since velocity **u** and density of momentum **g** were in the same direction this result would have been equal to zero. In fact we could have written in that case $g_y = \rho u_y$ and $g_x = \rho u_x$, which leads at once to the cancellation of the two terms in the bracket. In relativistic mechanics, nevertheless, we have already seen in equations (35.1) and (35.2) that we can have momentum at right angles to the direction of motion in the case of a stressed medium. Indeed, in relativistic mechanics the relation of the stress to the integrand in (38.17) is given by equation (38.14) already obtained above $(u_x g_y - u_y g_x) = (t_{xy} - t_{yx}).$

Hence in relativistic mechanics, owing to the lack of symmetry of the components of stress t_{ij}, we must conclude that the angular momentum of a stressed body can in general be changing with the time, even when it is in a steady state of motion in a straight line; and an external turning moment can be necessary in order to produce this change in angular momentum and maintain the body in its steady state of motion.

(c) **The right-angled lever as an example.** The apparently paradoxical case of the stressed right-angled lever affords an interesting

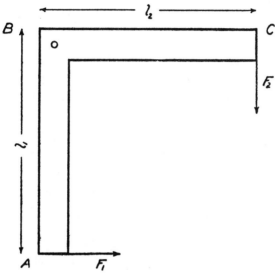

FIG. 3

example of the above conclusion that a turning moment may be necessary to maintain a stressed body in a uniform state of translatory motion.† Consider a right-angled lever as shown in Figure 3, with a pivot at the corner B and opposing forces F_1 and F_2 at the two ends A and C. Let the lever be stationary with respect to a system of proper coordinates S^0, the two lever arms being equal in this system of coordinates

$$l_1^0 = l_2^0$$

and the two forces also being equal

$$F_1^0 = F_2^0.$$

Let us now consider the lever as it appears using a new system of coordinates S with respect to which the lever is moving in the x-direction with the velocity V. Referred to this new system of coordinates

† Laue, *Verh. d. Deutsch. Phys. Ges.* **13**, 513 (1911).

the length l_1 of the arm which lies parallel to the y-axis will evidently be the same as in system S^0

$$l_1 = l_1^0,$$

but the other arm which lies parallel to the direction of motion will have the shorter length in accordance with the Lorentz contraction

$$l_2 = l_2^0 \sqrt{(1 - V^2/c^2)}.$$

Furthermore, in accordance with our transformation equations for force (25.3), we shall have for the forces acting at the two ends

$$F_1 = F_1^0$$

and $$F_2 = F_2^0 \sqrt{(1 - V^2/c^2)}.$$

With the help of these values for the forces and lever arms we can now calculate for the turning moment acting on the lever around the pivot B

$$F_1 l_1 - F_2 l_2 = F_1^0 l_1^0 - F_2^0 l_2^0 (1 - V^2/c^2) = F_1^0 l_1^0 V^2/c^2 = F_1 l_1 V^2/c^2.$$

Since the lever is obviously not rotating about the pivot B when looked at either from the point of view of system S^0 or S, we are thus led to a simple example of a stressed body in uniform translatory motion which nevertheless needs a turning moment to maintain this state of motion.

This result is, however, in entire agreement with the conclusions reached in the preceding section, since we can easily show that the angular momentum of the system is indeed actually being increased by a flow of energy into it at exactly the rate demanded by this turning moment. Since the force F_1 is doing the work $F_1 V$ per second at the point A, a stream of energy of this amount is evidently continuously entering the system at A and flowing out through the pivot at B, where an equal and opposite force is acting. In accordance then with our ideas as to the relation of mass and energy, we hence have the mass $F_1 V/c^2$ per second entering the system at A and are thus increasing its angular momentum at the rate

$$F_1 \frac{V}{c^2} V l_1 = F_1 l_1 \frac{V^2}{c^2}.$$

This, however, is the very result which was found above for the turning moment acting on the lever and we thus have the entire resolution of any apparent paradox.

(d) The complete static system. The complications, which can arise in the case of the uniform translatory motion of a stressed body, as just illustrated by the right-angled lever which was acted on by a turning moment and yet did not turn, are much reduced if we consider the complete static system involved. In the above example this would mean the consideration of the stressed lever together with the housing which carries its pivot and carries the supports for the springs that can be thought of as exerting the forces F_1 and F_2 on its two ends. And in general we shall understand by a complete static system, an entire structure which can remain in a permanent state of rest with respect to a set of proper coordinates S^0, without the necessity for any forces from the outside.

In the first place, since such systems will evidently remain in a state of uniform translatory motion with respect to any system of Lorentz coordinates S, without the application of forces from outside, it is evident that no turning moment is necessary for their steady motion and in accordance with (38.16) that their angular momentum as a whole is not changing with the time.

In the second place, we can demonstrate with the help of a certain amount of calculation that the expressions for the mass and momentum of such complete systems reduce to a very simple form. This we shall now show.

With respect to a set of proper coordinates S^0, we can write the equations of motion (36.6) for the system in the form

$$\frac{\partial p_{ij}^0}{\partial x_j} + \frac{\partial g_i^0}{\partial t} = 0, \tag{38.18}$$

and since the velocity of all parts of the system is zero in these coordinates, the densities of momentum g_i^0 will everywhere have the constant value zero in accordance with (36.5), so that we can rewrite this in the form

$$\frac{\partial p_{ij}^0}{\partial x_j} = \frac{\partial p_{ix}^0}{\partial x} + \frac{\partial p_{iy}^0}{\partial y} + \frac{\partial p_{iz}^0}{\partial z} = 0. \tag{38.19}$$

Let us now integrate this expression over a volume which is bounded on one side by a plane perpendicular to the x-axis, that cuts through the system at some arbitrary point, say x', and is completed by any surface whatever which lies entirely outside the system. We then have

$$\iiint \left(\frac{\partial p_{ix}^0}{\partial x} + \frac{\partial p_{iy}^0}{\partial y} + \frac{\partial p_{iz}^0}{\partial z} \right) dx\,dy\,dz$$

$$= \iint |p_{ix}^0|_x^{x'} \, dy\,dz + \iint |p_{iy}^0|_y^{y'} \, dx\,dz + \iint |p_{iz}^0|_z^{z'} \, dx\,dy$$

$$= 0, \tag{38.20}$$

where the limits of that part of the integration which has been performed are denoted by x, x', etc. Since, however, all these limits except x' lie outside the system itself, the 'stresses' p_{ij}^0 will all be zero at the limits except for p_{ix}^0 on the surface $x = x'$, and (38.20) will then reduce to

$$\iint p_{ix}^0 \, dydz = 0 \tag{38.21}$$

over a plane which cuts through the system at any arbitrary point perpendicular to the x-axis. Multiplying this by dx and integrating over the whole system we then obtain the useful result

$$\iiint p_{ix}^0 \, dxdydz = 0,$$

or in general

$$\iiint p_{ij}^0 \, dxdydz = 0, \tag{38.22}$$

since the plane cutting the system can be taken perpendicular to any one of the three axes.

This result now permits us to obtain the desired simple expressions for the mass and momentum of the system as a whole when referred to any set of coordinates S. For simplicity let us take the system as moving with respect to S in the x-direction with the velocity u. We can then arrive at its total mass by integrating the expression for the density given by (36.4) over the whole volume. We obtain

$$m = \int \rho \, dv = \int \frac{\rho_{00} + p_{xx}^0 u^2/c^4}{1 - u^2/c^2} \sqrt{(1 - u^2/c^2)} \, dv^0, \tag{38.23}$$

where we have substituted for the element of volume dv in terms of the element of proper volume dv^0, with the help of the Lorentz contraction. Noting (38.22) this evidently gives us the simple result

$$m = \int \frac{\rho_{00}}{\sqrt{(1 - u^2/c^2)}} \, dv^0 = \frac{m_0}{\sqrt{(1 - u^2/c^2)}}, \tag{38.24}$$

where m_0 is the rest-mass of the system. And carrying out similar integrations, using the expressions for momentum density given by (36.5), we at once obtain for the total momentum of the system the components

$$G_x = \frac{m_0}{\sqrt{(1 - u^2/c^2)}} u \qquad G_y = G_z = 0. \tag{38.25}$$

Hence for a complete static system the expressions for mass and momentum reduce to the same simple form that we originally found for a particle, and there are no complexities which would result from components of momentum not parallel to the velocity. Moreover, if we act on such systems with external forces which exert no turning moment, but merely produce what may be called a quasi-stationary adiabatic acceleration that leaves the internal condition unchanged from the point of view of a local observer travelling with the system, we can then apply the previous dynamics of a particle to the phenomena.

Vice versa this last result now permits us to regard the mechanics of a particle acted on by external forces as a special case of the mechanics of a continuous medium, if we treat the particle as a complete static system too small to be acted on by a turning moment and with an internal state which remains unchanged from the point of view of a local observer. This conclusion is satisfying as an indication of the logical coherence of our whole system of mechanics.

It may also be noted in closing that an extension of the conclusions as to the behaviour of a complete static system to include electrical as well as mechanical phenomena may be regarded as explaining the result obtained in the well-known Trouton-Noble experiments, which demonstrated that there is no tendency for a charged condenser moving with the earth to turn about its axis. Calculation shows that the field of a charged condenser which is in motion should exert a turning moment on the material parts of which the system is constructed, and a somewhat lengthy and complicated computation is necessary to show that this turning moment is just sufficient to account for an increase in the angular momentum of these stressed material parts which is occurring, even when the condenser is not turning, on account of a transverse energy flow. If, however, we regard the field and the material parts of the condenser taken together as forming a complete static system we can at once conclude, in accordance with the previous discussion, that the system as a whole can be in uniform translatory motion without exhibiting any tendency to turn about its axis. This conclusion and the experimental findings are of course both what would be directly demanded by the first postulate of the special theory of relativity.

IV

SPECIAL RELATIVITY AND ELECTRODYNAMICS

Part I. ELECTRON THEORY

39. The Maxwell-Lorentz field equations

In the present chapter we shall consider the relations between special relativity and electrodynamics, basing the treatment in Part I on the point of view of the Lorentz electron theory and in Part II on a macroscopic point of view.

Since the classical Newtonian mechanics was developed on the tacit basis of the simple Galilean transformation equations (8.7) for space and time, while the relativistic mechanics of Einstein, on the other hand, was explicitly developed on the basis of the Lorentz transformation (8.1), there are very conspicuous differences in the nature of the two resulting theories, as has been shown in the preceding chapter. In the case of electrodynamics, however, the introduction of special relativity produced a much smaller change in theory, since the development of electrodynamics in the hands of Lorentz had already actually led the way to the transformation equations now associated with his name.

On account of this small divergence between classical and relativistic electromagnetic theory, we can now take as a postulatory starting-point for relativistic electrodynamics the well-known Maxwell-Lorentz field equations which we write in the vector form

$$\operatorname{div} \mathbf{E} = \rho, \tag{39.1}$$

$$\operatorname{div} \mathbf{H} = 0, \tag{39.2}$$

$$\operatorname{curl} \mathbf{E} = -\frac{1}{c}\frac{\partial \mathbf{H}}{\partial t}, \tag{39.3}$$

$$\operatorname{curl} \mathbf{H} = \frac{1}{c}\frac{\partial \mathbf{E}}{\partial t} + \rho\frac{\mathbf{u}}{c}, \tag{39.4}$$

where $\mathbf{E}$ and $\mathbf{H}$ are the electric and magnetic field strengths, ρ is the density of electric charge, $\mathbf{u}$ the velocity with which it is moving, and c the velocity of light.

These equations were proposed by Lorentz as a basis for electrodynamics long before the modern development of quantum mechanics, and assume possibilities of attaching significance to microscopic quantities not in accord with present modes of thought. Thus the quantities ρ and u occuring in these equations were regarded as giving the microscopic density of charge and velocity, specified when necessary even for points within the electron, and E and H were taken as the forces on unit charge and unit pole even under conditions where no actual experiments with existing test charges and poles could be conceived for measuring their values.

This difficulty with the Lorentz axioms, arising from the classical microscopic point of view adopted in their development, is unfortunate. Nevertheless, at the time of writing no completely satisfactory quantum electrodynamics appears to have been developed, and in any case we can be sure that the conclusions drawn from the Lorentz starting-point will have much in common with later developments. Furthermore in Part II of the present chapter we shall give attention to a phenomenological treatment of the electrodynamics of ponderable bodies, which will be more closely analogous to the entirely macroscopic treatment which we were able to give to mechanics in the preceding chapter.

Several well-known conclusions that are customarily drawn directly from the Lorentz field equations may be mentioned before proceeding.

If we take the divergence of (39.4) and introduce (39.1) we at once obtain

$$\frac{\partial \rho}{\partial t} + \operatorname{div}(\rho \mathbf{u}) = 0, \tag{39.5}$$

since div curl H is of course equal to zero. This equation of continuity for the density of electric charge is an expression of the fact of the conservation of total charge.

Secondly, if we consider the field equations (39.1–4) for the case of free space with $\rho = 0$, we can easily transform them with the help of equation (13) in Appendix II into the form

$$\frac{\partial^2 \mathbf{E}}{\partial x^2} + \frac{\partial^2 \mathbf{E}}{\partial y^2} + \frac{\partial^2 \mathbf{E}}{\partial z^2} - \frac{1}{c^2}\frac{\partial^2 \mathbf{E}}{\partial t^2} = 0 \tag{39.6}$$

and

$$\frac{\partial^2 \mathbf{H}}{\partial x^2} + \frac{\partial^2 \mathbf{H}}{\partial y^2} + \frac{\partial^2 \mathbf{H}}{\partial z^2} - \frac{1}{c^2}\frac{\partial^2 \mathbf{H}}{\partial t^2} = 0, \tag{39.7}$$

which are the well-known wave equations for the propagation of electromagnetic disturbances in free space with the velocity c.

Finally, if we introduce the so-called scalar potential ϕ and vector potential **A** with the help of the equations

$$E = -\operatorname{grad}\phi - \frac{1}{c}\frac{\partial A}{\partial t} \tag{39.8}$$

and

$$H = \operatorname{curl} A, \tag{39.9}$$

it can be shown that these new quantities can be taken so as to satisfy the differential equations

$$\frac{\partial^2\phi}{\partial x^2} + \frac{\partial^2\phi}{\partial y^2} + \frac{\partial^2\phi}{\partial z^2} - \frac{1}{c^2}\frac{\partial^2\phi}{\partial t^2} = -\rho \tag{39.10}$$

and

$$\frac{\partial^2 A}{\partial x^2} + \frac{\partial^2 A}{\partial y^2} + \frac{\partial^2 A}{\partial z^2} - \frac{1}{c^2}\frac{\partial^2 A}{\partial t^2} = -\rho\frac{u}{c}. \tag{39.11}$$

These then have the well-known solutions

$$\phi = \frac{1}{4\pi}\int \frac{[\rho]}{r}\,dv \tag{39.12}$$

and

$$A = \frac{1}{4\pi c}\int \frac{[\rho u]}{r}\,dv, \tag{39.13}$$

where the integration is to be carried out over the whole of space, r is the distance from the point of interest to the element of volume dv, and the square brackets signify that the value of the quantity inside is to be taken at a time r/c earlier than that of the instant of interest. These results taken with (39.8, 9) thus provide a complete solution for the field in terms of the distribution of charge and current.

40. The transformation equations for E, H, and ρ

The postulatory basis for electrodynamics provided by the field equations (39.1–4), which apply in the first instance to some particular set of coordinates, must now be extended in such a way as to include the essential ideas of the special theory of relativity. To do this we must require in accordance with the two postulates of relativity firstly that equations of exactly the same form as the above shall correctly describe electromagnetic phenomena in all sets of coordinates in uniform relative motion, secondly that the transformation equations for the kinematical quantities occurring in the above equations shall be those already provided by the Lorentz transformation, and thirdly that the equations for transforming the newly introduced electromagnetic quantities **E**, **H**, and ρ from one set of coordinates to a second set, moving relative thereto, shall be entirely symmetrical with those for the reverse transformation except for the sign of the relative velocity between the two sets of coordinates.

Considering the transformation from an original set of coordinates S to a second set S' moving relative thereto with the velocity V, taken for simplicity in the x-direction, the foregoing conditions can, as a matter of fact, all be satisfied by taking the Lorentz transformation equations (8.1) for the coordinates x, y, z, and t, together with (10.1) for the components u_x, u_y, and u_z of the velocity u, and by taking for the transformation of the electromagnetic quantities the equations proposed by Einstein

$$E_x = E'_x$$

$$E_y = \frac{E'_y + \frac{V}{c}H'_z}{\sqrt{(1 - V^2/c^2)}}$$ (40.1)

$$E_z = \frac{E'_z - \frac{V}{c}H'_y}{\sqrt{(1 - V^2/c^2)}},$$

$$H_x = H'_x$$

$$H_y = \frac{H'_y - \frac{V}{c}E'_z}{\sqrt{(1 - V^2/c^2)}}$$ (40.2)

$$H_z = \frac{H'_z + \frac{V}{c}E'_y}{\sqrt{(1 - V^2/c^2)}},$$

$$\rho = \frac{\rho'(1 + u'_x V/c^2)}{\sqrt{(1 - V^2/c^2)}}.$$ (40.3)

By substituting these transformation equations together with those for coordinates and velocities, the field equations (39.1–4) are indeed found by a somewhat lengthy calculation to be unchanged in form when expressed in the primed instead of in the unprimed variables. Furthermore, on solving the above equations for the primed in terms of unprimed quantities, the equations for the reverse transformation are also found to be of the same form except for the sign of V. Hence these equations do satisfy the requirements of relativity.

As a result of the combined appearance of components of E' and H' in the transformation equations (40.1, 2) for both E and H, it is to be noted that the separation of an electromagnetic field into electric and magnetic portions is dependent on the state of motion of the set of coordinates which is being used. A field which would be regarded as purely electrostatic in system S' would have magnetic components in system S.

It should also be noted that the transformation equation for electric density (40.3) is such as to make total electric charge an invariant, since by the introduction of (11.1) this equation can be rewritten in the form

$$\frac{\rho}{\rho'} = \frac{\sqrt{(1-u'^2/c^2)}}{\sqrt{(1-u^2/c^2)}}. \tag{40.4}$$

This shows that the measurements of electric density in the two systems are inversely proportional to the factors which determine the Lorentz contraction, and hence that the measurements of a total charge e will agree giving us

$$e = e'. \tag{40.5}$$

This is in accordance with the idea that electric charge is essentially a quantity having discrete magnitude to be determined by counting numbers of electrons and protons, and hence necessarily invariant for different observers. Conversely, of course, this invariance could be used to establish (40.3).

41. The force on a moving charge

In addition to the four field equations, the Lorentz electromagnetic theory as originally developed contained as part of its postulatory basis a fifth equation for the force acting on a moving charge of electricity. It is, however, a gratifying result of the present method of development that this fifth fundamental equation can be derived with the help of the transformation equations for force obtained from the mechanics of a particle, and hence does not have to be taken as a separate axiom.†

Consider a charge e moving with respect to system S with any given velocity V, and for simplicity choose the direction of axes in S in such a way that the motion is in the x-direction, giving us

$$u_x = V \quad u_y = 0 \quad u_z = 0. \tag{41.1}$$

To calculate the force in system S acting on this moving charge, let us now first consider the force acting on it in a second system of coordinates S', which is itself moving relative to S with the same velocity V as the charge e. Since the charge is at rest in this new system S' the force acting on it in these coordinates will be immediately given as the product of charge by electric field strength,

† Tolman, *Phil. Mag.* **21**, 296 (1911).

owing to the definition of this quantity as the force acting on unit stationary charge, so that we can write at once

$$F'_x = e'E'_x$$
$$F'_y = e'E'_y \qquad (41.2)$$
$$F'_z = e'E'_z.$$

Introducing the transformation equations for force (25.3), for the invariance of electric charge (40.5), and for the components of electric field strength (40.1) and noting the values for u_x, u_y, and u_z given by (41.1) we then immediately obtain

$$F_x = eE_x$$
$$F_y = e\left(E_y - \frac{u_x H_z}{c}\right) \qquad (41.3)$$
$$F_z = e\left(E_z + \frac{u_x H_y}{c}\right).$$

Removing now the specialization involved in choosing the x-axis parallel to the motion, we can then write in general for the force acting on a charge e moving with the velocity $\mathbf{u}$ the vector expression

$$\mathbf{F} = e\left(\mathbf{E} + \frac{1}{c}[\mathbf{u} \times \mathbf{H}]\right), \qquad (41.4)$$

or for the force per unit charge

$$\mathbf{F} = \mathbf{E} + \frac{1}{c}[\mathbf{u} \times \mathbf{H}], \qquad (41.5)$$

which is the desired fifth fundamental equation of the Maxwell-Lorentz theory of electromagnetism.

42. The energy and momentum of the electromagnetic field

With the help of the above equation and the four field equations, expressions can be obtained by well-known methods for the energy and momentum resident in an electromagnetic field. The results are so important for the theory of relativity as to warrant the presentation of the calculations by which they are obtained.

In accordance with equation (41.5) the component of force acting on a moving charge due to the magnetic field lies at right angles to the direction of motion. It hence does no work on the charge, and we can write, for the total rate at which the electromagnetic field is doing work on the charged material inside a given boundary, the expression

$$dW/dt = \int (\rho\mathbf{u} \cdot \mathbf{E})\, dv, \qquad (42.1)$$

where ρ is the density of electric charge, and we shall regard the integration as taken over a definite *fixed volume in space*.

This result can be changed, however, for our present purposes with the help of the field equations, by substituting for ρu the value given by (39.4), and by adding to the integrand a quantity which is evidently zero in accordance with (39.3). We can then rewrite (42.1) in the form

$$\frac{dW}{dt} = -\int \frac{\partial}{\partial t}\left(\frac{E^2+H^2}{2}\right) dv - c \int (\mathbf{H}\cdot\operatorname{curl}\mathbf{E}-\mathbf{E}\cdot\operatorname{curl}\mathbf{H})\, dv, \quad (42.2)$$

and by well-known relations of vector analysis (equation 17, Appendix II) this can be rewritten in the form

$$\frac{dW}{dt} = -\int \frac{\partial}{\partial t}\left(\frac{E^2+H^2}{2}\right) dv - c \int [\mathbf{E}\times\mathbf{H}]_n\, d\sigma, \quad (42.3)$$

where the last term is integrated over the surface surrounding the volume under consideration and the subscript n denotes the outward normal component of the vector in question.

This last equation has a simple interpretation if we include the idea of the conservation of energy as part of our postulatory basis. We must then regard the rate at which work is being done on the material within the boundary as equal to the rate at which energy is being abstracted from the electromagnetic field. The first term on the right-hand side of (42.3) can hence be interpreted as the rate of change in the energy of the electromagnetic field lying inside the volume under consideration, and the second term can be interpreted as the rate of flow of electromagnetic energy across the boundary of that volume. For the density of electromagnetic energy we can then evidently take the well-known expression

$$\rho = \frac{E^2+H^2}{2} \quad (42.4)$$

and for the density of energy flow, or Poynting vector,

$$\mathbf{s} = c[\mathbf{E}\times\mathbf{H}]. \quad (42.5)$$

Or, making use of the relations between mass, energy, and momentum discussed in § 27, we can write for the density of electromagnetic mass

$$\rho = \frac{1}{c^2}\frac{E^2+H^2}{2}, \quad (42.6)$$

and for the density of momentum

$$\mathbf{g} = \frac{1}{c}[\mathbf{E}\times\mathbf{H}]. \quad (42.7)$$

43. The electromagnetic stresses

With the help of our fundamental equations we can in addition obtain the known expressions of Maxwell for the stresses in the field, which will also be important for the theory of relativity. To do this we consider the effect of the electromagnetic field in changing the momentum of electrically charged matter instead of its energy as was done in the last section.

In accordance with equation (41.5) we can write for the rate at which the electromagnetic field is changing the momentum of the charged material inside a given boundary, the expression

$$\frac{d\mathbf{G}}{dt} = \int \rho\left(\mathbf{E} + \frac{1}{c}[\mathbf{u}\times\mathbf{H}]\right) dv, \tag{43.1}$$

where we shall again regard the integration as taken over a definite *fixed volume* in space. And substituting for ρ and $\rho\mathbf{u}/c$ the values given by the field equations (39.1) and (39.4), this can be rewritten as

$$\frac{d\mathbf{G}}{dt} = \int \left(\mathbf{E}\operatorname{div}\mathbf{E} + (\operatorname{curl}\mathbf{H})\times\mathbf{H} - \frac{1}{c}\frac{\partial\mathbf{E}}{\partial t}\times\mathbf{H}\right) dv$$

$$= \int \left(\mathbf{E}\operatorname{div}\mathbf{E} + (\operatorname{curl}\mathbf{H})\times\mathbf{H} + \frac{1}{c}\mathbf{E}\times\frac{\partial\mathbf{H}}{\partial t} - \frac{1}{c}\frac{\partial}{\partial t}[\mathbf{E}\times\mathbf{H}]\right) dv$$

which by (39.3) and (42.7) becomes

$$\frac{d\mathbf{G}}{dt} = \int \left(\mathbf{E}\operatorname{div}\mathbf{E} + (\operatorname{curl}\mathbf{H})\times\mathbf{H} + (\operatorname{curl}\mathbf{E})\times\mathbf{E} - \frac{\partial\mathbf{g}}{\partial t}\right) dv, \tag{43.2}$$

where $\mathbf{g}$ is the density of electromagnetic momentum in the field. Or considering for specificity the component of momentum in the x-direction, and writing out in detail the values for the x-components of the vectors involved, this will give us after a somewhat long but straightforward transformation

$$\frac{dG_x}{dt} = \int \left[\frac{1}{2}\frac{\partial}{\partial x}(E_x^2 - E_y^2 - E_z^2 + H_x^2 - H_y^2 - H_z^2) + \frac{\partial}{\partial y}(E_x E_y + H_x H_y) + \right.$$

$$\left. + \frac{\partial}{\partial z}(E_x E_z + H_x H_z) - \frac{\partial g_x}{\partial t}\right] dv. \tag{43.3}$$

This result has a simple interpretation, however, if we now define the stresses in the field as given in general by the symmetrical expressions

$$p_{ii} = -\tfrac{1}{2}(E_i^2 - E_j^2 - E_k^2 + H_i^2 - H_j^2 - H_k^2)$$

$$p_{ij} = -(E_i E_j + H_i H_j), \tag{43.4}$$

which will permit us to rewrite (43.3) in the general form

$$\frac{dG_i}{dt} + \int \frac{\partial g_i}{\partial t}\, dv = -\iiint \left(\frac{\partial p_{ix}}{\partial x} + \frac{\partial p_{iy}}{\partial y} + \frac{\partial p_{iz}}{\partial z}\right) dx\,dy\,dz, \quad (43.5)$$

or, by performing a part of the integration, in the form

$$\frac{dG_i}{dt} + \int \frac{\partial g_i}{\partial t}\, dv$$

$$= -\iint |p_{ix}|_x^{x'}\, dy\,dz - \iint |p_{iy}|_y^{y'}\, dx\,dz - \iint |p_{iz}|_z^{z'}\, dx\,dy, \quad (43.6)$$

where the limits of integration are denoted by x, x', etc. These equations show that the change in the total momentum, mechanical plus that of the electromagnetic field, inside the boundary may be calculated from the electromagnetic stresses at the surface as defined by (43.4). Furthermore the appropriateness of the name electromagnetic stresses for the quantities p_{ij}' is now evident from the form of these equations.

44. Transformation equations for electromagnetic densities and stresses

We have thus obtained expressions in the last two sections for quantities which may be fittingly regarded as the density of electromagnetic mass, density of electromagnetic momentum, and the components of electromagnetic stress. And since these quantities are all defined with the help of equations (42.6), (42.7), and (43.4) in terms of the electromagnetic field strengths **E** and **H**, we can evidently obtain transformation equations from one set of coordinates to another with the help of the transformations (40.1) and (40.2) for the components of these two vectors. The calculations for doing this are somewhat tedious but perfectly straightforward. We obtain for the transformation from a set of coordinates S to a new set S' moving with the relative velocity V parallel to the x-axis the expressions

$$\rho = \frac{\rho' + p_{xx}' V^2/c^4 + 2g_x' V/c^2}{1 - V^2/c^2} \tag{44.1}$$

$$g_x = \frac{(c^2\rho' + p_{xx}')V/c^2 + (1 + V^2/c^2)g_x'}{1 - V^2/c^2} \qquad g_y = \frac{p_{xy}' V/c^2 + g_y'}{\sqrt{(1 - V^2/c^2)}} \tag{44.2}$$

$$p_{xx} = \frac{p_{xx}' + \rho' V^2 + 2g_x' V}{1 - V^2/c^2} \qquad p_{yy} = p_{yy}'$$

$$p_{xy} = \frac{p_{xy}' + g_y' V}{\sqrt{(1 - V^2/c^2)}} \qquad p_{yz} = p_{yz}' \tag{44.3}$$

and the transformation equations for the remaining components can be obtained from the above on account of the symmetry in y and z and of p_{ij}.

It should be specially noted that these equations reduce to the same form as equations (36.3–5) for the analogous mechanical quantities if the new set of coordinates S' are specially chosen with a velocity $V = u$ and the components of momentum g'_x, g'_y, and g'_z are set equal to zero, as was the case for the proper coordinates used in the mechanical case, and can be shown to be of exactly the same form as the mechanical equations for the more general transformation here considered.

45. Combined result of mechanical and electromagnetic actions

In the preceding chapter we obtained in § 38 expressions for the effect of mechanical actions in changing the mass (or energy) and momentum inside a given fixed spatial volume, and in the present chapter have obtained in §§ 42, 43 analogous expressions for the effect of electromagnetic actions in changing these same quantities. We may now consider the comparison and combination of the two kinds of effects. In doing so we shall distinguish between mechanical and electromagnetic quantities with the help of brackets carrying the subscripts (me) and (em) respectively, and for brevity we shall let the double occurrence of a subscript (j) in a given term denote as previously a summation over the three spatial coordinates x, y, and z.

We can then write in accordance with equations (38.1) and (38.3) as expressions for the effects of *mechanical action* on the mass and momentum in a given spatial volume

$$\int \frac{\partial}{\partial t} [\rho]_{me} \, dv = - \int \frac{\partial}{\partial x_j} [g_j]_{me} \, dv \qquad (45.1)$$

and

$$\int \frac{\partial}{\partial t} [g_i]_{me} \, dv = - \int \frac{\partial}{\partial x_j} [p_{ij}]_{me} \, dv. \qquad (45.2)$$

On the other hand, noting that the rate at which work is being done on matter is c^2 times the rate at which its mass is being increased, we can write with the help of the equations given in §§ 42, 43 as expressions for the effects of *electromagnetic action* again on the *mechanical* mass and momentum in a given spatial volume

$$\int \frac{\partial}{\partial t}[\rho]_{me}\, dv = -\int \left\{\frac{\partial}{\partial t}[\rho]_{em} + \frac{\partial}{\partial x_j}[g_i]_{cm}\right\} dv \qquad (45.3)$$

and

$$\int \frac{\partial}{\partial t}[g_i]_{me}\, dv = -\int \left\{\frac{\partial}{\partial t}[g_i]_{em} + \frac{\partial}{\partial x_j}[p_{ij}]_{em}\right\} dv. \qquad (45.4)$$

Since in general there will be a simultaneous mechanical and electromagnetic action on mass and momentum, we are now tempted to combine these equations by addition to obtain the total rate of change of these quantities. Before doing this, however, we must emphasize the very different character of the considerations by which we were led to the mechanical and to the electromagnetic equations.

The mechanical equations were obtained from a macroscopic phenomenological point of view. The quantities ρ, g, and p_{ij} occurring in them are all macroscopic in character and are defined with the help of equations (36.3–5) in terms of macroscopic quantities which could be directly measured by a local observer moving with the material under consideration. Furthermore, the theoretical treatment for obtaining the equations of mechanics included no microscopic considerations, but depended on natural, although perhaps not always inevitable, extensions of conclusions drawn from actual macroscopic experiments on the conservation of mass and momentum, on the postulates of relativity and on the phenomenological behaviour of elastic bodies when at rest. Hence we can expect the treatment to be relatively unaffected by the development of quantum mechanics.

On the other hand, our development of electromagnetic theory has so far been based on the microscopic point of view adopted in the classical electron theory of Lorentz. The quantities $[\rho]_{me}$, $[\rho]_{em}$, $[g]_{em}$, and $[p_{ij}]_{em}$ occurring in (45.3, 4) are microscopic in character and are regarded as referring to an exact point in space and instant in time even under conditions when no conceptual experiment can be devised for determining their values. Hence we must expect the treatment that we have given to be altered by the development of a satisfactory quantum electrodynamics, although many of the results when applied to macroscopic phenomena will certainly be unchanged.

The macroscopic character of the quantities in (45.1, 2) and the microscopic character of those in (45.3, 4) makes the immediate addition of the two kinds of action on mechanical density and momentum logically unsound. If, however, we assume that a correct process of averaging to obtain macroscopic instead of microscopic densities would leave the electromagnetic equations (45.3, 4) unaltered in form, we should then feel justified in adding the two kinds of action. With some change in order we should thus obtain

$$\int \left\{ \frac{\partial}{\partial t}[\rho]_{me} + \frac{\partial}{\partial t}[\rho]_{em} + \frac{\partial}{\partial x_j}[g_j]_{me} + \frac{\partial}{\partial x_j}[g_j]_{em} \right\} dv = 0 \qquad (45.5)$$

and

$$\int \left\{ \frac{\partial}{\partial t}[g_i]_{me} + \frac{\partial}{\partial t}[g_i]_{em} + \frac{\partial}{\partial x_j}[p_{ij}]_{me} + \frac{\partial}{\partial x_j}[p_{ij}]_{em} \right\} dv = 0. \qquad (45.6)$$

Or, combining mechanical and electromagnetic quantities which are of the same nature, and changing to the differential form, we could write

$$\frac{\partial \rho}{\partial t} + \frac{\partial g_j}{\partial x_j} = 0 \qquad (45.7)$$

and

$$\frac{\partial g_i}{\partial t} + \frac{\partial p_{ij}}{\partial x_j} = 0 \qquad (45.8)$$

for the dependence of the total densities of mass and momentum on the time. These expressions are of exactly the same form as our original equations of continuity and motion (36.6, 7) for the purely mechanical case.

46. Four-dimensional expression of the electron theory

(a) The field equations. The field equations (39.1–4) on which the electron theory is based can readily be expressed in the four-dimensional language which we have previously used for the space-time continuum. This is often very advantageous.

To obtain such an expression we shall again make use of the Galilean coordinates originally given by (20.2)

$$x^1 = x \quad x^2 = y \quad x^3 = z \quad x^4 = ct, \qquad (46.1)$$

corresponding to the simple formula for interval

$$ds^2 = -(dx^1)^2 - (dx^2)^2 - (dx^3)^2 + (dx^4)^2, \qquad (46.2)$$

which is always possible in the flat space-time of special relativity, and shall introduce two vectors on which the analysis will be made to depend.

The first of these is the so-called current vector J^μ, which can be defined in general for any system of coordinates by the expression

$$J^\mu = \rho_0 \, dx^\mu/ds, \qquad (46.3)$$

where ρ_0 is the proper density of electric charge at the point of interest as measured by a local observer and dx^μ/ds is its generalized velocity. In the special coordinates (46.1), the components of the current vector evidently become

$$
\begin{aligned}
J^\mu &= \left(\rho_0 \frac{dx^1}{ds}, \quad \rho_0 \frac{dx^2}{ds}, \quad \rho_0 \frac{dx^3}{ds}, \quad \rho_0 \frac{dx^4}{ds} \right) \\
&= \left(\rho_0 \frac{dx^4}{ds} \frac{dx^1}{dx^4}, \quad \rho_0 \frac{dx^4}{ds} \frac{dx^2}{dx^4}, \quad \rho_0 \frac{dx^4}{ds} \frac{dx^3}{dx^4}, \quad \rho_0 \frac{dx^4}{ds} \right) \\
&= \left(\rho \frac{u_x}{c}, \quad \rho \frac{u_y}{c}, \quad \rho \frac{u_z}{c}, \quad \rho \right),
\end{aligned}
\tag{46.4}
$$

since
$$ ds/dx^4 = \sqrt{(1-u^2/c^2)} \tag{46.5} $$

is the factor for the Lorentz contraction of the moving charge.

The second vector to be introduced is the so-called *generalized potential* ϕ^μ, which can be defined by taking its components in the coordinates (46.1) as given by

$$ \phi^\mu = (A_x, A_y, A_z, \phi) \tag{46.6} $$

in terms of the ordinary vector potential **A** and scalar potential ϕ previously introduced by equations (39.8–11). By applying the Lorentz transformation to the components of this vector as given by (39.12, 13), it can be shown that ϕ^μ will depend on **A** and ϕ in the way given in all systems of coordinates of the type (46.1).

With the help of the *covariant* associate ϕ_μ of this generalized potential we may also define the so-called *electromagnetic tensor* $F_{\mu\nu}$ by the tensor equation

$$ F_{\mu\nu} = \frac{\partial \phi_\mu}{\partial x^\nu} - \frac{\partial \phi_\nu}{\partial x^\mu}. \tag{46.7} $$

In the special coordinates (46.1), the covariant expression for the potential will have, in accordance with (20.7) the components

$$ \phi_\mu = (-A_x, -A_y, -A_z, \phi) \tag{46.8} $$

and the components of $F_{\mu\nu}$ in terms, of the electric and magnetic field strengths **E** and **H** are easily calculated from (39.8, 9) and found to have the values

$$
F_{\mu\nu} = \begin{array}{cccc}
0 & H_z & -H_y & E_x \\
-H_z & 0 & H_x & E_y \\
H_y & -H_x & 0 & E_z \\
-E_x & -E_y & -E_z & 0,
\end{array}
\tag{46.9}
$$

while in accordance with (20.8) its contravariant associate has the components

$$F^{\mu\nu} = \begin{array}{cccc} 0 & H_z & -H_y & -E_x \\ -H_z & 0 & H_x & -E_y \\ H_y & -H_x & 0 & -E_z \\ E_x & E_y & E_z & 0. \end{array} \qquad (46.10)$$

With the help of the quantities which we have thus defined, the content of the Lorentz field equations can now be expressed in very simple form by the two equations

$$\frac{\partial F_{\mu\nu}}{\partial x^\sigma} + \frac{\partial F_{\nu\sigma}}{\partial x^\mu} + \frac{\partial F_{\sigma\mu}}{\partial x^\nu} = 0 \qquad (46.11)$$

and

$$\partial F^{\mu\nu}/\partial x^\nu = J^\mu. \qquad (46.12)$$

The first of these equations is easily seen from the definition of $F_{\mu\nu}$ given by (46.7) to be an identity. It is a tensor equation true in all systems of coordinates if true in one [see equation (41), Appendix III]. The second equation must be regarded as an independent postulate and, since $\partial F^{\mu\nu}/\partial x^\nu$ is the form assumed in our present simple coordinates by the contracted covariant derivative of $F^{\mu\nu}$, may also be regarded as expressing a tensor relation.

Assigning to μ, ν, and σ the different values 1, 2, 3, 4, and introducing the components of J^μ, $F_{\mu\nu}$, and $F^{\mu\nu}$ which are given above by (46.4, 9, 10), it is readily shown that the first of these equations (46.11) is equivalent to the set of equations

$$\frac{\partial H_x}{\partial x} + \frac{\partial H_y}{\partial y} + \frac{\partial H_z}{\partial z} = 0 \qquad (46.13)$$

and

$$\frac{\partial E_z}{\partial y} - \frac{\partial E_y}{\partial z} = -\frac{1}{c}\frac{\partial H_x}{\partial t},$$

$$\frac{\partial E_x}{\partial z} - \frac{\partial E_z}{\partial x} = -\frac{1}{c}\frac{\partial H_y}{\partial t}, \qquad (46.14)$$

$$\frac{\partial E_y}{\partial x} - \frac{\partial E_x}{\partial y} = -\frac{1}{c}\frac{\partial H_z}{\partial t};$$

while the second equation (46.12) is equivalent to

$$\frac{\partial E_x}{\partial x} + \frac{\partial E_y}{\partial y} + \frac{\partial E_z}{\partial z} = \rho \qquad (46.15)$$

and

$$\frac{\partial H_z}{\partial y} - \frac{\partial H_y}{\partial z} = \frac{1}{c}\frac{\partial E_x}{\partial t} + \rho\frac{u_x}{c},$$

$$\frac{\partial H_x}{\partial z} - \frac{\partial H_z}{\partial x} = \frac{1}{c}\frac{\partial E_y}{\partial t} + \rho\frac{u_y}{c}, \qquad (46.16)$$

$$\frac{\partial H_y}{\partial x} - \frac{\partial H_x}{\partial y} = \frac{1}{c}\frac{\partial E_z}{\partial t} + \rho\frac{u_z}{c}.$$

These equations are, however—in scalar form—the four field equations (39.1–4) on which the Lorentz electron theory is founded. The two equations (46.11, 12) thus furnish an extremely satisfactory starting-point for electromagnetic investigations.

(b) **Four-dimensional expression of force on moving charge.** In addition to this possibility of expressing the Lorentz field equations in four-dimensional language, it should also be noted that the expression for force (41.4), given by the fifth fundamental equation of the Lorentz theory, also has the correct four-dimensional character. This is most easily made evident by considering the possibility of constructing a contravariant vector F^μ with components which are related to the ordinary components of force and rate of energy change in accordance with the scheme

$$F^\mu = \left(\frac{F_x}{\sqrt{(1-u^2/c^2)}}, \quad \frac{F_y}{\sqrt{(1-u^2/c^2)}}, \quad \frac{F_z}{\sqrt{(1-u^2/c^2)}}, \quad \frac{1}{\sqrt{(1-u^2/c^2)}}\frac{dE}{c\,dt}\right).$$
$$(46.17)$$

As a matter of fact, substituting from the fifth fundamental equation (41.4), the values for the components of force F_x, F_y, and F_z acting on a charge e moving with the velocity u, and for dE/dt the work done by them, it is easily found with the help of our transformation equations (40.1, 2) and (40.5), that the quantities given by (46.17) do transform as the components of a contravariant vector. In accordance with our previous discussion of equation (28.12) this agrees, however, with a necessary property of forces of any origin.

Furthermore, substituting these values for the components of F^μ into our previous equation (28.10) for the force acting on a moving particle

$$F^\mu = c^2 \frac{d}{ds}\left(m_0\frac{dx^\mu}{ds}\right) \qquad (46.18)$$

we can readily obtain the expected results

$$\frac{d}{dt}\left(\frac{m_0 \mathbf{u}}{\sqrt{(1-u^2/c^2)}}\right) = e\left(\mathbf{E}+\frac{1}{c}[\mathbf{u}\times\mathbf{H}]\right)$$

and

$$\frac{d}{dt}\left(\frac{m_0 c^2}{\sqrt{(1-u^2/c^2)}}\right) = e(\mathbf{E}\cdot\mathbf{u}).$$

(46.19)

(c) **Four-dimensional expression of electromagnetic energy-momentum tensor.** Finally, it should be noted that we can evidently construct from the electromagnetic stresses p_{ij} and the densities of electromagnetic mass ρ and of momentum g_i, an *electromagnetic energy-momentum tensor* $[T^{\mu\nu}]_{em}$ entirely similar in form to our previous mechanical tensor $[T^{\mu\nu}]_{me}$, since the transformation equations (44.1–3) for the electromagnetic quantities are of exactly the same form as those for the corresponding mechanical quantities. This electromagnetic energy momentum tensor will then have the form

$$[T^{\mu\nu}]_{em} = \begin{array}{cccc} p_{xx} & p_{xy} & p_{xz} & [\mathbf{E}\times\mathbf{H}]_x \\[2mm] p_{yx} & p_{yy} & p_{yz} & [\mathbf{E}\times\mathbf{H}]_y \\[2mm] p_{zx} & p_{zy} & p_{zz} & [\mathbf{E}\times\mathbf{H}]_z \\[2mm] [\mathbf{E}\times\mathbf{H}]_x & [\mathbf{E}\times\mathbf{H}]_y & [\mathbf{E}\times\mathbf{H}]_z & \dfrac{E^2+H^2}{2}, \end{array}$$

(46.20)

where the Maxwell stresses have the values previously given by the formulae

$$p_{ii} = -\tfrac{1}{2}(E_i^2-E_j^2-E_k^2+H_i^2-H_j^2-H_k^2),$$

(46.21)

$$p_{ij} = -(E_i E_j+H_i H_j).$$

And if we permit the possibility discussed in § 45 of combining the corresponding mechanical and electrical quantities, it is evident that the equations of motion and continuity (45.7, 8) for a combined mechanical and electrical system could then be expressed by the four-dimensional equation

$$\frac{\partial T^{\mu\nu}}{\partial x^\nu} = \frac{\partial}{\partial x^\nu}\{[T^{\mu\nu}]_{me}+[T^{\mu\nu}]_{em}\} = 0.$$

(46.22)

47. Applications of the electron theory

The foregoing completes the discussion of the relations between special relativity and electrodynamics from the point of view of the electron theory, in so far as will be necessary for our further work. The treatment shows that the Lorentz electron theory can be taken over into special relativity with almost no alteration, beyond the very agreeable one that the fundamental equations can now be taken as valid with respect to all sets of axes in uniform motion, and not merely with respect to axes at rest in a suppositious ether. This makes it possible to continue to use most previous applications of the Lorentz theory, in so far as they are not modified by quantum theory, and we shall not give special consideration to them here.

One of the most important applications of the Lorentz theory lies in the possibility of deriving Maxwell's electromagnetic field equations for ponderable matter from a consideration of the average behaviour of the electrons which such matter may be assumed to contain. The essential point of the treatment consists in relating the 'macroscopic' quantities E, D, H, B, J, and ρ of the Maxwell theory to the appropriate averages which can be obtained from the 'microscopic' quantities E, H, u, and ρ of the electron theory, and then showing that the 'macroscopic' quantities do satisfy Maxwell's equations for matter. This was successfully carried out by Lorentz himself for the case considered by Maxwell of matter at rest, while for the case of matter in motion treatments agreeing with the special theory of relativity have been given by Born[†] and by Dällenbach.[‡]

For the purposes of this book, however, we shall omit any detailed consideration of this possibility of basing the electromagnetic equations for ponderable matter on those of the electron theory, and shall now turn to Part II of the present chapter in which these equations are treated from a more strictly phenomenological point of view. Such a treatment will be more in keeping with that which we were able to give to the mechanics of ponderable matter, and will avoid the uncertainties which still obscure the applications of relativity in the microscopic field of quantum mechanics and quantum electrodynamics.

[†] Born, *Math. Ann.* **68**, 526 (1910).
[‡] Dällenbach, *Ann. der Physik*, **58**, 523 (1919).

IV

SPECIAL RELATIVITY AND ELECTRODYNAMICS (*contd.*)

Part II. MACROSCOPIC THEORY

48. The field equations for stationary matter

In order to treat the gross electromagnetic behaviour of ponderable matter we shall follow the method of Minkowski,[†] by first assuming in accordance with available experimental information that the behaviour in the case of stationary matter is correctly described by Maxwell's theory, and then drawing conclusions as to the behaviour of moving matter with the help of the theory of relativity. Since the results of any electromagnetic experiment made on stationary matter should be describable in terms of the spatial and temporal behaviour of identifiable objects, and the special theory of relativity has provided a unique method for translating the description of such behaviour to a new system of coordinates in which the matter would have any desired uniform velocity, the proposed method of attack should lead unambiguously to an electromagnetic theory applicable to any body in a state of uniform motion. In addition we shall find that the theory we obtain would also be rigorously applicable to a system of bodies with different uniform velocities provided they are separated by free space. For more complicated kinds of motion the theory is presumably at least a first approximation.

In accordance with Maxwell's theory, the field equations for *stationary matter* connecting the electric field strength **E**, electric displacement **D**, magnetic field strength **H**, and magnetic induction **B**, with the densities of charge ρ and conduction current **J** are given by the vector expressions

$$\text{div}^0 \mathbf{D}^0 = \rho^0, \tag{48.1}$$

$$\text{div}^0 \mathbf{B}^0 = 0, \tag{48.2}$$

$$\text{curl}^0 \mathbf{E}^0 = -\frac{1}{c}\frac{\partial \mathbf{B}^0}{\partial t^0}, \tag{48.3}$$

$$\text{curl}^0 \mathbf{H}^0 = \frac{1}{c}\left(\frac{\partial \mathbf{D}^0}{\partial t^0} + \mathbf{J}^0\right), \tag{48.4}$$

where we have attached the superscript ^{0} to the quantities involved

† Minkowski, *Math. Ann.* **68**, 472 (1910).

in order to indicate that they are measured with respect to a proper system of coordinates S^0 in which the matter is at rest.

The quantities occurring in these equations are to be regarded as macroscopically determinable. Thus E^0, D^0, H^0, and B^0 are the forces per unit of electric charge and per unit of magnetic pole strength as they would be determined by considering the limit approached in ideal but nevertheless conceivable macroscopic experiments to be carried out with obtainable test charges and poles inserted at the point of interest into prescribed longitudinal or transverse crevasses cut in the matter, and ρ^0 and J^0 are to be regarded as the macroscopic densities of charge and current for volume elements that can be treated as infinitesimals although large compared with intermolecular distances. Hence in so far as we restrict ourselves to problems which do not involve too small intervals of space or time our present treatment will be unaffected by the quantum-mechanical considerations which should be introduced into a correct microscopic treatment.

49. The constitutive equations for stationary matter

The above field equations are not sufficient in number to give a complete determination of electromagnetic phenomena but must be supplemented by further relations connecting the quantities employed with the constitution of the material involved. In order not to introduce too great complication, we shall take these further relations as given by the familiar simple equations of Maxwell

$$\mathbf{D}^0 = \epsilon \mathbf{E}^0 \qquad \mathbf{B}^0 = \mu \mathbf{H}^0 \qquad \mathbf{J}^0 = \sigma \mathbf{E}^0, \qquad (49.1)$$

where the dielectric constant ϵ, the magnetic permeability μ, and the electrical conductivity σ are to be regarded as known functions of the position and time.

Although the field equations (48.1–4) are regarded as holding in general for inhomogeneous and anisotropic bodies, the further results which depend on these particular constitutive equations will be limited to isotropic matter in the absence of so-called impressed electrical forces of an extraneous—for example thermal or chemical—nature.

50. The field equations in four-dimensional language

In our previous applications of the special theory of relativity to mechanics and to electron theory, we have first developed the treatment in terms of the coordinates (x, y, z) and (t) corresponding to three-dimensional space and a separate one-dimensional time, and then followed this by a translation or parallel treatment in terms of the coordinates (x^1, x^2, x^3, x^4) corresponding to a four-dimensional space-time continuum. The simplicity and appropriateness of the four-dimensional method has thereby been made evident. In our present application to the complicated electrodynamics of ponderable matter, we shall from the start take advantage of this powerful method, by first formulating the fundamental equations in four-dimensional language, and then translating results back into the older language when desirable.

To carry this out we now return to our earlier space-time coordinates (20.2)

$$x^1 = x \quad x^2 = y \quad x^3 = z \quad x^4 = ct, \tag{50.1}$$

corresponding to the simple form (20.1) for the element of interval

$$ds^2 = -(dx^1)^2 - (dx^2)^2 - (dx^3)^2 + (dx^4)^2, \tag{50.2}$$

and introduce for treating the electrodynamics of ponderable matter two anti-symmetric *electromagnetic tensors* $F^{\mu\nu}$ and $H^{\mu\nu}$, which we define by stating that their components in *proper* coordinates $(x_0^1, x_0^2, x_0^3, x_0^4)$, with respect to which the material is at rest, are given in terms of the quantities appearing in the Maxwell field equations (§ 48) by the expressions

$$F^{\mu\nu} = \begin{matrix} 0 & B_z^0 & -B_y^0 & -E_x^0 \\ -B_z^0 & 0 & B_x^0 & -E_y^0 \\ B_y^0 & -B_x^0 & 0 & -E_z^0 \\ E_x^0 & E_y^0 & E_z^0 & 0 \end{matrix} \tag{50.3}$$

and

$$H^{\mu\nu} = \begin{matrix} 0 & H_z^0 & -H_y^0 & -D_x^0 \\ -H_z^0 & 0 & H_x^0 & -D_y^0 \\ H_y^0 & -H_x^0 & 0 & -D_z^0 \\ D_x^0 & D_y^0 & D_z^0 & 0 \end{matrix} \tag{50.4}$$

together with the *current vector* J^μ whose components in proper coordinates are given by

$$J^\mu = \left(\frac{J_x^0}{c}, \quad \frac{J_y^0}{c}, \quad \frac{J_z^0}{c}, \quad \rho^0 \right). \tag{50.5}$$

With the help of the tensors, which we have thus defined, we can now express the content of the originally postulated Maxwell field equations by the simple tensor equations

$$\frac{\partial F_{\mu\nu}}{\partial x^\sigma} + \frac{\partial F_{\nu\sigma}}{\partial x^\mu} + \frac{\partial F_{\sigma\mu}}{\partial x^\nu} = 0 \qquad (50.6)$$

and $$\partial H^{\mu\nu}/\partial x^\nu = J^\mu. \qquad (50.7)$$

Since these are tensor equations, in the form corresponding to any set of rectangular coordinates of the type (50.1), they will be true in all such sets of coordinates if true in one. In proper coordinates, however, their identity with Maxwell's equations can readily be verified from the definitions we have given for the tensors involved, (50.6) being equivalent to (48.2) and (48.3), and (50.7) to (48.1) and (48.4). Hence we have thus obtained a simple expression for the field equations in a form valid for matter in uniform motion as well as for matter at rest.

These equations can also be regarded as applying to the case of several bodies moving with different uniform velocities and separated by free space, since the tensors $F^{\mu\nu}$, $H^{\mu\nu}$, and J^μ can be taken inside each of these bodies as reducing to the forms (50.3–5) when referred to proper coordinates moving with the material, thus guaranteeing that Maxwell's equations for stationary matter will hold for each body.

51. The constitutive equations in four-dimensional language

The constitutive equations (49.1) connecting displacement, magnetization, and current with the properties of the material can also easily be expressed in four-dimensional language with the help of the tensor equations†

$$H_{\alpha\beta} \frac{dx^\alpha}{ds} = \epsilon F_{\alpha\beta} \frac{dx^\alpha}{ds} \qquad (51.1)$$

$$(g_{\alpha\beta} F_{\gamma\delta} + g_{\alpha\gamma} F_{\delta\beta} + g_{\alpha\delta} F_{\beta\gamma}) \frac{dx^\alpha}{ds}$$
$$= \mu(g_{\alpha\beta} H_{\gamma\delta} + g_{\alpha\gamma} H_{\delta\beta} + g_{\alpha\delta} H_{\beta\gamma}) \frac{dx^\alpha}{ds} \qquad (51.2)$$

and $$J^\alpha - J_\beta \frac{dx^\beta}{ds} \frac{dx^\alpha}{ds} = \frac{\sigma}{c} g_{\beta\gamma} F^{\gamma\alpha} \frac{dx^\beta}{ds}, \qquad (51.3)$$

where ϵ, μ, and σ are the dielectric constant, magnetic permeability, and conductivity of the material as measured by a local observer moving with it, and dx^α/ds, dx^β/ds are components of the 'velocity' of the matter at the point of interest.

Noting that in proper coordinates dx^α/ds will be unity when

† Weyl, *Raum, Zeit, Materie*, Berlin, fourth edition, 1921, p. 174.

$\alpha = 4$ and otherwise zero, it is easy to verify for proper coordinates the equivalence of (51.1–3) to the original constitutive equations (49.1). Hence these new expressions for the constitutive equations, being valid in one set of coordinates, are valid in all sets of coordinates owing to their tensor character. They too can be applied in the case of several bodies in uniform motion separated by free space.

52. The field equations for moving matter in ordinary vector language

We have thus obtained, in the two preceding sections, expressions both for the field equations and for the constitutive equations in a four-dimensional form which can be used for matter in a state of uniform motion as well as for matter at rest. We have hence provided a complete basis for the macroscopic electrodynamics of moving matter, and no new content can be added to this basis by re-expressing it in other forms. Nevertheless, we can perhaps gain some further insight into the physical nature of the theory if we now translate the results obtained back into ordinary vector language.

To do this, let us now return to the tensors $F^{\mu\nu}$, $H^{\mu\nu}$, and J^{μ} defined above by their components in proper coordinates, and *as a matter of convention* use the same symbols without the superscript ^{0} to designate their components in any system of coordinates of the type (50.1). In agreement with the possibility of expressing the original Maxwell equations in the form (50.6, 7), we can then evidently write the field equations in general, for matter moving with a uniform velocity as well as for stationary matter, in the original Maxwellian form (48.1–4):

$$\operatorname{div} \mathbf{D} = \rho, \tag{52.1}$$

$$\operatorname{div} \mathbf{B} = 0, \tag{52.2}$$

$$\operatorname{curl} \mathbf{E} = -\frac{1}{c}\frac{\partial \mathbf{B}}{\partial t}, \tag{52.3}$$

$$\operatorname{curl} \mathbf{H} = \frac{1}{c}\left(\frac{\partial \mathbf{D}}{\partial t} + \mathbf{J}\right). \tag{52.4}$$

Furthermore, making use of the rules of tensor transformation (19.10), and the convention above by which the quantities occurring in these equations were defined, we can readily obtain as the equations for the transformation of these quantities from a given system of coordinates $S\,(x, y, z, t)$ to a new system $S'\,(x', y', z', t')$, corresponding to new axes moving in the x-direction with the velocity V relative to the old, the expressions

$$E_x = E'_x$$

$$E_y = \frac{E'_y + \dfrac{V}{c} B'_z}{\sqrt{(1 - V^2/c^2)}}$$ (52.5)

$$E_z = \frac{E'_z - \dfrac{V}{c} B'_y}{\sqrt{(1 - V^2/c^2)}},$$

$$B_x = B'_x$$

$$B_y = \frac{B'_y - \dfrac{V}{c} E'_z}{\sqrt{(1 - V^2/c^2)}}$$ (52.6)

$$B_z = \frac{B'_z + \dfrac{V}{c} E'_y}{\sqrt{(1 - V^2/c^2)}},$$

$$D_x = D'_x$$

$$D_y = \frac{D'_y + \dfrac{V}{c} H'_z}{\sqrt{(1 - V^2/c^2)}}$$ (52.7)

$$D_z = \frac{D'_z - \dfrac{V}{c} H'_y}{\sqrt{(1 - V^2/c^2)}},$$

$$H_x = H'_x$$

$$H_y = \frac{H'_y - \dfrac{V}{c} D'_z}{\sqrt{(1 - V^2/c^2)}}$$ (52.8)

$$H_z = \frac{H'_z + \dfrac{V}{c} D'_y}{\sqrt{(1 - V^2/c^2)}},$$

$$J_x = \frac{J'_x + \rho' V}{\sqrt{(1 - V^2/c^2)}}$$

$$J_y = J'_y$$ (52.9)

$$J_z = J'_z,$$

and

$$\rho = \frac{\rho' + J'_x V/c^2}{\sqrt{(1 - V^2/c^2)}}.$$ (52.10)

Several points of interest with respect to these results may be mentioned.

In the case of free space, it should be noted that we shall have $\mathbf{D} = \mathbf{E}$, $\mathbf{B} = \mathbf{H}$, $\mathbf{J} = 0$, and $\rho = 0$ and the above field equations and transformation equations will then reduce to the same form (§§ 39 and 40) as the corresponding equations for the electron theory in the absence of matter. This result is of course satisfactory from the point of view of the consistency of the two methods of attack.

With regard to the physical significance of the quantities occurring in our new formulation of the field equations (52.1–4), we can assign them no immediate meaning beyond what is determined by their definition as names for the components of certain tensors as provided by a previous paragraph. With the help of the transformation equations (52.5) to (52.10), however, we can relate the values of these quantities in any desired system of coordinates S with the values of the corresponding quantities as directly measured by a local observer, using proper coordinates S^0 in which the material is at rest.

Making use of this possibility together with the transformation equations for force (25.3), it can readily be shown that the forces $\mathbf{E}^0$ and $\mathbf{H}^0$ which would be found by a local observer to act on unit charge and unit pole moving with the matter, would lead to the relations

$$\mathbf{E}^* = \mathbf{E} + \left[\frac{\mathbf{u}}{c} \times \mathbf{B}\right] \tag{52.11}$$

and

$$\mathbf{H}^* = \mathbf{H} - \left[\frac{\mathbf{u}}{c} \times \mathbf{D}\right] \tag{52.12}$$

as expressions in terms of the variables of system S for the forces acting on unit charge and unit pole which are moving with the matter with the velocity u. These expressions will perhaps give a feeling for the physical nature of the quantities involved.

It is also of physical interest to consider the separation of the total current $\mathbf{J}$ into conduction current $\mathbf{C}$ and convection current $\rho\mathbf{u}$, in accordance with the equation of definition

$$\mathbf{J} = \mathbf{C} + \rho\mathbf{u}. \tag{52.13}$$

Making this separation, we can readily obtain from the transformation equations (52.9, 10) the following expressions for the conduction current and charge in matter moving parallel to the x-axis with the velocity V:

$$C_x = C_x^0 \sqrt{(1 - V^2/c^2)}$$
$$C_y = C_y^0$$
$$C_z = C_z^0$$

(52.14)

and

$$\rho = \frac{\rho_0}{\sqrt{(1 - V^2/c^2)}} + \frac{C_x^0 V/c^2}{\sqrt{(1 - V^2/c^2)}}.$$

(52.15)

In the case of a charged insulator moving with the velocity V in the x-direction, this would give for the total value of J to be substituted into the field equation (52.4), the purely convective term

$$J = \rho V = \frac{\rho_0 V}{\sqrt{(1 - V^2/c^2)}}$$

(52.16)

in agreement with Rowland's celebrated discovery of the existence of a convection current.

In more general cases the separation of total current into conduction and convection current is dependent on the system of coordinates being used. Even when the charge density is zero in proper coordinates, charge density and convection current can appear in other coordinates, in accordance with (52.15), provided the conduction current is not zero in proper coordinates. Looked at from a microscopic point of view, this possibility—that different observers may disagree as to the relative number of positive and negative electrons in a given volume element of the material—can be shown in detail to arise only when there is relative motion between the two kinds of electrons and to depend in an entirely expected manner on the lack of agreement as to simultaneity provided by the theory of relativity.†

53. The constitutive equations for moving matter in ordinary vector language

The constitutive equations connecting displacement, magnetization, and current with the properties of the material, which were given for proper coordinates in § 49 and in tensor form in § 51, can also easily be found with the help of the equations in the preceding section to be expressible in terms of the vectors that we are now using.

Defining in analogy with (52.11) and (52.12) two new vectors by the equations

$$\mathbf{D}^* = \mathbf{D} + \left[\frac{\mathbf{u}}{c} \times \mathbf{H} \right]$$

(53.1)

and

$$\mathbf{B}^* = \mathbf{B} - \left[\frac{\mathbf{u}}{c} \times \mathbf{E} \right],$$

(53.2)

† See Laue, *Das Relativitätsprinzip*, Braunschweig, second edition, 1913, p. 145.

the first two of the constitutive equations can be written in the simple form

$$\mathbf{D}^* = \epsilon\mathbf{E}^* \quad \text{and} \quad \mathbf{B}^* = \mu\mathbf{H}^*; \tag{53.3}$$

and using the expressions for conduction current given by (52.14) the third constitutive equation can be written in the form

$$C_x = \sigma\sqrt{(1-u^2/c^2)}E_x^*$$

$$C_y = \frac{\sigma}{\sqrt{(1-u^2/c^2)}}E_y^* \tag{53.4}$$

$$C_z = \frac{\sigma}{\sqrt{(1-u^2/c^2)}}E_z^*,$$

where the velocity u of the moving material is taken as parallel to the x-axis. This latter result may be regarded as the analogue of Ohm's law for moving material.

54. Applications of the macroscopic theory

The foregoing has given a complete statement of the underlying basis for macroscopic electrodynamics, expressed both in four-dimensional language and in the language of ordinary vector theory. It will be seen that this basis is but little altered from that which has usually been employed in electromagnetic considerations and we shall now merely wish to consider certain consequences of the theory which are specially illuminating or of importance for our later work.

(a) **The conservation of electric charge.** Making use of the equation

$$\partial H^{\mu\nu}/\partial x^\nu = J^\mu, \tag{54.1}$$

which is the second of our two fundamental equations (50.6, 7) for the macroscopic theory, we can immediately obtain the principle of the conservation of electric charge. Differentiating (54.1) with respect to x^μ we can evidently write

$$\frac{\partial J^\mu}{\partial x^\mu} = \frac{\partial^2 H^{\mu\nu}}{\partial x^\mu \partial x^\nu} = 0, \tag{54.2}$$

where the value zero arises owing to the antisymmetry of $H^{\mu\nu}$. Writing out the expression for the first term in (54.2) in full, however, this gives

$$\frac{\partial J^1}{\partial x^1} + \frac{\partial J^2}{\partial x^2} + \frac{\partial J^3}{\partial x^3} + \frac{\partial J^4}{\partial x^4} = 0, \tag{54.3}$$

or in terms of our usual coordinates as given by (50.1), and the expressions for the components of J^μ in terms of ordinary vector language set up by the conventions introduced in § 52, we can write this in the form

$$\frac{\partial J_x}{\partial x}+\frac{\partial J_y}{\partial y}+\frac{\partial J_z}{\partial z}+\frac{\partial \rho}{\partial t} = 0. \tag{54.4}$$

Since J is to be regarded as the sum of the convection and conduction currents, this can be taken as an equation of continuity guaranteeing the conservation of total electric charge.

(b) **Boundary conditions.** Boundary conditions at the surfaces between media of different properties, quite similar to those familiar in Maxwell's theory for stationary matter, can be derived† from the field equations given in § 52. In the case of moving matter we encounter a certain complication, however, since the field equations contain the expressions $\partial B/\partial t$ and $\partial D/\partial t$, for the rate of change of the vectors involved at a given point in space, and these vectors will in general be changing discontinuously at a point which is momentarily coincident with the moving boundary. If, nevertheless, we consider the rate of change with time at a point moving with the same velocity u as the material itself, as given by the operator

$$\frac{d}{dt} = \frac{\partial}{\partial t}+(\mathbf{u}\nabla) \tag{54.5}$$

we may expect to obtain finite rates of change even within the boundary layer.

With this in mind we may now rewrite our field equation (52.4) in the form

$$\mathrm{curl}\,\mathbf{H}+\left(\frac{\mathbf{u}}{c}\nabla\right)\mathbf{D}-\frac{\mathbf{J}}{c} = \frac{1}{c}\frac{\partial \mathbf{D}}{\partial t}+\left(\frac{\mathbf{u}}{c}\nabla\right)\mathbf{D}, \tag{54.6}$$

and conclude that the quantity on the left-hand side of this equation will everywhere be finite on account of the form of the right-hand side. Or, writing out in full the three components corresponding to the left-hand side and expressing the current J as the sum of the conduction current **C** and convection current ρu, we can conclude that the following three quantities are finite

$$\frac{\partial H_z}{\partial y}-\frac{\partial H_y}{\partial z}+\frac{u_x}{c}\frac{\partial D_x}{\partial x}+\frac{u_y}{c}\frac{\partial D_x}{\partial y}+\frac{u_z}{c}\frac{\partial D_x}{\partial z}-\rho\frac{u_x}{c}-\frac{C_x}{c}$$

$$\frac{\partial H_x}{\partial z}-\frac{\partial H_z}{\partial x}+\frac{u_x}{c}\frac{\partial D_y}{\partial x}+\frac{u_y}{c}\frac{\partial D_y}{\partial y}+\frac{u_z}{c}\frac{\partial D_y}{\partial z}-\rho\frac{u_y}{c}-\frac{C_y}{c} \tag{54.7}$$

$$\frac{\partial H_y}{\partial x}-\frac{\partial H_x}{\partial y}+\frac{u_x}{c}\frac{\partial D_z}{\partial x}+\frac{u_y}{c}\frac{\partial D_z}{\partial y}+\frac{u_z}{c}\frac{\partial D_z}{\partial z}-\rho\frac{u_z}{c}-\frac{C_z}{c}.$$

† Einstein and Laub, *Ann. der Physik*, **28**, 445 (1909).

To make use of these results let us now for simplicity choose coordinates such that the boundary surface between the two media will be parallel to the yz-plane at the point of interest. Since differentiation with respect to y and z will then evidently lead to finite results, and since the conduction current **C** may be taken as finite by hypothesis, we can evidently discard some of the terms occurring in (54.7), and take as finite the simpler quantities

$$\frac{u_x}{c}\left(\frac{\partial D_x}{\partial x}-\rho\right),$$

$$-\frac{\partial H_z}{\partial x}+\frac{u_x}{c}\frac{\partial D_y}{\partial x}-\rho\frac{u_y}{c}, \qquad (54.8)$$

$$\frac{\partial H_y}{\partial x}+\frac{u_x}{c}\frac{\partial D_z}{\partial x}-\rho\frac{u_z}{c}.$$

The first of these expressions gives us

$$\frac{\partial D_x}{\partial x}-\rho \quad \text{finite} \qquad (54.9)$$

or
$$\Delta D_x = \omega, \qquad (54.10)$$

where ΔD_x is the discontinuous change in the normal component of electric displacement on passing through the boundary due to any charge of surface density ω which may be present. The remaining two expressions may be combined with (54.9) to give

$$\frac{\partial H_y}{\partial x}-\left(\frac{u_z}{c}\frac{\partial D_x}{\partial x}-\frac{u_x}{c}\frac{\partial D_z}{\partial x}\right)$$

and
$$\frac{\partial H_z}{\partial x}-\left(\frac{u_x}{c}\frac{\partial D_y}{\partial x}-\frac{u_y}{c}\frac{\partial D_x}{\partial x}\right)$$

as finite quantities, or since the components of the velocity **u** are constant we can conclude that

$$H_y-\left(\frac{u_z}{c}D_x-\frac{u_x}{c}D_z\right)$$

and
$$H_z-\left(\frac{u_x}{c}D_y-\frac{u_y}{c}D_x\right) \qquad (54.11)$$

will vary continuously in passing across the boundary. These latter quantities, however, are the tangential components of the vector

$$\mathbf{H^*} = \mathbf{H}-\left[\frac{\mathbf{u}}{c}\times\mathbf{D}\right]$$

as previously defined by (52.12).

Applying similar methods to the field equation (52.3), we may now state in general

$$\Delta D_n = \omega \quad \Delta B_n = \Delta E_t^* = \Delta H_t^* = 0 \qquad (54.12)$$

as expressions for the presence or absence of any discontinuous change that would take place in the normal or tangential components of the vectors indicated on passing across the boundary between two media. The expressions also apply at the boundary between matter and free space where the velocity of the matter **u** is to be substituted into the expressions for **E*** and **H***.

(c) **The Joule heating effect.** If we consider the alteration which will be produced in the energy content of a small element of matter due to the action of an electromagnetic field in which it is immersed, it is evident that this can result either from the mechanical work done by the ponderomotive forces arising from the field or from the (Joule) heat generated within the element by electromagnetic action. Hence we can write for a small increment δE in the energy of the element, in terms of work done δW and heat generated δQ, the expression

$$\delta E = \delta W + \delta Q. \qquad (54.13)$$

And if we restrict ourselves for simplicity to cases, where the only work done is that corresponding to the motion of the element as a whole in the field of force, we can then write for the Joule heat developed

$$\delta Q = -\delta W + \delta E$$
$$= -F_x\,\delta x - F_y\,\delta y - F_z\,\delta z + \frac{dE}{c\,dt}\,c\,\delta t, \qquad (54.14)$$

where F_x, F_y, and F_z are the components of force of electromagnetic origin acting on the element as a whole.

In accordance with the discussion of Chapter III, however, all forces of whatever origin must obey the same transformation laws, and hence in agreement with (28.12) we can take the quantities

$$F_\mu = \left(\frac{-F_x}{\sqrt{(1-u^2/c^2)}}, \quad \frac{-F_y}{\sqrt{(1-u^2/c^2)}}, \quad \frac{-F_z}{\sqrt{(1-u^2/c^2)}}, \quad \frac{1}{\sqrt{(1-u^2/c^2)}}\frac{dE}{c\,dt} \right), \qquad (54.15)$$

where u is the ordinary velocity of the element, as being the components of a covariant vector in the space-time coordinates

$$x^1 = x \quad x^2 = y \quad x^3 = z \quad x^4 = ct. \qquad (54.16)$$

By combining (54.14) and (54.15) we can now write

$$\delta Q = \sqrt{(1 - u^2/c^2)}F_\mu \, \delta x^\mu \qquad (54.17)$$

as an expression for the increment of heat.

By the rules of tensor analysis, however, $F_\mu \, \delta x^\mu$ must be a scalar invariant, having the same value in all systems of coordinates, so that we now obtain

$$\delta Q = \sqrt{(1 - u^2/c^2)}\delta Q_0 \qquad (54.18)$$

as a general expression connecting the increment of heat δQ and velocity of the element of matter u, as measured in any given system of coordinates, with the increment of heat δQ_0 as measured in proper coordinates by a local observer. The result is of special interest because of its agreement with the transformation equation for heat which we shall obtain in our development of *relativistic thermodynamics*.

(d) **Electromagnetic energy and momentum.** With the help of the field equations we can readily obtain expressions which will permit a calculation of the rates at which the energy and momentum are changing inside a *boundary which lies in the free space* surrounding a material body acted on by electromagnetic force.

Taking the inner product of the field equation (52.3) with **H** and subtracting the inner product of (52.4) with **E**, we obtain the result

$$\mathbf{E} \cdot \frac{\partial \mathbf{D}}{\partial t} + \mathbf{H} \cdot \frac{\partial \mathbf{B}}{\partial t} + \mathbf{E} \cdot \mathbf{J} + c(\mathbf{H} \cdot \operatorname{curl} \mathbf{E} - \mathbf{E} \cdot \operatorname{curl} \mathbf{H}) = 0. \qquad (54.19)$$

Integrating this over the region inside a stationary boundary which encloses the system of interest we obtain with the help of a well-known relation of vector analysis [equation (17), Appendix II]

$$\int \left(\mathbf{E} \cdot \frac{\partial \mathbf{D}}{\partial t} + \mathbf{H} \cdot \frac{\partial \mathbf{B}}{\partial t} + \mathbf{E} \cdot \mathbf{J} \right) dv = -c \int [\mathbf{E} \times \mathbf{H}]_n \, d\sigma, \qquad (54.20)$$

where the right-hand term is integrated over the surface surrounding the volume under consideration and the subscript n denotes the outward normal component of the vector in question.

For a volume containing no ponderable matter this evidently reduces to the familiar equation

$$\int \frac{\partial}{\partial t} \left(\frac{E^2 + H^2}{2} \right) dv = -c \int [\mathbf{E} \times \mathbf{H}]_n \, d\sigma, \qquad (54.21)$$

where the left-hand side gives the known expression for the rate at which energy is increasing, and the right-hand side must then give the rate of energy flow through the boundary.

Hence returning to the more general equation, the left-hand side of (54.20) must be an expression for the rate of energy increase even when ponderable matter is present, since the boundary was by hypothesis located in the free space surrounding the matter. Equation (54.20) can hence be used to calculate the rate at which the energy is changing inside a fixed boundary located in the free space surrounding a material system by an integration extending over the volume involved. The equation is analogous to (42.3) in the electron theory but does not furnish unique expressions for the densities of electromagnetic energy and momentum *inside of matter*.

Similarly, we can obtain an expression for the rate at which the momentum inside the boundary is changing with time from the field equations. Taking the product of the field equation (52.1) with **E** and adding the outer product of (52.4) with **B** we obtain the result

$$\mathbf{E} \operatorname{div} \mathbf{D} - \mathbf{E}\rho + (\operatorname{curl} \mathbf{H}) \times \mathbf{B} - \frac{1}{c}\frac{\partial \mathbf{D}}{\partial t} \times \mathbf{B} - \frac{\mathbf{J}}{c} \times \mathbf{B} = 0. \quad (54.22)$$

Changing signs, separating **J** into convection current $\rho\mathbf{u}$ and conduction current **C**, and making use of the field equation (52.3) this can be written in the form

$$\rho\left(\mathbf{E} + \frac{\mathbf{u}}{c} \times \mathbf{B}\right) + \frac{\mathbf{C}}{c} \times \mathbf{B} + \frac{1}{c}\frac{\partial}{\partial t}[\mathbf{D} \times \mathbf{B}] - \mathbf{E} \operatorname{div} \mathbf{D} -$$

$$- (\operatorname{curl} \mathbf{H}) \times \mathbf{B} - (\operatorname{curl} \mathbf{E}) \times \mathbf{D} = 0. \quad (54.23)$$

Or, introducing the electric and magnetic polarizations defined by

$$\mathbf{P} = \mathbf{D} - \mathbf{E} \quad (54.24)$$

and

$$\mathbf{M} = \mathbf{B} - \mathbf{H}, \quad (54.25)$$

we can write

$$\rho\left(\mathbf{E} + \frac{\mathbf{u}}{c} \times \mathbf{B}\right) + \frac{\mathbf{C}}{c} \times \mathbf{B} - \mathbf{E} \operatorname{div} \mathbf{P} + \mathbf{P} \times \operatorname{curl} \mathbf{E} + \mathbf{M} \times \operatorname{curl} \mathbf{H} + \frac{1}{c}\frac{\partial}{\partial t}[\mathbf{D} \times \mathbf{B}]$$

$$= \mathbf{E} \operatorname{div} \mathbf{E} + (\operatorname{curl} \mathbf{E}) \times \mathbf{E} + (\operatorname{curl} \mathbf{H}) \times \mathbf{H}. \quad (54.26)$$

Considering for specificity the x-component and integrating over the region inside a fixed boundary located in the free space surrounding the system of interest, this will give us with help of a somewhat long but straightforward transformation

$$\int \left\{ \rho\left(\mathbf{E}+\frac{\mathbf{u}}{c}\times\mathbf{B}\right)+\frac{\mathbf{C}}{c}\times\mathbf{B}- \right.$$

$$\left. -\mathbf{E}\,\mathrm{div}\,\mathbf{P}+\mathbf{P}\times\mathrm{curl}\,\mathbf{E}+\mathbf{M}\times\mathrm{curl}\,\mathbf{H}+\frac{1}{c}\frac{\partial}{\partial t}[\mathbf{D}\times\mathbf{B}]\right\}_x dv$$

$$= \int \left\{ \frac{1}{2}\frac{\partial}{\partial x}(E_x^2-E_y^2-E_z^2+H_x^2-H_y^2-H_z^2)+\frac{\partial}{\partial y}(E_xE_y+H_xH_y)+ \right.$$

$$\left. +\frac{\partial}{\partial z}(E_xE_z+H_xH_z)\right\} dv. \quad (54.27)$$

This equation is analogous to equation (43.3) of the electron theory and the quantities whose partial differentials appear on the right-hand side give the known expressions for the Maxwell electromagnetic stresses p_{xx}, p_{xy}, and p_{xz}. In case the boundary encloses no matter the equation reduces to the familiar form

$$\int \frac{1}{c}\frac{\partial}{\partial t}[\mathbf{E}\times\mathbf{H}]_x dv = \int -\left(\frac{\partial p_{xx}}{\partial x}+\frac{\partial p_{xy}}{\partial y}+\frac{\partial p_{xz}}{\partial z}\right) dv, \quad (54.28)$$

where the left-hand side gives the known expression for the rate at which the x-component of momentum is increasing, and the right-hand side which can evidently be replaced by a surface integral must then give the rate of momentum flow through the boundary.

Hence, returning to the more general equation, the left-hand side of (54.27) must be an expression for the rate of momentum increase even when ponderable matter is present, since the boundary was by hypothesis located in the free space surrounding the matter. The equation can hence be used for calculating the rate of momentum increase by integrating over the volume involved but does not furnish unique expressions for the electromagnetic stresses *inside of matter*.

(e) **The energy-momentum tensor.** Since the macroscopic theory has not led to unique expressions for the densities of electromagnetic energy and momentum and for the electromagnetic stresses inside of matter, the construction of an electromagnetic energy-momentum tensor $[T^{\mu\nu}]_{em}$ cannot be carried out in an unambiguous manner, and several different proposals for such a tensor have actually been considered without the attainment of universal agreement.†

In accordance with our general ideas as to the relation between

† See Pauli, 'Relativitätstheorie', *Encyclopädie der math. Wiss.*, Band v. 2, Heft 4, Leipzig, 1921, § 35.

densities of momentum and energy flow as discussed in § 27, it appears that such a tensor should be symmetrical with respect to the components $T^{4\mu}$ and $T^{\mu 4}$, and presumably in the other components since the microscopic treatment of the electron theory led to a completely symmetrical tensor. With no matter present the macroscopic theory leads unambiguously to an energy-momentum tensor of the same form as that obtained in the Lorentz theory.

Even in the absence of unambiguous values for the components of such a tensor in terms of the variable in the field equations, it seems reasonable to assume the possibility of using a combined equation of the form

$$\frac{\partial T^{\mu\nu}}{\partial x^{\nu}} = \frac{\partial}{\partial x^{\nu}}\{[T^{\mu\nu}]_{me} + [T^{\mu\nu}]_{em}\} = 0 \qquad (54.29)$$

for treating the macroscopic behaviour of a combined mechanical and electromagnetic system, where $[T^{\mu\nu}]_{em}$ is presumably a symmetrical tensor.

(*f*) Applications to experimental observations. As mentioned at the beginning of this section (§ 54), the macroscopic electromagnetic theory which has been developed with the help of special relativity does not differ greatly from those usually employed in electromagnetic considerations; indeed, for the cases of free space and of stationary bodies it is identical with that of Maxwell. For this reason we can be assured of its agreement with a great mass of experimental fact.

In the case of moving bodies, the agreement of the present electrodynamics with Rowland's discovery of the convection current has already been pointed out in connexion with equation (52.16), and the result that the conduction current as given by (53.4) is proportional —except for terms of the order u^2/c^2—to the vector $\mathbf{E}^* = \mathbf{E} + [\mathbf{u}/c \times \mathbf{B}]$ is in satisfactory agreement with experiment.

The theory can also be shown to give satisfactory explanations of the Roentgen-Eichenwald experiment on the magnetic field produced by the rotation of a dielectric in an electric field, and of the H. A. Wilson experiment on the surface charge produced by the rotation of a dielectric in a magnetic field. These experiments were not satisfactorily explained by the Hertz theory of moving dielectrics, although the results were accounted for by the Lorentz theory. In addition, special attention should be called to the later experiments of M. Wilson and H. A. Wilson,[†] who repeated the original Wilson experi-

† M. Wilson and H. A. Wilson, *Proc. Roy. Soc.* (A) **89**, 99 (1914).

ment with an artificially constructed dielectric which had an appreciable magnetic permeability. In this case there is a disagreement between the prediction to which the Lorentz theory had seemed to lead, and that obtained without ambiguity from the macroscopic electrodynamics developed in this chapter. The experimental results agreed with the latter.

V
SPECIAL RELATIVITY AND THERMODYNAMICS
Part I. THE THERMODYNAMICS OF STATIONARY SYSTEMS

55. Introduction

In the present chapter we shall discuss the relations of special relativity to thermodynamics. These relations are found to be of two different sorts.

On the one hand, the special theory of relativity has provided us with a simple equation connecting mass and energy which permits us to calculate the change in the energy content of a thermodynamic system from its change in mass. This new relation proves to be of thermodynamic importance—without reference to the state of motion of the system considered—since it permits the calculation of thermodynamic equilibria for certain conceivable processes where our only possibility of knowing the energy changes that would accompany the process must at present be based on a knowledge of the changes in mass that would take place.

On the other hand, the special theory of relativity has provided us, through the Lorentz transformation, with a possible method of translating the experimental findings obtained by a local observer, who is stationary with respect to a thermodynamic system, into terms which would express the results for an observer with respect to whom the system is in motion. This hence leads, as first shown by Planck and by Einstein, to a thermodynamic theory for moving systems.

In Part I of the present chapter we shall consider the thermodynamics of stationary systems, first developing some well-known portions of the classical theory which will be specially useful to us later, and then exhibiting the application of the mass-energy relation of special relativity to thermodynamics by calculations of the equilibria between hydrogen and helium and between matter and radiation, assuming the possibility of their interconversion.

In Part II we shall consider the Lorentz transformation for thermodynamic quantities and the thermodynamics of moving systems. The work will include a four-dimensional formulation of thermodynamic principles which will be of particular interest when we later undertake the extension of thermodynamics to general relativity.

Before proceeding to these tasks, it is important to emphasize the macroscopic and phenomenological character of thermodynamic considerations. The principles of thermodynamics can, to be sure, be based with a certain degree of success on the microscopic considerations of statistical mechanics. Nevertheless, both on account of their historical origin and essential content, it is most satisfactory to regard the two laws of classical thermodynamics as a generalized formulation of observations made in the actual performance of numerous macroscopic experiments. Even the so-called third law of thermodynamics, although its content is greatly illuminated by the statistical mechanical interpretation of entropy, was originally formulated without the aid of this interpretation, and is now justified as a satisfactory principle by its dependence on a great mass of actual experimental data. The phenomenological character of thermodynamic considerations and the extended basis of experimental verification give us great confidence in thermodynamic predictions even when applied to quite new situations.

Since the considerations of the theory of relativity are also—for the present at least—primarily macroscopic in character, the construction of a relativistic thermodynamics seems a natural and evident extension to undertake. The construction of a fundamentally satisfactory relativistic statistical mechanics would be in any case a complicated business and at present a somewhat dubious undertaking. Nevertheless, some progress in this direction has already been made using classical rather than quantum-mechanical statistics as a starting-point.†

In connexion with the phenomenological character of thermodynamics it is also of interest to emphasize once more the phenomenological character of relativistic considerations. Indeed, the formulation of the first postulate of relativity, as a generalization of failures to detect the motion of the earth through a suppositious ether, has an interesting parallelism with the formulation of the second law of thermodynamics as a generalization of failures to construct perpetual motion machines of the so-called second kind. And the formulation of the second postulate of relativity as expressing the results of measurements on the velocity of light from moving sources, has something in common with the formulation of the first law of thermodynamics as expressing the results obtained in measurements such as those on the mechanical equivalent of heat.

† Jüttner, *Ann. d. Physik*, **34**, 856 (1911). Tolman, *Phil. Mag.* **28**, 583 (1914).

The experimental basis for the special theory of relativity is perhaps less extended than that for thermodynamics. In addition, there is often a more complicated chain of deductive reasoning involved in obtaining conclusions from the theory of relativity and more introduction of subsidiary hypotheses. Nevertheless, it does not appear that the mere process of combining the two theories should of itself involve any increase in uncertainty, and the main principles of relativistic thermodynamics can certainly be accepted with considerable confidence. Those applications which involve most in the way of subsidiary hypotheses must of course be regarded with the most suspicion. It is, however, one of the main functions of theoretical science, not merely to describe in complicated fashion those facts that are already known, but to extrapolate as wisely as may be into regions yet unexplored but pregnant with human interest.

56. The first law of thermodynamics and the zero point of energy content

In accordance with the ideas underlying the science of thermodynamics, the energy contained in a system is a definite function of its state and can only be changed when the state of the system is itself altered. When such a change in state takes place, it is important for the purposes of thermodynamics to distinguish two different modes of transfer by which the energy content may be affected, namely the flow of heat and the performance of work.

Recognizing these two possibilities, the first law of thermodynamics then states the principle of the conservation of energy in the form

$$\Delta E = Q - W, \qquad (56.1)$$

where ΔE is the increase in energy content corresponding to some given change in state, and Q and W are respectively the heat flow into the system from the surroundings and the work done by the system on the surroundings when a particular process takes place that leads to the given change in state. The equation may be regarded as an expression of the principle of the conservation of energy, since it excludes the possibility of creation or destruction of energy within any region by equating the change in its energy content to a transfer through the boundary in the form of heat or work.

The special theory of relativity has in no way destroyed our ideas as to the conservation and localization of energy,† nor modified our

---

† Indeed the relativistic association of mass with energy has fortified our concepts as to the localization of energy.

ideas as to the possibility of distinguishing between heat and work. Hence in the extension of thermodynamics to special relativity the first law can evidently be taken over unaltered in the form given by (56.1).

The theory of relativity has, however, provided an important *supplement* to the above equation by giving us the additional relation connecting energy with mass as discussed in §§ 27 and 29 (d). In accordance with this new relation we can also express the increase in the energy of a system ΔE in terms of its increase in mass Δm, by the equation,

$$\Delta E = c^2 \Delta m, \qquad (56.2)$$

where c is the velocity of light. And this equation will make it possible to apply the calculations of thermodynamics to processes where a change in mass furnishes the only information as to energy content.

Furthermore, although the first law equation (56.1) gives information only as to changes in energy content and provides no unique zero point of energy content, it may be noted that our previous generalization of the relativistic relation (56.2) to the form

$$E = c^2 m \qquad (56.3)$$

suggests that the absence of all mass can rationally be taken as the zero point of energy content.

57. The second law of thermodynamics and the starting-point for entropy content

In addition to its energy E, thermodynamics also recognizes the entropy S of a system as a definite function of its state. Furthermore, just as the first law relates the energy change in a system to the heat absorbed and work done when some process occurs which changes the state of the system, so too the second law of thermodynamics relates the change in entropy content of the system to the character of the process by which the change in state is brought about.

In order to obtain a definition of entropy and a statement of the second law it is first necessary to distinguish between irreversible and reversible processes, the former being actual processes by which the state of a system may be changed without presenting the possibility of restoring both the system and its surroundings to their original condition, and the latter being ideal processes—approached as a limit by actual processes as they are made more efficient—of such a nature that the system and its surroundings could *both* be returned to their original condition.

With the help of this distinction, we can now define the change in the entropy content of a system, which accompanies a change in its state, by the equation

$$\Delta S = \int_{\text{rev}} \frac{dQ}{T}, \tag{57.1}$$

where T is the temperature for each element of heat dQ transferred across the boundary from the surroundings into the system, and the integration—as indicated by the subscript [rev]—is to be taken for the case of an ideal reversible process by which the given change in state could be thought of as brought about. Moreover, in the light of this definition, we can then state the second law of thermodynamics in the general form

$$\Delta S \geqslant \int \frac{dQ}{T}, \tag{57.2}$$

where the integral can now be taken for any process under consideration by which the system goes from its initial to its final state, and the sign 'is greater than' is to be used unless the process actually is reversible.

In accordance with these expressions, and our previous statement that the entropy content of a system is a definite function of its state, it is evident that the quantity $\int dQ/T$ will have the same maximum value for all reversible processes that result in a given change in the state of a system and a smaller value for all irreversible processes that result in the same change of state. We are thus provided by the second law with a criterion for distinguishing between reversible and irreversible processes and at least a partial description of their character which will be found to lead to specific conclusions of interest and importance.

To complete our consideration as to the nature of entropy we must also inquire as to the total entropy content of a system. Just as the statement of the first law furnished no unique zero point for energy contents, so the above two expressions which give the substance of the second law, are merely statements as to changes in entropy content and furnish no unique zero point for entropy contents. In the case of energy we have seen that a rational zero point of energy could be provided by the mass-energy relationship of the theory of relativity. In the case of entropy a zero point—or more strictly a useful starting-point—is provided by the so-called third law of thermodynamics as discovered and formulated by Nernst and Planck.

In accordance with the third law of thermodynamics, there is no change in entropy content

$$\Delta S_{T=0} = 0 \qquad (57.3)$$

for any change—at the absolute zero of temperature—in the state of a system which is composed both initially and finally of pure crystalline substances. As a result of this principle it becomes specially convenient to take the value zero

$$S_{T=0} = 0, \qquad (57.4)$$

for the entropy of all pure crystalline substances at the absolute zero, and with this as a starting-point then take as their entropy under other conditions the increase which occurs in changing the substance to the particular condition of interest. With this convention it is evident that the entropy increase accompanying any change of state can then be obtained by subtracting the sum of the entropies assigned to the initial substances under the conditions in question from that for the final substances.

As mentioned above in § 55, a deeper insight into the nature of the third law of thermodynamics can be obtained with the help of the statistical-mechanical interpretation of entropy, which shows— speaking somewhat loosely—that the assignment of zero entropy to a pure crystal at the absolute zero corresponds to the complete lack of disorder in the atomic arrangement of such a crystal. Considerations of this microscopic kind can be specially important in criticizing the application of the third law in cases which involve internuclear reactions or the complete transformation of matter into radiation as will be undertaken in §§ 66 and 67. From our present point of view, however, since we desire to remain as far as possible on the macroscopic phenomenological level, it is perhaps most important to emphasize that the third law of thermodynamics can now be regarded—at least for the case of ordinary chemical reactions—as an empirical principle which is well supported by a mass of data obtained particularly under the direction and leadership of Nernst and of Lewis.

58. Heat content, free energy, and thermodynamic potential

In addition to the fundamental thermodynamic quantities energy and entropy, it also proves useful to introduce three further defined quantities H, A, and F which may be called for convenience by the usual but somewhat misleading names—heat content, free energy, and thermodynamic potential.

Restricting ourselves to systems which have the same pressure and temperature in all parts, these quantities may then be defined by the equations

$$H = E+pv, \tag{58.1}$$

$$A = E-TS, \tag{58.2}$$

$$F = E-TS+pv = H-TS, \tag{58.3}$$

where E, S, T, p, and v are respectively the energy, entropy, temperature, pressure, and volume of the system under consideration.

It will be noted from the above expressions that H, A, and F are respectively the χ, ψ, and ζ of Gibbs.[†] Furthermore, A is the free energy as originally defined by Helmholtz,[‡] and F is the quantity usually called free energy by chemists.[§] The nature of the three new quantities and reasons for the names by which they are denoted can be seen from the following considerations.

If we consider a system which is kept under *constant pressure* in such a manner that the only work it can do on the surroundings will be due to change in volume against this pressure, we can write from equation (58.1)

$$\Delta H = \Delta E+p\Delta v$$
$$= \Delta E+W,$$

or in accordance with the first law equation (56.1)

$$\Delta H = Q. \tag{58.4}$$

Hence for such processes the heat absorbed is equal to the increase in the quantity called heat-content. The designation heat-content is, nevertheless, not a happy one since the above simple relation is not true for processes in general. In addition, the designation heat-content unfortunately suggests—in agreement with the abandoned caloric theory—the incorrect use of the term heat to characterize a portion of the energy actually contained within a system, instead of its correct use to characterize a portion of the energy being transferred across the boundary separating the system from its surroundings.

Turning next to a system which is kept at *constant temperature*, we can write from equation (58.2)

$$\Delta A = \Delta E-T\Delta S,$$

† Gibbs, 'On the Equilibrium of Heterogeneous Substances', *Collected Works*, vol. i, p. 87, New York, 1928.   ‡ Helmholtz, *Berl. Ber.* **1**, 22 (1882).
§ Lewis, *Journ. Amer. Chem. Soc.* **35**, 14 (1913).

and, making use of the first and second laws (56.1) and (57.2), we can substitute

$$\Delta E = Q - W$$
$$T\Delta S \geqslant Q,$$

(58.5)

which leads to the result

$$W \leqslant -\Delta A.$$

(58.6)

In accordance with this relation, the work which can be done on its surroundings by a system maintained at constant temperature cannot be greater than the decrease in the quantity which has been called its free energy. This makes the reason for the name obvious, although it must be emphasized that the result obtained applies, of course, only to isothermal processes.

Finally considering a system which is kept both at *constant temperature* and under *constant pressure*, we can write from equation (58.3)

$$\Delta F = \Delta E - T\Delta S + p\Delta v,$$

and again substituting the results of the first and second laws as given by (58.5) we obtain the expression

$$W - p\Delta v \leqslant -\Delta F.$$

(58.7)

In accordance with this relation, for a system maintained under the conditions specified, the total work which can be done by the system on its surroundings diminished by that done against the pressure under which it is maintained cannot be greater than the decrease in its thermodynamic potential F. Since the excess work, over and above that which must be expended in any case against the external pressure, is often the portion of special interest on account of its availability for accomplishing desired results, the thermodynamic potential F is also often called—in particular by chemists—the free energy of the system. The relation given by (58.7) applies of course only to isothermal isopiestic processes.

59. General conditions for thermodynamic change and equilibrium

With the help of the foregoing we can now investigate the conditions which are necessary if a thermodynamic system is to undergo change or to be in a state of equilibrium. This we shall do first for isolated systems which cannot interact with the surroundings, then for systems which are maintained at constant volume and temperature, and finally for systems maintained at constant pressure and temperature.

In the case of an *isolated* system, the heat Q absorbed in any change of state will necessarily be zero owing to the postulated lack of any interaction with the surroundings. Hence, substituting into the second law expression (57.2) we obtain

$$\Delta S \geqslant 0 \qquad (59.1)$$

as a necessary condition for any change that takes place in the state of an isolated system. In accordance with this result the entropy of an isolated system cannot decrease but will increase with the time if irreversible processes take place, or at the limit remain constant if reversible processes take place. Moreover, if the system is in a state of maximum possible entropy, such that variations in its condition cannot lead to further increase in entropy, as denoted by the formulation

$$\delta S = 0, \qquad (59.2)$$

the system will evidently be in a condition of thermodynamic equilibrium where further changes will be impossible. In applying this condition to simple homogeneous systems, it is to be noted that holding the energy and volume constant will be sufficient to secure the necessary lack of interaction with the surroundings.

Turning next to the case of systems subject to external constraints which maintain *constant volume and temperature,* we can evidently write

$$\Delta E = Q$$

in accordance with the first law equation (56.1), since the external work will be zero on account of the constancy of volume. Combining this result with the second law expression (57.2), and making use of the constancy of temperature, we then obtain

$$T\Delta S \geqslant \Delta E,$$

or introducing the definition of free energy (58.2)

$$-\Delta A \geqslant 0 \qquad (59.3)$$

as a necessary condition for any change of state at constant volume and temperature. In accordance with this result, the free energy of a system maintained under these conditions can only decrease or remain constant with the time, and the condition for thermodynamic equilibrium will be that for a minimum value of the free energy, as denoted by the formulation

$$\delta A = 0. \qquad (59.4)$$

Similarly, in the case of systems subject to external constraints which maintain *constant pressure and temperature*, we can write

$$\Delta E = Q - p\Delta v$$

in accordance with the first law, and combining with the second law obtain a decrease in thermodynamic potential

$$-\Delta F \geqslant 0 \qquad (59.5)$$

as a necessary condition for any change in state, and a minimum of thermodynamic potential as denoted by

$$\delta F = 0 \qquad (59.6)$$

as the condition for thermodynamic equilibrium.

The foregoing conditions for thermodynamic change and equilibrium prove very useful in predicting the behaviour of physical-chemical systems. The first pair of conditions as given by (59.1) and (59.2), for the case of an isolated system subject to no external constraints, seem perhaps the most fundamental, since by the inclusion of a sufficient region within the boundary of the system to be considered—or indeed if allowable by considering the universe as a whole—we can undertake the treatment of any situation of interest. The third pair of conditions as given by (59.5) and (59.6), for the case of a system maintained at constant pressure and temperature, is often the most useful on account of our frequent interest in the equilibrium of a system at some specified pressure and temperature—for example atmospheric pressure and room temperature.

60. Conditions for change and equilibrium in homogeneous systems

In order to apply the foregoing conditions for thermodynamic change and equilibrium to determine the behaviour of any given system, we should have to know the dependence of its entropy, free energy, or thermodynamic potential on the variables used for the description of its state. The form of this dependence must, of course, be worked out for the particular system under consideration, and in the present section we shall investigate this form for the case of simple homogeneous systems. This can be done with the help of an equation which combines the requirements of the first and second laws.

Consider a simple system—having uniform pressure, temperature, and composition throughout—whose state can be completely specified by the energy E, volume v, and number of mols $n_1, n_2, ..., n_n$ of the different substances which it contains. Since the entropy S of a system is a definite function of its state, we can then evidently write in accordance with the principles of the differential calculus

$$dS = \frac{\partial S}{\partial E} dE + \frac{\partial S}{\partial v} dv + \frac{\partial S}{\partial n_1} dn_1 + ... + \frac{\partial S}{\partial n_n} dn_n \qquad (60.1)$$

as an expression for the dependence of the entropy of this system on the variables determining its state.

For an infinitesimal reversible change solely in energy and volume we can evidently write, however, in accordance with the first and second laws

$$dS = \frac{dQ}{T} = \frac{dE + p\, dv}{T}, \qquad (60.2)$$

which gives us for the partial differentials with respect to energy and volume the well-known expressions

$$\frac{\partial S}{\partial E} = \frac{1}{T} \quad \text{and} \quad \frac{\partial S}{\partial v} = \frac{p}{T}, \qquad (60.3)$$

where p and T are the pressure and temperature. Substituting these expressions, equation (60.1) can now be written in the more useful form

$$dS = \frac{1}{T} dE + \frac{p}{T} dv + \frac{\partial S}{\partial n_1} dn_1 + ... + \frac{\partial S}{\partial n_n} dn_n. \qquad (60.4)$$

With the help of this equation and our previous definitions of free energy and thermodynamic potential, it is also possible to derive useful expressions for the dependence of these latter quantities on the variables which determine the state of a system. To carry this out we have only to differentiate the equations (58.2, 3) by which free energy and thermodynamic potential were defined, which will give us

$$dA = dE - S\, dT - T\, dS \qquad (60.5)$$

and

$$dF = dE - S\, dT - T\, dS + v\, dp + p\, dv, \qquad (60.6)$$

and then substitute the expression for dS given by (60.4).

Doing this we can then write the three parallel expressions for the dependence of entropy, free energy, and thermodynamic potential on the state of the system

$$dS = \frac{1}{T} dE + \frac{p}{T} dv + \left(\frac{\partial S}{\partial n_1}\right)_{E,v} dn_1 + ... + \left(\frac{\partial S}{\partial n_n}\right)_{E,v} dn_n \qquad (60.7)$$

$$dA = -S\, dT - p\, dv - T\left(\frac{\partial S}{\partial n_1}\right)_{E,v} dn_1 - ... - T\left(\frac{\partial S}{\partial n_n}\right)_{E,v} dn_n \qquad (60.8)$$

$$dF = -S\, dT + v\, dp - T\left(\frac{\partial S}{\partial n_1}\right)_{E,v} dn_1 - ... - T\left(\frac{\partial S}{\partial n_n}\right)_{E,v} dn_n, \qquad (60.9)$$

where subscripts have been introduced to prevent mistake as to the variables held constant when differentiating the entropy with respect to the variables determining the composition.

These equations are of such a form that in using them it is evident that we are to treat entropy as a function of energy, volume, and composition—free energy as a function of temperature, volume, and composition—and thermodynamic potential as a function of temperature, pressure, and composition. Doing so, we can now write for the partial derivatives with respect to composition the useful relations

$$-T\left(\frac{\partial S}{\partial n_i}\right)_{E,v} = \left(\frac{\partial A}{\partial n_i}\right)_{T,v} = \left(\frac{\partial F}{\partial n_i}\right)_{T,p}, \qquad (60.10)$$

where the subscripts indicate the variables in addition to those giving the composition which are taken as regarded as determining the quantities S, A, and F.

With the help of the above considerations, we may now easily investigate the possibilities for change in the composition of a homogeneous system by chemical reactions involving the substances present. To do this let us consider any possible chemical reaction which might be written down for these substances, and denote by $\delta n_1, \delta n_2, ..., \delta n_i, ..., \delta n_n$ the changes that would occur in composition if this reaction should proceed to an infinitesimal extent. We can then write for the infinitesimal changes in entropy, free energy, or thermodynamic potential that would accompany such an infinitesimal reaction proceeding under the respective conditions of constant E and v, constant T and v, or constant T and p, the expressions

$$(\delta S)_{E,v} = \sum_i \left(\frac{\partial S}{\partial n_i}\right)_{E,v} \delta n_i,$$

$$(\delta A)_{T,v} = \sum_i \left(\frac{\partial A}{\partial n_i}\right)_{T,v} \delta n_i, \qquad (60.11)$$

$$(\delta F)_{T,p} = \sum_i \left(\frac{\partial F}{\partial n_i}\right)_{T,p} \delta n_i,$$

where the summation is to be taken over all the substances involved in the reaction.

In accordance with the preceding section (§ 59), however, we can take the occurrence of a maximum of entropy, a minimum of free energy, or a minimum of thermodynamic potential, under the respective conditions specified, as a criterion of thermodynamic equilibrium. Hence, noting the relation given by (60.10), we can now use as the criterion of chemical stability any one of the three following expressions which proves most convenient for the particular problem

$$\sum_i \left(\frac{\partial S}{\partial n_i}\right)_{E,v} \delta n_i = 0, \tag{60.12}$$

$$\sum_i \left(\frac{\partial A}{\partial n_i}\right)_{T,v} \delta n_i = 0, \tag{60.13}$$

$$\sum_i \left(\frac{\partial F}{\partial n_i}\right)_{T,p} \delta n_i = 0, \tag{60.14}$$

and these expressions will hold for each reaction which has no thermodynamic tendency to proceed.

In case the system is not in a state of chemical equilibrium, the quantities given above will not be equal to zero for all possible reactions that might occur. Thus if we have a homogeneous isolated system of constant energy and volume, and there is a reaction for which the quantity given in (60.12) is greater than zero, the progress of this reaction would lead to an increase in entropy, and we can expect it to take place and continue until the values of the coefficients $(\partial S/\partial N_i)$ become such that equilibrium is reached (see § 63).

Although the criteria for chemical stability given above have been obtained from a consideration of the possibility of chemical reaction in a finite homogeneous system subject to specified external restraints, they can be applied in general since the tendency for a chemical reaction to take place is determined solely by conditions at the point of interest. Thus if ϕ is the density of entropy at any particular point, there will be no tendency for chemical reaction at that point, provided we have in agreement with (60.12) the relation

$$\sum_i \left(\frac{\partial \phi}{\partial c_i}\right)_{\rho,\bar{v}} \delta c_i = 0, \tag{60.15}$$

where the quantities $(\partial \phi/\partial c_i)_{\rho,\bar{v}}$ denote rates of change in entropy density with concentration at constant energy density ρ and specific volume $\bar{v}$, and the quantities δc_i denote the infinitesimal changes in concentration of the different reacting substances which would accompany the progress of the reaction under those conditions.

61. Uniformity of temperature at thermal equilibrium

We have now developed sufficient apparatus for thermodynamic considerations so that we can proceed to develop consequences of interest. In the present section and the two following ones we shall consider three well-known principles, commonly employed in the classical thermodynamics, in accordance with which (i) a state of thermal equilibrium would necessarily be characterized by uniformity of temperature; (ii) thermodynamic processes taking place at a finite rate would necessarily be irreversible; and (iii) the final state of an isolated system would necessarily be one of maximum entropy where further change would be impossible.

These three principles have been obtained in the past with the help of the ideas of the classical thermodynamics by such simple and direct methods as to seem inescapable, and have frequently been made the basis for philosophic reflection on the nature of the universe as a whole. Nevertheless, when we consider the extension of thermodynamics to general relativity in the later parts of this book we shall find that all three of these principles must be regarded as subject to exception.

To investigate the distribution of temperature at thermal equilibrium, let us consider the transfer of a small amount of heat dQ from one part of an isolated system at temperature T_1 to a second part at the *lower* temperature T_2. In accordance with the expression of the second law of thermodynamics given by (57.2), we can evidently write for the increase in the entropies of the two parts of the system

$$dS_1 \geqslant -\frac{dQ}{T_1} \quad \text{and} \quad dS_2 \geqslant \frac{dQ}{T_2}$$

and hence by addition for the change in entropy of the whole system

$$dS \geqslant -\frac{dQ}{T_1} + \frac{dQ}{T_2} > 0, \tag{61.1}$$

the value being greater than zero since T_1 is greater than T_2 by hypothesis.

In accordance with this result, an isolated system having parts at different temperatures would not be in a state of thermodynamic equilibrium since a process could occur which would lead to an increase in entropy, in contradiction to the criterion for equilibrium given by (59.2). In the classical thermodynamics we are thus led to the general conclusion that there is a tendency for heat to flow from regions of higher to those of lower temperature, and that uniform temperature throughout is a necessary condition for thermal equilibrium.

In our later development of relativistic thermodynamics, however, we shall find a necessity for modifying this conclusion when different portions of the system under consideration are at different gravitational potentials (see § 129). Roughly speaking, the reason for the modification can be said to lie in the fact that heat must be regarded as having weight. Hence on the transfer of heat from a place of higher to a place of lower gravitational potential, the quantity abstracted at the upper level is less than that added at the lower level and the analysis that led to (61.1) is no longer valid. Defining temperature as that which would be measured by a local observer using proper coordinates, the result obtained in the relativistic treatment will actually show the necessity for a definite temperature gradient at thermal equilibrium to prevent the flow of heat from places of higher to those of lower gravitational potential.

62. Irreversibility and rate of change

The second familiar principle used in the classical thermodynamics, to which we wish to draw attention, is the conclusion that thermodynamic processes which take place at a finite rate are necessarily irreversible. The common reason for belief in this principle lies in the general idea that thermodynamic processes would necessarily have to be carried out at an infinitesimally slow rate in order to secure that maximum efficiency which would be needed for reversibility. A detailed analysis of the application of this idea to a specific typical example will make the reasons for the principle clearer, and will indicate the possibility for later modification when we treat the extension of thermodynamics to general relativity.

As a thermodynamic system sufficiently typical to illustrate the different kinds of processes that must be considered, let us take a mixture of gases, enclosed in a cylinder provided with a movable piston. Any change in the thermodynamic state of this system must involve either the transfer of heat, or of work, between the system and its surroundings, or be due solely to a change in internal conditions. We may consider these three possibilities seriatim.

First of all, it is evident from our knowledge of the phenomena of heat conduction that the transfer of heat between the system and its surroundings at a finite rate could only occur as the accompaniment of a finite temperature gradient. It would thus involve the transfer of heat from regions of higher to those of lower temperature, and hence in accordance with (61.1) of the preceding section there would be an increase in entropy for the system and its surroundings taken as a whole. Since this whole, however, could itself be regarded as an isolated system, such an increase in entropy would involve irreversibility in accordance with our treatment of isolated systems in § 59, and hence we can allow no transfer of heat into the system at a finite rate if the process is to be reversible.

Turning next to the exchange of energy between the system and its surroundings by the performance of work, this could be accomplished in the case of the system mentioned by an expansion of the gases which would force the piston out in such a way as to do work on some suitable external mechanism for storing potential energy. In order to carry this out reversibly, however, it is evident that the force exerted on the external mechanism during the expansion could not be less than the force necessary to recompress the gases on reversal of the direction of motion. It is evident, nevertheless, that this could not be accomplished with a finite rate of expansion—in the first place because of the friction that would accompany a finite velocity of the piston, and in the second place because the flow of gases necessary to fill in the space left by the moving piston would not take place rapidly enough to maintain as great a gas pressure on the piston during expansion as would be present during compression. Since similar considerations could be applied to other modes of doing work, we are led to the general conclusion that the system can do no work on its surroundings at a finite rate and still maintain reversibility.

Since the system cannot interact reversibly with its surroundings either by the transfer of heat or work at a finite rate, we must now inquire into the possibility of internal processes. Furthermore, since the system can have no interaction with its surroundings it may now be treated as isolated, and these internal processes in accordance with (59.1) must lead to no increase in entropy if we are to maintain reversibility.

For the system considered, the possible internal processes could be the transfer of heat from one portion of the gases to another inside the cylinder, the performance of work by one portion of the gases on another on account of pressure differences, the diffusion of one of the component gases from a place of higher to one of lower concentration, or the chemical reaction of the gases among themselves. None of these processes, however, could take place at a finite rate without increase in entropy and hence irreversibility. The transfer of heat and work inside the system at a finite rate would of course be just as irreversible as the transfer between system and surroundings considered above, and we must conclude that the temperature and pressure would have to be uniform throughout the contents of the cylinder. Furthermore, the diffusion of a mol of gas from concentration c_1 to c_2 at constant temperature would be accompanied by the increase in entropy $R \ln (c_1/c_2)$, without reference to the rate at which it took place. And finally, in accordance with our knowledge of chemical kinetics and the criterion for change in an isolated system given by (59.1), if any chemical reaction were possible which did not lead to increase in entropy it would on the average take place as often in the forward and reverse directions without resulting change in composition.

Hence for the simple system considered we are led to the conclusion that no processes, with or without interaction between system and surroundings, could take place *both* reversibly *and* at a finite rate. Furthermore, the system treated is sufficiently typical to illustrate the line of thought by which this result has come to be regarded as a general principle for use in connexion with the classical thermodynamics.

It remains to point out a reason for exceptions to this principle which we shall later find resident in the extension of thermodynamics to general relativity. This will be found to lie in the possibility for changes in the proper volume of an element of matter—as measured by a local observer—due to changes in gravitational potentials which are neglected in the classical theory. We were led above to the conclusion that a reversible increase in the volume of the gas could not take place at a finite rate because of friction that would develop and because of a falling off in outward pressure that would accompany the flow of gas to fill in the space left by the moving piston. In relativistic mechanics, however, we shall find possibilities for a change in *proper* volume without friction and with a complete balance between internal and external pressures, and hence shall be led to different thermodynamic conclusions.

63. Final state of an isolated system

The third principle of the classical thermodynamics which we wish to consider is the conclusion that the final state of an isolated system would necessarily be one of maximum entropy where further change would be impossible. The justification for this principle in the classical thermodynamics is found to depend on the first and second laws in a relatively simple manner.

In accordance with the first law of classical thermodynamics the energy content of an isolated system must remain constant, and in accordance with the second law—see §§ 59 and 62—its entropy must increase with the time as a result of any actual thermodynamic changes that take place in it. Hence if there is an upper limit, giving the maximum possible entropy of the system, this will determine the final state of the system, where in accordance with (59.2) further change will be impossible.

The proof that a maximum upper limit of entropy would exist can be carried out in detail by the methods of the classical thermodynamics for any specified isolated system chosen as typical. More generally it is evident that the entropy of a system can be regarded as a function of its energy, volume, and sufficient further variables to determine its internal configuration and constitution, and since the energy of an isolated system will be constant we need only to consider the dependence of entropy on the volume and internal variables. In the case of unconfined gases, however, this dependence is such that with constant energy content a final state of infinite dilution and complete dissociation into atoms would be one of maximum entropy. And in the case of systems held together by their own coherence a final state of maximum entropy would be obtained when the internal variables have adjusted themselves to the most favourable values possible in the restricted range permitted by the fixed value of the energy.

For example, in the case of a homogeneous system of constant energy and volume, the considerations of § 60 have shown that the condition for a given chemical reaction to take place would be given by the expression (see 60.12)

$$\sum_i \left(\frac{\partial S}{\partial n_i}\right)_{E,v} \delta n_i > 0, \tag{63.1}$$

where $\delta n_1, \delta n_2, ..., \delta n_i, ...$ denote the changes in the number of mols of the different interacting substances which would occur if the reaction in question should proceed to an infinitesimal extent. And since the value of any individual differential coefficient $(\partial S/\partial n_i)$ is actually found to decrease with increasing values of n_i it is evident that continued reaction would ultimately lead to a condition of maximum entropy.

By considerations such as these the classical thermodynamics has been led to the belief that isolated systems would approach a final state of maximum possible entropy where further change would not take place. In our later extension of thermodynamics to general relativity, however, this conclusion will be modified by the fact that relativistic mechanics does not require a constant value for what may be called the total proper energy of an isolated system, and this removes the restriction on the adjustment of variables to secure increased entropy imposed in the classical thermodynamics by the principle of the conservation of energy.

64. Energy and entropy of a perfect monatomic gas

As a preparation for later applications we may now treat several matters of a more specific nature. In the present section we shall give expressions for the energy and entropy of a perfect monatomic gas.

For the relation between the pressure, volume, and temperature of such a gas we can take the perfect gas laws in the form

$$pv = NkT, \tag{64.1}$$

where N is the number of molecules present, and Boltzmann's constant k is the ordinary gas law constant R for one mol of the gas divided by the number of molecules in a mol (Avogadro's number A)

$$k = R/A. \tag{64.2}$$

Furthermore, in accordance with experiment and the simplest considerations of the kinetic theory, we can write

$$C_v = \tfrac{3}{2}Nk \quad \text{and} \quad C_p = \tfrac{5}{2}Nk \tag{64.3}$$

for the heat capacities of such a gas at constant volume and constant pressure respectively, and in addition can take†

$$E_{\text{kin}} = \tfrac{3}{2}NkT \tag{64.4}$$

as an expression for the translational kinetic energy of the molecules at temperature T.

---

† For sufficiently light molecules at sufficiently high temperatures the expression becomes
$$E = 3NkT.$$
See Jüttner and Tolman, loc. cit.

Making use of the fundamental starting-point for energy content provided by the mass-energy relationship of Einstein, as discussed in § 56, we can then take for the total energy of such a gas the sum of the energy due to the *rest-mass m* of the particles themselves, and the above value for the kinetic energy, in accordance with the expression

$$E = Nmc^2 + \tfrac{3}{2}NkT. \tag{64.5}$$

This result will be of importance when we desire to consider the possible transformation of matter into radiation, or the possible transformation of one kind of atoms into another as in the formation of helium out of hydrogen, since the store of internal energy Nmc^2 can then be drawn upon.

For the dependence of the entropy of a perfect gas on temperature and volume, we can evidently write

$$dS = C_v \frac{dT}{T} + \frac{p\,dv}{T}, \tag{64.6}$$

since $(C_v\,dT + p\,dv)$ would be the heat absorbed in a reversible change of temperature and volume. And substituting the values of C_v and v given by (64.1) and (64.3) and integrating we obtain

$$S = \tfrac{5}{2}Nk \log T - Nk \log p + \text{const.} \tag{64.7}$$

as an expression for the entropy of N molecules of perfect monatomic gas at temperature T and pressure p. Or introducing the concentration of the gas c as given by the gas laws in the form

$$p = \frac{N}{v}kT = ckT \tag{64.8}$$

we can also rewrite the above expression for entropy in the equivalent form

$$S = \tfrac{3}{2}Nk \log T - Nk \log c + \text{const.} \tag{64.9}$$

The value of the constant of integration occurring in equation (64.7) can be taken proportional to the number of molecules N, but is of course otherwise undetermined, until we choose some specific zero point for entropy contents. Taking the zero of entropy—in accordance with the third law of thermodynamics (§ 57)—to be that for the substance in the form of a pure crystal at the absolute zero, we can then determine the constant from a knowledge of the reversible heat of evaporation from the crystalline to the gaseous form. This can readily be done theoretically† and leads to the well-known

† See, for example, Tolman, *Statistical Mechanics*, New York, 1927.

Sackur-Tetrode equation for the entropy of a monatomic gas

$$S = \tfrac{5}{2}Nk\log T - Nk\log p + Nk\log\frac{(2\pi m)^{\frac{3}{2}}(ke)^{\frac{5}{2}}}{h^3}, \qquad (64.10)$$

or in terms of concentration

$$S = \tfrac{3}{2}Nk\log T - Nk\log c + Nk\log\frac{(2\pi mk)^{\frac{3}{2}}e^{\frac{5}{2}}}{h^3}, \qquad (64.11)$$

where the additive constant is seen to depend on the mass m per molecule for the particular gas in question, and on certain universal constants, the base of the natural system of logarithms e, Boltzmann's constant k, and Planck's constant h. The actual dependence of the entropy of monatomic gases on these quantities in the way stated may now be regarded as a satisfactorily tested empirical fact.

The quantity given by equations (64.10, 11) is often spoken of as the absolute entropy of the gas. Since such a designation might be misleading, however, it is well to emphasize that this quantity is in any case—both theoretically and experimentally—the increase in entropy that would accompany a change in state of the substance considered from the form of a pure crystal at the absolute zero to that of a perfect monatomic gas under the conditions specified.

For practical calculations equations (64.10, 11) can be written in the following forms for the entropy per mol of gas at a given concentration c or at a given pressure p:

$$S = \tfrac{3}{2}R\log T - R\log c + \tfrac{3}{2}R\log M + S_0 \qquad (64.12)$$

and

$$S = \tfrac{5}{2}R\log T - R\log p + \tfrac{3}{2}R\log M + S_0', \qquad (64.13)$$

where the logarithms are to the base e, the entropy S and gas constant R are in calories per mol per degree centigrade, T is in degrees centigrade absolute, M is the molecular weight of the gas in grammes, c is in mols per cubic centimetre, p is in normal atmospheres, and the constants have the values†

$$S_0 = -11 \cdot 0533 \qquad (64.14)$$

and

$$S_0' = -2 \cdot 29852 \qquad (64.15)$$

calories per mol per degree centigrade. The expressions are, of course, for monatomic gases.

Expressions for the energy and entropy of gases composed of more complicated molecules, where allowance must be made for the rota-

† Birge, *Phys. Rev.* Supplement, **1**, 1 (1929).

tion of the molecule as a whole and if necessary also for the oscillation of the atoms within the molecule, can also be obtained, but will not be necessary for the applications that will be undertaken.

65. Energy and entropy of black-body radiation

In the present section we shall give the well-known expressions for the energy and entropy of black-body radiation, for use in our later applications.

As shown by the work of Stefan and Boltzmann, the energy density u for radiation in equilibrium with the walls of a hollow enclosure at temperature T is given by the formula

$$u = aT^4, \tag{65.1}$$

where Stefan's constant a has the value†

$$a = 7 \cdot 6237 \times 10^{-15} \tag{65.2}$$

in ergs per cubic centimetre per degree centigrade to the fourth power. Furthermore, the pressure of this radiation is given by

$$p = \frac{a}{3} T^4. \tag{65.3}$$

In accordance with (65.1) we may then write for the total energy of the radiation present at equilibrium in a hollow enclosure of volume v at temperature T

$$E = avT^4. \tag{65.4}$$

Furthermore, in accordance with the above expressions, we can evidently write for the heat absorbed when the volume of the enclosure is increased by a reversible isothermal expansion

$$\begin{aligned} dQ &= dE + dW \\ &= aT^4 \, dv + \tfrac{1}{3}aT^4 \, dv \\ &= \tfrac{4}{3}aT^4 \, dv, \end{aligned}$$

and hence for the increase in the entropy content of the enclosure

$$dS = \tfrac{4}{3}aT^3 \, dv.$$

This expression can now be integrated, however, to give the total entropy increase corresponding to an increase in the volume from zero to v. We thus obtain, for the total entropy of the radiation at equilibrium in a hollow enclosure of volume v at temperature T, the expression

$$S = \tfrac{4}{3}avT^3. \tag{65.5}$$

Moreover, this quantity could be strictly spoken of as the absolute

† Birge, loc. cit.

entropy of the radiation, since it is the total entropy increase accompanying the actual introduction of the radiation into the space created inside the enclosure.

As a further important characteristic of black-body radiation we may also note that the energy distribution among the different frequencies is given at equilibrium by the Planck law

$$du = 8\pi \frac{h\nu^3}{c^3} \frac{1}{e^{h\nu/kT}-1} \, d\nu. \tag{65.6}$$

66. The equilibrium between hydrogen and helium

As an interesting thermodynamic application of the relation between mass and energy provided by the theory of relativity, we may now consider the possible formation of helium out of hydrogen in accordance with a quasi-chemical reaction which we can write in the form

$$4H = He. \tag{66.1}$$

If the hydrogen atom does consist of one proton and one electron and the helium atom of a nucleus containing four protons and two electrons surrounded by two external electrons—as it seems reasonable to believe—such an internuclear reaction should be entirely possible. Furthermore, since the mass of the helium atom is considerably less than that of four hydrogen atoms there should be a great evolution of heat accompanying this process and hence in accordance with the qualitatively correct principle of Berthelot a great tendency for the reaction to occur. In the present section we shall calculate the conditions of equilibrium for this postulated process.†

To do this it will be most convenient to take the criterion for chemical equilibrium in the form given by our previous equation (60.14)

$$\sum_i \left(\frac{\partial F}{\partial n_i}\right)_{T,p} \delta n_i = 0, \tag{66.2}$$

where the quantities δn_i are the changes in number of mols of the different substances present which would occur if the reaction under test should proceed to an infinitesimal extent, and the quantities $(\partial F/\partial n_i)$ are the rates of change in the thermodynamic potential of the system at constant temperature and pressure per mol of the substance indicated. Applying this criterion to the reaction between

† Tolman, *Journ. Amer. Chem. Soc.* **44**, 1902 (1922).

hydrogen and helium as given by (66.1), and using the subscripts 1 and 2 to refer to hydrogen and helium respectively we can write the requirement for equilibrium in the form

$$\left(\frac{\partial F}{\partial n_1}\right)_{T,p} \delta n_1 + \left(\frac{\partial F}{\partial n_2}\right)_{T,p} \delta n_2 = 0,$$

or since we shall necessarily have four hydrogen atoms used for each helium atom formed we can substitute

$$-\delta n_1 = 4\,\delta n_2$$

and obtain

$$-4\left(\frac{\partial F}{\partial n_1}\right)_{T,p} + \left(\frac{\partial F}{\partial n_2}\right)_{T,p} = 0 \qquad (66.3)$$

as a relation which must hold at equilibrium.

To use this equation in our present problem, we may assume the hydrogen and helium both sufficiently dilute, so that they can be treated as perfect monatomic gases at their partial pressures and the temperature of the mixture. The thermodynamic potential per mol of hydrogen or helium produced in the mixture can then be taken as equal to the actual thermodynamic potential for one mol of that gas in a pure state at the temperature and pressure thus given. And since these thermodynamic potentials will themselves be calculable in terms of the energy E, pressure p, volume v, and temperature T of the gas in question from the equation of definition (58.3)

$$F = E + pv - TS$$

the condition of equilibrium (66.3) can now be rewritten in the form

$$(E_2 + p_2 v - TS_2) - 4(E_1 + p_1 v - TS_1) = 0, \qquad (66.4)$$

where the subscripts 1 and 2 again refer to hydrogen and helium, E_1 and E_2 being the energies and S_1 and S_2 the entropies of a mol of pure hydrogen or helium at the temperature T and the pressures p_1 and p_2 respectively, which are the partial pressures in the equilibrium mixture.

For the energy difference between a mol of helium and 4 mols of hydrogen, allowing for the relativistic relation between mass and energy, we may write in accordance with (64.5)

$$E_2 - 4E_1 = (M_2 - 4M_1)c^2 - \tfrac{9}{2}RT, \qquad (66.5)$$

where M_1 and M_2 are the molecular weights of hydrogen and helium, thus placing the energy change equal to the change in internal energy plus the change in the kinetic energy of the molecules. Furthermore, for the pressure-volume products we may write in accordance with the gas laws

$$p_2 v - 4 p_1 v = -3RT. \tag{66.6}$$

Finally, for the entropies we can use the values provided by the Sackur-Tetrode equation in the form (64.13), which will give us

$$-TS_2 + 4TS_1$$
$$= \tfrac{15}{2} RT \log T - RT \log \frac{p_1^4}{p_2} + \tfrac{3}{2} RT \log \frac{M_1^4}{M_2} + 3TS_0'. \tag{66.7}$$

Substituting these expressions into the equilibrium condition (66.4) and solving, we then obtain as an expression for the equilibrium constant for the reaction

$$\log \frac{p_2}{p_1^4} = -\frac{(M_2 - 4M_1)}{RT} c^2 - \tfrac{15}{2} \log T + \tfrac{3}{2} \log \frac{M_2}{M_1^4} - \frac{3S_0'}{R} + \frac{15}{2}, \tag{66.8}$$

where p_1 and p_2 are the equilibrium pressures of hydrogen and helium.

To obtain numerical results with the help of this equation, we may take for the molecular weights of monatomic hydrogen and helium the 1932 atomic weights 1·0078 and 4·002 gm., thus neglecting the small fraction of the isotope of hydrogen of approximate weight 2 recently discovered. For the other quantities we may use the values of Birge,[†] for the velocity of light $c = 2 \cdot 99796 \times 10^{10}$ cm. sec.$^{-1}$, for the gas constant $R = 8 \cdot 31360 \times 10^7$ erg. deg.$^{-1}$ mol^{-1} or $1 \cdot 98643$ cal. deg.$^{-1}$ mol^{-1}, and for $S_0' = -2 \cdot 29852$ cal. deg.$^{-1}$ mol^{-1} corresponding to taking the pressures in atmospheres. Substituting these values and changing to logarithms to the base 10, we then obtain with sufficient accuracy for our present purposes

$$\log \frac{p_2}{p_1^4} = \frac{1 \cdot 371 \times 10^{11}}{T} - 7 \cdot 5 \log T + 5 \cdot 648, \tag{66.9}$$

where the equilibrium pressures p_1 and p_2 of hydrogen and helium are in atmospheres. Or denoting the total pressure of the mixture by

$$p = p_1 + p_2 \tag{66.10}$$

and letting α be the fraction of helium which would be dissociated into hydrogen at equilibrium we can also easily rewrite this equation in the form

$$\log \frac{256 p^3 \alpha^4}{(1-\alpha)(1+3\alpha)^3} = -\frac{1 \cdot 371 \times 10^{11}}{T} + 7 \cdot 5 \log T - 5 \cdot 648. \tag{66.11}$$

† Birge, loc. cit.

In accordance with these results, we see that there would be an extremely great thermodynamic tendency for hydrogen to change over into helium unless we should go to extremely high temperatures and low pressures. This tendency arises because of the great evolution of energy accompanying the formation of helium from hydrogen corresponding to a loss in mass which is not large in grammes per mol formed, but very large in ergs on account of the appearance of the square of the velocity of light as the conversion factor. Thus in accordance with (66.9) we should have

$$\frac{p_2}{p_1^4} = 1 \quad \text{at} \quad T \simeq 2 \times 10^9 \,^\circ\text{C}. \tag{66.12}$$

and hence should need a temperature over 10^9 degrees absolute in order to have monatomic hydrogen at one atmosphere in equilibrium with helium at that same pressure. And in accordance with (66.11), even at a temperature of a million degrees and a pressure as low as 10^{-100} atmospheres, the fraction of helium α dissociated into monatomic hydrogen would only be

$$\alpha = 10^{-30,000} \begin{cases} T = 10^6 \,^\circ\text{C}. \\ p = 10^{-100} \text{ atm}. \end{cases} \tag{66.13}$$

These calculations have been made for the equilibrium between unionized helium and unionized monatomic hydrogen. Nevertheless, the free energy changes, accompanying such processes as ionization or ordinary chemical reaction, are so small compared with that for the internuclear reaction as not to change the general conclusion that hydrogen would in any case be almost completely transformed into helium at equilibrium under all but the most extreme conditions of high temperature and low pressure.

This result must now be compared with the known facts concerning the presence of hydrogen and helium on the earth, and in the sun and other stars. On the earth it is well known that hydrogen shows as yet no discovered tendency to go over into helium. Hydrogen at terrestrial temperatures and pressures is of course largely in the diatomic form or combined with oxygen to form water. As noted above, however, this would not appreciably diminish the thermodynamic tendency to form helium. Furthermore, it is well known from spectral data that hydrogen in the unionized monatomic form and unionized helium are both found in appreciable amounts in the chromosphere of the sun and in stars of a number of classes at temperatures ranging say from 6,000 to 20,000 degrees absolute and at pressures enormously greater than the 10^{-100} atmospheres mentioned above. We must hence conclude in general that the observed concentrations of hydrogen and helium in the universe are very far from correspondence with the amounts calculated for equilibrium.

Similar conclusions, as to a discrepancy between relative amounts calculated for thermodynamic equilibrium and actually observed, have been obtained by Urey and Bradley[†] for the case of a considerable number of isotopes which could be conceivably transformed into each other by internuclear reactions, of the type given by the example

$$C^{12} + O^{17} = C^{13} + O^{16}.$$

The method of calculation is similar to that employed above for the equilibrium between hydrogen and helium. The calculations have the advantage that the effect of deviations from the perfect gas laws —which might arise under the actual conditions obtaining when equilibrium is established—would tend to cancel out by affecting both reactants and products in the same way. The calculations have the disadvantage, however, of smaller and at least in some cases less certain mass changes, and of less general information as to the relative abundance in the universe as a whole of the substances involved. The actual relative abundances of the isotopes in terrestrial material was found not to be in agreement with the relative abundances that would be calculated for equilibrium conditions at any assigned temperature.

To account for the discrepancy, between the observed concentration of hydrogen on the earth or in the sun and stars and that calculated for equilibrium, three general types of explanation present themselves. First, it is possible that helium cannot be formed out of the constituents of hydrogen as assumed; secondly, the theoretical basis for the calculations may not be justified at some point; and thirdly, hydrogen may actually have a thermodynamic tendency to go over into helium but the reaction be so slow that equilibrium has not been attained.

The reasons for believing in the possible formation of helium and the other higher elements from the constituents of hydrogen lie in the approximate whole number relations between atomic weights, in the observed emission of electrons and protons where certain nuclei are decomposed either artificially or by radioactive disintegration, and

[†] Urey and Bradley, *Phys. Rev.* 38, 718 (1931).

in the general simplification and reduction in number of necessary independent assumptions obtained with the help of this hypothesis. These reasons are strong but certainly not conclusive evidence that helium can be formed from hydrogen, since our present knowledge as to the number and nature of fundamental particles, such as positive and negative electrons, protons, and neutrons, is not complete. If helium cannot be produced solely from the constituents of hydrogen the treatment given is of course not applicable.

The theoretical basis for the calculations might be wrong either on account of the expression used for the energy change that would accompany the reaction, or on account of the expression used for the entropy change. The energy change was calculated with the help of Einstein's mass-energy relationship. This principle forms such a simple and integral part of the theory of relativity that we should be loth to abandon it. The entropy change was calculated with the help of the Sackur-Tetrode expression for the entropy of monatomic gases. This expression undoubtedly gives correctly that part of the entropy which is associated with the unordered spatial arrangement and motion of the atoms as a whole, but neglects any possible disorder within the nucleus. Nevertheless, we should have to assume an enormous increase in the internal disorder within protons or electrons themselves in going from helium to hydrogen in order to change the nature of the conclusions in the direction of higher concentrations of hydrogen.

The most probable explanation for the high concentration of hydrogen in the observable portion of the universe appears to lie, hence, in the assumption of an exceedingly slow rate for the reaction by which hydrogen would go over into helium, coupled with the hypothesis that even larger amounts of hydrogen were present in the universe in the past.

The assumption of great slowness for the reaction is in itself entirely reasonable. If the reaction took place in accordance with the simple mechanism

$$4H \rightarrow He$$

it would be of a very high order—the fourth—and in addition the nuclei of the hydrogen atoms would then be hindered in coming into intimate contact by the presence of the valency electrons. And if the reaction took place in accordance with the mechanism

$$4H^+ + 2E^- \rightarrow He^{++}$$

after ionization of the valency electrons, it would be of even higher order, and intimate contact would in this case be hindered by a net electrostatic repulsion. Hence, except under extreme conditions of pressure and temperature, we should in any case expect the reaction to be very slow.

On the other hand, the assumption of even larger amounts of hydrogen present in the universe in the past, at once introduces us to difficulties of the well-known kind, always encountered in the application of the second law over long time-intervals. In the case at hand, even if one assume a very small rate of reaction, we can still ask why the equilibrium concentration of hydrogen has not been reached in the infinite past time presumably available for transformation into helium. The consideration of such difficulties will form an important part of our later work. For the time being we may content ourselves with pointing out that the high concentration of hydrogen, the lack of equilibrium ratios for the relative amounts of the isotopes, and the presence of still undisintegrated radioactive substances are all phenomena of a similar kind, which indicate the possibility that the present composition of the matter in this portion of the universe results from a past history that involved exceedingly high temperatures.

67. The equilibrium between matter and radiation

As a second thermodynamic application of the relativistic relation between mass and energy, we may now consider the possible transformation of matter into radiation. This process, which is often called the annihilation of matter, would occur if negative electrons and protons or negative and positive electrons should be able to combine, with a resulting mutual neutralization of electric charge, and a change of the energy corresponding to their total mass into the form of electromagnetic radiation. We have, of course, at the present time no direct evidence that such a process ever does occur, although at least in the case of negative and positive electrons it seems highly probable. Nevertheless the annihilation of matter, together with the transformation of hydrogen into helium, have both of them seemed attractive hypotheses to astrophysicists in order to account for the long life during which energy emission has taken place from the sun and other stars.

Assuming for the time being the possibility of such a transformation, we shall investigate in the present section the conditions for thermodynamic equilibrium between matter and radiation. Such an investigation was first made by Stern,† and later with a somewhat altered point of view by the present writer.‡

To carry out the investigation we shall find it most convenient to employ the criterion for equilibrium given by equation (59.2)

$$(\delta S)_{E,v} = 0, \tag{67.1}$$

in accordance with which the entropy of a system maintained at constant energy and volume would be at a maximum. In order to use this criterion we shall need expressions both for the energy and entropy of a system containing an interacting mixture of matter and radiation. To obtain these expressions, we shall take the matter as being in the form of a perfect monatomic gas, and shall assume matter and radiation both sufficiently dilute so that we shall be justified in neglecting interaction and regarding the total energy and entropy as the sum of the usual expressions for the energies and entropies of the two constituents.

For the total energy of a system containing N molecules of mass m in volume v and at temperature T, we may then write in accordance with equations (64.5) and (65.4)

$$E = Nmc^2 + \tfrac{3}{2}NkT + avT^4, \tag{67.2}$$

where the first term allows for the internal energy associated with the mass of the molecules, as is necessary if we are to contemplate the transformation of matter into radiation, and the other two terms give the kinetic energy of the monatomic molecules and the energy of the radiation.

For the entropy of the gas contained in the system we shall find it most convenient to use the expression in terms of concentration as given by (64.9)

$$S = \tfrac{3}{2}Nk \log T - Nk \log c + \text{const.}, \tag{67.3}$$

and for the entropy of the radiation we can use the expression given by (65.5)

$$S = \tfrac{4}{3}avT^3. \tag{67.4}$$

The additive constant in (67.3) must be taken proportional to the number of molecules N, but will otherwise be determined by the choice of starting-point for entropy values, and in combining the two expressions for entropy we must use the same starting-point for the entropy of matter and radiation if their inter-conversion is to be

† Stern, *Zeits. Elektrochem.* **31**, 448 (1925).
‡ Tolman, *Proc. Nat. Acad.* **12**, 670 (1926).

considered. In order to do this let us write $Nk \log be^{\frac{5}{2}}$ as an expression for the constant, where the particular form is taken merely in the interests of simplicity in the final formula, and b is a quantity whose value by hypothesis shall be such as to secure the necessary identity of starting-point. Adding the two expressions we can then write for the total entropy of our mixture

$$S = \tfrac{3}{2}Nk \log T - Nk \log(N/v) + Nk \log be^{\frac{5}{2}} + \tfrac{4}{3}avT^3. \qquad (67.5)$$

In accordance with our criterion for equilibrium, this quantity is to be a maximum at constant energy and volume. Taking the variation with respect to N and T, keeping v constant, we can then write

$$\delta S = \{\tfrac{3}{2}k \log T - k \log(N/v) - k + k \log be^{\frac{5}{2}}\}\, \delta N +$$
$$+ (\tfrac{3}{2}Nk/T + 4avT^2)\, \delta T = 0, \qquad (67.6)$$

together with the subsidiary equation for the constancy of the energy

$$\delta E = (mc^2 + \tfrac{3}{2}kT)\, \delta N + (\tfrac{3}{2}Nk + 4avT^3)\, \delta T = 0 \qquad (67.7)$$

as the necessary conditions for equilibrium. And by combining these two equations and solving, we easily obtain as the desired expression for the concentration of monatomic gas in equilibrium with radiation at temperature T the simple expression

$$N/v = bT^{\frac{3}{2}}e^{-mc^2/kT}, \qquad (67.8)$$

which shows that the equilibrium concentration of matter would increase with rise in temperature.

In order to obtain specific values from this equation for the equilibrium concentration of matter at any given temperature we need to have a value for the constant b. Empirically we have of course no direct knowledge as to what this value should be. Theoretically, however, we know that its value must be such as to give the same starting-point to the entropy of matter and radiation, and the treatment given to the problem by Stern was equivalent to assuming that the entropies of a hollow enclosure containing radiation, and of matter in the form of a pure crystal, would both approach zero on cooling down to the absolute zero of temperature. The justification for the first part of this assumption seems reasonable, since there would be no radiation at all left in the enclosure at absolute zero. The justification for the second part of the assumption seems less certain, however, since it takes for the total entropy of matter only that part which would be connected with the disordered positions and motions of the component atoms and neglects other possibilities. Nevertheless, this assumption is probably the best that we can make at present.

Assuming that we are justified in taking as the total entropy of matter that which it has over and above the entropy of a pure crystal at the absolute zero, we can easily proceed since it is evident that the term in (67.5), containing the constant b, can then be set equal to the last term in our previous expression (64.11) for the entropy increase in going from a crystal at the absolute zero to the form of gas. This will give us

$$Nk \log b e^{\frac{5}{2}} = Nk \log \frac{(2\pi m k)^{\frac{3}{2}} e^{\frac{5}{2}}}{h^3},$$

and on solving for b and substituting into (67.8), we obtain Stern's expression for the concentration of monatomic gas in equilibrium with radiation at the temperature T

$$\frac{N}{v} = \left(\frac{2\pi m k}{h^2}\right)^{\frac{3}{2}} T^{\frac{3}{2}} e^{-mc^2/kT}. \tag{67.9}$$

For the purposes of practical calculation, by substituting values for the universal constants, this can be rewritten in the form

$$c = 3 \cdot 143 \times 10^{-4} M^{\frac{3}{2}} T^{\frac{3}{2}} e^{-1 \cdot 081 \times 10^{13} M/T}, \tag{67.10}$$

where c is the concentration in mols per cubic centimetre and M is the molecular weight in grams.

In accordance with these expressions it is immediately evident that the calculated equilibrium concentration of matter would be exceedingly low except at enormously high temperatures on account of the great effect of the negative exponent $-mc^2/kT$. Thus for a gas of molecular weight one composed of simple neutral particles (neutron gas) whose mass could be directly transformed into radiation, the calculated equilibrium concentration even at 10^9 degrees centigrade would only be

$$c = 10^{-4686} \frac{\text{mols}}{\text{cm.}^3} \quad \text{or} \quad N = 6 \cdot 06 \times 10^{-4663} \frac{\text{molecules}}{\text{cm.}^3}. \tag{67.11}$$

Instead of considering the equilibrium between radiation and a gas composed of simple neutral particles, it might seem more in correspondence with actuality to consider the equilibrium between radiation and a mixture of negative electrons and protons of masses m_1 and m_2, or of negative and positive electrons, produced from radiation in equal numbers in order to maintain electrical neutrality. The treatment of this case can also easily be carried out by the methods employed above and leads in place of (67.9) to the expression

$$\frac{N}{v} = \left(\frac{2\pi \sqrt{(m_1 m_2)}k}{h^2}\right)^{\frac{3}{2}} T^{\frac{3}{2}}e^{-(m_1+m_2)c^2/2kT} \qquad (67.12)$$

for the equal concentrations N/v of the two kinds of particle, where the average mass of the two particles $(m_1+m_2)/2$ now appears in the negative exponent. The result still leads, of course, to excessively low equilibrium concentrations of matter.

To account for the enormous discrepancy between the observed concentrations of matter and radiation in the universe, and what would be calculated for thermodynamic equilibrium, similar considerations present themselves as in the previous case of the possible transformation of hydrogen into helium. Thus it is possible, first that matter and radiation are not interconvertible as assumed, secondly that the theoretical basis for the calculations is not justified at some point, and thirdly that matter does have a great tendency to go over into radiation, but that the change is so slow that equilibrium has not been attained.

In the case of the present problem, it is felt that the actual occurrence of combinations between negative and positive electrons to form radiation is highly probable,[†] but the general question of the possible transformation of all kinds of matter, including neutrons and protons, into radiation is less certain. To justify such an assumption we have little to go on except the fact that matter and radiation both have mass, and hence—accepting the principle of the conservation of mass—we can tell how much radiation would be formed from a given amount of matter. It seems entirely possible, however, that some forms of matter contain entities which cannot be changed into radiation.

With regard to the theoretical basis for the calculations, special emphasis must be laid on the uncertainties involved in obtaining a value for the constant b in equation (67.8) by taking the entropy of matter in the crystalline form at the absolute zero as the correct starting-point. This is certainly an appropriate procedure in considering the transformation of one kind of matter into another by ordinary chemical reaction, but is much more dubious for processes which would actually involve the destruction of fundamental particles, since it neglects the possibility of entropy resident within their structure. It is to be noted, nevertheless, that we have no theory of such entropy inside the particles, and perfectly enormous stores of such entropy would be necessary to overcome the great effect of the negative exponent $-mc^2/kT$ in leading to low concentrations of matter.

With regard to the rate of reaction by which radiation might be formed from matter, theoretical computation has indicated a very high rate for the mutual annihilation of positive and negative electrons,[†] and this is perhaps in agreement with the apparent lack of accumulation of the positive electrons which we now know to be continuously driven out from terrestrial matter by the bombardment of cosmic rays. In the case of other processes we have no information. The mutual annihilation of one proton and one electron would be a complicated process involving the production of two light quanta under just the necessary conditions to satisfy the conservation laws and might well have a low *a priori* probability of occurrence. Nevertheless, it is possible that the rate of annihilation would have to be exceedingly small if we should desire to account both for the total concentration of matter and the relative concentrations of its different forms by assuming general equilibrium at a very high temperature at some time in the past.

In spite of the uncertainties which attend the foregoing treatment of the equilibrium between matter and radiation, and to a lesser extent that for the equilibrium between hydrogen and helium, it is believed that the methods of calculation employed are instructive, and the results obtained are at least of some interest in our present state of knowledge. It is perhaps specially interesting to note that thermodynamic calculations can be made which are logical and consequent, even at a time when our ignorance of necessary facts precludes a definite assertion as to their actual applicability to the physical situation proposed. Such an experimentation with 'if' 'then' considerations can be of great importance both to the intellect and the imagination.

† Oppenheimer, *Phys. Rev.* **35**, 939 (1930).

V

SPECIAL RELATIVITY AND THERMODYNAMICS (*contd.*)

Part II. THE THERMODYNAMICS OF MOVING SYSTEMS

68. The two laws of thermodynamics for a moving system

In Part I of this chapter we have considered the classical thermodynamics of stationary systems, and have investigated the effect of
the theory of relativity only in so far as it has provided a new means
for determining the energy content of a system. We must now turn
to the more far-reaching effects of relativity in providing, as first
shown by Planck† and by Einstein,‡ a satisfactory theory for the
treatment of thermodynamic systems which are in motion relative
to the set of axes which are being used by the observer.

As a basis for the theory we shall find it possible to use the two laws
of thermodynamics written in exactly their previous forms:

$$\Delta E = Q - W \qquad (68.1)$$

for the energy change of a system in terms of heat absorbed and work
done, and

$$\Delta S \geqslant \int \frac{dQ}{T} \qquad (68.2)$$

for the entropy change in terms of heat absorbed and temperature.
In applying these expressions, however, it will now be understood that
the quantities, energy, entropy, heat, work, and temperature which
appear therein are to be assigned the values which are appropriate to
the particular set of axes which is being used, with reference to which
the thermodynamic system under consideration is not necessarily at
rest but may be in a state of uniform translatory motion.

The justification for using the above expressions (68.1) and (68.2),
as giving the content of the first and second laws of thermodynamics
when applied to systems in a state of uniform motion, will depend
on the fact that the transformation equations for the quantities
involved will be such as to make the validity of these expressions, in
a set of coordinates with respect to which a thermodynamic system
is in motion, equivalent to their validity in proper coordinates with
respect to which the system is at rest. In these latter coordinates,

† Planck, *Berl. Ber.* 1907, p. 542; *Ann. der Physik,* **26,** 1 (1908).
‡ Einstein, *Jahrb. der Radioaktivität und Elektronik,* **4,** 411 (1907).

however, these expressions are merely a statement of the classical first and second laws for which we assume that there is adequate empirical justification.

We may now turn to a consideration of the Lorentz transformation equations for the quantities involved.

69. The Lorentz transformation for thermodynamic quantities

For most of our purposes it will be sufficient if we limit our treatment to simple systems containing a thermodynamic fluid which can exert an equal pressure in all directions but cannot withstand shear, and whose state can be specified by two variables such as energy and volume or temperature and pressure. Such a limitation is familiar in thermodynamic discussions and its introduction will make it sufficient for the present to consider the Lorentz transformation only for the quantities—volume, pressure, energy, work, heat, entropy, and temperature.

The first four of these quantities are of a mechanical nature, and the equations for their transformation have already been given or implied in what has preceded and will not be subject to alteration on account of thermodynamic considerations. Nevertheless, in order to unify our treatment we shall also give here, on the basis of earlier principles, a discussion of the Lorentz transformation of these quantities, especially as some simplification is introduced by the limitation which we have placed on the kind of stress which the fluid can withstand. The transformation equations for the new quantities heat, entropy, and temperature must be—in accordance with our previous remarks— such as to make the validity of the two laws of thermodynamics (68.1) and (68.2) in any given set of coordinates equivalent to their validity in proper coordinates, with respect to which the thermodynamic system is at rest. This requirement with one acceptable addition, which will appear in obtaining the transformation equation for entropy, is sufficient to lead to a unique solution.

It will prove most convenient to have our transformation equations in a form which relates the quantity of interest in a given set of coordinates S with quantities as measured in proper coordinates S^0 by a local observer moving with the thermodynamic system in question. We now proceed to obtain such equations.

(a) **Volume and pressure.** For the volume v of a thermodynamic system moving with the uniform velocity u we can immediately write in accordance with our previous consideration of the Lorentz contraction

$$v = v_0\sqrt{(1-u^2/c^2)}, \tag{69.1}$$

where v_0 is the volume as measured in proper coordinates.

For the pressure p we can base the Lorentz transformation on the definition of pressure as force per unit area and on the known transformation equations for force. To do this let us temporarily use for simplicity axes chosen in such a way that the velocity u of the system of interest will be parallel to the x-axis. For the forces F_x, F_y, and F_z acting on surfaces of the system which lie perpendicular to the indicated axes we can then evidently write in accordance with the transformation equations for force (25.3)

$$\begin{aligned}
F_x &= F_x^0, \\
F_y &= F_y^0\sqrt{(1-u^2/c^2)}, \\
F_z &= F_z^0\sqrt{(1-u^2/c^2)},
\end{aligned} \tag{69.2}$$

where F_x^0, F_y^0, and F_z^0 are the forces acting on these same surfaces as measured in proper coordinates S^0. Hence, since an area perpendicular to the x-axis will not be affected by the Lorentz contraction, while areas perpendicular to the other two axes will be contracted in the ratio of $\sqrt{(1-u^2/c^2)} : 1$, we at once obtain the simple result

$$p = p_0, \tag{69.3}$$

as the transformation equation for pressure. This result will be seen to be merely a specialization of the general transformation equations for the components of stress (34.5), for the case now being considered, in which the stresses reduce to a hydrostatic pressure

$$p = t_{xx} = t_{yy} = t_{zz} \qquad t_{ij} = 0 \quad (i \neq j). \tag{69.4}$$

(b) **Energy.** To obtain an expression for the energy of our moving system we shall start with the system in a state of rest, in the internal condition which is to be considered, and then determine the work necessary to bring it to the velocity of interest by a quasi-stationary adiabatic acceleration which will not disturb that internal condition—as measured by a local observer—which we desire to consider. To carry this out we shall first need to obtain an expression for the force acting during the process of acceleration.

In accordance with our previous discussions of the relation between density of momentum and density of energy flow, see §§ 27 and 35, it is evident that we can write, in the case of a fluid of density ρ and pressure p, moving with the velocity $\mathbf{u}$,

$$\mathbf{g} = \rho\mathbf{u} + \frac{p\mathbf{u}}{c^2}, \tag{69.5}$$

as an expression for the density of momentum, where the first term is the density of momentum associated with the mass motion of the fluid, and the second term allows for the additional momentum that is associated with the flow of energy resulting from work done on the moving fluid by the pressure acting on it. And introducing the relation between mass and energy we can then write

$$\mathbf{G} = \frac{E + pv}{c^2}\mathbf{u}, \tag{69.6}$$

as an expression for the total momentum of the fluid in volume v, in entire agreement with our previous more general equations (35.2). This then gives us

$$\mathbf{F} = \frac{d}{dt}\left(\frac{E + pv}{c^2}\mathbf{u}\right) \tag{69.7}$$

as the desired expression for the external force that will accompany the acceleration of the system.

We are now ready to calculate the work done and energy increase associated with a change in velocity. This will evidently be the sum of the work done by the external force $\mathbf{F}$, and by the action of the pressure p on the changing volume v of the system, so that we can put for the rate of change in energy

$$\frac{dE}{dt} = \mathbf{F}\cdot\mathbf{u} - p\,\frac{dv}{dt}. \tag{69.8}$$

In applying this result to the process to be considered we can take in accordance with (69.1) and (69.3)

$$p = p_0 \quad \text{and} \quad v = v_0\sqrt{(1 - u^2/c^2)}, \tag{69.9}$$

where p_0 and v_0 will be constants, since we desire to carry out the acceleration in such a way as to leave p_0 and v_0 for the state of the system as measured by a local observer with the unchanged values which are of interest to us. Making use of the constancy of p thus provided, and substituting (69.7) for $\mathbf{F}$, we can then write

$$\frac{dE}{dt} = \frac{dE}{dt}\frac{u^2}{c^2} + p\frac{u^2}{c^2}\frac{dv}{dt} + \frac{E+pv}{c^2}u\frac{du}{dt} - p\frac{dv}{dt},$$

or by transposing,

$$\left(1 - \frac{u^2}{c^2}\right)\frac{d}{dt}(E+pv) = \frac{E+pv}{c^2}u\frac{du}{dt},$$

which can easily be integrated to give us

$$E+pv = \frac{\text{const.}}{\sqrt{(1-u^2/c^2)}}.$$

Evaluating the constant by considering the value of $E+pv$ at $u=0$, we can then write

$$E+pv = \frac{E_0+p_0 v_0}{\sqrt{(1-u^2/c^2)}}, \tag{69.10}$$

or in accordance with (69.9)

$$E = \frac{E_0+p_0 v_0 u^2/c^2}{\sqrt{(1-u^2/c^2)}} \tag{69.11}$$

as the desired transformation equations for energy. The result will readily be seen to be a specialization of our previous equation (35.7) for the case that the stress reduces to a simple hydrostatic pressure.

(c) **Work.** To obtain an expression for the work done when the internal state of the system is changed, *keeping the velocity u constant*, we must remember in accordance with the principles of relativistic mechanics that the momentum of a system can change even at constant velocity if its energy changes, and hence an external force which does work may be necessary to maintain the constant velocity. For the work dW accompanying a change in internal state at constant velocity we can then write

$$dW = p\,dv - \mathbf{u}\cdot d\mathbf{G}, \tag{69.12}$$

where the first term gives the work done against the pressure and the second term takes care of the work associated with the external force necessary to maintain constant velocity. Or substituting from (69.6) for the case of constant velocity, we can put

$$dW = p\,dv - \frac{u^2}{c^2}\,d(E+pv). \tag{69.13}$$

Introducing (69.1), (69.3), and (69.10) this gives us

$$dW = \sqrt{(1-u^2/c^2)}p_0\,dv_0 - \frac{u^2/c^2}{\sqrt{(1-u^2/c^2)}}\,d(E_0+p_0 v_0)$$

or $\qquad dW = \sqrt{(1-u^2/c^2)}\,dW_0 - \frac{u^2/c^2}{\sqrt{(1-u^2/c^2)}}\,d(E_0+p_0 v_0) \qquad (69.14)$

as an expression for the work dW in terms of quantities which are measured in proper coordinates.

(d) **Heat.** The quantities whose transformation equations have just been considered were of a mechanical nature and no new principles beyond those of our previous mechanics were introduced into their treatment. Turning now to the first of the non-mechanical quantities heat, we can obtain its transformation equation from the requirement that the first law of thermodynamics as given by (68.1) is to hold both in the set of coordinates S that are being employed and in proper coordinates S^0.

In accordance with (68.1) we can write

$$dQ = dE + dW \tag{69.15}$$

for a small element of heat absorbed; and substituting for dE and dW from (69.11) and (69.14), we obtain for the case of a change in internal state without change in velocity

$$dQ = \frac{dE_0 + d(p_0 v_0)u^2/c^2}{\sqrt{(1-u^2/c^2)}} + \sqrt{(1-u^2/c^2)}\, dW_0 - \frac{dE_0 + d(p_0 v_0)}{\sqrt{(1-u^2/c^2)}}\frac{u^2}{c^2}$$

or

$$dQ = \sqrt{(1-u^2/c^2)}(dE_0 + dW_0).$$

Since the first law of thermodynamics, however, must certainly hold in proper coordinates we may put

$$dQ_0 = dE_0 + dW_0, \tag{69.16}$$

and are thus uniquely led to the transformation equation for heat

$$dQ = \sqrt{(1-u^2/c^2)}\, dQ_0 \tag{69.17}$$

or

$$Q = \sqrt{(1-u^2/c^2)}Q_0.$$

With this transformation equation for heat, the validity of the first law in any given coordinate system S is then seen to be equivalent to its validity in proper coordinates S^0, which agrees with the justification proposed in § 68 for our choice of fundamental principles. The transformation equation (69.17) will also be seen to be in complete agreement with equation (54.18) obtained in our previous investigation of the Joule heating effect.

(e) **Entropy.** In order to obtain the transformation equation for entropy we shall add to our thermodynamic requirements for systems in a state of rest or uniform motion, the requirement that the entropy of a system would be unaltered by a reversible adiabatic change in velocity without absorption of heat. This addition is evidently in acceptable agreement with our ideas as to reversible processes and as to the significance of entropy.

Considering now a thermodynamic system in some internal state of interest and originally at rest with the entropy S_0, we can then accelerate it to the velocity u reversibly and adiabatically without change in its internal state, and hence without change either in its proper entropy S_0 or entropy S with respect to the coordinate system actually being used. We are thus led to the simple transformation equation for entropy

$$S = S_0. \qquad (69.18)$$

In further justification it may also be noted that this result is in agreement with the statistical mechanical interpretation of entropy in terms of probability, since the probability of finding a system in a given state should evidently be independent of the velocity of the observer relative to it.

(f) **Temperature.** Finally with the help of the second law of thermodynamics (68.2), and the transformation equations for heat and entropy just obtained, the transformation for temperature becomes immediately evident. In accordance with the second law we have

$$\Delta S \geqslant \int \frac{dQ}{T},$$

and substituting (69.17) and (69.18) we obtain

$$\Delta S_0 \geqslant \int \frac{\sqrt{(1-u^2/c^2)}\, dQ_0}{T}.$$

In proper coordinates, however, the second law must certainly hold in the simple classical form

$$\Delta S_0 \geqslant \int \frac{dQ_0}{T_0},$$

so that we can at once take

$$T = \sqrt{(1-u^2/c^2)}T_0 \qquad (69.19)$$

as the transformation equation for temperature.

The transformation equations for all three of the non-mechanical quantities Q, S, and T have thus been taken so that the validity of the two laws of thermodynamics in a given coordinate system S is equivalent to their validity in proper coordinates S^0, which was the justification proposed in § 68 for our choice of fundamental principles.

To conclude the section we may now collect into one place the transformation equations for thermodynamic quantities, for convenience of future reference.

$$v = v_0 \sqrt{(1-u^2/c^2)},$$

$$p = p_0,$$

$$E = \frac{E_0 + p_0 v_0 u^2/c^2}{\sqrt{(1-u^2/c^2)}},$$

$$dW = \sqrt{(1-u^2/c^2)}\, dW_0 - \frac{u^2/c^2}{\sqrt{(1-u^2/c^2)}}\, d(E_0+p_0 v_0), \quad (69.20)$$

$$dQ = \sqrt{(1-u^2/c^2)}\, dQ_0,$$

$$S = S_0,$$

$$T = \sqrt{(1-u^2/c^2)}\, T_0.$$

70. Thermodynamic applications

From the foregoing equations it is immediately evident that the thermodynamic equations for moving systems differ from those for stationary systems only in terms of the second order or higher in u/c. Hence the direct empirical verification of the applications of this extension of thermodynamics is hardly to be expected. There are, however, two simple conceptual applications, which we may now develop as illustrating the internal consistency of the theory.

(*a*) **Carnot cycle involving change in velocity.** Our first application will be the consideration of a simple reversible cycle involving the transfer of heat from a stationary to a moving heat reservoir. The process may be regarded as analogous to the Carnot cycle of ordinary thermodynamics, and the result obtained will illustrate the consistency of our transformation equation for temperature.

Consider a simple system S (the engine), containing a fluid which will be kept at the constant pressure $p = p_0$ throughout the cycle, and two heat reservoirs R_1 being at temperature T_1 and at rest, and R_2 being at temperature T_2 and moving with the velocity u. In the initial state (*a*) of the system let it be at rest with the same temperature $T_a = T_1$ as the reservoir R_1, and having the energy content and volume E_a and v_a; and let the *first step* of the cycle consist in a change to state (*b*) by the reversible isopiestic absorption of heat from the reservoir R_1. For the heat Q_1 absorbed from the reservoir and the work W_1 done by the system we can evidently write

$$Q_1 = E_b - E_a + p(v_b - v_a) \qquad (70.1)$$

and
$$W_1 = p(v_b - v_a). \qquad (70.2)$$

In the *second step* of the cycle let us change to state (*c*) by a reversible adiabatic acceleration to the same velocity u as that of the reservoir R_2, keeping the internal condition of the system unaltered as measured by a local observer moving therewith. There will be no heat change in this process and the work done will be

$$W_2 = E_b - E_c. \qquad (70.3)$$

By hypothesis let the temperature of the system in state (c) be the same as that of the reservoir R_2, and let the *third step* of the cycle consist in the reversible transfer to the reservoir of a certain amount of heat $-Q_2$ whose value will be determined later. The work W_3 done by the system during this process will be in part due to a change in volume under the constant pressure p, and in part due to a change in momentum even at the constant velocity u as previously discussed. Making use of equation (69.13) we can then evidently write for the heat absorbed and work done during this step

$$Q_2 = Q_2 \tag{70.4}$$

and $$W_3 = p(v_d - v_c) - (u^2/c^2)\{E_d - E_c + p(v_d - v_c)\}. \tag{70.5}$$

Finally, by hypothesis let the heat transferred in the above step be just sufficient so that the system can be returned to its original state (a) by a reversible deceleration which leaves the internal condition unaltered as measured by a local observer. There will be no heat change in this process and the work done will be

$$W_4 = E_d - E_a. \tag{70.6}$$

We have now completed the cycle and are ready to apply the two laws of thermodynamics.

In accordance with the first law of thermodynamics, the total heat absorbed by the system in this cycle must be equal to the total work done, since the system finally arrives in its initial state with its original energy content. Hence we can evidently write

$$Q_1 + Q_2 = W_1 + W_2 + W_3 + W_4, \tag{70.7}$$

and on solving for Q_2 and substituting the above values for the other quantities, this is found to lead to the result

$$Q_2 = \{(E_d + pv_d) - (E_c + pv_c)\}\{1 - u^2/c^2\}. \tag{70.8}$$

In accordance with (69.10), however, we can evidently put

$$E_d + pv_d = \frac{E_a + pv_a}{\sqrt{(1 - u^2/c^2)}} \quad \text{and} \quad E_c + pv_c = \frac{E_b + pv_b}{\sqrt{(1 - u^2/c^2)}},$$

since the cycle was carried out in such a way that the internal condition of the system as measured by a local observer moving therewith was the same in states (d) and (a), and in states (c) and (b). Substituting in (70.8) this then gives us

$$Q_2 = \{E_a - E_b + p(v_a - v_b)\}\sqrt{(1 - u^2/c^2)},$$

or in accordance with (70.1)

$$Q_2 = -Q_1\sqrt{(1 - u^2/c^2)}. \tag{70.9}$$

On the other hand, in accordance with the second law of thermo-dynamics, we can evidently write

$$\frac{Q_1}{T_1} + \frac{Q_2}{T_2} = 0, \tag{70.10}$$

since the total entropy change of the system will be zero for the cycle. And substituting (70.9) in (70.10) we obtain the result

$$T_2 = T_1\sqrt{(1-u^2/c^2)}. \tag{70.11}$$

In accordance with the process described, the quantity T_1 occurring in this expression is the temperature of the system S (the engine) when at rest and the quantity T_2 is the temperature to which it falls when its velocity is raised to u by a process which does not change its internal condition as measured by a local observer moving therewith. The result is in complete agreement with the transformation equation for temperature (69.19) obtained above by somewhat different considerations.

(b) **The dynamics of thermal radiation.** As a second application of the considerations developed in this chapter we may consider the dynamics of a hollow enclosure filled with black-body radiation, and moving with the velocity u.

In accordance with (65.3) and (65.4) the energy and pressure for such a system will have the values

$$E_0 = av_0 T_0^4 \tag{70.12}$$

and $$p_0 = \tfrac{1}{3}aT_0^4 \tag{70.13}$$

when measured by a local observer who is moving with the same velocity as the enclosure. Hence, making use of (69.11), we can write for the energy with reference to coordinates such that the system has the velocity u

$$E = \frac{av_0 T_0^4 + \tfrac{1}{3}av_0 T_0^4 u^2/c^2}{\sqrt{(1-u^2/c^2)}}$$
$$= E_0 \frac{1+\tfrac{1}{3}u^2/c^2}{\sqrt{(1-u^2/c^2)}}, \tag{70.14}$$

and in accordance with (69.6) and (69.10) we can write for the momentum of the system

$$\mathbf{G} = \frac{av_0 T_0^4 + \tfrac{1}{3}av_0 T_0^4}{\sqrt{(1-u^2/c^2)}} \frac{\mathbf{u}}{c^2}$$
$$= \frac{4}{3} \frac{E_0}{\sqrt{(1-u^2/c^2)}} \frac{\mathbf{u}}{c^2}. \tag{70.15}$$

These expressions for the energy and momentum of a moving 'Hohlraum' are of interest because of their agreement with the conclusions obtained by Mosengeil† directly from the electromagnetic theory of radiation without the explicit use of relativity.

71. Use of four-dimensional language in thermodynamics

In our development of the dynamics of a mechanical medium it was found possible to express the content of the laws for the conservation of mass, energy, and momentum in four-dimensional language by a single equation (37.9) having the simple form

$$\partial T^{\mu\nu}/\partial x^{\nu} = 0, \tag{71.1}$$

where the components of the energy-momentum tensor $T^{\mu\nu}$ are related to densities of mass, energy, and momentum and to the stresses in the manner given by the table of components (37.8). And this equation can be employed to investigate the energy changes within a mechanical medium for use in connexion with the first law of thermodynamics.

In the present section we shall show the possibility of expressing the second law of thermodynamics in a four-dimensional form. This will be important for our later extension of thermodynamics to general relativity.

To obtain the desired result let us start with the second law of thermodynamics in its original form (68.2), and consider a small element of any given thermodynamic fluid or medium as the system to which we apply it. If δv is the volume of this element and ϕ is the density of entropy at the point where the element is located, the entropy content of the element will be $\phi \, \delta v$, and we can evidently write in accordance with the second law

$$\frac{d}{dt}(\phi \, \delta v)\delta t \geqslant \frac{\delta Q}{T} \tag{71.2}$$

as an expression which relates the change in this entropy content in the infinitesimal time δt with the heat δQ which flows into the element during that time interval and the temperature T at the point in question.

Expanding the left-hand side of this expression we obtain

$$\left(\frac{d\phi}{dt}\delta v + \phi\frac{d\,\delta v}{dt}\right)\delta t \geqslant \frac{\delta Q}{T},$$

---

† Mosengeil, *Ann. der Physik*, **22**, 867 (1907).

and, substituting evident expressions in terms of partial derivatives for the two total derivatives with respect to the time, this becomes

$$\left(u_x \frac{\partial \phi}{\partial x} + u_y \frac{\partial \phi}{\partial y} + u_z \frac{\partial \phi}{\partial z} + \frac{\partial \phi}{\partial t}\right) \delta v \delta t + \left(\phi \frac{\partial u_x}{\partial x} + \phi \frac{\partial u_y}{\partial y} + \phi \frac{\partial u_z}{\partial z}\right) \delta v \delta t \geqslant \frac{\delta Q}{T},$$

where u_x, u_y, and u_z are the components of the velocity of the fluid at the point under consideration. Combining terms this result can be written in the simpler form

$$\left[\frac{\partial}{\partial x}(\phi u_x) + \frac{\partial}{\partial y}(\phi u_y) + \frac{\partial}{\partial z}(\phi u_z) + \frac{\partial \phi}{\partial t}\right] \delta v \delta t \geqslant \frac{\delta Q}{T}, \tag{71.3}$$

or, introducing expressions for u_x, u_y, u_z, and δv in terms of the coordinates x, y, z, and t, in the form

$$\left[\frac{\partial}{\partial x}\left(\phi \frac{dx}{dt}\right) + \frac{\partial}{\partial y}\left(\phi \frac{dy}{dt}\right) + \frac{\partial}{\partial z}\left(\phi \frac{dz}{dt}\right) + \frac{\partial \phi}{\partial t}\right] \delta x \delta y \delta z \delta t \geqslant \frac{\delta Q}{T}. \tag{71.4}$$

In order to re-express this result in four-dimensional language, let us now return to our fundamental idea of a four-dimensional space-time continuum characterized by the formula for interval (20.1)

$$ds^2 = -(dx^1)^2 - (dx^2)^2 - (dx^3)^2 + (dx^4)^2 \tag{71.5}$$

in terms of the space-time coordinates (x^1, x^2, x^3, x^4) where

$$x^1 = x \quad x^2 = y \quad x^3 = z \quad x^4 = ct. \tag{71.6}$$

Introducing these new coordinates we can then evidently rewrite (71.4) in the form

$$\left[\frac{\partial}{\partial x_1}\left(\phi \frac{dx^1}{dx^4}\right) + \frac{\partial}{\partial x^2}\left(\phi \frac{dx^2}{dx^4}\right) + \frac{\partial}{\partial x^3}\left(\phi \frac{dx^3}{dx^4}\right) + \frac{\partial \phi}{\partial x^4}\right] \delta x^1 \delta x^2 \delta x^3 \delta x^4 \geqslant \frac{\delta Q}{T},$$

or, by an obvious substitution, in the form

$$\left[\frac{\partial}{\partial x^1}\left(\phi \frac{ds}{dx^4} \frac{dx^1}{ds}\right) + \frac{\partial}{\partial x^2}\left(\phi \frac{ds}{dx^4} \frac{dx^2}{ds}\right) + \frac{\partial}{\partial x^3}\left(\phi \frac{ds}{dx^4} \frac{dx^3}{ds}\right) + \right.$$
$$\left. + \frac{\partial}{\partial x^4}\left(\phi \frac{ds}{dx^4} \frac{dx^4}{ds}\right)\right] \delta x^1 \delta x^2 \delta x^3 \delta x^4 \geqslant \frac{\delta Q}{T}. \tag{71.7}$$

In accordance with the formula for interval (71.5), however, it is evident that ds/dx^4 is equal to the factor which gives the Lorentz contraction

$$ds/dx^4 = \sqrt{(1 - u^2/c^2)} \tag{71.8}$$

for matter moving with the velocity at the point in question. Furthermore, since entropy in accordance with (69.18) is an invariant for the Lorentz transformation, we can evidently write as the transformation equation for entropy density

$$\phi = \frac{\phi_0}{\sqrt{(1-u^2/c^2)}}. \tag{71.9}$$

Hence by combining (71.8) and (71.9) we shall be able to put

$$\phi \frac{ds}{dx^4} = \phi_0. \tag{71.10}$$

In addition, in accordance with the transformation equations for heat and temperature (69.17) and (69.19) we can put

$$\frac{\delta Q}{T} = \frac{\delta Q_0}{T_0}, \tag{71.11}$$

where T_0 and δQ_0 are the temperature and the heat that enters the element as measured by a local observer moving therewith.

Substituting (71.10) and (71.11), our expression (71.7) for the requirements of the second law can then be written in the symmetrical form

$$\left[\frac{\partial}{\partial x^1}\left(\phi_0 \frac{dx^1}{ds}\right) + \frac{\partial}{\partial x^2}\left(\phi_0 \frac{dx^2}{ds}\right) + \frac{\partial}{\partial x^3}\left(\phi_0 \frac{dx^3}{ds}\right) + \right.$$
$$\left. + \frac{\partial}{\partial x^4}\left(\phi_0 \frac{dx^4}{ds}\right)\right] \delta x^1 \delta x^2 \delta x^3 \delta x^4 \geqslant \frac{\delta Q_0}{T_0}. \tag{71.12}$$

Introducing the summation convention this can be written in the shorter form

$$\frac{\partial}{\partial x^\mu}\left(\phi_0 \frac{dx^\mu}{ds}\right) \delta x^1 \delta x^2 \delta x^3 \delta x^4 \geqslant \frac{\delta Q_0}{T_0}. \tag{71.13}$$

Or defining the *entropy vector* S^μ, in terms of proper entropy density ϕ_0 and generalized velocity of the fluid dx^μ/ds, by the equation

$$S^\mu = \phi_0 \frac{dx^\mu}{ds}, \tag{71.14}$$

we may finally write the very simple equation

$$\frac{\partial S^\mu}{\partial x^\mu} \delta x^1 \delta x^2 \delta x^3 \delta x^4 \geqslant \frac{\delta Q_0}{T_0}. \tag{71.15}$$

The foregoing equations (71.12), (71.13), and (71.15) express the requirements of the second law of thermodynamics in the desired four-dimensional form which will be valid for any space-time coordinates of the type (71.6). They are of the form assumed by tensor equations of rank zero in 'rectangular' coordinates of this type, and can be written by a slight modification in a general tensor form valid in 'curvilinear' coordinates as well. This latter form will provide the basis for our later extension of thermodynamics to general relativity.

THE GENERAL THEORY OF RELA⌐

72. Introduction

Einstein's theory of relativity may be regarded as based on the fundamental idea of the relativity of all motion. In accordance with this idea we can detect and measure the motion of a given body relative to other bodies, but cannot assign any meaning to its absolute motion.

The special theory of relativity makes only a restricted use of this general idea, since it merely assumes the relativity of uniform translatory motion in a region of free space where gravitational effects can be neglected. As a result of this assumption we are led to the conclusion that the laws of physics for the description of phenomena in free space must be independent of the velocity of the particular observer who makes measurements for their determination, and must hence have the same form and content when referred to different sets of Cartesian axes which are in uniform relative translatory motion. Making use of this conclusion, the special theory of relativity then guides us in determining the necessary form of the laws of physics when expressed in the coordinates corresponding to any desired set of unaccelerated Cartesian axes, assuming that the effects of gravitation can be neglected. The special theory of relativity, however, makes no hypothesis as to the relativity of all kinds of motion, gives no discussion of the form of the laws of physics when referred to more general coordinates corresponding, for example, to spatial axes in non-uniform motion, and provides no treatment of gravitational action.

The general theory of relativity, to which we now turn, attempts, on the other hand, to make full use of the general idea of the relativity of all kinds of motion. In the first place, this immediately leads to a consideration of the laws of physics when referred to any kind of space-time coordinates, and to the conclusion that these laws must be expressible in a form which is independent of the particular space-time coordinates chosen, since otherwise the difference in form could provide a criterion for judging the absolute motion of the spatial framework used in the construction of different systems of dinates. In the second place the programme thus initiated is also found to involve a consideration of the effects of gravitational action. This arises from the fact that the expression of the equations of physics in a *form* which is independent of the coordinate system does not in general prevent a change in their *numerical content* when we change from one system of coordinates to another, and it is only by relating such changes in numerical content to conceivable changes in gravitational field that we are able to eliminate criteria for absolute motion and to preserve the idea of the relativity of all kinds of motion. This, however, is found to lead to a complete theory of gravitational action. Hence by a natural extension of the fundamental basis, the general theory of relativity leads to a satisfactory solution of the two obvious problems which were left untouched by the special theory of relativity.

The assumption that the laws of physics can be expressed in a form which is independent of the coordinate system is called the *principle of covariance*, and the actual hypothesis by which gravitational considerations are introduced into the development has been named the *principle of equivalence*, for reasons which will appear later. We may now undertake the detailed consideration of these two principles and their more immediate consequences.

73. The principle of covariance

In accordance with the principle of covariance the general laws of physics can be expressed in a form which is independent of the choice of space-time coordinates. In the present section we shall first discuss the justification for the introduction of this hypothesis, the theoretical and practical nature of the consequences that could follow its adoption, and the methods by which it is to be used. We shall then consider two simple and important examples of the employment of the principle, which are furnished by the covariant expression of the formula for space-time interval and by the covariant expression for the equations of motion for free particles and light rays.

(*a*) **Justification for the principle of covariance.** As already indicated in the preceding section, our primary motive in introducing the principle of covariance can be regarded as residing in our desire to make full use of the idea of the relativity of all kinds of motion. If the general laws of physics could not be expressed in a form which is the same for all space-time coordinate systems we could take the differences in form for different coordinate systems as an evidence of differences in the absolute motion of the spatial frameworks used in setting up the space-time coordinate systems. This we avoid by the introduction of the principle of covariance, even though we shall later find—as already mentioned—that invariance of form alone is still not sufficient to preserve the relativity of all kinds of motion.

Although our original motive for introducing the principle of covariance may thus be thought of as furnished by the idea of the relativity of motion, there is an even more immediate justification for believing in its validity. As emphasized by Einstein the laws of physics are to be regarded as a codification of the results of experimental observations, and these consist in the last analysis in the determination of space-time coincidences.† The recording of such space-time coincidences is, of course, conveniently carried out with the help of some system of space-time coordinates. Nevertheless, the actual physical behaviour can be in no way affected by the coordinate system used, which may be introduced by the experimenter in any arbitrary way which suits his convenience or fancy. As a result of this independence of physical reality and coordinate system, we are then led to the conviction that the laws of physics—whatever they may be—can be expressed in a form which makes no reference to any particular coordinate system, and we are further strengthened in this conviction by the great success which the mathematician has already had in devising language—in particular that of the tensor calculus—for the covariant expression of geometrical and physical relations. We thus come to regard the principle of covariance as in any case an inescapable axiom, and to regard it as merely a task—possibly difficult but theoretically possible—for the mathematician to find a form, invariant to coordinate transformation, for the expression of any desired physical law.

(*b*) **Consequences of the principle of covariance.** The full appreciation of this inescapable character of the principle of covariance has an immediate effect on our estimate of the theoretical consequences that could follow from the adoption of the principle. If the laws of physics—whatever they might be—could in any case be expressed in invariant form, given sufficient ingenuity on the part of the investigator, it becomes at once evident that the adoption of the principle imposes no necessary restriction on the nature of these laws. Hence the very reasoning that leads to our certainty of belief

† As elsewhere in the book, we are considering macroscopic phenomena.

in the validity of the principle of covariance, has at the same time robbed the principle of any absolutely necessary consequences, a conclusion first presented by Kretschmann[†] and concurred in by Einstein.[‡]

Nevertheless, as further emphasized by Einstein, the explicit use of the principle of covariance does have important actual consequences in our investigation of the axioms of physics. In searching for the appropriate axioms, we shall wish to eliminate unsuspected assumptions that could arise from the use of any particular coordinate system. Hence from the very start we shall desire to express our axioms by covariant equations that make no use of a particular coordinate system. This, however, has important actual consequences, since we are then led to adopt as axioms, not such principles as appear simple when we use some special coordinate system, but such principles as can be simply expressed by covariant equations that are independent of the coordinates. There is, moreover, a certain theoretical justification for this mode of procedure, since even without any belief in the necessary simplicity of nature it is evident that our progress in understanding must lie in a process of successive approximation that starts with the provisional use of simple expressions, and as stated above to eliminate unsuspected assumptions these must be stated in covariant language. Hence the adoption and use of the principle does have great heuristic value, as illustrated, for example, by the fact that it would certainly be practically impossible to take the Newtonian law of gravitation as an appropriate axiom, since its expression in covariant language would undoubtedly be too complicated either for comprehension or use.

(c) **Method of obtaining covariant expressions.** In the actual employment of the principle of covariance, we are enormously assisted in our task of expressing the fundamental axioms or principles of physics in covariant form by the use of the tensor calculus, developed by Ricci and Levi-Civita, since as we have already seen in § 19 the expression of a physical law by a tensor equation has exactly the same form in all systems of space-time coordinates. Hence in the development of general relativity we are at once led to seek expressions for the fundamental postulates of physics in the form of tensor equations, and are greatly helped in this task by the fact that we have

† Kretschmann, *Ann. der Physik*, **53**, 575 (1917).
‡ Einstein, ibid. **55**, 241 (1918).

already found tensor equations for many of the principles of the special theory of relativity.

Although the tensor analysis is thus of the greatest importance for the development of general relativity, it would be wrong to assume, as has sometimes been done with unfortunate results in the past, that we must limit ourselves to the use of tensor equations in investigating the fundamental principles of physics. The fact that all tensor equations are necessarily covariant equations does not, of course, eliminate the possibility of covariant equations which are not tensor equations. Indeed the frequent use of covariant equations connecting tensor densities—instead of tensors—is a specially simple and familiar example to the contrary. In addition Einstein's development of the equations of relativistic mechanics in a covariant form containing the pseudo-tensor density of potential energy and momentum has been of great importance in obtaining an insight into mechanics, and we shall not hesitate to employ it in this book.

(*d*) **Covariant expression for interval.** In the development of the special theory of relativity we have found that the principles of physics can be treated with great effectiveness with the help of a four-dimensional space-time geometry, characterized by the formula for the element of interval

$$ds^2 = -dx^2 - dy^2 - dz^2 + c^2\,dt^2, \tag{73.1}$$

where x, y, z, and t are our usual spatial and temporal variables. In the development of the general theory of relativity we shall find the use of the idea of a four-dimensional space-time continuum even more necessary, and in accordance with the principle of covariance shall need a covariant expression for the formula for interval by which the geometry can be characterized. As a preliminary step in this direction let us first examine the possibility of re-expressing this special relativity formula for interval in covariant form.

The expression for interval given by (73.1) is not a completely covariant one, since it retains unaltered form only for the limited class of transformations discussed in § 17. These include the Lorentz transformation to a new set of variables x', y', z', t', corresponding to new Cartesian axes moving with uniform velocity relative to the old, but with more general transformations, corresponding, for example, to a change to accelerated axes or even to the mere change to spatial polar coordinates, the form will not be unaltered.

It is easily possible, however, to re-express the formula for interval in covariant language, since we immediately recognize (73.1) as the simplified expression in 'rectangular' coordinates for the general tensor relation

$$ds^2 = g_{\mu\nu} \, dx^\mu dx^\nu, \tag{73.2}$$

which is valid in any coordinates, using the appropriate values for the components $g_{\mu\nu}$ of the metrical tensor and summing as indicated over all values of $\mu, \nu = 1, 2, 3, 4$.

To demonstrate the existence of this possibility in detail, we have merely to note in the first place that our covariant expression (73.2) is indeed equivalent to the original expression for interval (73.1) with the specially simple values for the components of the metrical tensor

$$g_{11} = g_{22} = g_{33} = -1 \qquad g_{44} = c^2$$
$$g_{\mu\nu} = 0 \quad (\mu \neq \nu), \tag{73.3}$$

and then to show in the second place that by any arbitrary transformation to new coordinates the formula for interval will still be left in the form (73.2).

To prove this, let us consider an arbitrary change to any desired new set of general (curvilinear) coordinates x^1, x^2, x^3, x^4 which are related to the original (rectangular) coordinates x, y, z, t in any way

$$x^\mu = x^\mu(x, y, z, t), \tag{73.4}$$

which is consistent with the necessary conditions of continuity and unambiguity. Making use of the relations between the two systems of coordinates we can then write

$$dx = \frac{\partial x}{\partial x^1} \, dx^1 + \frac{\partial x}{\partial x^2} \, dx^2 + \frac{\partial x}{\partial x^3} \, dx^3 + \frac{\partial x}{\partial x^4} \, dx^4$$
$$\cdot \quad \cdot \quad \cdot \quad \cdot \quad \cdot \quad \cdot \quad \cdot \quad \cdot \quad \cdot \quad \cdot \quad \cdot \quad \cdot \quad \cdot \tag{73.5}$$
$$\cdot \quad \cdot \quad \cdot \quad \cdot \quad \cdot \quad \cdot \quad \cdot \quad \cdot \quad \cdot \quad \cdot \quad \cdot \quad \cdot \quad \cdot$$
$$dt = \frac{\partial t}{\partial x^1} \, dx^1 + \frac{\partial t}{\partial x^2} \, dx^2 + \frac{\partial t}{\partial x^3} \, dx^3 + \frac{\partial t}{\partial x^4} \, dx^4$$

for the differentials of the old variables in terms of the new. By squaring these and introducing into (73.1) we then obtain

$$ds^2 = \left[-\left(\frac{\partial x}{\partial x^1}\right)^2 - \left(\frac{\partial y}{\partial x^1}\right)^2 - \left(\frac{\partial z}{\partial x^1}\right)^2 + c^2\left(\frac{\partial t}{\partial x^1}\right)^2 \right](dx^1)^2 \; +$$
$$\cdot \quad \cdot \quad \cdot \quad \cdot \quad \cdot \quad \cdot \quad \cdot \quad \cdot \quad \cdot \quad \cdot \quad \cdot \quad \cdot \quad \cdot$$
$$\cdot \quad \cdot \quad \cdot \quad \cdot \quad \cdot \quad \cdot \quad \cdot \quad \cdot \quad \cdot \quad \cdot \quad \cdot \quad \cdot$$

$$+\left[-\left(\frac{\partial x}{\partial x^4}\right)^2-\left(\frac{\partial y}{\partial x^4}\right)^2-\left(\frac{\partial z}{\partial x^4}\right)^2+c^2\left(\frac{\partial t}{\partial x^4}\right)^2\right](dx^4)^2$$

$$+2\left[-\frac{\partial x}{\partial x^1}\frac{\partial x}{\partial x^2}-\frac{\partial y}{\partial x^1}\frac{\partial y}{\partial x^2}-\frac{\partial z}{\partial x^1}\frac{\partial z}{\partial x^2}+c^2\frac{\partial t}{\partial x^1}\frac{\partial t}{\partial x^2}\right]dx^1dx^2+$$

$$\cdots\cdots\cdots\cdots\cdots\cdots$$ (73.6)

$$\cdots\cdots\cdots\cdots\cdots\cdots$$

which is seen to be still in the form (73.2), as was to be proved.

It should be noted from the form of (73.6), that the $g_{\mu\nu}$ will always be symmetrical in μ and ν. It should also be remarked that the tensor character of $g_{\mu\nu}$ is immediately evident, since in accordance with the postulated invariance of interval we can write for any pair of coordinate systems x'^μ and x^μ the equivalent expressions

$$ds^2 = g'_{\mu\nu}\,dx'^\mu dx'^\nu = g_{\alpha\beta}\,dx^\alpha dx^\beta,$$ (73.7)

which will evidently give us

$$g'_{\mu\nu} = \frac{\partial x^\alpha}{\partial x'^\mu}\frac{\partial x^\beta}{\partial x'^\nu}g_{\alpha\beta}$$ (73.8)

as the transformation equation for the $g_{\mu\nu}$, and this is in agreement with the general equation (19.10) that we have given for the definition of tensors.

The generally covariant tensor expression (73.2) which we have thus obtained for the element of interval now makes it possible to treat the facts of special relativity using not only our usual coordinates x, y, z, t, but also using any set of general coordinates x^1, x^2, x^3, x^4 which we may desire to introduce. At the present stage of the argument we have only demonstrated the justice of using this covariant formula in the absence of gravitational action when the principles of the special theory are actually valid. Nevertheless, we shall show in the next section [(§ 74 (e)] with the help of the principle of equivalence that we shall also have a measure of justification for using this same formula in the more general case when gravitational action is involved, when no coordinates can be found which would make it possible to express the formula for interval throughout the whole of space-time in the original simple form (73.1).

(e) **Covariant expression for the trajectories of free particles and light rays.** As a second example of the introduction of the idea of covariance we may now consider the covariant expression of the equations governing the motion of free particles and light rays. The possibility of obtaining such a covariant expression has already been shown in § 28 in our discussion of the four-dimensional treatment of the mechanics of a particle as given by the special theory of relativity. A somewhat more complete treatment from our present point of view will, however, be useful.

In accordance with the special theory of relativity the behaviour of a free particle would be governed by Newton's first law of motion, so that the particle would move in a straight line with constant components of velocity

$$\frac{dx}{dt} = u_x \quad \frac{dy}{dt} = u_y \quad \frac{dz}{dt} = u_z, \tag{73.9}$$

where x, y, z, and t are our usual spatial and temporal variables. And by combining these expressions with the formula for space-time interval (73.1), we can re-express them in the four-dimensional form

$$\begin{aligned}
\frac{dx}{ds} &= \frac{u_x}{\sqrt{(c^2 - u^2)}} \\[2mm]
\frac{dy}{ds} &= \frac{u_y}{\sqrt{(c^2 - u^2)}} \\[2mm]
\frac{dz}{ds} &= \frac{u_z}{\sqrt{(c^2 - u^2)}} \\[2mm]
\frac{dt}{ds} &= \frac{1}{\sqrt{(c^2 - u^2)}}.
\end{aligned} \tag{73.10}$$

Interpreting this result, we see that the four-dimensional 'velocity' of a free particle would be a vector with constant components, and that its four-dimensional trajectory would hence be a straight line.

By differentiating these expressions a second time with respect to the element of interval, we can re-express the conditions for the four-dimensional trajectory in the form

$$\frac{d^2x}{ds^2} = \frac{d^2y}{ds^2} = \frac{d^2z}{ds^2} = \frac{d^2t}{ds^2} = 0. \tag{73.11}$$

Furthermore, these conditions can also be expressed in accordance with the known properties of the straight line by the single equation

$$\delta \int ds = 0, \tag{73.12}$$

which states that the total interval along the trajectory shall be an extremum for small variations which vanish at the two limits of integration. This final form of expression is, however, a tensor (scalar) equation which makes no reference to any particular coordinates and would lead to the same results in all systems of coordinates.

We thus have no difficulty in finding a covariant expression for the motion which could be assumed by a free particle in accordance with the special theory of relativity. Moreover, just as in the case of the covariant expression for interval, it is to be emphasized that we shall later find (§ 74 e) a certain justification, with the help of the principle of equivalence, for taking our covariant expression (73.12) for the trajectory as also valid in the presence of gravitational fields when no coordinates are possible such that the formula for interval could be written throughout in the simple form (73.1). In this case equation (73.12) is the general condition for a geodesic, of which the straight line is a special case.

In making practical use of the condition for a geodesic it is usually convenient to replace (73.12) by the equivalent equations

$$\frac{d^2x^\sigma}{ds^2} + \{\mu\nu, \sigma\}\frac{dx^\mu}{ds}\frac{dx^\nu}{ds} = 0, \tag{73.13}$$

which can easily be obtained from (73.12) by familiar methods by substituting the general formula for interval (73.2). The Christoffel three-index symbols with 'curly' brackets occurring in (73.13) are defined by

$$\{\mu\nu, \sigma\} = \tfrac{1}{2}g^{\sigma\lambda}\left(\frac{\partial g_{\mu\lambda}}{\partial x^\nu} + \frac{\partial g_{\nu\lambda}}{\partial x^\mu} - \frac{\partial g_{\mu\nu}}{\partial x^\lambda}\right) \tag{73.14}$$

while three-index symbols with 'square' brackets are defined by

$$[\mu\nu, \sigma] = \tfrac{1}{2}\left(\frac{\partial g_{\mu\sigma}}{\partial x^\nu} + \frac{\partial g_{\nu\sigma}}{\partial x^\mu} - \frac{\partial g_{\mu\nu}}{\partial x^\sigma}\right), \tag{73.15}$$

neither of these quantities being tensors. It will be immediately seen from (73.13) that the general conditions for a geodesic do reduce in the case of the 'flat' space-time of special relativity to the simple form (73.11), since for the coordinates x, y, z, t the components of the metrical tensor $g_{\mu\nu}$ will then be constants and the three-index symbols will vanish.

Turning to the case of light rays, the equations of motion will be the same as for the case of particles with the additional restriction

$$ds = 0 \tag{73.16}$$

already discussed in § 21, corresponding to the fixed value for the velocity of light in free space.

The foregoing discussions of the covariant expression for interval and of the covariant expressions for the motion of particles and light rays give typical examples of the possibility of re-expressing the principles of the special theory of relativity in a covariant form which permits the use of any desired 'curvilinear' coordinates x^1, x^2, x^3, x^4 instead of the usual 'rectilinear' coordinates x, y, z, t. In both cases we shall later obtain with the help of the principle of equivalence (§ 74 e) a measure of justification for taking these covariant expressions as valid not only in the case of the 'flat' space-time of special relativity, but also in the case of the 'curved' space-time which we shall find to be associated with the presence of permanent gravitational fields.

74. The principle of equivalence

We may now turn to an examination of the principle of equivalence, which furnishes the second main element in the general theory of relativity and the one which leads to the necessary introduction of gravitational fields and 'curved' space-time into the considerations. We must first discuss the method of formulating this principle which is a somewhat more involved matter than in the case of the principle of covariance.

(a) **Formulation of the principle of equivalence. Metric and gravitation.** The principle of equivalence gives specific expression to the correspondence between the results which would be obtained by an observer who makes measurements in a gravitational field using a frame of reference which is held stationary, and the results obtained by a second observer who makes measurements in the absence of gravitational field but using an accelerated frame of reference. In a qualitative way it is immediately evident that some measure of correspondence between the two sets of measurements should exist, since both observers would find an acceleration with respect to their frames of reference for all free particles left to their own motion.

To obtain a precise expression of the principle, we may first consider the hypothetical limiting case of a non-accelerated observer in a perfectly uniform gravitational field, as contrasted with a uniformly accelerated observer in a region of free space where the gravitational field can be neglected. In this case the principle of equivalence makes the definite assertion that the results obtained by the two observers in performing any given physical experiment will be precisely identical, provided of course that the observer in free space is given an acceleration, relative to the non-accelerated axes of the special theory of relativity, which is equal and opposite to the gravitational acceleration found by the other observer.

An alternative expression of the principle of equivalence can also be given which is often more convenient for use. Having asserted the complete equivalence between the two observers, we appreciate that the equivalence will have to persist when we make analogous changes in their states of motion. Thus if the observer in the field is himself allowed to fall freely with the natural acceleration due to gravity, and the forced acceleration given to the observer in free space is reduced to zero, they must still obtain identical results in any given experiment that they may perform. In other words, for a freely falling observer in a uniform gravitational field the effects of gravitation would be abolished. Hence the principle of equivalence can also be taken as the assertion that it is always possible in the case of a uniform gravitational field to transform to space-time coordinates such that the effects of gravity will not appear.

In the general case of non-uniform fields, the statement of the principle of equivalence has to be modified since the natural acceleration due to gravity would be different in different parts of the field. Nevertheless, we may still maintain for a sufficiently small region that the effects of gravitation could be removed by the use of freely falling axes, having the natural acceleration due to gravity for that region. This is illustrated, for example, by the temporary and limited abolition of gravitational action which would be obtained inside a freely falling lift at the surface of the earth. Hence the principle of equivalence may be finally formulated by the statement that it is always possible at any space-time point of interest to transform to coordinates such that the effects of gravity will disappear over a differential region in the neighbourhood of that point, which is taken small enough so that the spatial and temporal variation of gravity within the region may be neglected.

The recognition, which the principle of equivalence thus gives to the possibility of permanent gravitational fields which cannot be completely transformed away by choice of coordinate system, leads at once to an intimate relation between metric and gravitation. In accordance with the special theory of relativity, coordinates x, y, z, t can be chosen such that throughout space-time the formula for interval can be written in the simple form

$$ds^2 = -dx^2 - dy^2 - dz^2 + c^2\, dt^2, \qquad (74.1)$$

where the metrical tensor has the *constant* values

$$g_{11} = g_{22} = g_{33} = -1 \qquad g_{44} = c^2$$
$$g_{\mu\nu} = 0 \quad (\mu \neq \nu). \qquad (74.2)$$

In accordance with the above, however, it will not be possible in the case of permanent gravitational fields to find coordinates such that the components of the metrical tensor assume these values except in the neighbourhood of some selected point, and we shall find it necessary to use the more general formula for interval

$$ds^2 = g_{\mu\nu}\, dx^\mu dx^\nu, \qquad (74.3)$$

where the components of the metrical tensor may be any function of the coordinates

$$g_{\mu\nu} = g_{\mu\nu}(x^1, x^2, x^3, x^4). \qquad (74.4)$$

There is thus an intimate relation between metric and gravitation to be more precisely investigated as we proceed. Using the language of our four-dimensional geometry, however, we can already say that the absence of gravitational field corresponds to the metric for a 'flat' space-time and the presence of any given permanent gravitational field corresponds to the metric for some particular kind of 'curved' space-time.

(b) **Principle of equivalence and relativity of motion.** We may next consider the relation of the principle of equivalence to the fundamental idea of the relativity of all kinds of motion.

The first step towards the preservation of this idea lay in the introduction of the principle of covariance, in accordance with which the equations of physics can be expressed in a form the same for all coordinate systems, thus removing the possibility of using essential differences in form as a criterion of absolute differences in motion. As already mentioned, nevertheless, this alone is not necessarily sufficient to preserve the idea of the relativity of all kinds of motion, since equations of the same form can exhibit essential differences in numerical content which could be used as possible criteria of absolute motion. At this point, however, the introduction of the principle of equivalence can be regarded as the second step in preserving the idea of the relativity of all kinds of motion, since with the help of this principle it proves possible, if so desired, to interpret the essential changes in numerical content which are actually found to accompany changes in coordinate system as due to changes in gravitational field rather than to changes in the absolute state of motion of the axes of reference.

To illustrate this by a specific example we may concentrate our attention on the simple case of two observers in free space, the first being regarded as in a state of rest or uniform motion and the second as accelerated. In spite of the principle of covariance which allows the two observers to treat, for example, the motion of free particles by equations of exactly the same form, it is evident that they must find essential differences in numerical content, since the first observer would find no motion relative to himself for a free particle which he places in his immediate neighbourhood and the second observer would find a definite motion for such a test particle owing to his own state of acceleration. And this difference could be interpreted by the second observer as a definite criterion for the absolute character of his acceleration, if the principle of equivalence did not enter at this point and permit him—as we have seen above—to ascribe the acceleration of free particles with equal justice as being due to the presence of a gravitational field.

To treat this same example somewhat more mathematically, let us consider that the unaccelerated observer uses a coordinate system corresponding to our usual spatial and temporal variables x, y, z, and t and to the formula for interval

$$ds^2 = -dx^2 - dy^2 - dz^2 + c^2\, dt^2. \qquad (74.5)$$

On the other hand, let us assume that the second observer, who can be taken as moving relative to the first with the acceleration a in the x-direction, uses the coordinates x', y', z', and t' as given by

$$x' = x - \tfrac{1}{2}at^2 \qquad y' = y \qquad z' = z \qquad t' = t \qquad (74.6)$$

in accordance with the usual transformation to accelerated axes, which we may certainly regard as a reasonable change at least at low velocities. Substituting from (74.6) into (74.5) we then find as the formula for interval for the second observer the expression

$$ds^2 = -dx'^2 - dy'^2 - dz'^2 + (c^2 - a^2 t'^2)dt'^2 - 2at'\, dx'dy'. \quad (74.7)$$

Examining the two formulae for interval (74.5) and (74.7), we immediately appreciate their essential difference in content, in spite of the fact that both formulae are in agreement with the generally covariant expression for interval

$$ds^2 = g_{\mu\nu} \, dx^\mu dx^\nu,$$

since in the one case the components of the metrical tensor $g_{\mu\nu}$ have the simple constant values $-1, c^2, 0$, and in the other case are considerably more complicated. This difference, moreover, is immediately reflected in a difference in the experimental results which the two observers obtain. Thus although both observers can use the same covariant equations (73.13) previously given

$$\frac{d^2x^\sigma}{ds^2} + \{\mu\nu, \sigma\} \frac{dx^\mu}{ds} \frac{dx^\nu}{ds} = 0$$

to describe the motion of free particles, the first observer will find that the application of this equation leads to the general result

$$\frac{d^2x}{ds^2} = \frac{d^2y}{ds^2} = \frac{d^2z}{ds^2} = \frac{d^2t}{ds^2} = 0, \qquad (74.8)$$

while the second observer will obtain a more complicated result which reduces for the case of particles having negligible velocity to

$$\frac{d^2x'}{ds^2} = \frac{-a}{c^2 - a^2 t^2} \qquad \frac{d^2y'}{ds^2} = \frac{d^2z'}{ds^2} = \frac{d^2t'}{ds^2} = 0. \qquad (74.9)$$

In accordance with the principle of equivalence, nevertheless, the second observer is permitted to interpret this difference in experimental results—arising from the changed values of the components of the metrical tensor—as due to the presence of a gravitational field rather than to any absolute quality in his state of motion.

As a result of the foregoing discussion, we now appreciate in general that the principle of equivalence will permit us, if we so desire, to interpret the change in the content of the equations of physics when we change to a new coordinate system as due to a change in gravitational field rather than to a change in the absolute motion of the spatial framework. This, however, is sufficient to preserve the idea of the relativity of all motion. Thus the changed results, that we ordinarily describe as being due to a change in our reference framework from a state of rest to a state of accelerated motion, can also be described as due to the changed gravitational field which results when the reference system is left at rest and the remainder of the universe is accelerated in the opposite direction. Acceleration as well as velocity thus partakes in the quality of relativity. Similarly the effects accompanying the change, ordinarily described as that from stationary to rotating axes, can only be regarded as due to the relative rotation of the axes and the gravitating bodies in the rest of the universe.

We shall later have further examples of the validity of the idea of the relativity of all kinds of motion, a very instructive one being given by the discussion in § 79 (c) of the so-called clock paradox. In general, the possibility of contradictions to this idea may now be regarded as satisfactorily removed by the introduction of the principle of equivalence.

(c) Justification for the principle of equivalence. Although the general idea of the relativity of all kinds of motion thus provides a strong motive for the acceptance of the principle of equivalence, our justification for the introduction of this principle can also be based on more immediate grounds. Unlike the principle of covariance, the principle of equivalence cannot be regarded as a necessarily inescapable axiom of physics, since it makes perfectly definite statements as to the interrelated character of coordinate systems and gravitational fields which might or might not be true. Hence it is also unlike the principle of covariance in demanding necessary physical consequences, and our final justification for the introduction of the principle must depend on the comparison of these predicted consequences with the results of observation and experiment.

The simplest of these consequences is the conclusion that the gravitational acceleration of all bodies would have to be the same when tested in the same gravitational field, since the presence and amount of this acceleration would be solely a function of the coordinate system used. Hence the far-reaching discovery of Galileo that all bodies fall at the same rate, and the precise tests of this law furnished in the case of ordinary materials by the exhaustive investigations of Eötvös and in the case of radioactive material by the work of Southerns, can be regarded as furnishing immediate support for the principle of equivalence.

In addition to this simple, but nevertheless very general and well tested, consequence of the principle of equivalence, we shall see in § 80 that the general theory of relativity leads to the Newtonian theory of gravitation as a first and very close approximation. Hence the accurately confirmed laws of celestial mechanics can also be regarded as furnishing support for the building-stones upon which the theory is based.

Finally, moreover, our belief in the principles of the general theory of relativity receives the compelling sanction provided by the three so-called crucial tests, § 83, which distinguish between the predictions of the approximate Newtonian theory and those of the more precise Einstein theory. So that we must regard all the postulates of the theory as being very satisfactorily chosen.

In addition to these observational verifications, which justify the introduction of the principle of equivalence, we must also assign a high importance to our intuitive appreciation of the rationality of assuming the abolition of gravitational effects for a freely falling observer, and to our intellectual appreciation of the simplicity, clarity, and effectiveness of the postulate that we thus obtain. These qualities of intuitive rationality and of intellectual simplicity, clarity, and effectiveness, which bespeak so unmistakably the insight and genius of Einstein, furnish of themselves of course no evidence of correspondence with experimental and observational fact. They are, nevertheless, necessary qualities for those principles which the human mind is willing to use as the fundamental postulates for science, and their presence must hence be regarded as also furnishing important justification for the acceptance of the principle of equivalence.

(*d*) **Use of the principle of equivalence in generalizing the principles of special relativity. Natural and proper coordinates.** In accordance with the principle of equivalence we can always choose coordinates so as to make the effects of gravity disappear in the immediate neighbourhood of any space-time point of interest, over a differential region taken small enough so that we can neglect the spatial and temporal variations of gravity for the range involved. In the absence of gravity, however, the principles of the special theory of relativity can be regarded as being valid. Hence the principle of equivalence can also be understood as requiring the possibility of choosing coordinates such that the general statements of the laws of physics will reduce in the immediate neighbourhood of any desired point to forms previously given by the special theory of relativity in terms of our usual spatial and temporal variables x, y, z, and t, or more simply in terms of the so-called Galilean coordinates introduced in § 20:

$$x^1 = x, \qquad x^2 = y, \qquad x^3 = z, \qquad x^4 = ct.$$

Such coordinates may be called *natural coordinates* for the point in question. In these coordinates, in accordance with the special relativity formula for interval, the components of the metrical tensor $g_{\mu\nu}$ will assume, at the chosen point of interest, their previous simple values -1, $+1$, and 0, and the first differential coefficients of the $g_{\mu\nu}$ with respect to these coordinates will be zero at that point. In general, however, the second differential coefficients will not be zero except for the special case of space-time that actually is flat. It will thus be seen that the assumption of approximate correctness for the special theory of relativity in the immediate neighbourhood of any desired space-time point is analogous to the approximate replacement of a curved surface by its tangent plane at a given point of interest, made use of in geometrical considerations.

For any given space-time point there will be an infinite number of different possible systems of natural coordinates, which can be obtained by rotating the spatial axes to different orientations, and by making the Lorentz transformation to different velocities of the origin. Among these different systems we shall often be specially interested in coordinates which are so chosen that some particular observer with his measuring instruments or some particular thing such as a given particle of matter will be at least momentarily at rest with respect to the spatial axes. Such systems may be called *proper coordinates* for the observer or thing in question, and by § 18 a transformation to such coordinates can of course always be made.

The possibility thus furnished by the principle of equivalence of using natural coordinates gives us a powerful instrument for use in determining the general laws of physics, since we now require that they must in any case be of such form, when expressed in natural coordinates, that they will reduce at the selected point to their previously obtained special relativity forms. This provides a procedure for testing covariant expressions which may be proposed as general laws of physics, and eliminating those which do not agree with the principle of equivalence. The method of course does not lead to necessarily unique results, since more than one generalization of the principles of special relativity having the required property can be possible. Nevertheless, in many cases the simplest possible generalizations which present themselves are found to provide satisfactory principles.

(e) Interval and trajectory in the presence of gravitational fields. In our discussion of the principle of covariance [see §§ 73 (d) and 73 (e)], it has already been intimated that we shall adopt the covariant forms, in which the special relativity formulae for the element of interval and for the trajectories of free particles and light rays can be expressed, as also valid in the 'curved' space-time associated with permanent gravitational fields. We must now show that this does agree with the requirements of the principle of equivalence as just discussed above.

To show this in the case of the formula for interval

$$ds^2 = g_{\mu\nu}\, dx^\mu dx^\nu \tag{74.10}$$

we must first prove that a transformation of coordinates is always possible such that the first differential coefficients of the components of the metrical tensor $g_{\mu\nu}$ will become zero at any selected point. This, however, is a well-known theorem of differential geometry which demonstrates that it is always possible to reduce to such 'geodesic coordinates'. To do this we first transfer the origin of coordinates to the point of interest and then change from unprimed to primed variables by the substitution

$$x^\epsilon = g_\mu^\epsilon x'^\mu - \tfrac{1}{2}\{\alpha\beta,\, \epsilon\}_0\, g_\mu^\alpha g_\nu^\beta x'^\mu x'^\nu, \tag{74.11}$$

where $\{\alpha\beta,\, \epsilon\}_0$ is the value of the three-index symbol at the origin. It is then readily proved that we shall have the relations†

$$\frac{\partial g'_{\mu\nu}}{\partial x'^\alpha} = 0 \tag{74.12}$$

holding at the origin in the new coordinates. Having secured the desired constancy at the origin for the components of the metrical tensor, the further transformation to coordinates such that the components $g_{\mu\nu}$ will assume at that point the prescribed values -1, c^2, 0 or ± 1, 0 can then be secured by familiar methods. Hence the choice of the covariant expression (74.10) as the general relativity formula for interval, in the presence as well as in the absence of gravitational fields, is in agreement with the requirements of the principle of equivalence.

Turning next to the covariant expression

$$\frac{d^2 x^\sigma}{ds^2} + \{\mu\nu,\, \sigma\}\, \frac{dx^\mu}{ds}\, \frac{dx^\nu}{ds} = 0 \tag{74.13}$$

as given by the special theory of relativity (73.13) for the trajectory of a free particle or light ray, we see at once that this is also suitable to take as a postulate which will be applicable in general relativity in the presence of gravitational fields, since in natural coordinates the

† See for example Eddington, *The Mathematical Theory of Relativity*, Cambridge, 1923, p. 77.

Christoffel three-index symbols (73.14) will evidently vanish at the point of interest in accordance with (74.12), and the general formula (74.13) will reduce to the special relativity form (73.11)

$$\frac{d^2x^\sigma}{ds^2} = 0. \tag{74.14}$$

Furthermore, the additional limitation

$$ds = 0 \tag{74.15}$$

which must be used in the case of light rays can evidently be appropriately taken as a general condition also in the presence of gravitational fields. The conditions for a geodesic as given by (74.13) can thus be postulated as also applying to the motion of particles or light rays in a gravitational field. This gives an enormous step forward in the direction of securing a complete theory of gravitation, a step which of course must be justified, when the time comes, by comparison with the observational data of astronomy.

The fundamental tensor $g_{\mu\nu}$ occurs both in the formula for interval (74.10) and in the formula for trajectory (74.13). In the formula for interval it appears as a set of metrical quantities which determine the nature of the space-time geometry by relating the measured values of different intervals to the corresponding coordinate differences. In the formula for trajectory the first derivatives of the $g_{\mu\nu}$ with respect to the coordinates appear in the Christoffel three-index symbol in a certain analogy with the appearance of the derivatives of the Newtonian gravitational potential in the older expressions for trajectory. This dual character of the fundamental tensor may be recognized by referring to the ten independent quantities $g_{\mu\nu}$ either as the components of the *metrical tensor* or as the *gravitational potentials* in the Einstein theory of gravitation. The dependence of the geometry of space-time and hence also of space itself on gravitation, arising from this dual character of the fundamental tensor, is a noteworthy result of the general theory of relativity.

The rather abstract quality of the formula for interval (74.10) must not be allowed to obscure the fact of its reference to matters which are completely observational in character. Any interval expressed by the formula (74.10) will be either space-like, time-like, or singular in character according as ds^2 is negative, positive, or zero. By transformation to suitably chosen proper coordinates x, y, z, t, the expression for any space-like interval can be thrown into the form

$$-ds^2 = dx^2 + dy^2 + dz^2, \qquad (74.16)$$

and the expression for any time-like interval into the form

$$ds^2 = c^2\, dt^2, \qquad (74.17)$$

which makes the corresponding *proper length* or *proper time* immediately determinable from the readings of suitably taken metre sticks or clocks.

Similarly the formula for trajectory (74.13) refers to observational situations, since the time-like interval ds is then the proper time for a local observer moving with the particle in question, and the rate of change of the coordinates of the particle with this quantity can be observationally determined.

75. The dependence of gravitational field and metric on the distribution of matter and energy. Principle of Mach

In addition to the principles of covariance and equivalence we must evidently introduce some further element into the theory of gravitation. With the help of the two foregoing principles we have learned to interpret the fundamental tensor $g_{\mu\nu}$ in its metrical aspect as determining the nature of space-time geometry and in its gravitational aspect as determining the motion of particles and light rays. We have so far, nevertheless, no laws for the actual dependence of the values of the $g_{\mu\nu}$ on the coordinates, beyond that provided by the very general notion that 'flat' space-time corresponds to the absence of intrinsic gravitational action and that 'curved' space-time corresponds to the presence of permanent gravitational fields. Hence we must now seek, as the third element in the relativistic theory of gravitation, a precise statement of the laws giving the dependence of the metrical and gravitational field on space-time position, which will permit the calculation of gravitational effects in the presence of any given distribution of matter and energy.

In accordance with the Newtonian theory of gravitation the action of gravity at any point in space at a given instant is determined by the location of the surrounding matter, and this general idea with suitable modifications must evidently be taken over into the relativistic theory of gravitation since the Newtonian theory is in any case an exceedingly close first approximation. In Newtonian theory the dependence of the gravitational potential ψ on the distribution of matter of density ρ is given by Poisson's equation

$$\frac{\partial^2\psi}{\partial x^2}+\frac{\partial^2\psi}{\partial y^2}+\frac{\partial^2\psi}{\partial z^2} = 4\pi k\rho, \qquad (75.1)$$

where k is the gravitational constant. In the relativistic theory of gravitation modifications will be necessary, in the first place since we shall need to calculate the ten components of the metrical tensor or gravitational potentials $g_{\mu\nu}$ instead of the single gravitational potential ψ of the Newtonian theory, and in the second place because the special theory of relativity has provided us with relations between mass, energy, and momentum which indicate that covariant expressions are to be obtained by making use of all ten components of the energy-momentum tensor $T_{\mu\nu}$, rather than by singling out some single quantity which we could call *the* density of matter.

Our general aim, hence, will be to obtain a covariant equation connecting the $g_{\mu\nu}$ with the $T_{\mu\nu}$ which will be the analogue of Poisson's equation, and which will lead to the same results as the Newtonian theory to a first approximation. Before proceeding to the complete solution of this task, however, it will first be profitable to consider two special cases, that of the field corresponding to the special theory of relativity, and that of a field in empty space but in the neighbourhood of gravitating bodies.

The general hypothesis that the metrical field is determined by the distribution of matter and energy may be called the principle of Mach.†

76. The field corresponding to the special theory of relativity. The Riemann-Christoffel tensor

The special theory of relativity can be regarded as developed on the assumption of a 'flat' space-time, which neglects the presence of intrinsic gravitational fields, and the results obtained may be regarded as approximately correct in what may be called the *free* space at great distances from gravitating bodies. We may now inquire into the covariant expression of the conditions necessary for the 'flat' space-time corresponding to the special theory.

---

† This hypothesis was designated as the principle of Mach by Einstein, *Ann. der Physik*, **55**, 241 (1918), since he took it to be a generalization of Mach's requirement that inertia must be regarded as based on the interaction of bodies. At the time Einstein believed that the Mach principle necessitated the introduction of the cosmological Λ-term into the field equations. This term, however, no longer appears necessary.

To obtain this we must first consider the Riemann-Christoffel tensor, which can be expressed in terms of the Christoffel three-index symbols by the equation

$$R^{\tau}_{\mu\nu\sigma} = \{\mu\sigma, \alpha\}\{\alpha\nu, \tau\} - \{\mu\nu, \alpha\}\{\alpha\sigma, \tau\} + \frac{\partial}{\partial x^{\nu}}\{\mu\sigma, \tau\} - \frac{\partial}{\partial x^{\sigma}}\{\mu\nu, \tau\}. \quad (76.1)$$

The tensor character of this expression can be readily demonstrated. In accordance with the form given and the definition of the three-index symbols (73.14), it will be seen that the tensor is constructed solely from the components of the metrical tensor $g_{\mu\nu}$ and their first and second derivatives with respect to the coordinates. It can be shown, moreover, that the only tensors which can be thus constructed solely from the fundamental tensor without going beyond the second derivatives are themselves functions of the $g_{\mu\nu}$ and $R^{\tau}_{\mu\nu\sigma}$.

The conditions for 'flat' space-time can now be expressed by setting the Riemann-Christoffel tensor equal to zero, giving us the covariant equation

$$R^{\tau}_{\mu\nu\sigma} = 0. \quad (76.2)$$

This equation is evidently a *necessary* condition, since in the case of 'flat' space-time we know that it is possible to choose coordinates which will make the components of $g_{\mu\nu}$ constants, and thus give all the Christoffel three-index symbols the value zero. The proof can also be given that the vanishing of the Riemann-Christoffel tensor is a *sufficient* condition for the possibility of choosing coordinates which will make all the components of $g_{\mu\nu}$ constants, as first shown by Lipschitz.†

The tensor equation (76.2) thus expresses the conditions necessary for the validity of the special theory of relativity and for the absence of permanent gravitational fields, which cannot be transformed away by a suitable choice of coordinates. In the actual universe the density of matter is found to be approximately uniform as far as the Mount Wilson 100-inch telescope can penetrate, at least to over 10^8 light years; and there is no reason to believe that there is any region in the universe where the gravitational field could actually be completely transformed away. Indeed the mere presence of physical measuring instruments would be accompanied by an irreducible gravitational field. In other words, that which we have hitherto designated as the *free* space in which the special theory of

† Lipschitz, *Crelle's Journ.*, **70**, 71 (1869).

relativity would be exactly true presumably does not exist in the actual universe except as an idealization. Nevertheless, it is evident that the principles of the special theory of relativity are approximately true even in the permanent gravitational field at the surface of the earth, and would be valid to an extremely high degree of approximation in internebular space. The use of the special theory of relativity as an abstract idealization thus seems entirely legitimate.

77. The gravitational field in empty space. The contracted Riemann-Christoffel tensor

Since the equation obtained by setting the Riemann-Christoffel tensor equal to zero would eliminate the possibility of permanent gravitational fields, it is evident that we must seek some less stringent relation for the gravitational field in the *empty* space in the neighbourhood of gravitating bodies.

We can arrive at such a less stringent relation with the help of the contracted Riemann-Christoffel tensor which is obtained by setting $\sigma = \tau$ in (76.1) and summing. This gives us the tensor

$$R_{\mu\nu} = \{\mu\sigma, \alpha\}\{\alpha\nu, \sigma\} - \{\mu\nu, \alpha\}\{\alpha\sigma, \sigma\} + \frac{\partial}{\partial x^\nu}\{\mu\sigma, \sigma\} - \frac{\partial}{\partial x^\sigma}\{\mu\nu, \sigma\}, \quad (77.1)$$

which, with the help of equation (37) in Appendix III and a change in order and in dummy suffixes, can also be written in the simplified form

$$R_{\mu\nu} =$$

$$-\frac{\partial}{\partial x^\alpha}\{\mu\nu, \alpha\} + \{\mu\alpha, \beta\}\{\nu\beta, \alpha\} + \frac{\partial^2}{\partial x^\mu dx^\nu}\log\sqrt{-g} - \{\mu\nu, \alpha\}\frac{\partial}{\partial x^\alpha}\log\sqrt{-g}.$$

$$(77.2)$$

As the field equations in empty space but nevertheless in the neighbourhood of gravitating masses, Einstein has proposed the relation

$$R_{\mu\nu} = 0, \quad (77.3)$$

which would evidently be true when the condition for 'flat' space-time (76.2) is satisfied, but could also be true under less stringent conditions.

The theoretical justification for the choice of this equation will become apparent in the next section, where it will be possible to regard it as a limiting case of the more general equation for gravitational field in the presence of matter; and the very exact observational justification provided by the observed motions of the planets will be considered in § 83.

78. The gravitational field in the presence of matter and energy

We must now turn to the complete solution of the fundamental problem outlined in § 75, of obtaining a covariant relation between the gravitational potentials $g_{\mu\nu}$ and components of the energy-momentum tensor $T_{\mu\nu}$ which will be the appropriate relativistic analogue of Poisson's equation

$$\frac{\partial^2\psi}{\partial x^2}+\frac{\partial^2\psi}{\partial y^2}+\frac{\partial^2\psi}{\partial z^2} = 4\pi k\rho, \tag{78.1}$$

which in the Newtonian theory of gravitation connects the single gravitational potential ψ with the density of matter ρ and the gravitational constant k.

In obtaining a solution of this fundamental problem Einstein had several kinds of consideration to assist him. In the first place, in agreement with the preliminary outline of the problem given in § 75, we may expect that the relativistic analogue of Poisson's equation will be a relation connecting all ten gravitational potentials $g_{\mu\nu}$ with the distribution of matter and energy as given by the ten components of the energy-momentum tensor $T_{\mu\nu}$. In the second place, in accordance with the principle of covariance, we shall desire to express this result in covariant form, which will suggest the search for a tensor of the second rank, constructed from the $g_{\mu\nu}$ and their derivatives with respect to the coordinates, which can be equated to the energy-momentum tensor $T_{\mu\nu}$. In the third place, since Poisson's equation does not involve higher derivatives of the Newtonian potential than the second, it will be natural to assume that it will not be necessary—at least in first approximation—to use a tensor containing higher derivatives of the $g_{\mu\nu}$ than the second. Finally, in accordance with the principle of equivalence, we know that the energy-momentum tensor is a quantity whose ordinary divergence can be made to vanish at any selected point by the use of natural coordinates, since in the special theory of relativity we have already obtained, in Galilean coordinates, the relation (37.9)

$$\frac{\partial T^{\mu\nu}}{\partial x^\nu} = 0 \tag{78.2}$$

in the treatment given in § 37 to the mechanics of a continuous medium.

These considerations were sufficient to suggest as the relativistic analogue of Poisson's equation, the tensor equation first proposed by Einstein

$$R_{\mu\nu} - \tfrac{1}{2}Rg_{\mu\nu} + \Lambda g_{\mu\nu} = -\kappa T_{\mu\nu}, \tag{78.3}$$

where $R_{\mu\nu}$ is the contracted Riemann-Christoffel tensor, R is the invariant obtained by the further contraction of this tensor, Λ is a constant, the so-called cosmological constant whose significance will be considered in more detail later, κ is a constant which is related to the ordinary constant of gravitation by a factor to be obtained in § 80 when Poisson's equation is obtained as a first approximation, and $T_{\mu\nu}$ is the energy-momentum tensor which is defined for the purpose of general relativity by assigning to its components in proper coordinates—and hence in any set of natural coordinates—the values which it would have in accordance with the special theory of relativity.

This relation satisfactorily fulfils all the conditions mentioned above. It reduces in the case of weak gravitational fields to Poisson's equation as a first approximation as will be shown in § 80. It connects the ten gravitational potentials $g_{\mu\nu}$ and their derivatives with the components of the energy-momentum tensor $T_{\mu\nu}$. It satisfies the principle of covariance by being a tensor equation valid in all systems of coordinates if valid in one. And it contains no derivatives of the $g_{\mu\nu}$ higher than the second.

Furthermore, it is to be noted that it secures the validity of (78.2) in natural coordinates in a very fundamental manner, since it can readily be shown from the definition of the Riemann-Christoffel tensor that the relation

$$(R^{\mu\nu} - \tfrac{1}{2}Rg^{\mu\nu} + \Lambda g^{\mu\nu})_\nu = 0 \tag{78.4}$$

is a necessary identity with any constant value for Λ; and this will give as the fundamental equation of mechanics in any system of coordinates

$$(T^{\mu\nu})_\nu = \frac{\partial T^{\mu\nu}}{\partial x^\nu} + \{\alpha\nu, \mu\}T^{\alpha\nu} + \{\alpha\nu, \nu\}T^{\mu\alpha} = 0, \tag{78.5}$$

or

$$\frac{\partial T^{\mu\nu}}{\partial x^\nu} = 0, \tag{78.6}$$

in the special case of natural coordinates where the Christoffel three-index symbols vanish. Moreover, it is to be emphasized in this connexion, in the first place that $(R^{\mu\nu}-\frac{1}{2}Rg^{\mu\nu}+\Lambda g^{\mu\nu})$, *with Λ an arbitrary constant*, can be shown to be the most general tensor of the second rank constructed solely from the $g_{\mu\nu}$ and their first and second derivatives whose contracted covariant derivative (78.4) would be identically equal to zero, and in the second place that four identical relations must necessarily be present since otherwise the solution of the ten field equations (78.3) for the ten $g_{\mu\nu}$ would not permit the four-fold transformation of coordinates which must necessarily be possible.

We are thus impelled with considerable force at least to the tentative acceptance of Einstein's relation (78.3) as the appropriate field equations for the relativistic theory of gravitation. The complete justification for accepting this relation must of course depend on the correspondence between the predictions which it provides and the results of observation. To test this correspondence we can make use of the field equations (78.3) with any given distribution of matter and energy to predict the dependence of the tensor $g_{\mu\nu}$ on the coordinates used, and then compare the predicted values of the $g_{\mu\nu}$ with observed values of the $g_{\mu\nu}$. Theoretically, these observed values could of course be obtained from the direct measurement of space-like and time-like intervals with the help of the formula for interval $ds^2 = g_{\mu\nu}\,dx^\mu dx^\nu$. Practically, however, such direct measurements cannot be carried out with sufficient accuracy even to distinguish between 'flat' and 'curved' space-time, and our accurate determinations of the $g_{\mu\nu}$ depend on the measurement of astronomical motions, followed by the calculation of the $g_{\mu\nu}$ with the help of the formula for trajectory

$$\frac{d^2x^\sigma}{ds^2} + \{\mu\nu, \sigma\}\frac{dx^\mu}{ds}\frac{dx^\nu}{ds} = 0. \tag{78.7}$$

By raising indices, the field equations (78.3) can be written in the different forms

$$-\kappa T_{\mu\nu} = R_{\mu\nu}-\tfrac{1}{2}Rg_{\mu\nu}+\Lambda g_{\mu\nu} \tag{78.8}$$

$$-\kappa T_\mu^\nu = R_\mu^\nu-\tfrac{1}{2}Rg_\mu^\nu+\Lambda g_\mu^\nu \tag{78.9}$$

$$-\kappa T^{\mu\nu} = R^{\mu\nu}-\tfrac{1}{2}Rg^{\mu\nu}+\Lambda g^{\mu\nu}, \tag{78.10}$$

and by the contraction of (78.9) we can evidently write

$$\kappa T = R-4\Lambda. \tag{78.11}$$

In empty space with all the components of the energy-momentum tensor equal to zero we see, by combining (78.8) and (78.11), that the field equations will then reduce to the simple form

$$R_{\mu\nu} = \Lambda g_{\mu\nu}. \tag{78.12}$$

Nevertheless, as already stated in the preceding section § 77, we shall actually find in empty space that the motions of the planets correspond with great precision to the even simpler field equations

$$R_{\mu\nu} = 0. \tag{78.13}$$

Therefore, since we shall find later [see, for example, equation (82.10)] that the effects of the Λ-term would increase with the size of the region considered, we may conclude that the unspecified constant Λ, introduced above in order to obtain the most general expression with a vanishing covariant derivative (78.4), is either actually equal to zero or in any case small enough so that its effects are inappreciable within a region of the size of the solar system. Hence, in many of our considerations at the very least, we shall be justified in setting Λ equal to zero, and taking the field equations in the simpler form

$$-\kappa T_{\mu\nu} = R_{\mu\nu} - \tfrac{1}{2} R g_{\mu\nu}, \tag{78.14}$$

together with $\qquad\qquad \kappa T = R \tag{78.15}$

as the result of contraction.

For regions of great size, on the other hand, it can be shown that effects could result even from a very small value of Λ. Hence for cosmological considerations we shall retain the possibility that the quantity Λ, customarily known as the *cosmological constant*, may not necessarily be exactly equal to zero.

We are now ready to consider a number of the simpler applications of general relativity, some of which will be specially important in illustrating the correspondence between theory and observation.

VI

THE GENERAL THEORY OF RELATIVITY (*contd.*)

Part II. ELEMENTARY APPLICATIONS OF GENERAL RELATIVITY

79. Simple consequences of the principle of equivalence

As already noted, unlike the principle of covariance, the principle of equivalence cannot be regarded as a necessarily inescapable axiom of physics, but must be considered as a definite postulate whose consequences are to be tested by comparison with observation and experiment. We may now consider certain simple qualitative and semi-quantitative consequences that can be drawn directly from this principle without the full apparatus of the general theory of relativity.

(*a*) **The proportionality of weight and mass.** The most important of these simple consequences of the principle of equivalence is the conclusion, already mentioned in § 74 (*c*), that the gravitational acceleration of all bodies would have to be the same when tested in the same gravitational field, since the presence and amount of this acceleration would by this principle be solely a function of the coordinate system used. The result is in immediate agreement with the fundamental discovery of Galileo that different bodies do fall at the same rate in the earth's gravitational field.

Since the gravitational acceleration g of a body at the surface of the earth is connected, in the language of Newton's second law of motion, with its mass m and the gravitational force acting on it or weight W by the equation

$$W = mg, \tag{79.1}$$

the above result can also be stated as requiring a constant proportionality between weight and mass for different bodies. The precise and exhaustive tests of this proportionality made on ordinary materials by Eötvös,[†] and the similar experiments on radioactive materials made by Southerns[‡] are in complete agreement with the theoretical conclusion.

(*b*) **Effect of gravitational potential on the rate of a clock.** In accordance with the principle of equivalence there should be an agreement between the results obtained by a uniformly accelerated

---

[†] Eötvös, *Math. und Naturw. Ber. aus Ungarn,* **8,** 65 (1890).
[‡] Southerns, *Proc. Roy. Soc.* **84A,** 325 (1910).

observer who makes measurements in the absence of any intrinsic gravitational field, and those obtained in similar experiments by a stationary observer in the presence of a uniform gravitational field. Since we can easily make approximate calculations as to the nature of the results obtained by the accelerated observer, this provides a simple method for investigating certain of the effects of gravity.

This method can be readily applied to determine the effect of differences in gravitational potential on the observed rate of clocks. Let us first consider an observer in the absence of any intrinsic gravitational field who is subject to the constant acceleration g, and is provided with two identically constructed clocks placed one ahead of the other on a line parallel to the direction of acceleration at a distance apart h. Let the clocks have the natural period τ_0, and let light signals be sent at the end of each period from one clock to the other to permit a comparison of their observed rates.

Since the time necessary for a signal to pass between the two clocks will be approximately

$$t = \frac{h}{c},$$

where c is the velocity of light, the forward clock will acquire the added velocity in the direction of motion

$$v = gt = g\frac{h}{c}$$

in the interval necessary for light to pass from the rear to the forward clock. Hence by the ordinary Doppler effect when the rates of the clocks are compared, the period of the rear clock, when measured in terms of that of the forward clock with the help of the arriving light signals, will be found to be approximately

$$\tau = \tau_0\left(1 + \frac{v}{c}\right) = \tau_0\left(1 + \frac{gh}{c^2}\right). \tag{79.2}$$

With the help of the principle of equivalence, however, this result can be at once reinterpreted as also applying to the analogous situation of two stationary clocks separated by a distance h in the direction of a uniform gravitational field of intensity g, so that we may immediately write as a consequence of (79.2)

$$\tau_2 = \tau_1\left(1 + \frac{\Delta\psi}{c^2}\right) \tag{79.3}$$

for the relation connecting the periods τ_2 and τ_1 of the two identically constructed clocks with their difference in gravitational potential $\Delta\psi = gh$, the clock at the lower potential having the longer observed period.

Furthermore, since time measurements could be made with the help of the period of light corresponding to any given spectral line, we can evidently regard different atoms of the same substance as furnishing the identically constructed clocks necessary for the validity of the above relation. Hence, making use of (79.3) together with the relation between the period and wave-length of light, we are at once led to the conclusion that there should be an observed shift $\delta\lambda$ of the approximate amount

$$\delta\lambda = \lambda\frac{\Delta\psi}{c^2} \tag{79.4}$$

in the wave-length λ of light which passes through a difference in gravitational potential of amount $\Delta\psi$, in travelling from the point of origin to that where the observation is made. The observational verification of this result will be more particularly mentioned in § 83 (c), in connexion with the three so-called crucial tests of relativity.

(c) **The clock paradox.** The foregoing relation between the rate of a clock and its gravitational potential has also been found to furnish the solution for a well-known paradox, which can arise when the behaviour of clocks is treated in accordance with the principles of special relativity without making due allowance for the principles of the general theory.

Consider two identically constructed clocks A and B, originally together and at rest, and let a force F_1 be then applied for a short time to clock B giving it the velocity u with which it then travels away from A at a constant rate for a time which is long compared with that necessary for the acceleration. At the end of this time let a second force F_2 be applied in the reverse direction which brings B to rest and starts it back towards A with the reversed velocity $-u$. And finally, when it has returned to the neighbourhood of A, let the clock B be brought once more to rest by the action of a third force F_3.

Since by hypothesis the time intervals necessary for the acceleration and deceleration of clock B are made negligibly short compared with the time of travel at the constant velocity u, we can then write, in accordance with the decreased rate of a moving clock given by the special theory of relativity (see § 9),

$$\Delta t_A = \frac{\Delta t_B}{\sqrt{(1-u^2/c^2)}} = \Delta t_B\left(1 + \frac{1}{2}\frac{u^2}{c^2} + ...\right) \qquad (79.5)$$

as an expression connecting the measurements Δt_A and Δt_B—on the two different clocks—of the elapsed time necessary for the clock B to move out from A and return. In accordance with this expression we are thus led to the definite conclusion that clock B would register a smaller number of divisions than clock A at the end of the indicated experiment.

At first sight, nevertheless, this conclusion—obtained quite correctly from the special theory of relativity—appears incompatible with the idea of the relativity of all motion, since it should then be equally as acceptable to regard B as the clock which remains at rest and consider A as moving away with the velocity $-u$ and returning with the velocity $+u$. And taking A as the moving clock, it then seems as if A should be the clock that registers the smaller number of divisions.

The apparent paradox is, however, readily solved with the help of the general theory of relativity, if we do not neglect the actual lack of symmetry between the treatment given to the clock A which was at no time subjected to any force, and that given to the clock B which was subjected to the successive forces F_1, F_2, and F_3 when the relative motion of the clocks was changed. To preserve this same state of affairs in a valid description of the experiment, taking A as the moving clock and B as the one which remains at rest, we may assume that the changes in the relative motion of the two clocks are produced by the temporary introduction of homogeneous gravitational fields, which are allowed to act freely on A in such a way as to produce the desired changes in velocity without A experiencing any force, and in such a way as to necessitate the action of the same forces on B as before in order to maintain it at rest. This then gives us a valid description of the *identical* experiment in the new language, and we can easily calculate the relation which would now be expected between the two time measurements Δt_A and Δt_B.

To do this, let us first put

$$\Delta t_A = \tau_A + \tau_A' + \tau_A'' + \tau_A''' \qquad (79.6)$$

and
$$\Delta t_B = \tau_B + \tau_B' + \tau_B'' + \tau_B''', \qquad (79.7)$$

where τ_A and τ_B are the time measurements referred to the two clocks during which the clock A is now regarded as having the uniform velocity u, and τ_A', τ_A'', τ_A''' and τ_B', τ_B'', τ_B''' are the times needed for the three changes in the velocity of A which are brought about at the beginning, middle, and end of the experiment by the temporary introduction of an appropriate gravitational field as mentioned above. And let us take these latter intervals as very short compared with the time during which A is in uniform motion, in correspondence with the previous description of the experiment.

Since the clock A is now the one which moves, we can in the first place write in accordance with the special theory of relativity to the desired order of precision

$$\tau_A = \tau_B\left(1 - \frac{1}{2}\frac{u^2}{c^2} + \dots\right), \tag{79.8}$$

in contrast to the previous relation (79.5) where B was taken as the moving clock. Furthermore, since the two clocks will be at practically the same potential when the gravitational fields are introduced at the beginning and end of the experiment, we can evidently write with sufficient precision

$$\tau_A' = \tau_B' \quad \text{and} \quad \tau_A''' = \tau_B'''. \tag{79.9}$$

On the other hand, when the gravitational field is introduced at the middle of the experiment to produce the necessary reversal in the motion of A, the two clocks will be at a great distance from each other, and we must evidently write in accordance with our previous treatment

$$\tau_A'' = \tau_B''\left(1 + \frac{\Delta\psi}{c^2}\right), \tag{79.10}$$

where $\Delta\psi$ is their difference in gravitational potential at that time.

This difference in potential, however, is given in terms of the distance between the two clocks h and the gravitational acceleration g by the simple expression

$$\Delta\psi = hg.$$

Furthermore, we can evidently put

$$h = \tfrac{1}{2}u\tau_B,$$

since $2h$ is the total distance travelled at the speed u, and can write

$$g = \frac{2u}{\tau_B''}$$

since $2u$ is the total change in velocity in the time τ_B''. Substituting these three expressions we can then write (79.10) in the more useful form

$$\tau_A'' = \tau_B'' + \tau_B \frac{u^2}{c^2},$$ (79.11)

and combining this equation with the previous equations (79.6–9) we obtain

$$\Delta t_A = \tau_B\left(1 - \frac{1}{2}\frac{u^2}{c^2} + \ldots\right) + \tau_B' + \tau_B'' + \tau_B\frac{u^2}{c^2} + \tau_B'''$$

$$= \tau_B\left(1 + \frac{1}{2}\frac{u^2}{c^2} + \ldots\right) + \tau_B' + \tau_B'' + \tau_B''',$$

or to our order of approximation, since the primed quantities are very short compared with τ_B,·

$$\Delta t_A = \Delta t_B\left(1 + \frac{1}{2}\frac{u^2}{c^2}\right).$$ (79.12)

Comparing this result with the earlier equation (79.5), we now see that whether we consider A or B to be the clock which moves we obtain the same expression for the relative readings of the two clocks, to the order of approximation that has been employed. The treatment of the problem without approximation would involve the full apparatus of the general theory of relativity.

The solution thus provided for the well-known clock paradox of the special theory gives a specially illuminating example of the justification for regarding all kinds of motion as relative, that has been made possible by the adoption of the general theory of relativity.

A similar treatment can also be given with entire success to the difference in rate between a clock placed at the centre of a rotating platform and a second clock fixed to the periphery of the platform. If the platform is taken as rotating, the peripheral clock will be regarded as having a slower rate than the central clock because of its velocity of motion. On the other hand, if the platform is taken as at rest and the remainder of the universe as rotating in the opposite direction, the slower rate of the peripheral clock will be ascribed to its position of lower gravitational potential corresponding to the gravitational interpretation which would then be given to centrifugal action. The general idea of the relativity of all kinds of motion will thus again be preserved, since we can with equal success treat the platform or the remainder of the universe as subject to the rotation.

80. Newton's theory as a first approximation

As our next application of the general theory, we may now show that Newton's theory of gravitation can be regarded as giving a first approximation to the more rigorous results of the general theory of relativity, with the quantity g_{44} of the general theory closely related to the gravitational potential ψ of the Newtonian theory. To demonstrate this, we must show in the first place that the Newtonian motion of a free particle would agree in first approximation with that predicted from the relativistic equations of motion, and in the second place that the field equation of Poisson can be regarded as a first approximation to the more general field equations of Einstein. In doing so, we may restrict our considerations to very weak static fields and to test particles having very small velocities compared with that of light, since the Newtonian theory was only developed to cover such cases. Hence we may take the line element as differing only slightly from the special relativity form, as expressed in Galilean coordinates,

$$ds^2 = -(dx^1)^2 - (dx^2)^2 - (dx^3)^2 + (dx^4)^2 \qquad (80.1)$$

with components of $g_{\mu\nu}$ which have very closely the special relativity values

$$g_{11} \simeq g_{22} \simeq g_{33} \simeq -1 \qquad g_{44} \simeq 1 \qquad (80.2)$$

$$g_{\mu\nu} \simeq 0 \quad (\mu \neq \nu)$$

and which are independent of the time

$$\frac{\partial g_{\mu\nu}}{\partial x^4} = 0. \qquad (80.3)$$

Furthermore, we may take the components for the generalized velocity of our test particles as having the approximate values

$$\frac{dx^1}{ds} \simeq \frac{dx^2}{ds} \simeq \frac{dx^3}{ds} \simeq 0 \qquad \frac{dx^4}{ds} \simeq 1. \qquad (80.4)$$

(a) **Motion of free particle in a weak gravitational field.** We are now ready to consider the motion of a free test particle in such a weak field.

In accordance with the theory of relativity [see § 74 (e)], the trajectory of a free particle will be given in general by the equations for a geodesic (74.13)

$$\frac{d^2x^\sigma}{ds^2} + \{\mu\nu, \sigma\} \frac{dx^\mu}{ds} \frac{dx^\nu}{ds} = 0,$$

and in the present simplified case these will reduce for $\sigma = 1, 2, 3$ as a result of equations (80.4), to the approximate form

$$\frac{d^2x^\sigma}{d(x^4)^2} + \{44, \sigma\} = 0. \tag{80.5}$$

Furthermore, in accordance with the definition for the three-index symbols (73.14) and the approximate values for the $g_{\mu\nu}$ given by (80.2), we can write

$$\{44, \sigma\} = \tfrac{1}{2}g^{\sigma\sigma}\left(\frac{\partial g_{4\sigma}}{\partial x^4} + \frac{\partial g_{4\sigma}}{\partial x^4} - \frac{\partial g_{44}}{\partial x^\sigma}\right)$$

where the summation convention is suspended; and on account of the static nature of the field expressed by (80.3), this leads to the simplified result

$$\{44, \sigma\} = \frac{1}{2}\frac{\partial g_{44}}{\partial x^\sigma}, \tag{80.6}$$

which on substitution into (80.5) gives

$$\frac{d^2x^\sigma}{d(x^4)^2} = -\frac{1}{2}\frac{\partial g_{44}}{\partial x^\sigma}. \tag{80.7}$$

This result, however, can easily be rewritten in a form familiar in the Newtonian theory of gravitation if we introduce our usual spatial and temporal variables by their relation with Galilean coordinates

$$x^1 = x \qquad x^2 = y \qquad x^3 = z \qquad x^4 = ct \tag{80.8}$$

and define the Newtonian potential ψ in terms of g_{44} by the expression

$$\frac{\psi}{c^2} = \frac{g_{44}}{2} + \text{const.}$$

or

$$\frac{\psi}{c^2} = \frac{g_{44}-1}{2} \qquad g_{44} = 1 + \frac{2\psi}{c^2}, \tag{80.9}$$

where the additive constant has been so chosen as to make the potential ψ approach the value zero in the free space at great distances from gravitating bodies where g_{44} approaches its special relativity value unity. Substituting (80.8) and (80.9) in (80.7), we can then write our result in the well-known Newtonian form

$$\frac{d^2x}{dt^2} = -\frac{\partial\psi}{\partial x} \qquad \frac{d^2y}{dt^2} = -\frac{\partial\psi}{\partial y} \qquad \frac{d^2z}{dt^2} = -\frac{\partial\psi}{\partial z}. \tag{80.10}$$

(b) Poisson's equation as an approximation for Einstein's field equations. To complete the justification for this interpretation of the Newtonian potential ψ, we must also show that Einstein's relativistic field equations will give us as a first approximation the same dependence of ψ on the distribution of matter as Poisson's equation in the Newtonian theory of gravitation.

To do this we note, in accordance with the expression for the energy-momentum tensor (37.8) given by the special theory of relativity in the coordinates (x^1, x^2, x^3, x^4), that the components of the energy-momentum tensor will in the present case all be approximately zero except for the component

$$T^{44} = c^2\rho,$$

provided we do have a weak static field as has been assumed, and provided the mechanical stresses p_{ij} are negligible compared with the density of energy as is true in ordinary applications. Hence on account of the specially simple values for the components of the metrical tensor $g_{\mu\nu}$ given by (80.2) which will be involved in the raising and lowering of suffixes, we can write for the application in hand

$$T^{44} = T^4_4 = T_{44} = T = c^2\rho, \qquad (80.11)$$

with all other components of the energy-momentum tensor $T_{\mu\nu}$ equal to zero.

Combining (80.11), moreover, with equations (78.14) and (78.15) we see for the case in hand that the relativistic field equations will now provide the simple result

$$-\kappa c^2\rho = R_{44} - \tfrac{1}{2}\kappa c^2\rho\, g_{44}$$

or

$$R_{44} = -\frac{\kappa c^2\rho}{2}. \qquad (80.12)$$

And examining the expression for $R_{\mu\nu}$ given by (77.1)—since products of the Christoffel three-index symbols can be neglected on account of the weakness of the field and derivatives with respect to the time are zero in a static field—we can rewrite this in the form

$$\frac{\partial}{\partial x^\sigma}\{44, \sigma\} = \frac{\kappa c^2\rho}{2},$$

or finally by the introduction of (80.6) in the form of the desired equation

$$\frac{\partial^2}{\partial x^2}\left(\frac{g_{44}}{2}\right) + \frac{\partial^2}{\partial y^2}\left(\frac{g_{44}}{2}\right) + \frac{\partial^2}{\partial z^2}\left(\frac{g_{44}}{2}\right) = \frac{\kappa c^2\rho}{2}. \qquad (80.13)$$

This equation, however, is in complete agreement with Poisson's equation (75.1) in the Newtonian theory of gravitation, provided we again take

$$\frac{\psi}{c^2} = \frac{g_{44}}{2} + \text{const.}$$

as in (80.9), and assign to the constant κ the value

$$\kappa = \frac{8\pi k}{c^4}, \tag{80.14}$$

where k is the ordinary constant of gravitation and c is the velocity of light.

We thus complete the demonstration that Newton's theory of gravitation can be regarded as a first approximation for the more complete theory of gravitation furnished by the general theory of relativity. Furthermore, since it can be shown to be an exceedingly close approximation—for fields of the strength encountered in the more usual applications of gravitational theory—we can now regard all the well-tested results of celestial mechanics as furnishing important support for the relativistic theory of gravitation.

81. Units to be used in relativistic calculations

Equation (80.14) provides a definite value, in terms of the usual constant of gravity k and the velocity of light c, for the constant κ in Einstein's field equations (78.3), and this makes it possible to obtain numerical conclusions from these equations for comparison with observational results. Substituting the values†

$$c = 2{\cdot}99796 \times 10^{10}\,\text{cm. sec.}^{-1} \quad \text{and} \quad k = 6{\cdot}664 \times 10^{-8}\,\text{cm.}^3\,\text{gm.}^{-1}\,\text{sec.}^{-2} \tag{81.1}$$

we obtain

$$\kappa = \frac{8\pi k}{c^4} = 2{\cdot}073 \times 10^{-48}\,\text{cm.}^{-1}\text{gm.}^{-1}\,\text{sec.}^2 \tag{81.2}$$

This value of κ is dependent in the first place on the fact that we are using the centimetre-gramme-second system of units, and in the second place on the fact that we have taken the components of the energy-momentum tensor $T^{\mu\nu}$ when referred to Galilean coordinates (37.8) as having the dimensions of energy density instead of mass density as is sometimes done.

By changing at this point, however, to a new system of units, some simplification can be introduced into the writing of relativistic equations and the effect of any arbitrary convention as to the dimensions of $T^{\mu\nu}$ can be eliminated. To do this we may retain the centimetre as the unit of length and then choose the units of time and mass so as to give the velocity of light in free space c, and the constant of

† Birge, *Phys. Rev.* Supplement, **1**, 1 (1929).

gravitation k the values unity

$$c = 1 \qquad k = 1. \tag{81.3}$$

With this choice of units, the energy and mass of a given system will have in accordance with (27.4) the same numerical value

$$E = m, \tag{81.4}$$

so that the two different proposals mentioned above for setting up the energy-momentum tensor will lead to the same numerical result. Furthermore, the constant κ will now have the simple value

$$\kappa = 8\pi, \tag{81.5}$$

and the relativistic field equations (78.8) can be written in the form

$$-8\pi T_{\mu\nu} = R_{\mu\nu} - \tfrac{1}{2} R g_{\mu\nu} + \Lambda g_{\mu\nu}. \tag{81.6}$$

In any computations which follow we shall assume the use of these units. The results of the computations, however, can easily be translated into c.g.s. units with the help of the following relations connecting the values for lengths, times, and masses L, T, M in c.g.s. units with their values l, t, m in the new units.

$$L = l \text{ cm.}$$

$$T = \frac{t}{2 \cdot 998 \times 10^{10}} = 3 \cdot 335 \times 10^{-11} t \text{ sec.} \tag{81.7}$$

$$M = \frac{(2 \cdot 998 \times 10^{10})^2}{6 \cdot 664 \times 10^{-8}} m = 1 \cdot 349 \times 10^{28} m \text{ gm.}$$

82. The Schwarzschild line element

As a specially important application of the general theory of relativity, we may next consider the problem of obtaining an expression for the line element or formula for interval in the empty space surrounding a gravitating point particle. The complete solution of this problem was first obtained by Schwarzschild† and is of great significance, since it provides a treatment of the gravitational field surrounding the sun for use in discussing the three crucial tests, that distinguish between the predictions of the Newtonian theory of gravitation and the more exact predictions of the theory of relativity.

The methods of solving this problem are so well known that it will be sufficient for our purposes to indicate the derivation. In accordance with the static and spherically symmetrical nature of the field which would surround an attracting point particle, it can be shown

† Schwarzschild, *Berl. Ber.* 1916, p. 189.

necessarily possible (see § 95) to choose coordinates r, θ, ϕ, and t such that the line element will be of the simple form

$$ds^2 = -e^\lambda\, dr^2 - r^2\, d\theta^2 - r^2\sin^2\theta\, d\phi^2 + e^\nu\, dt^2, \tag{82.1}$$

where λ and ν are functions of r alone. Furthermore, the components of the energy-momentum tensor T^ν_μ corresponding to this formula for interval [see the later equations (95.3)] are known to have the values

$$8\pi T^1_1 = -e^{-\lambda}\left(\frac{\nu'}{r} + \frac{1}{r^2}\right) + \frac{1}{r^2}$$

$$8\pi T^2_2 = 8\pi T^3_3 = -e^{-\lambda}\left(\frac{\nu''}{2} - \frac{\lambda'\nu'}{4} + \frac{\nu'^2}{4} + \frac{\nu'-\lambda'}{2r}\right)$$

$$8\pi T^4_4 = e^{-\lambda}\left(\frac{\lambda'}{r} - \frac{1}{r^2}\right) + \frac{1}{r^2} \tag{82.2}$$

$$8\pi T^\nu_\mu = 0 \quad (\mu \neq \nu),$$

where the primes indicate differentiation with respect to r, and the cosmological constant Λ has been taken as zero.

In the empty space surrounding our particle, nevertheless, all the components of the energy-momentum tensor will evidently be equal to zero. Combining the first and third of these equations, this then leads to the result

$$\lambda' = -\nu', \tag{82.3}$$

and, combining with the second of the above equations, we obtain

$$\nu'' + \nu'^2 + \frac{2\nu'}{r} = 0, \tag{82.4}$$

which is easily seen to have the solution

$$e^\nu = a + \frac{b}{r}, \tag{82.5}$$

where a and b are constants of integration.

At great distances from the particle, however, where r approaches infinity, we must expect the line element to approach the special relativity form

$$ds^2 = -dr^2 - r^2\, d\theta^2 - r^2\sin^2\theta\, d\phi^2 + dt^2 \tag{82.6}$$

with $e^\lambda = e^\nu = 1$, in the units adopted in the preceding section. As a result, equation (82.5) can then be rewritten in the form

$$e^\nu = 1 - \frac{2m}{r}, \tag{82.7}$$

where the constant a has been set equal to unity and the constant b has been called $-2m$ to correspond to the later physical interpretation of m as the mass of the particle. And as a further result, in accordance with (82.3), we must also evidently have

$$e^{-\lambda} = 1 - \frac{2m}{r}, \qquad (82.8)$$

and these expressions when introduced into the first and third of equations (82.2) will also secure the values zero for T_1^1 and T_4^4.

Hence, substituting (82.7) and (82.8) into the general expression (82.1), we can now write, for the line element in the neighbourhood of an attracting point particle, the Schwarzschild solution

$$ds^2 = - \frac{dr^2}{1-2m/r} - r^2 \, d\theta^2 - r^2 \sin^2\theta \, d\phi^2 + \left(1 - \frac{2m}{r}\right) dt^2. \qquad (82.9)$$

Since this result has been obtained with the help of expressions (82.2) for the components of the energy-momentum tensor T_μ^ν in which the cosmological constant Λ has been set equal to zero, the solution corresponds, in accordance with (78.12) and (78.13), to Einstein's original field equations for the case of empty space

$$R_{\mu\nu} = 0.$$

It is also easily possible, however, to employ the complete expressions for the energy-momentum tensor (see 95.3) without omitting the cosmological term, and this is found to lead to the result

$$ds^2 = - \frac{dr^2}{1 - \frac{2m}{r} - \frac{\Lambda}{3}r^2} - r^2 \, d\theta^2 - r^2 \sin^2\theta \, d\phi^2 + \left(1 - \frac{2m}{r} - \frac{\Lambda}{3}r^2\right) dt^2, \qquad (82.10)$$

which corresponds to the more general possibility for the field equations in empty space $\quad R_{\mu\nu} = \Lambda g_{\mu\nu}.$

Comparing the two expressions for the line element (82.9) and (82.10), we now see as already remarked in § 78 that the effect of the Λ term on the field surrounding an attracting point particle would increase with the size of the region considered. Hence, since the motions of the planets are actually given with great accuracy by (82.9), we can conclude that Λ is in any case small enough not to produce appreciable effects within a region of the order of size of the solar system.

The particular form for the Schwarzschild line element given by (82.9) is of course dependent on the coordinate system which is being used, and the forms which it assumes in other coordinate systems are sometimes more useful. Substituting $\bar{r}$ in place of r with the help of the relation

$$r = \left(1 + \frac{m}{2\bar{r}}\right)^2 \bar{r} \tag{82.11}$$

we readily obtain the Schwarzschild line element in the form

$$ds^2 = -\left(1 + \frac{m}{2\bar{r}}\right)^4 (d\bar{r}^2 + \bar{r}^2\, d\theta^2 + \bar{r}^2 \sin^2\theta\, d\phi^2) + \frac{(1 - m/2\bar{r})^2}{(1 + m/2\bar{r})^2}\, dt^2, \tag{82.12}$$

and by substituting 'rectangular coordinates'

$$x = \bar{r}\sin\theta\cos\phi \qquad y = \bar{r}\sin\theta\sin\phi \qquad z = \bar{r}\cos\theta, \tag{82.13}$$

this can be written in the form

$$ds^2 = -\left(1 + \frac{m}{2r}\right)^4 (dx^2 + dy^2 + dz^2) + \frac{(1 - m/2r)^2}{(1 + m/2r)^2}\, dt^2, \tag{82.14}$$

where we now have $r = \sqrt{(x^2 + y^2 + z^2)}$.

These new coordinates may be called isotropic, since the formula for interval is symmetrical in x, y, and z. At great distances from the central particle where terms of the order of $(m/r)^2$ and higher can be neglected in comparison with unity, this expression for the Schwarzschild line element reduces to the approximate form

$$ds^2 = -\left(1 + \frac{2m}{r}\right)(dx^2 + dy^2 + dz^2) + \left(1 - \frac{2m}{r}\right) dt^2 \tag{82.15}$$

with $\qquad\qquad\qquad r = \sqrt{(x^2 + y^2 + z^2)}.$

83. The three crucial tests of relativity

We must now turn to the actual correspondence between the Schwarzschild expression for the line element surrounding an attracting point particle and the observational facts of astronomy. The methods of investigating this correspondence are so well known that it will be sufficient for our purposes merely to indicate the treatment.†

We may first consider the motion of the planets in the gravitational field of the sun. Since the planets can be regarded as free particles, their space-time trajectories will be given in accordance with the theory of relativity [see § 74 (e)] by the equations for a geodesic (74.13)

$$\frac{d^2 x^\sigma}{ds^2} + \{\mu\nu, \sigma\}\frac{dx^\mu}{ds}\frac{dx^\nu}{ds} = 0. \tag{83.1}$$

Since the field surrounding the sun can be regarded as that due to an attracting point particle, the values of the Christoffel three-index

† We follow the treatment of Eddington, *The Mathematical Theory of Relativity*, Cambridge, 1923.

symbols to be used in (83.1) will be those for the Schwarzschild line element, which we shall use in the form in which it was derived (82.9) and these—in so far as they do not vanish—are known to be given (see 95.2) by the expressions

$$\{11, 1\} = \frac{1}{2}\frac{d\lambda}{dr} \quad \{21, 2\} = \frac{1}{r} \qquad \{31, 3\} = \frac{1}{r} \qquad\qquad \{41, 4\} = \frac{1}{2}\frac{d\nu}{dr}$$

$$\{12, 2\} = \frac{1}{r} \qquad \{22, 1\} = -re^{-\lambda} \quad \{32, 3\} = \cot\theta \qquad \{44, 1\} = \frac{e^{\nu-\lambda}}{2}\frac{d\nu}{dr}$$

$$\{13, 3\} = \frac{1}{r} \qquad \{23, 3\} = \cot\theta \quad \{33, 1\} = -r\sin^2\theta\, e^{-\lambda}$$

$$\{14, 4\} = \frac{1}{2}\frac{d\nu}{dr} \qquad\qquad\qquad\qquad \{33, 2\} = -\sin\theta\cos\theta, \qquad (83.2)$$

provided we substitute for λ and ν the values for the Schwarzschild line element given by (82.7) and (82.8).

Introducing these expressions for the Christoffel symbols into (83.1), we then obtain for the four possible cases $\sigma = 1, 2, 3, 4$

$$\frac{d^2r}{ds^2} + \frac{1}{2}\frac{d\lambda}{dr}\left(\frac{dr}{ds}\right)^2 - re^{-\lambda}\left(\frac{d\theta}{ds}\right)^2 - r\sin^2\theta\, e^{-\lambda}\left(\frac{d\phi}{ds}\right)^2 + \frac{e^{\nu-\lambda}}{2}\frac{d\nu}{dr}\left(\frac{dt}{ds}\right)^2 = 0, \quad (83.3)$$

$$\frac{d^2\theta}{ds^2} + \frac{2}{r}\frac{dr}{ds}\frac{d\theta}{ds} - \sin\theta\cos\theta\left(\frac{d\phi}{ds}\right)^2 = 0, \qquad (83.4)$$

$$\frac{d^2\phi}{ds^2} + \frac{2}{r}\frac{dr}{ds}\frac{d\phi}{ds} + 2\cot\theta\frac{d\phi}{ds}\frac{d\theta}{ds} = 0, \qquad (83.5)$$

$$\frac{d^2t}{ds^2} + \frac{d\nu}{dr}\frac{dr}{ds}\frac{dt}{ds} = 0, \qquad (83.6)$$

as the equations which would govern the motion of a planet. These equations can be readily simplified, however, by choosing coordinates such that the planet is originally moving in the plane $\theta = \frac{1}{2}\pi$. This will make $d\theta/ds$ and $\cos\theta$ both initially equal to zero and hence in accordance with (83.4) permanently equal to zero, so that the equations will reduce to the simpler form

$$\frac{d^2r}{ds^2} + \frac{1}{2}\frac{d\lambda}{dr}\left(\frac{dr}{ds}\right)^2 - re^{-\lambda}\left(\frac{d\phi}{ds}\right)^2 + \frac{e^{\nu-\lambda}}{2}\frac{d\nu}{dr}\left(\frac{dt}{ds}\right)^2 = 0, \qquad (83.7)$$

$$\frac{d^2\phi}{ds^2} + \frac{2}{r}\frac{dr}{ds}\frac{d\phi}{ds} = 0, \qquad (83.8)$$

$$\frac{d^2t}{ds^2} + \frac{d\nu}{ds}\frac{dt}{ds} = 0. \qquad (83.9)$$

These equations can now easily be handled since the original expression (82.1) for the line element itself provides one integral and two of the equations, (83.8) and (83.9), can readily be integrated by inspection. We thus obtain

$$e^\lambda\left(\frac{dr}{ds}\right)^2+r^2\left(\frac{d\phi}{ds}\right)^2-e^\nu\left(\frac{dt}{ds}\right)^2+1 = 0,$$

$$\frac{d\phi}{ds} = \frac{h}{r^2},$$

$$\frac{dt}{ds} = ke^{-\nu}$$

as the first integrals of the above equations, where h and k are constants of integration. And by combining the first and third of these equations, and substituting the values for λ and ν given by (82.7) and (82.8), we obtain as the relativistic equations for the motion of a planet

$$\left(\frac{dr}{ds}\right)^2+r^2\left(\frac{d\phi}{ds}\right)^2-\frac{2m}{r}\left(1+r^2\frac{d\phi^2}{ds^2}\right) = k^2-1 \qquad (83.10)$$

$$r^2\frac{d\phi}{ds} = h, \qquad (83.11)$$

where r and ϕ are the spatial coordinates originally introduced, m and k are constants, and ds is an element of proper time as measured by a local clock moving with the planet.

This puts the relativistic equations for the orbit of a planet in a form suitable for comparison with the two Newtonian equations, resulting from the application of the ordinary laws for the conservation of energy and of angular momentum,

$$\left(\frac{dr}{dt}\right)^2+r^2\left(\frac{d\phi}{dt}\right)^2-\frac{2m}{r} = \text{const.} \qquad (83.12)$$

$$r^2\frac{d\phi}{dt} = \text{const.}, \qquad (83.13)$$

where m is the mass of the sun expressed in the units of § 81, and where we must now regard r and ϕ as ordinary polar coordinates and dt as an ordinary time interval as used in prerelativistic considerations which neglected the possibility of effects of motion and of curvature on spatial and temporal measurements.

Since these effects of motion and of curvature would be extremely small for the slow velocities of the planets and in the nearly 'flat' space-time surrounding the sun,† and since the added term $r^2d\phi^2/ds^2$ occurring in (83.10) would be very small compared with unity, being as it is the square of the transverse velocity of the planet divided by the square of the velocity of light, the reasons can now be appreciated for the high order of accuracy obtained in the application of the Newtonian theory of gravitation to the field of celestial mechanics.

There are, nevertheless, three consequences which can be obtained from the Schwarzschild line element which can be used to distinguish between the relativistic and Newtonian theories of gravitation. To these we must now turn our attention.

(a) **The advance of perihelion.** The first of these three crucial tests of relativity is made possible by the fact that the added term in the relativistic equation (83.10), as compared with the analogous Newtonian equation (83.12), leads to planetary orbits with a slow rotation of perihelion instead of to the perfectly closed elliptical orbits of the older theory.

Substituting (83.11) into (83.10), differentiating with respect to ϕ, and for simplicity putting

$$u = \frac{1}{r}, \tag{83.14}$$

we can easily obtain

$$\frac{d^2u}{d\phi^2} + u = \frac{m}{h^2} + 3mu^2 \tag{83.15}$$

---

† In accordance with the Schwarzschild line element, the spatial geometry around the sun would be characterized by the formula for interval

$$du^2 = \frac{dr^2}{1-2m/r} + r^2\,d\theta^2 + r^2\sin^2\theta\,d\phi^2$$

instead of by the usual formula for flat space

$$du^2 = dr^2 + r^2\,d\theta^2 + r^2\sin^2\theta\,d\phi^2.$$

Even at the surface of the sun, however, the term $2m/r$ would be only about 4×10^{-6} and at the distance of the earth would drop to about 2×10^{-8}. Hence the space around the sun is sufficiently flat so that the coordinates r, θ, and ϕ for the position of a planet could not be distinguished from the values assigned on the basis of considerations that neglect spatial curvature.

Furthermore, in accordance with the form of the line element, the relation between increments in proper time ds as measured on the planet and in coordinate time dt would be given by

$$\frac{ds^2}{dt^2} = 1 - \left(\frac{1}{1-2m/r}\frac{dr^2}{dt^2} + r^2\frac{d\theta^2}{dt^2} + r^2\sin^2\theta\frac{d\phi^2}{dt^2} + \frac{2m}{r}\right),$$

where the second term would be very small compared with unity, being for example about 3×10^{-8} in the case of the earth. Hence the two kinds of time could also not be distinguished in describing the orbit of a planet.

Our present considerations give a concrete illustration of the fact that deviations of the $g_{\mu\nu}$ from their Galilean values, which are small from the metrical point of view, can be very important from the gravitational point of view.

as a relativistic equation for a planetary orbit, to be compared with the analogous Newtonian equation

$$\frac{d^2u}{d\phi^2} + u = \frac{m}{h^2}. \tag{83.16}$$

Since the added term 3^-mu^2 on the right-hand side of (83.15) is easily seen with the help of (83.11) to be very small compared with m/h^2, the difference between the relativistic and Newtonian equations is only slight. Hence in solving the relativistic equation (83.15) we may take as a first approximation the well-known solution of the Newtonian equation (83.16)

$$u = \frac{m}{h^2}\{1 + e\cos(\phi - \omega)\}, \tag{83.17}$$

where e is the eccentricity of the orbit and ω the longitude of perihelion. By substituting this solution back into (83.15) it then becomes possible to obtain

$$u = \frac{m}{h^2}\left\{1 + e\cos\left(\phi - \omega - \frac{3m^2}{h^2}\phi\right)\right\} \tag{83.18}$$

as a satisfactory second approximation. This result can then be interpreted by assigning per revolution of the planet an advance in the longitude of its perihelion of the amount

$$\delta\omega = \frac{6\pi m^2}{h^2}. \tag{83.19}$$

Mercury is the only one of the solar planets for which the predicted advance is sufficient to be observationally determinable with certainty. The predicted advance in the longitude of perihelion amounts in the case of Mercury to 42·9 seconds of arc per century and the observational advance to 43·5 seconds.† The agreement must be regarded as satisfactory.

(b) **The gravitational deflexion of light.** The second of the three crucial tests of relativity is furnished by the deflexion of light in passing through the gravitational field in the neighbourhood of the sun.

In accordance with the general theory of relativity [see § 74 (e)], the trajectory of a light ray as well as that of a free particle should be governed by the equations for a geodesic, with the added condition $ds = 0$ for the interval associated with the motion. Hence, by

† Chazy, *Comptes Rendus,* **182,** 1134 (1926).

introducing this further condition, our previous equation for the orbit of a planet (83.15) should also be applicable to the path of light rays in the field of an attracting point particle. In accordance with (83.11), it is evident, moreover, that the effect of this further condition can be obtained by setting $h = \infty$ in (83.15) and writing

$$\frac{d^2u}{d\phi^2} + u = 3mu^2 \qquad (83.20)$$

with
$$u = \frac{1}{r} \qquad (83.21)$$

as an equation for the path of a ray of light in the neighbourhood of an attracting point particle of mass m.

In the absence of the disturbing term $3mu^2$, the solution for (83.20) could be taken as the equation

$$r\cos\phi = R \qquad (83.22)$$

for a straight line which passes the attracting point at the distance R. And by substituting (83.22) back into (83.20) it is possible to obtain

$$r\cos\phi = R - \frac{m}{R}(r\cos^2\phi + 2r\sin^2\phi) \qquad (83.23)$$

as a satisfactory second approximation. Changing to Cartesian coordinates, which can be taken as approximately valid in the nearly Euclidean space surrounding the sun, this can be rewritten in the form

$$x = R - \frac{m}{R}\frac{x^2 + 2y^2}{\sqrt{(x^2 + y^2)}}. \qquad (83.24)$$

For large values of y this gives us

$$x = R - \frac{m}{R}(\pm 2y),$$

where the upper sign is to be used for y positive and the lower sign for y negative. Hence for the angle between the asymptotic directions of the ray we obtain

$$\theta = \frac{4m}{R}. \qquad (83.25)$$

For a ray of light which grazes the sun's limb this leads to an angle of deflexion of 1·75 seconds of arc. This prediction can be tested by observations made at times of total eclipse on the apparent positions of stars whose light has passed close to the limb of the sun. The results must be regarded as in exceedingly satisfactory correspondence with theory. The first and quite good checks on the relativistic theory were obtained by the British eclipse expeditions of 1919, and the most satisfactory data at present are presumably those of Campbell and Trumpler† who obtained the results $1 \cdot 72'' \pm 0 \cdot 11''$ and $1 \cdot 82'' \pm 0 \cdot 15''$ with two different sizes of cameras in the 1922 expedition of the Lick Observatory.

It is interesting to point out that the relativistic expression for the deflexion of light passing a mass m, as given by (83.25), is twice as great as would be calculated from the *simple* Newtonian theory for a particle travelling with the velocity of light.

To obtain the Newtonian result we may consider the particle to be travelling approximately parallel to the y-axis and to pass the mass m at the distance $x = R$. For the acceleration in the x-direction we can then write

$$\frac{d^2x}{dt^2} = -\frac{mx}{(x^2+y^2)^{\frac{3}{2}}},$$

or with sufficient approximation for our present purposes

$$\frac{d^2x}{dy^2} = -\frac{mR}{(R^2+y^2)^{\frac{3}{2}}},$$

solving and choosing the constants of integration so as to make $dx/dy = 0$ and $x = R$ at $y = 0$, we easily obtain as the approximate trajectory for large values of y

$$x = R - \frac{m}{R}(\pm y),$$

giving for the angle between the asymptotic directions

$$\theta = \frac{2m}{R},$$

which is half the previous result (83.25).

This large difference between the relativistic and quasi-Newtonian results makes the observational check of the former all the more significant.

(c) **Gravitational shift in spectral lines.** As the third crucial test of the general theory of relativity, we have the dependence of the wavelength of light on the gravitational potential of its source, already approximately treated in § 79 (b) with the help of the principle of equivalence. Making use of the Schwarzschild line element, we may now investigate somewhat more in detail the shift that would be

† Campbell and Trumpler, *Lick Observatory Bull.* **11**, 41 (1923); **13**, 130 (1928).

expected in the period of spectral lines originating at the surface of the sun or other star. This can be done very easily.

On the one hand, in accordance with the Schwarzschild line element (82.9) and the relation $ds = 0$ for the trajectory of light, we note that the velocity of light originating at the surface of the star would be given in terms of the coordinates r and t by the expression

$$\frac{dr}{dt} = \left(1 - \frac{2m}{r}\right), \tag{83.26}$$

which is seen to be independent of the coordinate t. We may hence conclude that successive light impulses which are separated by the coordinate period δt when they originate at the surface of the star would still be separated by this coordinate period on reaching a stationary observer.

On the other hand, we note in accordance with the Schwarzschild line element that the proper period δs for a stationary atom and its coordinate period δt would be connected by the relation

$$\delta s = \sqrt{\left(1 - \frac{2m}{r}\right)}\, \delta t. \tag{83.27}$$

Hence since the proper period of an atom should be independent of its location, and since we have seen above that the coordinate period of light is in the present case unaltered by transmission, we can now write

$$\frac{\lambda + \delta\lambda}{\lambda} = \frac{\delta t}{\delta s} = \frac{1}{\sqrt{\{1 - (2m/r)\}}} \simeq 1 + \frac{m}{r} \tag{83.28}$$

for the ratio of the observed wave-lengths of light corresponding to a given spectral line which originates in the one case at the surface of the star at r and in the other case at a great distance from the star where the observer himself is located.

In the case of light originating on the surface of the sun this should lead to a very small shift towards the red to the extent

$$\frac{\delta\lambda}{\lambda} = 2{\cdot}12 \times 10^{-6}. \tag{83.29}$$

In the case of the very dense companion to Sirius, however, the shift should be about thirty times as great. In both cases the agreement between theory and observation is now satisfactory as a result of the work of St. John[†] and of Adams.[‡]

[†] St. John, *Astrophysical Journ.* **67**, 195 (1928).
[‡] Adams, *Proc. Nat. Acad.* **11**, 382 (1925).

The satisfactory verification of the general theory of relativity provided by these three crucial tests may well be emphasized. The verification is all the more significant, since the advance in the perihelion of Mercury was the only one of the three phenomena in question which was actually known at the time when Einstein's theory was developed, and the effects of gravitation both in determining the path and wave-length of light had not even been observed as qualitative phenomena prior to their prediction by the theory of relativity.

It is also remarkable that Einstein's development of relativity was in no sense the result of a mere attempt to account for a small known difference between the observed orbit of Mercury and that predicted by Newtonian theory, but was the full flowering of a complicated theoretical structure, growing from fundamental principles whose main justification seemed to lie in their inherent qualities of reasonableness and generality. The extraordinary success of a theory, obtained by those methods of intellectualistic approach, whose dangers have been so evident since the time of Galileo, well bespeaks the genius of the founder.

The observational verification which the theory of relativity has already received must make us regard it as a distinct advance over Newtonian theory, and can encourage us to proceed now to the consideration of further developments where the possibilities for observational verification are not always immediately present.

VII

RELATIVISTIC MECHANICS

Part I. SOME GENERAL MECHANICAL PRINCIPLES

84. The fundamental equations of relativistic mechanics

In the present chapter we shall undertake a somewhat detailed development of certain consequences of relativistic mechanics that are needed for our further work or that appear to be especially illuminating. These consequences are all implicitly contained in the field equation of Einstein

$$-8\pi T^{\mu\nu} = R^{\mu\nu} - \tfrac{1}{2}Rg^{\mu\nu} + \Lambda g^{\mu\nu}, \tag{84.1}$$

which connects the distribution of matter and energy with the geometry of space-time, by relating the energy-momentum tensor $T_{\mu\nu}$ to the fundamental metrical tensor $g_{\mu\nu}$ and its derivatives. And it is the business of relativistic mechanics to investigate with the help of this equation the principles which govern the energy-momentum tensor, and hence determine the behaviour of matter and energy.

For many purposes the full import of the above equation will not be necessary. The right-hand side of (84.1) gives a quantity whose tensor divergence is known to be identically equal to zero. Hence, we may write as an immediate consequence of (84.1)

$$(T^{\mu\nu})_{\nu} = 0, \tag{84.2}$$

and from this simple equation alone we can then draw many important conclusions as to the behaviour of matter and energy. Indeed, since this equation reduces in natural coordinates to the form

$$\frac{\partial T^{\mu\nu}}{\partial x^{\nu}} = 0, \tag{84.3}$$

which we took in § 37 as an expression for the fundamental equations of mechanics in special relativity, it will now be natural to refer to (84.2) as the general relativity expression for the fundamental equations of mechanics.

Expanding this expression in accordance with the rules for covariant differentiation, the fundamental equations of mechanics can also be written in the form

$$\frac{\partial T^{\mu\nu}}{\partial x^{\nu}} + \{\alpha\nu, \mu\}T^{\alpha\nu} + \{\alpha\nu, \nu\}T^{\mu\alpha} = 0, \tag{84.4}$$

and by lowering the suffix μ in the form

$$\frac{\partial T_\mu^\nu}{\partial x^\nu} - \{\mu\nu, \alpha\}T_\alpha^\nu + \{\alpha\nu, \nu\}T_\mu^\alpha = 0. \tag{84.5}$$

Furthermore, by introducing in place of the energy-momentum tensor the corresponding tensor-density

$$\mathfrak{T}_\mu^\nu = T_\mu^\nu \sqrt{-g} \tag{84.6}$$

this can be rewritten in accordance with well-known transformations (equation (47), Appendix III) in the simpler forms

$$\frac{\partial \mathfrak{T}_\mu^\nu}{\partial x^\nu} - \tfrac{1}{2}\mathfrak{T}^{\alpha\beta}\frac{\partial g_{\alpha\beta}}{\partial x^\mu} = 0 \tag{84.7}$$

and

$$\frac{\partial \mathfrak{T}_\mu^\nu}{\partial x^\nu} + \tfrac{1}{2}\mathfrak{T}_{\alpha\beta}\frac{\partial g^{\alpha\beta}}{\partial x^\mu} = 0. \tag{84.8}$$

85. The nature of the energy-momentum tensor. General expression in the case of a perfect fluid

In order to obtain physical conclusions from these fundamental equations of mechanics, we must of course apply them to some particular kind of physical medium for which we actually know the dependence of the energy-momentum tensor on observable properties of the medium. We shall hence desire explicit expressions for the energy-momentum tensor $T^{\mu\nu}$ in terms of quantities which can be determined by ordinary methods of measurement. In accordance with the principle of equivalence, we can obtain such expressions by the covariant generalization of expressions for the energy-momentum tensor which have already been provided in the special theory of relativity.

In the case of a purely mechanical medium, whose state at any point can be specified by the mechanical stresses p_{ij}^0 and density ρ_{00} as measured by a local observer, we have already found in the special theory of relativity that the energy-momentum tensor can be defined, by taking†

$$T_0^{\alpha\beta} = \begin{matrix} p_{xx}^0 & p_{xy}^0 & p_{xz}^0 & 0 \\ p_{yx}^0 & p_{yy}^0 & p_{yz}^0 & 0 \\ p_{zx}^0 & p_{zy}^0 & p_{zz}^0 & 0 \\ 0 & 0 & 0 & \rho_{00} \end{matrix} \tag{85.1}$$

as the components of the energy-momentum tensor in a special set

† See § 37. Note that we have set $c^2 = 1$ in agreement with the units adopted in § 81.

of Galilean coordinates so chosen that the material is at rest in these coordinates at the position and time of interest. Turning to the general theory of relativity, however, it is evident from the principle of equivalence that the energy-momentum tensor would also have to reduce to this same array in proper coordinates $(x_0^1, x_0^2, x_0^3, x_0^4)$ for any given point of interest. Hence the same array also gives us a definition of the energy-momentum tensor for a mechanical medium in general relativity by stating its components in a chosen system of proper coordinates. To obtain its components in any other coordinate system (x^1, x^2, x^3, x^4), we have then merely to employ the rule for tensor transformation

$$T^{\mu\nu} = \frac{\partial x^\mu}{\partial x_0^\alpha} \frac{\partial x^\nu}{\partial x^\beta} T_0^{\alpha\beta}, \qquad (85.2)$$

which allows us to compute the desired components—with the help of the derivatives connecting the new system of coordinates with the original proper system—in terms of the proper density ρ_{00} and stresses p_{ij}^0 as measured by a local observer using ordinary physical methods. Vice versa, if we know the components of the energy-momentum in a given set of coordinates, the possibility is presented of calculating the proper stresses and density with the help of the reverse transformation.

Although the foregoing equation (85.2) gives us a general expression for the energy-momentum tensor of a mechanical medium in any desired system of coordinates, its actual content will be dependent on the derivatives connecting these coordinates with some set of proper coordinates. In the case of a perfect fluid, however, it is readily possible to introduce substitutions which will eliminate the explicit appearance of the proper coordinates, and give an expression which depends in a clearer way on the actual coordinate system which is being employed.

In the case of a *perfect fluid*, which we define as a mechanical medium incapable of exerting transverse stresses, the only components of stress for a local observer will be those corresponding to the proper hydrostatic pressure p_0, so that the energy-momentum tensor will then have in proper coordinates the simple set of components

$$T_0^{11} = T_0^{22} = T_0^{33} = p_0 \qquad T_0^{44} = \rho_{00}$$
$$T^{\alpha\beta} = 0 \quad (\alpha \neq \beta), \qquad (85.3)$$

and substituting these values into the general expression for the energy-momentum tensor given by (85.2), we can then write as an expression for it

$$T^{\mu\nu} = \frac{\partial x^{\mu}}{\partial x_0^1}\frac{\partial x^{\nu}}{\partial x_0^1}p_0 + \frac{\partial x^{\mu}}{\partial x_0^2}\frac{\partial x^{\nu}}{\partial x_0^2}p_0 + \frac{\partial x^{\mu}}{\partial x_0^3}\frac{\partial x^{\nu}}{\partial x_0^3}p_0 + \frac{\partial x^{\mu}}{\partial x_0^4}\frac{\partial x^{\nu}}{\partial x_0^4}\rho_{00}, \quad (85.4)$$

where $(x_0^1, x_0^2, x_0^3, x_0^4)$ are proper coordinates for the point under consideration and (x^1, x^2, x^3, x^4) are the coordinates of actual interest.

To simplify this expression for the energy-momentum tensor, we can write in the first place for the contravariant components of the metrical tensor in the desired coordinates in terms of their values in proper coordinates

$$g^{\mu\nu} = \frac{\partial x^{\mu}}{\partial x_0^{\alpha}}\frac{\partial x^{\nu}}{\partial x_0^{\beta}}g_0^{\alpha\beta},$$

which on substituting the simple values in proper coordinates gives us

$$g^{\mu\nu} = -\frac{\partial x^{\mu}}{\partial x_0^1}\frac{\partial x^{\nu}}{\partial x_0^1} - \frac{\partial x^{\mu}}{\partial x_0^2}\frac{\partial x^{\nu}}{\partial x_0^2} - \frac{\partial x^{\mu}}{\partial x_0^3}\frac{\partial x^{\nu}}{\partial x_0^3} + \frac{\partial x^{\mu}}{\partial x_0^4}\frac{\partial x^{\nu}}{\partial x_0^4}. \quad (85.5)$$

And in the second place, we can evidently write for the macroscopic velocity of the fluid with respect to the desired coordinates

$$\frac{dx^{\mu}}{ds} = \frac{\partial x^{\mu}}{\partial x_0^1}\frac{dx_0^1}{ds} + \frac{\partial x^{\mu}}{\partial x_0^2}\frac{dx_0^2}{ds} + \frac{\partial x^{\mu}}{\partial x_0^3}\frac{dx_0^3}{ds} + \frac{\partial x^{\mu}}{\partial x_0^4}\frac{dx_0^4}{ds},$$

which reduces to

$$\frac{dx^{\mu}}{ds} = \frac{\partial x^{\mu}}{\partial x_0^4}, \quad (85.6)$$

owing to the value zero for the spatial components of velocity and the value unity for its temporal component in proper coordinates.

Substituting (85.5) and (85.6) in (85.4), we can then express the energy-momentum tensor for a perfect fluid in the very useful and general form

$$T^{\mu\nu} = (\rho_{00}+p_0)\frac{dx^{\mu}}{ds}\frac{dx^{\nu}}{ds} - g^{\mu\nu}p_0, \quad (85.7)$$

where ρ_{00} and p_0 are the proper macroscopic density and pressure of the fluid and the quantities dx^{μ}/ds are the components of the macroscopic velocity of the fluid with respect to the actual coordinate system that is being used.

Since a disordered distribution of radiation can be regarded as a perfect fluid characterized by its density and pressure, with the specially simple relation

$$\rho_{00} = 3p_0 \quad (85.8)$$

connecting these two quantities, see § 65, it also proves possible to use the above equation (85.7) together with this additional restriction as the expression for the energy-momentum tensor for such a distribution of radiation, provided we now take dx^μ/ds as the velocity—with respect to the coordinate system that is being used—of an observer who himself finds on the average no net flow of energy (see the later work of § 109).

We shall find the expression for the energy-momentum tensor of a perfect fluid given by (85.7) extremely useful in our later work. For more complicated mechanical media which exert transverse stresses, and for fluids in which heat flow is taking place it is not applicable. Furthermore, for electromagnetic fields which are more complicated than a disordered distribution of radiation we should have to use the more general expression for the energy-momentum tensor, which will be developed in the next chapter. Nevertheless, many important problems can be investigated with the help of models composed of a perfect fluid.

86. The mechanical behaviour of a perfect fluid

To illustrate the physical significance of the fundamental equations of mechanics which were discussed in § 84, we may now apply them to the case of a perfect fluid with the help of the expression for its energy-momentum tensor which we have just obtained. For simplicity and to obtain insight into the physical nature of the results we shall express them in terms of proper coordinates for some particular point of interest.

Using proper coordinates, it is evident that the general equations of mechanics (84.4) will reduce to the form

$$\frac{\partial T^{\mu\nu}}{\partial x^\nu} = 0, \tag{86.1}$$

owing to the null value for the Christoffel three-index symbols in proper coordinates. Furthermore, in proper coordinates the components of the metrical tensor will assume their Galilean values and their first differential coefficients will vanish at the point of interest so that we can write

$$g_{\mu\nu} = g^{\mu\nu} = \pm 1, 0 \qquad \frac{\partial g_{\mu\nu}}{\partial x^\alpha} = \frac{\partial g^{\mu\nu}}{\partial x^\alpha} = 0. \tag{86.2}$$

In addition, the spatial and temporal components of the velocity of the fluid will have the values

$$\frac{dx}{ds} = \frac{dy}{ds} = \frac{dz}{ds} = 0 \qquad \frac{dt}{ds} = 1 \tag{86.3}$$

at the point of interest. Finally, since we can write as a result of the general form of the formula for interval

$$1 = g_{11}\frac{dx}{ds}\frac{dx}{ds} + 2g_{12}\frac{dx}{ds}\frac{dy}{ds} + \ldots + g_{44}\frac{dt}{ds}\frac{dt}{ds},$$

it is evident that we must have as a consequence of differentiating both sides of this expression the relation

$$\frac{\partial}{\partial x^\alpha}\left(\frac{dt}{ds}\right) = 0 \tag{86.4}$$

at the point of interest, since the differentiation of all terms in the above expression except the last, followed by substitution of (86.2) and (86.3), would evidently lead to null results. Hence in proper coordinates at the point of interest the derivatives of the temporal component of velocity will vanish, although the derivatives of the spatial components will not in general be zero even at that point.

The foregoing equations together with our expression for the energy-momentum tensor of a perfect fluid

$$T^{\mu\nu} = (\rho_{00} + p_0)\frac{dx^\mu}{ds}\frac{dx^\nu}{ds} - g^{\mu\nu}p_0 \tag{86.5}$$

are all that is necessary for the investigation.

Substituting into (86.1) for the case $\mu = 1$, we easily obtain

$$\frac{\partial p_0}{\partial x} + (\rho_{00} + p_0)\frac{\partial}{\partial t}\left(\frac{dx}{ds}\right) = 0$$

as the only terms that survive, and in accordance with (86.3) and (86.4) this can be rewritten in the form

$$\frac{\partial p_0}{\partial x} + (\rho_{00} + p_0)\frac{du_x}{dt} = 0, \tag{86.6}$$

where du_x/dt is the acceleration of the fluid parallel to the x-axis. Remembering, however, the contribution to momentum to be expected from the work done by mechanical forces such as the pressure, as discussed in § 35, and not forgetting that the velocity of the fluid is zero at the point of interest in the coordinates which we are using, it is at once evident that this result is what would be expected as a consequence of the usual relation between force and rate of change of momentum. Similar equations will of course be obtained for the cases $\mu = 2$ and 3.

For the case $\mu = 4$, we obtain by substitution into (86.1) as the only surviving terms

$$\frac{\partial}{\partial x}\left[(\rho_{00}+p_0)\frac{dx}{ds}\frac{dt}{ds}\right]+\frac{\partial}{\partial y}\left[(\rho_{00}+p_0)\frac{dy}{ds}\frac{dt}{ds}\right]+$$

$$+\frac{\partial}{\partial z}\left[(\rho_{00}+p_0)\frac{dz}{ds}\frac{dt}{ds}\right]+\frac{\partial}{\partial t}\left[(\rho_{00}+p_0)\left(\frac{dt}{ds}\right)^2\right]-\frac{\partial p_0}{\partial t}=0,$$

and in accordance with (86.3) and (86.4), this can be rewritten in the form

$$\rho_{00}\left(\frac{\partial u_x}{\partial x}+\frac{\partial u_y}{\partial y}+\frac{\partial u_z}{\partial z}\right)+\frac{\partial \rho_{00}}{\partial t}+p_0\left(\frac{\partial u_x}{\partial x}+\frac{\partial u_y}{\partial y}+\frac{\partial u_z}{\partial z}\right)=0$$

which on multiplication by

$$\delta v_0 = \delta x \delta y \delta z, \tag{86.7}$$

where δv_0 is the proper volume of an element of the fluid, gives us

$$\frac{d}{dt}(\rho_{00}\,\delta v_0)+p_0\frac{d}{dt}(\delta v_0)=0, \tag{86.8}$$

which states that for proper coordinates the rate of change in the energy of an element of the fluid can be calculated in the expected way from the rate at which work is being done against the external pressure.

Using proper coordinates, the application of the equations of relativistic mechanics to a perfect fluid thus leads to expressions which have an immediate interpretation in terms of physical measurements. Furthermore, for a local observer at rest in the fluid, who examines an element of the fluid small enough so that gravitational curvature can be neglected, we find the same laws of mechanical behaviour as we should predict from our previous knowledge of the energy-momentum principle. In addition, using natural coordinates in which the fluid is not at rest at the point of interest, results can readily be obtained which are in agreement with what would be expected from the more complicated expressions for the energy, momentum, and stress of a moving fluid which were developed for the special theory of relativity in Chapter III. In more general coordinates, however, the physical interpretation of the equations of mechanics will be less direct as we shall see in later sections.

As we have shown above, a local observer at rest in a fluid, who examines an element of the surrounding medium small enough so that gravitational curvature can be neglected, will find the same *mechanical* behaviour as we should have been inclined to predict from our previous knowledge of the energy-momentum principle. Furthermore, it may be emphasized that we can expect, as a result of the principle of equivalence, some similar familiar findings in the case of a local observer who measures the *electrodynamic* or *thermodynamic* behaviour of a small element of the fluid in his immediate neighbourhood. Nevertheless, it must not be forgotten, in the general theory of relativity, that proper coordinates can be expected to lead to simple relations only in the immediate neighbourhood of a selected point. The mistake must not be made, for example, of supposing that the energy-momentum principle in its special relativity form would hold in general relativity over a region of finite size.

This can be well illustrated with the help of the equation

$$\frac{d}{dt_0}(\rho_{00}\delta v_0) + p_0 \frac{d}{dt_0}(\delta v_0) = 0, \tag{86.9}$$

which we have just derived, connecting the rate of energy change which a local observer would find for a small element of fluid in his neighbourhood with the rate at which work is being done on the surroundings. The result for the individual element agrees with what we might expect from our usual ideas as to the conservation and transfer of energy. It should be noted, however, that this same equation can evidently be applied to each one of all the elements into which the total fluid of a finite system could be divided. And, with positive pressure throughout, we shall later find that this leads to possibilities for isolated systems in which the proper energy $(\rho_{00}\,\delta v_0)$ of *every* element of the fluid is decreasing when the system is expanding or increasing when the system is contracting.

This fact that the sum total of the proper energies of the elements of a fluid which make up an isolated system is not necessarily constant seems at first sight quite strange. It corresponds, however, in Newtonian gravitational theory to the necessity of ascribing potential energy to the gravitational field in order to preserve the principle of the conservation of energy. And in the next two sections we shall see how the analogous treatment of potential gravitational energy *and* momentum is to be carried out in the general theory of relativity.

The result that the sum total of the proper energies of the elements of a fluid is not necessarily conserved proves to be of great importance for relativistic thermodynamics. We shall later see (§ 131) that this removes restrictions on the possibilities for entropy increase in an isolated system which were imposed in the classical thermodynamics by the ordinary principle of the conservation of energy.

In conclusion, it should be specially noted in accordance with (86.8) that perfect fluids have been defined in such a way as to behave adiabatically when examined by a local observer—the proper energy of an element of the fluid being subject to change as a result of external work but not as the result of heat-flow. This circumstance should be kept in mind in using perfect fluids for the construction of conceptual models.

87. Re-expression of the equations of mechanics in the form of an ordinary divergence

Although mechanical principles can often be applied most readily with the help of the forms in which we have already expressed the fundamental equations of relativistic mechanics in § 84, it will be necessary to re-express these equations in the form of an ordinary divergence in order to obtain for finite systems the analogues of the classical principles for the conservation of energy and momentum. This can be done with the help of a somewhat lengthy but well-known consideration which we may now consider in outline.[†]

We first define the so-called Lagrangian function $\mathfrak{L}$ in terms of the Christoffel three-index symbols by the equation

$$\mathfrak{L} = \sqrt{-g}\, g^{\mu\nu}[\{\mu\alpha,\beta\}\{\nu\beta,\alpha\}-\{\mu\nu,\alpha\}\{\alpha\beta,\beta\}]. \tag{87.1}$$

Since the combination of Christoffel symbols appearing inside the square brackets is not itself a tensor, the quantity $\mathfrak{L}$ is not a scalar density. Nevertheless, since equation (87.1) is taken as defining $\mathfrak{L}$ in all systems of coordinates, we shall be able to find its value in any coordinates and can construct non-tensor but nevertheless covariant equations in which it occurs.

Taking a small variation in $\mathfrak{L}$ with respect to the quantities on which it depends, it is found possible after considerable simplification to write this in the form

$$\delta\mathfrak{L} = [-\{\mu\alpha,\beta\}\{\nu\beta,\alpha\}+\{\mu\nu,\alpha\}\{\alpha\beta,\beta\}]\,\delta(g^{\mu\nu}\sqrt{-g})+$$

$$+[-\{\mu\nu,\alpha\}+\tfrac{1}{2}g_\mu^\alpha\{\nu\beta,\beta\}+\tfrac{1}{2}g_\nu^\alpha\{\mu\beta,\beta\}]\,\delta\left\{\frac{\partial}{\partial x^\alpha}(g^{\mu\nu}\sqrt{-g})\right\}. \tag{87.2}$$

Hence if we now regard $\mathfrak{L}$ as a function of the two new quantities

---

† The treatment of this section follows Eddington, *The Mathematical Theory of Relativity*, Cambridge, 1923, §§ 58 and 59.

defined by

$$\mathfrak{g}^{\mu\nu} = g^{\mu\nu}\sqrt{-g} \quad \text{and} \quad \mathfrak{g}^{\mu\nu}_\alpha = \frac{\partial}{\partial x^\alpha}(g^{\mu\nu}\sqrt{-g}), \tag{87.3}$$

we can write

$$\frac{\partial\mathfrak{L}}{\partial\mathfrak{g}^{\mu\nu}} = -\{\mu\alpha, \beta\}\{\nu\beta, \alpha\} + \{\mu\nu, \alpha\}\{\alpha\beta, \beta\} \tag{87.4}$$

and

$$\frac{\partial\mathfrak{L}}{\partial\mathfrak{g}^{\mu\nu}_\alpha} = -\{\mu\nu, \alpha\} + \tfrac{1}{2}g^\alpha_\mu\{\nu\beta,\beta\} + \tfrac{1}{2}g^\alpha_\nu\{\mu\beta, \beta\}. \tag{87.5}$$

With the help of these two expressions for the dependence of $\mathfrak{L}$ on the variables $\mathfrak{g}^{\mu\nu}$ and $\mathfrak{g}^{\mu\nu}_\alpha$, we can now obtain several useful equations containing $\mathfrak{L}$ which will be needed in the present section or later. Comparing (87.4) and (87.5) with the expression for the contracted Riemann-Christoffel tensor given by (77.1), it will be seen on examination that we can write this in the form

$$\frac{\partial}{\partial x^\alpha}\frac{\partial\mathfrak{L}}{\partial\mathfrak{g}^{\mu\nu}_\alpha} - \frac{\partial\mathfrak{L}}{\partial\mathfrak{g}^{\mu\nu}} = R_{\mu\nu}. \tag{87.6}$$

This result shows a formal resemblance to the equations of motion in the classical Lagrangian form in agreement with the name that we have given to the function $\mathfrak{L}$. Furthermore, multiplying (87.4) and (87.5) by $\mathfrak{g}^{\mu\nu}$ and $\mathfrak{g}^{\mu\nu}_\alpha$ it can be shown after some simplification that we can write

$$\mathfrak{g}^{\mu\nu}\frac{\partial\mathfrak{L}}{\partial\mathfrak{g}^{\mu\nu}} = -\mathfrak{L} \tag{87.7}$$

and

$$\mathfrak{g}^{\mu\nu}_\alpha\frac{\partial\mathfrak{L}}{\partial\mathfrak{g}^{\mu\nu}_\alpha} = 2\mathfrak{L}. \tag{87.8}$$

For the scalar density $\mathfrak{R}$ we can then obtain the expresson

$$\begin{aligned}
\mathfrak{R} &= g^{\mu\nu}R_{\mu\nu}\sqrt{-g} = \mathfrak{g}^{\mu\nu}R_{\mu\nu} \\
&= \mathfrak{g}^{\mu\nu}\frac{\partial}{\partial x^\alpha}\frac{\partial\mathfrak{L}}{\partial\mathfrak{g}^{\mu\nu}_\alpha} - \mathfrak{g}^{\mu\nu}\frac{\partial\mathfrak{L}}{\partial\mathfrak{g}^{\mu\nu}} \\
&= \frac{\partial}{\partial x^\alpha}\left(\mathfrak{g}^{\mu\nu}\frac{\partial\mathfrak{L}}{\partial\mathfrak{g}^{\mu\nu}_\alpha}\right) - \mathfrak{g}^{\mu\nu}_\alpha\frac{\partial\mathfrak{L}}{\partial\mathfrak{g}^{\mu\nu}_\alpha} - \mathfrak{g}^{\mu\nu}\frac{\partial\mathfrak{L}}{\partial\mathfrak{g}^{\mu\nu}} \\
&= \frac{\partial}{\partial x^\alpha}\left(\mathfrak{g}^{\mu\nu}\frac{\partial\mathfrak{L}}{\partial\mathfrak{g}^{\mu\nu}_\alpha}\right) - \mathfrak{L}.
\end{aligned} \tag{87.9}$$

We are now ready to undertake the re-expression of the fundamental equations of mechanics in the form of an ordinary divergence. To do this we shall wish to transform the second term in the previous form for the equations of mechanics given by (84.8). In accordance with the original field equations (84.1), we can write

$$
\begin{aligned}
-8\pi\mathfrak{T}_{\mu\nu}\,dg^{\mu\nu} &= (R_{\mu\nu}-\tfrac{1}{2}Rg_{\mu\nu}+\Lambda g_{\mu\nu})\sqrt{-g}\,dg^{\mu\nu} \\
&= R_{\mu\nu}\sqrt{-g}\,dg^{\mu\nu}-\tfrac{1}{2}R\sqrt{-g}\,g_{\mu\nu}\,dg^{\mu\nu}+\Lambda\sqrt{-g}\,g_{\mu\nu}\,dg^{\mu\nu} \\
&= R_{\mu\nu}\sqrt{-g}\,dg^{\mu\nu}+R\,d\sqrt{-g}-2\Lambda\,d\sqrt{-g} \\
&= R_{\mu\nu}\sqrt{-g}\,dg^{\mu\nu}+g^{\mu\nu}R_{\mu\nu}\,d\sqrt{-g}-2\Lambda\,d\sqrt{-g} \\
&= R_{\mu\nu}\,d(g^{\mu\nu}\sqrt{-g})-2\Lambda\,d\sqrt{-g},
\end{aligned}
\tag{87.10}
$$

where we have made use of equation (39) in Appendix III, in going from the second to the third of the above expressions. Substituting the value for $R_{\mu\nu}$ in terms of the Lagrangian function given by (87.6), and the value of $\mathfrak{g}_\beta^{\mu\nu}$ corresponding to the equation of definition (87.3) we then obtain

$$
\begin{aligned}
-8\pi\mathfrak{T}_{\mu\nu}\frac{\partial g^{\mu\nu}}{\partial x^\beta} &= \mathfrak{g}_\beta^{\mu\nu}\left(\frac{\partial}{\partial x^\alpha}\frac{\partial\mathfrak{L}}{\partial\mathfrak{g}_\alpha^{\mu\nu}}-\frac{\partial\mathfrak{L}}{\partial\mathfrak{g}^{\mu\nu}}\right)-2\Lambda\,\frac{\partial\sqrt{-g}}{\partial x^\beta} \\
&= \frac{\partial}{\partial x^\alpha}\left(\mathfrak{g}_\beta^{\mu\nu}\frac{\partial\mathfrak{L}}{\partial\mathfrak{g}_\alpha^{\mu\nu}}\right)-\frac{\partial}{\partial x^\alpha}\mathfrak{g}_\beta^{\mu\nu}\frac{\partial\mathfrak{L}}{\partial\mathfrak{g}_\alpha^{\mu\nu}}-\mathfrak{g}_\beta^{\mu\nu}\frac{\partial\mathfrak{L}}{\partial\mathfrak{g}^{\mu\nu}}-2\Lambda\,\frac{\partial\sqrt{-g}}{\partial x^\beta} \\
&= \frac{\partial}{\partial x^\alpha}\left(\mathfrak{g}_\beta^{\mu\nu}\frac{\partial\mathfrak{L}}{\partial\mathfrak{g}_\alpha^{\mu\nu}}\right)-\frac{\partial\mathfrak{L}}{\partial\mathfrak{g}_\alpha^{\mu\nu}}\frac{\partial}{\partial x^\beta}\mathfrak{g}_\alpha^{\mu\nu}-\frac{\partial\mathfrak{L}}{\partial\mathfrak{g}^{\mu\nu}}\frac{\partial}{\partial x^\beta}\mathfrak{g}^{\mu\nu}-2\Lambda\,\frac{\partial\sqrt{-g}}{\partial x^\beta} \\
&= \frac{\partial}{\partial x^\alpha}\left(\mathfrak{g}_\beta^{\mu\nu}\frac{\partial\mathfrak{L}}{\partial\mathfrak{g}_\alpha^{\mu\nu}}\right)-\frac{\partial\mathfrak{L}}{\partial x^\beta}-2\Lambda\,\frac{\partial\sqrt{-g}}{\partial x^\beta} \\
&= \frac{\partial}{\partial x^\alpha}\left[\mathfrak{g}_\beta^{\mu\nu}\frac{\partial\mathfrak{L}}{\partial\mathfrak{g}_\alpha^{\mu\nu}}-g_\beta^\alpha\mathfrak{L}-2g_\beta^\alpha\Lambda\sqrt{-g}\right].
\end{aligned}
\tag{87.11}
$$

To make use of this result, let us now define a new quantity, which may be called the pseudo-tensor density of gravitational energy and momentum, by the equation

$$
\mathfrak{t}_\beta^\alpha = \frac{1}{16\pi}\left[-\mathfrak{g}_\beta^{\mu\nu}\frac{\partial\mathfrak{L}}{\partial\mathfrak{g}_\alpha^{\mu\nu}}+g_\beta^\alpha\mathfrak{L}+2g_\beta^\alpha\Lambda\sqrt{-g}\right].
\tag{87.12}
$$

In accordance with this definition together with (87.11) we shall evidently have

$$
\frac{\partial\mathfrak{t}_\mu^\nu}{\partial x^\nu} = \tfrac{1}{2}\mathfrak{T}_{\alpha\beta}\frac{\partial g^{\alpha\beta}}{\partial x^\mu}.
\tag{87.13}
$$

And substituting in (84.8), we may now write the equations of mechanics in the desired form of an ordinary divergence

$$
\frac{\partial}{\partial x^\nu}(\mathfrak{T}_\mu^\nu+\mathfrak{t}_\mu^\nu) = 0.
\tag{87.14}
$$

This equation is not a tensor equation, both because the quantity $\mathfrak{t}_\mu^\nu$ is not a true tensor density and because the expression is that for an ordinary divergence instead of a tensor divergence. Nevertheless, t_μ^ν is a quantity which is defined in all systems of coordinates by (87.12), and the equation is a covariant one valid in all systems of coordinates. Hence we may have no hesitation in using this very beautiful result of Einstein.

In accordance with the definition of t_μ^ν given by (87.12) and the values for the quantities occurring therein as given by previous equations of this section, it will be noted that the value of t_μ^ν at any point will be determined by the values for the components of the metrical tensor $g_{\alpha\beta}$ and their first derivatives $\partial g_{\alpha\beta}/\partial x^\gamma$ at that point. Furthermore, if we use *natural coordinates* for the particular point of interest it is seen that the expression for t_μ^ν will then reduce to

$$8\pi t_\mu^\nu = g_\mu^\nu \Lambda \sqrt{-g}, \tag{87.15}$$

which by combining with the expression for the energy-momentum tensor (84.1) will also give us in these coordinates

$$-8\pi(\mathfrak{T}_\mu^\nu + t_\mu^\nu) = (R_\mu^\nu - \tfrac{1}{2}Rg_\mu^\nu)\sqrt{-g}. \tag{87.16}$$

Since t_μ^ν is not a true tensor density, however, we shall not have these simple results in all coordinate systems.

88. The energy-momentum principle for finite systems

With the help of our new expression for the principles of mechanics, we may now obtain an important result which may be regarded as the relativistic analogue of the ordinary laws of the conservation of energy and momentum.

To do this, let us take x^1, x^2, and x^3 as being space-like coordinates and x^4 the time-like coordinate, and apply equation (87.14) to a given finite system of interest by multiplying by $dx^1 dx^2 dx^3$ and integrating at some given 'time' x^4 over the spatial region in question. Carrying this out, we at once obtain with some rearrangement of terms

$$\iiint \frac{\partial}{\partial x^4}(\mathfrak{T}_\mu^4 + t_\mu^4)\, dx^1 dx^2 dx^3$$
$$= -\iiint \left[\frac{\partial}{\partial x^1}(\mathfrak{T}_\mu^1 + t_\mu^1) + \frac{\partial}{\partial x^2}(\mathfrak{T}_\mu^2 + t_\mu^2) + \frac{\partial}{\partial x^3}(\mathfrak{T}_\mu^3 + t_\mu^3) \right] dx^1 dx^2 dx^3, \tag{88.1}$$

and taking the limits of integration corresponding to the region as being constants x^1 to x'^1, x^2 to x'^2, etc. independent of the 'time' x^4, this can be rewritten in the form

$$\frac{d}{dx^4} \int \int \int (\mathfrak{T}_\mu^4 + \mathfrak{t}_\mu^4)\, dx^1 dx^2 dx^3 = - \int \int |\mathfrak{T}_\mu^1 + \mathfrak{t}_\mu^1|_{x^1}^{x'^1}\, dx^2 dx^3 -$$

$$- \int \int |\mathfrak{T}_\mu^2 + \mathfrak{t}_\mu^2|_{x^2}^{x'^2}\, dx^1 dx^3 - \int \int |\mathfrak{T}_\mu^3 + \mathfrak{t}_\mu^3|_{x^3}^{x'^3}\, dx^1 dx^2 \quad (88.2)$$

by performing the indicated integrations on the right-hand side of (88.1).

Equation (88.2) as written is true in all sets of coordinates, owing to its immediate dependence on the covariant equation (87.14). The interpretation and use of the equation are often simplified, however, if we choose coordinates in such a way that the limits of integration x^1, x'^1, etc., which must be taken in order to include the region in question, actually lie on the boundary surface which separates the system from its surroundings. Thus, for example, quasi-Cartesian coordinates x, y, z with the limits of integration x to x', y to y', and z to z', lying on the actual boundary of the system, are preferable for our present purposes to polar coordinates r, θ, and ϕ with the limits of integration 0 to r, 0 to π, and 0 to 2π, in which case r would be the only limit actually lying on the boundary. The increased simplicity of the properly chosen coordinates arises from the fact that the right-hand side of (88.2) is then determined solely by the values assumed by $\mathfrak{T}_\mu^1$, $\mathfrak{t}_\mu^1$, etc. at the boundary of the system and is not dependent on their values within the system. Having chosen coordinates in the way suggested, equation (88.2) then states that the rate of change with the 'time' x^4 of the volume integral on the left-hand side is equal to the sum of the surface integrals on the right side, which has a value that is entirely determined by conditions prevailing at the boundary separating the system from its surroundings.

Equation (88.2) is thus of the proper form to be considered as the expression of a conservation law provided we regard the right-hand side as defining a flux through the boundary. Furthermore, if we consider the limiting case of 'flat' space-time where the special theory of relativity would be valid, equation (88.2) could then be rewritten using Galilean coordinates in the form

$$\frac{d}{dx^4} \int \int \int T_\mu^4\, dx^1 dx^2 dx^3$$

$$= - \int \int |T_\mu^1|_{x^1}^{x'^1}\, dx^2 dx^3 - \int \int |T_\mu^2|_{x^2}^{x'^2}\, dx^1 dx^3 - \int \int |T_\mu^3|_{x^3}^{x'^3}\, dx^1 dx^2,$$

$$(88.3)$$

since $\mathfrak{t}_\mu^\nu$ would then be zero, in accordance with (87.15) and the value $\Lambda = 0$ for the case of 'flat' space-time, and $\mathfrak{T}_\mu^\nu$ would equal T_μ^ν owing to the value unity for $\sqrt{-g}$ in these coordinates. Referring, however, to our previous equation (38.10) we see that (88.3) is entirely equivalent with $\mu = 1, 2, 3$ to the special relativity expression resulting from the law of the conservation of the three components of linear momentum, and with $\mu = 4$ to the result of the law of the conservation of energy.†

Hence we may now take (88.2) as the analogue in general relativity of the usual energy-momentum principle if we define

$$J_\mu = \iiint (\mathfrak{T}_\mu^4 + \mathfrak{t}_\mu^4)\, dx^1 dx^2 dx^3, \tag{88.4}$$

with $\mu = 1, 2, 3$ as the expressions for the components of momenta of the region, and with $\mu = 4$ as the expression for its energy. And in accordance with this definition we may now regard the quantities $\mathfrak{T}_\mu^4$ as the densities of material energy and momentum and the $\mathfrak{t}_\mu^4$ as densities of potential gravitational energy and momentum. This necessity of including potential energy and momentum in order to secure the analogue of the usual energy-momentum principle is in agreement with the possibilities for the sum total of the proper energy of an isolated system not to remain constant which were mentioned at the end of § 86.

As a result of our definition (88.4), the quantities J_μ which we may regard as representing the energy and momenta of a finite system are not the components of a true covariant vector. They are, however, defined by (88.4) for all systems of coordinates and the equations in which they appear are covariant equations true in all systems of coordinates.

The physical significance of the quantities J_μ can be most easily grasped in the case of an isolated system. Consider an isolated system enclosed by a boundary located in the surrounding empty space at a sufficient distance so that we are justified in neglecting the curvature of space-time for points on the boundary and beyond. The spatial region inside this boundary can then be regarded as generating a tube in a surrounding 'flat' space-time, and we can choose coordinates in such a way that they will go continuously over into some

---

† The lowered position of μ in (88.3) as compared with (38.10) is not important, since $(T^{\mu\nu})_\nu = 0$ implies $(g_{\alpha\mu}T^{\alpha\nu})_\nu = (T_\mu^\nu)_\nu = 0$ owing to the relation $(g_{\alpha\mu})_\nu = 0$. See Appendix III, equation (35).

particular set of special relativity Galilean coordinates in the region outside this tube.

Using these coordinates it is then evident that the general energy-momentum principle (88.2) will reduce for such an isolated system to the simple principles of the conservation of energy and momentum

$$\frac{dJ_\mu}{dx^4} = 0, \tag{88.5}$$

since the right-hand side of (88.2) will now be zero in accordance with (87.16), if we take the curvature of space-time as negligible at the boundary of the tube as assumed.

Furthermore, it may be shown that the quantities J_μ are independent of any changes that we may make in the coordinate system inside the tube, provided the changed coordinate system still coincides with the original Galilean system in regions outside the tube. To see this we merely have to note that a third auxiliary coordinate system could be introduced coinciding with the common Galilean coordinate system in regions outside the tube, and coinciding inside the tube for one value of the 'time' x^4 (as given outside the tube) with the original coordinate system and at a later 'time' x^4 with the changed coordinate system. Then, since in accordance with (88.5) the values of J_μ would be independent of x^4 in all three coordinate systems, we can conclude that the values would have to be identical for the three coordinate systems.

In addition, it can be shown that the quantities J_μ would transform like the components of a four-dimensional vector for the linear transformations which could be introduced to change to any desired new set of Galilean coordinates for the region outside the tube. The rigorous proof of this, nevertheless, is somewhat complicated and we need not include it here.†

As a result of the foregoing, we then see that the physical significance of the quantities J_μ in the case of an isolated system can be appreciated from the four properties of reducing at the limit to the quantities which we have already taken as energy and momenta in the special theory of relativity, of obeying a conservation law when we use coordinates that are Galilean in the flat space-time outside the system, of being independent of the choice of coordinates within the

† See Pauli, 'Relativitätstheorie', *Encyclopädie der math. Wiss.* Band V₂, Heft 4, Leipzig, 1921. § 21.

region of appreciable space-time curvature, and of depending on different Galilean systems of coordinates in the surrounding 'flat' space-time in an analogous manner to the quantities $m_0\,dx^\mu/ds$ which can be regarded (see 28.4) as determining the momenta and energy of a single particle in special relativity. This should assist in giving us a feeling for the physical nature of these quantities.

89. The densities of energy and momentum expressed as divergences

For some of our further applications of the energy-momentum principle, it will now be desirable to re-express the quantities $(\mathfrak{T}_\mu^\nu + \mathfrak{t}_\mu^\nu)$, which we regard as giving the densities of energy and momentum, themselves in the form of divergences. To do this, we may first combine the expressions for $\mathfrak{T}_\mu^\nu$ and $\mathfrak{t}_\mu^\nu$ given by (84.1) and (87.12) and write

$$8\pi(\mathfrak{T}_\mu^\nu + \mathfrak{t}_\mu^\nu) = -\mathfrak{R}_\mu^\nu + \tfrac{1}{2}g_\mu^\nu \mathfrak{R} - \tfrac{1}{2}\mathfrak{g}_\mu^{\alpha\beta}\frac{\partial \mathfrak{L}}{\partial \mathfrak{g}_\nu^{\alpha\beta}} + \tfrac{1}{2}g_\mu^\nu \mathfrak{L}, \qquad (89.1)$$

where it is interesting to note that the Λ-term cancels out even if the cosmological constant is not exactly equal to zero. Substituting from (87.6) and (87.9) we obtain

$$8\pi(\mathfrak{T}_\mu^\nu + \mathfrak{t}_\mu^\nu) = -\mathfrak{g}^{\alpha\nu}\frac{\partial}{\partial x^\gamma}\frac{\partial \mathfrak{L}}{\partial \mathfrak{g}_\gamma^{\mu\nu}} + \mathfrak{g}^{\alpha\nu}\frac{\partial \mathfrak{L}}{\partial \mathfrak{g}^{\mu\alpha}} + \tfrac{1}{2}g_\mu^\nu\frac{\partial}{\partial x^\gamma}\left(\mathfrak{g}^{\alpha\beta}\frac{\partial \mathfrak{L}}{\partial \mathfrak{g}_\gamma^{\alpha\beta}}\right) - \tfrac{1}{2}\mathfrak{g}_\mu^{\alpha\beta}\frac{\partial \mathfrak{L}}{\partial \mathfrak{g}_\nu^{\alpha\beta}},$$

and this can evidently be rewritten in the form

$$8\pi(\mathfrak{T}_\mu^\nu + \mathfrak{t}_\mu^\nu) = \frac{\partial}{\partial x^\gamma}\left(-\mathfrak{g}^{\alpha\nu}\frac{\partial \mathfrak{L}}{\partial \mathfrak{g}_\gamma^{\mu\alpha}} + \tfrac{1}{2}g_\mu^\nu\,\mathfrak{g}^{\alpha\beta}\frac{\partial \mathfrak{L}}{\partial \mathfrak{g}_\gamma^{\alpha\beta}}\right) +$$
$$+ \mathfrak{g}_\gamma^{\alpha\nu}\frac{\partial \mathfrak{L}}{\partial \mathfrak{g}_\gamma^{\mu\alpha}} + \mathfrak{g}^{\alpha\nu}\frac{\partial \mathfrak{L}}{\partial \mathfrak{g}^{\mu\alpha}} - \tfrac{1}{2}\mathfrak{g}_\mu^{\alpha\beta}\frac{\partial \mathfrak{L}}{\partial \mathfrak{g}_\nu^{\alpha\beta}}. \qquad (89.2)$$

From the definitions which we have given for the quantities entering into this expression, it can be shown, nevertheless, by a rather lengthy but straightforward computation† that the sum of the last three terms of the expression will be identically equal to zero. This then permits us to write the divergence

$$8\pi(\mathfrak{T}_\mu^\nu + \mathfrak{t}_\mu^\nu) = \frac{\partial}{\partial x^\gamma}\left(-\mathfrak{g}^{\alpha\nu}\frac{\partial \mathfrak{L}}{\partial \mathfrak{g}_\gamma^{\mu\alpha}} + \tfrac{1}{2}g_\mu^\nu\,\mathfrak{g}^{\alpha\beta}\frac{\partial \mathfrak{L}}{\partial \mathfrak{g}_\gamma^{\alpha\beta}}\right) \qquad (89.3)$$

as a useful relation for calculating the relativistic densities of energy and momentum.

† Tolman, *Phys. Rev.* **35**, 875 (1930).

90. Limiting values for certain quantities at a large distance from an isolated system

In the following section we shall wish to use equation (89.3) to obtain expressions for the total energy and momentum of an isolated material system. As a preliminary, we shall first calculate the limiting values which would be approached at large distances from the system by certain of the quantities which can occur on the right-hand side of this expression.

In carrying out the computations, we shall use a system of quasi-Galilean coordinates (x, y, z, t) which are chosen so that the material system of interest is permanently located in the neighbourhood of the origin $x = y = z = 0$, and so chosen that the line element will approach the Galilean form at very great distances from this origin. As a consequence we can then assign to the line element at sufficient distances from the origin the approximate Schwarzschild form given by (82.15)

$$ds^2 = -\left(1 + \frac{2m}{r}\right)(dx^2 + dy^2 + dz^2) + \left(1 - \frac{2m}{r}\right) dt^2 \qquad (90.1)$$

with $\qquad r = \sqrt{(x^2 + y^2 + z^2)} \quad$ and $\quad m = $ constant, $\qquad (90.2)$

since at sufficient distances the field will be spherically symmetrical owing to the location of the material system, and will be static owing to the isolation, which we shall regard as requiring the metric at these distances to be unaffected by changes taking place within the material system itself.

Neglecting terms of the order of (m/r) compared with unity, it is easily found that the Christoffel three-index symbols corresponding to this line element will be of the forms

$$\{\mu\mu, \mu\} = -\frac{m}{r^2}\frac{\partial r}{\partial x^\mu}$$

$$\{\mu\mu, \nu\} = \frac{m}{r^2}\frac{\partial r}{\partial x^\nu}$$

$$\{\nu\mu, \mu\} = \{\mu\nu, \mu\} = \pm\frac{m}{r^2}\frac{\partial r}{\partial x^\nu} \quad \begin{cases} \mu = 4 \\ \mu \neq 4 \end{cases}$$

$$\{\mu\nu, \sigma\} = 0, \qquad (90.3)$$

where μ, ν, and σ represent different indices. In using these expressions it will be noted in accordance with (90.2) that certain of these quantities will be zero owing to the independence of r and $x^4 = t$.

With the help of these expressions for the Christoffel symbols, and the expression for $\partial\mathfrak{L}/\partial\mathfrak{g}_\alpha^{\mu\nu}$ given by (87.5), we can now obtain explicit expressions for quantities which can occur on the right-hand side of (89.3) which will be needed in the next section. We calculate to the *same order of approximation* as the expressions for the Christoffel symbols

$$\mathfrak{g}^{\alpha 4}\frac{\partial\mathfrak{L}}{\partial\mathfrak{g}_1^{1\alpha}} = \frac{\partial\mathfrak{L}}{\partial\mathfrak{g}_1^{14}} = -\{14,1\}+\tfrac{1}{2}\{4\epsilon,\epsilon\} = 0$$

$$\mathfrak{g}^{\alpha 4}\frac{\partial\mathfrak{L}}{\partial\mathfrak{g}_2^{1\alpha}} = \frac{\partial\mathfrak{L}}{\partial\mathfrak{g}_2^{14}} = -\{14,2\} = 0$$

$$\mathfrak{g}^{\alpha 4}\frac{\partial\mathfrak{L}}{\partial\mathfrak{g}_1^{4\alpha}} = \frac{\partial\mathfrak{L}}{\partial\mathfrak{g}_1^{44}} = -\{44,1\} = -\frac{m}{r^2}\frac{\partial r}{\partial x}$$

$$\mathfrak{g}^{\alpha\beta}\frac{\partial\mathfrak{L}}{\partial\mathfrak{g}_1^{\alpha\beta}} = -\frac{\partial\mathfrak{L}}{\partial\mathfrak{g}_1^{11}}-\frac{\partial\mathfrak{L}}{\partial\mathfrak{g}_1^{22}}-\frac{\partial\mathfrak{L}}{\partial\mathfrak{g}_1^{33}}+\frac{\partial\mathfrak{L}}{\partial\mathfrak{g}_1^{44}}$$

$$= +\{11,1\}-\tfrac{1}{2}\{1\epsilon,\epsilon\}-\tfrac{1}{2}\{1\epsilon,\epsilon\}+\{22,1\}+\{33,1\}-\{44,1\}$$

$$= \frac{m}{r^2}\left\{-\frac{\partial r}{\partial x}-\left(-\frac{\partial r}{\partial x}-\frac{\partial r}{\partial x}-\frac{\partial r}{\partial x}+\frac{\partial r}{\partial x}\right)+\frac{\partial r}{\partial x}+\frac{\partial r}{\partial x}-\frac{\partial r}{\partial x}\right\}$$

$$= \frac{2m}{r^2}\frac{\partial r}{\partial x}.$$

Extending these results with the help of the symmetry in $x, y,$ and z, and replacing the derivatives of r with respect to the coordinates by the direction cosines for the radius vector, we can then write

$$\mathfrak{g}^{\alpha 4}\frac{\partial\mathfrak{L}}{\partial\mathfrak{g}_1^{1\alpha}} = \mathfrak{g}^{\alpha 4}\frac{\partial\mathfrak{L}}{\partial\mathfrak{g}_2^{1\alpha}} = \mathfrak{g}^{\alpha 4}\frac{\partial\mathfrak{L}}{\partial\mathfrak{g}_3^{1\alpha}} = 0 \qquad (90.4)$$

$$\mathfrak{g}^{\alpha 4}\frac{\partial\mathfrak{L}}{\partial\mathfrak{g}_1^{4\alpha}} = -\frac{m}{r^2}\cos(nx)$$

$$\mathfrak{g}^{\alpha 4}\frac{\partial\mathfrak{L}}{\partial\mathfrak{g}_2^{4\alpha}} = -\frac{m}{r^2}\cos(ny) \qquad (90.5)$$

$$\mathfrak{g}^{\alpha 4}\frac{\partial\mathfrak{L}}{\partial\mathfrak{g}_3^{4\alpha}} = -\frac{m}{r^2}\cos(nz)$$

$$\mathfrak{g}^{\alpha\beta}\frac{\partial\mathfrak{L}}{\partial\mathfrak{g}_1^{\alpha\beta}} = \frac{2m}{r^2}\cos(nx)$$

$$\mathfrak{g}^{\alpha\beta}\frac{\partial\mathfrak{L}}{\partial\mathfrak{g}_2^{\alpha\beta}} = \frac{2m}{r^2}\cos(ny) \qquad (90.6)$$

$$\mathfrak{g}^{\alpha\beta}\frac{\partial\mathfrak{L}}{\partial\mathfrak{g}_3^{\alpha\beta}} = \frac{2m}{r^2}\cos(nz),$$

as a list which gives the limiting values at large distances from the material system for those quantities which we shall need in the next section.

91. The mass, energy, and momentum of an isolated system

With the help of the foregoing values, we may now obtain expressions for the energy and components of momentum of an isolated system. In accordance with (88.4) and (89.3), we can write for the energy of the system

$$U = J_4 = \iiint (\mathfrak{T}_4^4 + \mathfrak{t}_4^4)\, dxdydz$$

$$= \frac{1}{8\pi} \iiint \frac{\partial}{\partial x^\gamma} \left(-\mathfrak{g}^{\alpha 4} \frac{\partial \mathfrak{L}}{\partial \mathfrak{g}_\gamma^{4\alpha}} + \tfrac{1}{2}\mathfrak{g}^{\alpha\beta} \frac{\partial \mathfrak{L}}{\partial \mathfrak{g}_\gamma^{\alpha\beta}} \right) dxdydz, \qquad (91.1)$$

provided we take the integration over a sufficient volume surrounding the system of interest. Taking this volume as a sphere of radius r around the origin, noting the summation implied by the double occurence of the dummy γ, using Gauss's theorem to transform the first three terms of the summation to a surface integral, and noting the values given by (90.5, 6) which will be assumed by quantities on the right-hand side of (91.1) at sufficient values of r, we then obtain

$$U = \frac{1}{8\pi} \iint \frac{2m}{r^2} \{\cos^2(nx) + \cos^2(ny) + \cos^2(nz)\}\, d\sigma +$$

$$+ \frac{1}{8\pi} \frac{\partial}{\partial t} \iiint \left(-\mathfrak{g}^{\alpha 4} \frac{\partial \mathfrak{L}}{\partial \mathfrak{g}_4^{4\alpha}} + \tfrac{1}{2}\mathfrak{g}^{\alpha\beta} \frac{\partial \mathfrak{L}}{\partial \mathfrak{g}_4^{\alpha\beta}} \right) dxdydz. \quad (91.2)$$

The first term in this expression is immediately seen to have the value m. The second term in the expression cannot be explicitly computed, however, since it involves an integration over the whole volume of the sphere which includes regions in the neighbourhood of the origin where we know nothing about the nature of the line element. Nevertheless, since U is a constant in accordance with the conservation of energy for an isolated system expressed by (88.5), and since m is a constant as by hypothesis it determines the static field at large distances, it is evident that the second term on the right-hand side of (91.2) must also be a constant; and hence indeed have the value zero owing to the impossibility for the integral involved to change permanently at a constant finite rate. We thus obtain for the energy of our isolated system Einstein's very satisfactory result

$$U = m. \qquad (91.3)$$

The momentum of the system with respect to the coordinates being used can be similarly determined. In accordance with (88.4) and (89.3) we can write for the component of momentum in the x-direction

$$J_1 = \iiint (\mathfrak{T}_1^4 + \mathfrak{t}_1^4)\, dxdydz$$

$$= \frac{1}{8\pi} \iiint \frac{\partial}{\partial x^\gamma}\left(-\mathfrak{g}^{\alpha 4} \frac{\partial \mathfrak{L}}{\partial \mathfrak{g}_\gamma^{1\alpha}}\right) dxdydz,$$

and treating this in the same way as we did (91.1), noting the zero values for terms in the integrand which will arise from (90.4), we obtain

$$J_1 = \frac{1}{8\pi} \frac{\partial}{\partial t} \iiint \left(-\mathfrak{g}^{\alpha 4} \frac{\partial \mathfrak{L}}{\partial \mathfrak{g}_4^{1\alpha}}\right) dxdydz = 0, \qquad (91.4)$$

where the value zero rises from reasoning similar to that given in the immediately preceding paragraph.

Summarizing, we may now write for the three components of momentum J_1, J_2, and J_3 and for the energy $J_4 = U$ of an isolated material system permanently located at the origin of a system of quasi-Galilean coordinates of the kind used

$$J_\mu = (0, 0, 0, m). \qquad (91.5)$$

The value zero for the three components of momentum arises of course from our particular choice of coordinates, having the system of interest at rest so to speak at the origin. The value m obtained for the energy of the system seems very appropriate, since it shows that the total energy of an isolated object is also the quantity, occurring in the approximate Schwarzschild expression, which determines the gravitational field at large distances from that object.

Making use of the possibility, already mentioned at the end of § 88, of showing that the components of J_μ would transform like those of a covariant vector, for linear transformations which change from one system of Galilean coordinates to another in the surrounding 'flat' space-time, we could also write in agreement with (91.5) the more general contravariant expression

$$J^\mu = m \frac{dx^\mu}{ds}, \qquad (91.6)$$

where dx^μ/ds may be regarded as corresponding roughly to the velocity of the system as a whole with respect to the particular set of coordinates in use.

92. The energy of a quasi-static isolated system expressed by an integral extending only over the occupied space

For certain purposes both of the expressions for the energy of an isolated system

$$U = \iiint (\mathfrak{T}_4^4 + \mathfrak{t}_4^4)\, dxdydz \quad \text{and} \quad U = m \tag{92.1}$$

may be unsatisfactory. The first of these expressions suffers from the fact that the indicated integration will in general have to be carried out over a volume which is large compared with the actual system of interest, since $\mathfrak{t}_4^4$ is not in general equal to zero in empty space. And the second expression suffers from the fact that it gives no method of computing the energy from the actual distribution of matter and radiation within the system. For a particular class of systems, which may be called quasi-static, another expression can be obtained that is sometimes more usable.

Substituting into the first of the two expressions (92.1) the value for the density of potential gravitational energy $\mathfrak{t}_4^4$ given by (87.12), we obtain

$$U = \iiint \left(\mathfrak{T}_4^4 + \frac{\mathfrak{L}}{16\pi} - \frac{1}{16\pi} \mathfrak{g}_4^{\alpha\beta}\, \frac{\partial \mathfrak{L}}{\partial \mathfrak{g}_4^{\alpha\beta}} \right) dxdydz,$$

where the cosmological term has been omitted since the application will be to small systems which can be regarded as surrounded by 'flat' space-time. Introducing the expression for $\mathfrak{L}$ given by (87.9) this becomes

$$U = \iiint \left[\mathfrak{T}_4^4 - \frac{\mathfrak{R}}{16\pi} + \frac{1}{16\pi} \frac{\partial}{\partial x^\gamma}\left(\mathfrak{g}^{\alpha\beta} \frac{\partial \mathfrak{L}}{\partial \mathfrak{g}_\gamma^{\alpha\beta}} \right) - \frac{1}{16\pi} \mathfrak{g}_4^{\alpha\beta} \frac{\partial \mathfrak{L}}{\partial \mathfrak{g}_4^{\alpha\beta}} \right] dxdydz.$$

Furthermore, substituting for $\mathfrak{R}$, in agreement with (78.11), the well-known expression

$$\mathfrak{R} = 8\pi\mathfrak{T} = 8\pi(\mathfrak{T}_1^1 + \mathfrak{T}_2^2 + \mathfrak{T}_3^3 + \mathfrak{T}_4^4),$$

expanding the third term of the integrand, and then combining with the fourth term, the expression for U can be rewritten in the form

$$U = \iiint \tfrac{1}{2}(\mathfrak{T}_4^4 - \mathfrak{T}_1^1 - \mathfrak{T}_2^2 - \mathfrak{T}_3^3)\, dxdydz +$$

$$+ \frac{1}{16\pi} \iiint \left[\frac{\partial}{\partial x}\left(\mathfrak{g}^{\alpha\beta} \frac{\partial \mathfrak{L}}{\partial \mathfrak{g}_1^{\alpha\beta}} \right) + \frac{\partial}{\partial y}\left(\mathfrak{g}^{\alpha\beta} \frac{\partial \mathfrak{L}}{\partial \mathfrak{g}_2^{\alpha\beta}} \right) + \frac{\partial}{\partial z}\left(\mathfrak{g}^{\alpha\beta} \frac{\partial \mathfrak{L}}{\partial \mathfrak{g}_3^{\alpha\beta}} \right) \right] dxdydz +$$

$$\tag{92.2}$$

$$+ \frac{1}{16\pi} \iiint \mathfrak{g}^{\alpha\beta} \frac{\partial}{\partial t}\left(\frac{\partial \mathfrak{L}}{\partial \mathfrak{g}_4^{\alpha\beta}} \right) dxdydz.$$

To proceed, let us now introduce the definite requirement that the coordinates (x, y, z, t) be chosen so as to be of the quasi-Galilean type used in the two preceding sections, with the physical system of interest permanently located at the origin. The second integral on the right-hand side of (92.2) can then be readily evaluated with the help of Gauss's theorem, by taking the region of integration as a sphere and introducing the values given by (90.6) at the distant boundary. We thus obtain for the second integral the value $\frac{1}{2}m$ which in accordance with (91.3) is also equal to $\frac{1}{2}U$. Substituting in (92.2) we then obtain

$$U = \iiint (\mathfrak{T}_4^4 - \mathfrak{T}_1^1 - \mathfrak{T}_2^2 - \mathfrak{T}_3^3)\, dx\, dy\, dz +$$
$$+ \frac{1}{8\pi} \iiint \mathfrak{g}^{\alpha\beta} \frac{\partial}{\partial t}\!\left(\frac{\partial \mathfrak{L}}{\partial \mathfrak{g}_4^{\alpha\beta}}\right) dx\, dy\, dz. \qquad (92.3)$$

Finally let us define a quasi-static system as one in which changes are taking place with the 'time' t slowly enough so that the second term on the right-hand side of (92.3) is negligible compared with the first. This will, of course, be strictly true when we are interested in quiescent states of temporary or permanent equilibrium. For such systems we can then use the simple expression for the energy†

$$U = \iiint (\mathfrak{T}_4^4 - \mathfrak{T}_1^1 - \mathfrak{T}_2^2 - \mathfrak{T}_3^3)\, dx\, dy\, dz. \qquad (92.4)$$

And this expression has the great advantage that it can be evaluated by integrating only over the region actually occupied by matter or electromagnetic energy, since the values of $\mathfrak{T}_\mu^\nu$ will be zero in empty space.

† Tolman, *Phys. Rev.* **35**, 875 (1930).

VII

RELATIVISTIC MECHANICS (*contd.*)

Part II. SOLUTIONS OF THE FIELD EQUATIONS

93. Einstein's general solution of the field equations in the case of weak fields

As mentioned at the beginning of this chapter, the principles of relativistic mechanics are all implicitly contained in Einstein's field equations
$$-8\pi T_\mu^\nu = R_\mu^\nu - \tfrac{1}{2}Rg_\mu^\nu + \Lambda g_\mu^\nu, \qquad (93.1)$$
which connects the energy-momentum tensor with the geometry of space-time. In the preceding part of the chapter we have investigated those mechanical conclusions which arise from the fact that the tensor divergence of the energy-momentum tensor T_μ^ν must be equal to zero, since the tensor divergence of the right-hand side of (93.1) is known to be identically equally to zero. In what follows we shall be interested in the more complete problem of obtaining solutions for the ten differential equations denoted by (93.1), which will permit us to correlate the components of the energy-momentum tensor T_μ^ν as directly as may be with the components of the metrical tensor $g_{\mu\nu}$.

In the case of weak enough fields this problem has been completely solved by Einstein's approximate solution of these field equations. In the case of strong fields we can obtain no general solution of the equations, but by introducing special assumptions as to the physical nature of the system under consideration, can obtain a number of simplified expressions relating the components of T_μ^ν to the components of $g_{\mu\nu}$ and its derivatives which prove to be useful in solving the equations in particular cases.

We may first consider Einstein's general approximate solution. To obtain this solution we shall assume the gravitational field so weak that we can employ coordinates which are nearly Galilean in character, and can hence represent the components of the metrical tensor by the expression
$$g_{\mu\nu} = \delta_{\mu\nu} + h_{\mu\nu}, \qquad (93.2)$$
where the $\delta_{\mu\nu}$ are the constant Galilean values for the $g_{\mu\nu}$, ± 1 and 0, and the $h_{\mu\nu}$ are small correction terms. The quantities $h_{\mu\nu}$ and their derivatives with respect to the coordinates will be regarded as terms of the first order whose squares may be neglected. We shall also find

it convenient to introduce the quantities

$$h_\mu^\lambda = \delta^{\lambda\alpha}h_{\mu\alpha}; \qquad h = h_\alpha^\alpha = \delta^{\sigma\lambda}h_{\sigma\lambda} \tag{93.3}$$

where the $\delta^{\mu\nu}$ are the Galilean values of the $g^{\mu\nu}$.

Turning now to the expression for the contracted Riemann-Christoffel tensor given by (77.1), it is evident by neglecting higher order terms that we can write correct to the first order

$$
\begin{aligned}
R_{\mu\nu} &= \frac{\partial}{\partial x^\nu}\{\mu\sigma,\sigma\} - \frac{\partial}{\partial x^\sigma}\{\mu\nu,\sigma\} \\
&= \frac{\partial}{\partial x^\nu}\left(\tfrac{1}{2}\delta^{\sigma\lambda}\left\{\frac{\partial h_{\mu\lambda}}{\partial x^\sigma} + \frac{\partial h_{\sigma\lambda}}{\partial x^\mu} - \frac{\partial h_{\mu\sigma}}{\partial x^\lambda}\right\}\right) - \\
&\qquad - \frac{\partial}{\partial x^\sigma}\left(\tfrac{1}{2}\delta^{\sigma\lambda}\left\{\frac{\partial h_{\mu\lambda}}{\partial x^\nu} + \frac{\partial h_{\nu\lambda}}{\partial x^\mu} - \frac{\partial h_{\mu\nu}}{\partial x^\lambda}\right\}\right) \\
&= \tfrac{1}{2}\delta^{\sigma\lambda}\left(\frac{\partial^2 h_{\sigma\lambda}}{\partial x^\mu \partial x^\nu} - \frac{\partial^2 h_{\mu\sigma}}{\partial x^\nu \partial x^\lambda} - \frac{\partial^2 h_{\nu\lambda}}{\partial x^\sigma \partial x^\mu} + \frac{\partial^2 h_{\mu\nu}}{\partial x^\sigma \partial x^\lambda}\right).
\end{aligned}
$$

Rearranging, introducing the quantities defined by (93.3), and changing dummy suffixes, this can be rewritten as

$$R_{\mu\nu} = \tfrac{1}{2}\delta^{\sigma\lambda}\frac{\partial^2 h_{\mu\nu}}{\partial x^\sigma \partial x^\lambda} + \frac{1}{2}\left(\frac{\partial^2 h}{\partial x^\mu \partial x^\nu} - \frac{\partial^2 h_\mu^\alpha}{\partial x^\nu \partial x^\alpha} - \frac{\partial^2 h_\nu^\alpha}{\partial x^\mu \partial x^\alpha}\right). \tag{93.4}$$

We shall now show the possibility of satisfying this relation by the two equations

$$R_{\mu\nu} = \tfrac{1}{2}\delta^{\sigma\lambda}\frac{\partial^2 h_{\mu\nu}}{\partial x^\sigma \partial x^\lambda} \tag{93.5}$$

and

$$\frac{\partial^2 h}{\partial x^\mu \partial x^\nu} - \frac{\partial^2 h_\mu^\alpha}{\partial x^\nu \partial x^\alpha} - \frac{\partial^2 h_\nu^\alpha}{\partial x^\mu \partial x^\alpha} = 0. \tag{93.6}$$

In accordance with (93.5), and our original expression for the energy-momentum tensor (93.1), omitting the cosmological term, we can evidently write

$$
\begin{aligned}
-16\pi T_\mu^\nu &= 2R_\mu^\nu - g_\mu^\nu R \\
&= \delta^{\sigma\lambda}\frac{\partial^2}{\partial x^\sigma \partial x^\lambda}(h_\mu^\nu - \tfrac{1}{2}\delta_\mu^\nu h) \\
&= \left(-\frac{\partial^2}{\partial x^2} - \frac{\partial^2}{\partial y^2} - \frac{\partial^2}{\partial z^2} + \frac{\partial^2}{\partial t^2}\right)(h_\mu^\nu - \tfrac{1}{2}\delta_\mu^\nu h).
\end{aligned}
\tag{93.7}
$$

This 'wave equation' has the well-known solution familiar in the theory of retarded potentials

$$(h_\mu^\nu - \tfrac{1}{2}\delta_\mu^\nu h) = \frac{1}{4\pi}\int \frac{[-16\pi T_\mu^\nu]}{r}\,dx\,dy\,dz, \tag{93.8}$$

where the integration is to be carried out over the whole spatial volume, r is the distance from the point of interest where the value of $(h_\mu^\nu - \tfrac{1}{2}\delta_\mu^\nu h)$ is desired to the particular element of volume $dxdydz$ under consideration, and the square brackets indicate that we are to use the value of T_μ^ν at a time earlier than that of interest by the interval needed for a signal to pass with unit velocity from the element $dxdydz$ under consideration to the point of interest.

To complete our justification for this solution we must show that it also secures the validity of (93.6). From the differentiation of (93.8), we can write

$$\frac{\partial}{\partial x^\alpha}(h_\mu^\alpha - \tfrac{1}{2}\delta_\mu^\alpha h) = -4 \int \frac{[(\partial/\partial x^\alpha)T_\mu^\alpha]}{r}\, dxdydz.$$

In accordance with (93.7), nevertheless, T_μ^α is a quantity of the first order, and hence in accordance with the fundamental relation for the divergence of the energy-momentum tensor (84.5), its divergence will be a quantity of the second order, and we can write to our order of approximation

$$\frac{\partial h_\mu^\alpha}{\partial x^\alpha} = \frac{1}{2}\frac{\partial h}{\partial x^\mu}.$$

Substituting this, however, together with the analogous expression in ν, we immediately see that this will secure the necessary validity of (93.6).

The approximate solution of Einstein's field equations which we have thus justified

$$(h_\mu^\nu - \tfrac{1}{2}\delta_\mu^\nu h) = -4 \int \frac{[T_\mu^\nu]}{r}\, dxdydz \qquad (93.9)$$

proves to be very useful in permitting for the case of weak fields a straightforward calculation of the small deviations $h_{\mu\nu}$ from the Galilean values for the metrical tensor, whenever the energy-momentum tensor is given as a function of the coordinates. Although the method of treatment provided by this solution is limited to weak fields, it should be specially noted that there is no limitation as to the velocity of the matter producing the field; and this provides a great step forward from the Newtonian level of treatment where there was great uncertainty as to the mechanism and velocity with which gravitational effects would be propagated.

In accordance with the interpretation which we have for the right-hand side of (93.9), it is evident that we must now think of gravitational effects as propagated in the present coordinates with unit velocity, that is, with the same velocity as light. Furthermore, in accordance with the 'wave equation' (93.7), it is evident that we may expect gravitational waves carrying energy and propagated with this velocity. The emission and absorption of such waves, which to be sure may be expected to carry only extremely small amounts of energy, have been investigated by Einstein.†

The solution (93.9) has been used by Thirring and Lense‡ to discuss the effect of the rotation of a central astronomical body on the surrounding gravitational field and hence on the motion of satellites. The effects of such rotations are too small to be of practical astronomical interest. The solution has also been used by Thirring§ to investigate the theoretical problem of the gravitational field inside a thin rotating shell of matter, with the interesting and clarifying result that the rotation of the shell leads as might be expected to analogues of the centrifugal and coriolis forces of ordinary mechanics. In the next chapter the solution will be used in investigating the gravitational field produced by pencils and pulses of light.

94. Line elements for systems with spherical symmetry

Although we have no general solution for Einstein's field equations, except in the above case of weak fields, we can nevertheless often proceed by assuming a form for the solution which corresponds to the nature of the physical problem under consideration, and then investigating the properties of this proposed form. Thus, for example, if the physical system of interest is such that we know that its structure is spatially spherically symmetrical, we can feel sure that coordinates can be chosen in such a way that the line element for the system will exhibit this symmetry.

As the most general form of line element exhibiting spherical symmetry we may evidently write

$$ds^2 = -e^\lambda\,dr^2 - e^\mu(r^2\,d\theta^2 + r^2\sin^2\theta\,d\phi^2) + e^\nu\,dt^2 + 2a\,dr dt, \quad (94.1)$$

where λ, μ, ν, and a are functions of r and t alone and the coefficients $-e^\lambda$, $-e^\mu$, and $+e^\nu$ have been chosen in the exponential form in order to distinguish clearly between the space-like coordinates r, θ, and ϕ and the time-like coordinate t.

This general form of spherically symmetrical line element can,

† Einstein, *Berl. Ber.* 1918, p. 154.
‡ Thirring and Lense, *Phys. Zeits.* 19, 156 (1918).
§ Thirring, *Phys. Zeits.* 19, 33 (1918); ibid. 22, 29 (1921).

however, be subjected to simplifying transformations. To do this we may first introduce a new variable r' in accordance with the equation

$$r'^2 = e^\mu r^2. \tag{94.2}$$

Making this substitution, and dropping primes, it is easily seen that the line element is then expressible in the form

$$ds^2 = -e^\lambda \, dr^2 - r^2 \, d\theta^2 - r^2\sin^2\theta \, d\phi^2 + e^\nu \, dt^2 + 2a \, drdt, \tag{94.3}$$

where λ, ν, and a are now new functions of the new r and of t.

A further simplification which will eliminate the cross product can now be made by substituting a new time variable t' in accordance with the equation

$$dt' = \eta(a \, dr + e^\nu \, dt), \tag{94.4}$$

where η is an integrating factor which will make the right-hand side a perfect differential. In accordance with (94.4) we shall have

$$e^\nu \, dt^2 + 2a \, drdt = \frac{dt'^2}{\eta^2 e^\nu} - \frac{a^2}{e^\nu} \, dr^2, \tag{94.5}$$

so that on substitution into (94.3), and dropping primes, we can then express the line element in the simple standard form

$$ds^2 = -e^\lambda \, dr^2 - r^2 \, d\theta^2 - r^2\sin^2\theta \, d\phi^2 + e^\nu \, dt^2$$
$$\lambda = \lambda(r,t) \qquad \nu = \nu(r,t), \tag{94.6}$$

where λ and ν are functions of the present r and t alone.†

The possibility of eliminating a single cross product by the use of an integrating factor as in (94.4) greatly simplifies the treatment of problems exhibiting spherical symmetry.

For some purposes a somewhat different form of the line element for cases of spherical symmetry is more convenient. This may be obtained from (94.6) by introducing a new variable r' in accordance with the equation

$$\frac{dr'}{r'} = e^{\frac{1}{2}\lambda} \frac{dr}{r}. \tag{94.7}$$

Making this substitution, and dropping primes, the line element can then be expressed in the form

$$ds^2 = -e^\mu(dr^2 + r^2 \, d\theta^2 + r^2\sin^2\theta \, d\phi^2) + e^\nu \, dt^2$$
$$\mu = \mu(r,t) \qquad \nu = \nu(r,t), \tag{94.8}$$

where μ and ν are now functions of the present r and t. And by an

† Lemaitre, *Monthly Notices*, **91**, 490 (1931).

obvious further substitution this can also be rewritten in the form

$$ds^2 = -e^\mu(dx^2 + dy^2 + dz^2) + e^\nu \, dt^2$$

$$\mu = \mu(r, t) \qquad \nu = \nu(r, t) \qquad r = \sqrt{(x^2 + y^2 + z^2)}. \qquad (94.9)$$

These two latter systems of coordinates may be called isotropic.

95. Static line element with spherical symmetry

We must now turn to a more detailed examination of the foregoing proposed forms of solution for the field equations. To assist in obtaining actual solutions from them, we shall desire explicit expressions for the Christoffel three-index symbols and for the components of the energy-momentum tensor in terms of the quantities used in expressing these proposed line elements.

We may first consider physical systems which are static as well as spherically symmetrical. In accordance with (94.6) we can then write our line element in the standard form

$$ds^2 = -e^\lambda \, dr^2 - r^2 \, d\theta^2 - r^2\sin^2\theta \, d\phi^2 + e^\nu \, dt^2$$

$$\lambda = \lambda(r) \qquad \nu = \nu(r). \qquad (95.1)$$

The Christoffel three-index symbols corresponding to this form of line element can easily be evaluated from the definition given by (73.14) and are well known to have the values

$$\{11, 1\} = \tfrac{1}{2}\lambda' \qquad\qquad\qquad \{21, 2\} = 1/r$$
$$\{12, 2\} = 1/r \qquad\qquad\qquad \{22, 1\} = -re^{-\lambda}$$
$$\{13, 3\} = 1/r \qquad\qquad\qquad \{23, 3\} = \cot\theta$$
$$\{14, 4\} = \tfrac{1}{2}\nu'$$

$$\hspace{10cm} (95.2)$$

$$\{31, 3\} = 1/r \qquad\qquad\qquad \{41, 4\} = \tfrac{1}{2}\nu'$$
$$\{32, 3\} = \cot\theta \qquad\qquad\quad \{44, 1\} = \tfrac{1}{2}e^{\nu-\lambda}\nu',$$
$$\{33, 1\} = -r\sin^2\theta \, e^{-\lambda}$$
$$\{33, 2\} = -\sin\theta\cos\theta$$

where accents denote differentiation with respect to r, and all further three-index symbols vanish.

Using these values of the three-index symbols the components $R_{\mu\nu}$ of the contracted Riemann-Christoffel tensor can then be computed with the help of (77.2), and the components $T_{\mu\nu}$ of the energy-momentum tensor obtained from (81.6). It is simplest to express these in the form of the mixed tensor. The only ones which do not vanish are found to be

$$8\pi T_1^1 = -e^{-\lambda}\left(\frac{\nu'}{r}+\frac{1}{r^2}\right)+\frac{1}{r^2}-\Lambda$$

$$8\pi T_2^2 = 8\pi T_3^3 = -e^{-\lambda}\left(\frac{\nu''}{2}-\frac{\lambda'\nu'}{4}+\frac{\nu'^2}{4}+\frac{\nu'-\lambda'}{2r}\right)-\Lambda \qquad (95.3)$$

$$8\pi T_4^4 = e^{-\lambda}\left(\frac{\lambda'}{r}-\frac{1}{r^2}\right)+\frac{1}{r^2}-\Lambda.$$

Instead of taking the line element in the form (94.6), we could of course also use it in the isotropic form (94.8)

$$ds^2 = -e^\mu(dr^2+r^2\,d\theta^2 +r^2\sin^2\theta\,d\phi^2)+e^\nu\,dt^2$$

$$\mu = \mu(r) \qquad \nu = \nu(r). \qquad (95.4)$$

The Christoffel symbols corresponding to this form in the case of a static system are

$$\{11, 1\} = \tfrac{1}{2}\mu' \qquad\qquad\qquad \{21, 2\} = 1/r+\tfrac{1}{2}\mu'$$
$$\{12, 2\} = 1/r+\tfrac{1}{2}\mu' \qquad\qquad \{22, 1\} = -(r+\tfrac{1}{2}r^2\mu')$$
$$\{13, 3\} = 1/r+\tfrac{1}{2}\mu' \qquad\qquad \{23, 3\} = \cot\theta$$
$$\{14, 4\} = \tfrac{1}{2}\nu' \qquad\qquad\qquad\qquad\qquad\qquad (95.5)$$
$$\{31, 3\} = 1/r+\tfrac{1}{2}\mu' \qquad\qquad \{41, 4\} = \tfrac{1}{2}\nu'$$
$$\{32, 3\} = \cot\theta \qquad\qquad\qquad \{44, 1\} = \tfrac{1}{2}e^{\nu-\mu}\nu',$$
$$\{33, 1\} = -(r+\tfrac{1}{2}r^2\mu')\sin^2\theta$$
$$\{33, 2\} = -\sin\theta\cos\theta$$

where accents again denote differentiation with respect to r, and all further three-index symbols vanish.

The non-vanishing components of the energy-momentum tensor corresponding to this form of line element are

$$8\pi T_1^1 = -e^{-\mu}\left(\frac{\mu'^2}{4}+\frac{\mu'\nu'}{2}+\frac{\mu'+\nu'}{r}\right)-\Lambda$$

$$8\pi T_2^2 = 8\pi T_3^3 = -e^{-\mu}\left(\frac{\mu''}{2}+\frac{\nu''}{2}+\frac{\nu'^2}{4}+\frac{\mu'+\nu'}{2r}\right)-\Lambda \qquad (95.6)$$

$$8\pi T_4^4 = -e^{-\mu}\left(\mu''+\frac{\mu'^2}{4}+\frac{2\mu'}{r}\right)-\Lambda.$$

In applying either of the above forms of line element to a system which consists of a *perfect fluid*, we shall have in accordance with

(85.7) for the energy-momentum tensor

$$T^{\mu\nu} = (\rho_{00}+p_0)\frac{dx^\mu}{ds}\frac{dx^\nu}{ds} - g^{\mu\nu}p_0, \tag{95.7}$$

or lowering the index μ

$$T^\nu_\mu = (\rho_{00}+p_0)g_{\alpha\mu}\frac{dx^\alpha}{ds}\frac{dx^\nu}{ds} - g^\nu_\mu p_0. \tag{95.8}$$

Since we are dealing with a static problem, however, we can evidently write, in the case of both of the above line elements, for the components of fluid 'velocity'

$$\frac{dr}{ds} = \frac{d\theta}{ds} = \frac{d\phi}{ds} = 0 \qquad \frac{dt}{ds} = e^{-\frac{1}{2}\nu}. \tag{95.9}$$

Introducing these values into (95.8) we then obtain for the components of the energy-momentum tensor

$$T^1_1 = T^2_2 = T^3_3 = -p_0 \qquad T^4_4 = \rho_{00}, \tag{95.10}$$

which may be substituted in the case of a perfect fluid into (95.3) and (95.6).

Furthermore, in the case of a perfect fluid, the equality between the radial stress T^1_1 and the transverse stresses $T^2_2 = T^3_3$ makes it possible to derive a very simple expression for pressure gradient. Thus equating the two expressions for T^1_1 and T^2_2 given by (95.3) we obtain

$$e^{-\lambda}\left(\frac{\nu''}{2} - \frac{\lambda'\nu'}{4} + \frac{\nu'^2}{4} + \frac{\nu'-\lambda'}{2r} - \frac{\nu'}{r} - \frac{1}{r^2}\right) + \frac{1}{r^2} = 0.$$

And multiplying through by $2/r$ and rearranging terms, this can be rewritten in the form

$$e^{-\lambda}\left(\frac{\nu''}{r} - \frac{\nu'}{r^2} - \frac{2}{r^3}\right) - e^{-\lambda}\lambda'\left(\frac{\nu'}{r} + \frac{1}{r^2}\right) + \frac{2}{r^3} + e^{-\lambda}\left(\frac{\lambda'}{r} + \frac{\nu'}{r}\right)\frac{\nu'}{2} = 0,$$

which on comparison with (95.3) and (95.10) is seen to be equivalent to

$$\frac{dp_0}{dr} + (\rho_{00}+p_0)\frac{\nu'}{2} = 0. \tag{95.11}$$

This is the relativistic analogue of the Newtonian expression for the dependence of pressure on gravitational potential

$$\frac{dp}{dr} + \rho\frac{d\psi}{dr} = 0.$$

A result of exactly the same form as (95.11) can also be obtained in the case of isotropic coordinates by equating the two expressions for

T^1_1 and T^2_2 given by (95.6), and again multiplying by $2/r$ and rearranging terms.

Hence in the case of a static system having spherical symmetry and consisting of a *perfect fluid* we can take the line element in the standard form

$$ds^2 = -e^\lambda\, dr^2 - r^2\, d\theta^2 - r^2\sin^2\theta\, d\phi^2 + e^\nu\, dt^2$$

$$\lambda = \lambda(r) \qquad \nu = \nu(r)$$

(95.12)

with

$$8\pi p_0 = e^{-\lambda}\left(\frac{\nu'}{r} + \frac{1}{r^2}\right) - \frac{1}{r^2} + \Lambda$$

$$8\pi p_0 = e^{-\lambda}\left(\frac{\nu''}{2} - \frac{\lambda'\nu'}{4} + \frac{\nu'^2}{4} + \frac{\nu'-\lambda'}{2r}\right) + \Lambda$$

$$8\pi \rho_{00} = e^{-\lambda}\left(\frac{\lambda'}{r} - \frac{1}{r^2}\right) + \frac{1}{r^2} - \Lambda$$

$$\frac{dp_0}{dr} = -\frac{(\rho_{00}+p_0)\nu'}{2};$$

(95.13)

or in the isotropic form

$$ds^2 = -e^\mu(dr^2 + r^2\, d\theta^2 + r^2\sin^2\theta\, d\phi^2) + e^\nu\, dt^2$$

$$\mu = \mu(r) \qquad \nu = \nu(r)$$

(95.14)

with

$$8\pi p_0 = e^{-\mu}\left(\frac{\mu'^2}{4} + \frac{\mu'\nu'}{2} + \frac{\mu'+\nu'}{r}\right) + \Lambda$$

$$8\pi p_0 = e^{-\mu}\left(\frac{\mu''}{2} + \frac{\nu''}{2} + \frac{\nu'^2}{4} + \frac{\mu'+\nu'}{2r}\right) + \Lambda$$

$$8\pi \rho_{00} = -e^{-\mu}\left(\mu'' + \frac{\mu'^2}{4} + \frac{2\mu'}{r}\right) - \Lambda$$

$$\frac{dp_0}{dr} = -\frac{(\rho_{00}+p_0)\nu'}{2}.$$

(95.15)

It should be noted, moreover, is using equations (95.13) or (95.15) to determine the form of line element in terms of the distribution of density and pressure that they only express three original conditions. In solving the equations, this then permits us when desired to substitute, in place of the more complicated of the two expressions for pressure, the much simpler and physically more illuminating expression for pressure gradient.

96. Schwarzschild's exterior and interior solutions

Before proceeding to more complicated line elements, we may now illustrate the method of using the relations given in the preceding section in order to obtain actual solutions for the form of line element. In § 82 we have already used the relations (95.3) to obtain the line element surrounding an attracting point particle. The solution obtained, however, is equally applicable to the empty space surrounding a finite static system having spherical symmetry, and we may call the result Schwarzschild's *exterior solution*.

To obtain this exterior solution, we may take the line element as being in the form (95.1) already discussed

$$ds^2 = -e^\lambda \, dr^2 - r^2 \, d\theta^2 - r^2 \sin^2\theta \, d\phi^2 + e^\nu \, dt^2 \qquad (96.1)$$

and then set all the components of the energy-momentum tensor which are given by (95.3) as equal to zero in the empty space surrounding the sphere of matter. This will provide us with the three differential equations

$$-e^{-\lambda}\left(\frac{\nu'}{r}+\frac{1}{r^2}\right)+\frac{1}{r^2}-\Lambda = 0$$

$$-e^{-\lambda}\left(\frac{\nu''}{2}-\frac{\lambda'\nu'}{4}+\frac{\nu'^2}{4}+\frac{\nu'-\lambda'}{2r}\right)-\Lambda = 0 \qquad (96.2)$$

$$e^{-\lambda}\left(\frac{\lambda'}{r}-\frac{1}{r^2}\right)+\frac{1}{r^2}-\Lambda = 0,$$

which are readily found to be satisfied by the solution previously given, corresponding to the line element

$$ds^2 = -\frac{dr^2}{1-\dfrac{2m}{r}-\dfrac{\Lambda r^2}{3}}-r^2 \, d\theta^2 - r^2\sin^2\theta \, d\phi^2 + \left(1-\frac{2m}{r}-\frac{\Lambda r^2}{3}\right) dt^2,$$
$$(96.3)$$

where $2m$ is a constant.

This form of solution is to be taken as valid everywhere in the empty space outside the sphere of matter, and must be continued inside the sphere by another form of solution which will depend on the properties of the matter composing the sphere. To obtain such an *interior solution* for a particular case, we may assume with Schwarzschild† that the material composing the sphere consists of an incompressible perfect fluid of constant proper density ρ_{00}. In accordance

† Schwarzschild, *Berl. Ber.* 1916, p. 424.

with (95.13) we can then write

$$8\pi p_0 = e^{-\lambda}\left(\frac{\nu'}{r}+\frac{1}{r^2}\right)-\frac{1}{r^2}+\Lambda \tag{96.4}$$

$$8\pi\rho_{00} = e^{-\lambda}\left(\frac{\lambda'}{r}-\frac{1}{r^2}\right)+\frac{1}{r^2}-\Lambda \tag{96.5}$$

$$\frac{dp_0}{dr} = -\frac{(\rho_{00}+p_0)\nu'}{2}, \tag{96.6}$$

as equations for the interior of the sphere, which are to be solved under the conditions that the pressure be zero at the boundary of the sphere, and that the density ρ_{00} be constant inside this boundary.

As a result of the constancy of ρ_{00} and Λ we can immediately integrate the second of these equations (96.5) and obtain

$$e^{-\lambda} = 1-\frac{\Lambda+8\pi\rho_{00}}{3}r^2+\frac{C}{r}$$

as can readily be verified by redifferentiation, C being a constant of integration; and to remove singularities at the origin, we shall assign the value zero to this constant and write the desired solution for λ in the form

$$e^{-\lambda} = 1-\frac{r^2}{R^2} \quad \text{with} \quad R^2 = \frac{3}{\Lambda+8\pi\rho_{00}}. \tag{96.7}$$

To obtain a solution for ν we may first integrate equation (96.6), which on account of the constancy of ρ_{00}, will give us the simple result

$$(\rho_{00}+p_0) = \text{const. } e^{-\frac{1}{2}\nu}.$$

Combining this with the expressions for p_0 and ρ_{00} given by (96.4) and (96.5) we obtain

$$e^{\frac{1}{2}\nu}e^{-\lambda}\left(\frac{\lambda'}{r}+\frac{\nu'}{r}\right) = \text{const.},$$

and, substituting the value for $e^{-\lambda}$ given by (96.7), this becomes

$$e^{\frac{1}{2}\nu}\left(\frac{2}{R^2}+\frac{\nu'}{r}-\frac{r\nu'}{R^2}\right) = \text{const.},$$

which will be seen to have the solution

$$e^{\frac{1}{2}\nu} = A-B\sqrt{(1-r^2/R^2)}, \tag{96.8}$$

where A and B are the two constants of integration.

In accordance with (96.7) and (96.8) we can then write Schwarz-schild's interior solution for a fluid sphere of constant density ρ_{00} in the form

$$ds^2 = -\frac{dr^2}{1-r^2/R^2} - r^2\,d\theta^2 - r^2\sin^2\theta\,d\phi^2 + [A - B\sqrt{(1-r^2/R^2)}]^2\,dt^2,$$

$$(96.9)$$

or substituting $$\sin\chi = \frac{r}{R}$$ $$(96.10)$$

we can also write it in the form

$$ds^2 = -R^2(d\chi^2 + \sin^2\chi\,d\theta^2 + \sin^2\chi\sin^2\theta\,d\phi^2) + (A - B\cos\chi)^2\,dt^2,$$

$$(96.11)$$

which shows that the spatial geometry inside the fluid is that for the 'surface' of a sphere in four dimensions.

The pressure corresponding to the line element (96.9) is found with the help of (96.4) to be given by

$$8\pi p_0 = \frac{1}{R^2}\left(\frac{3B\sqrt{(1-r^2/R^2)}-A}{A-B\sqrt{(1-r^2/R^2)}}\right) + \Lambda.$$

$$(96\ 12)$$

Neglecting terms containing Λ which can in any case only be of importance at great distances from the origin, we can then make the pressure equal to zero at the boundary of the sphere $r = r_1$, and make the interior solution (96.9) agree at this radius with the exterior solution (96.3) by assigning the following values to the constants in the expressions

$$R^2 = \frac{3}{8\pi\rho_{00}}, \qquad A = \frac{3}{2}\sqrt{\left(1-\frac{r_1^2}{R^2}\right)}, \qquad B = \tfrac{1}{2}, \qquad m = \frac{4\pi}{3}\rho_{00}r_1^3,$$

$$(96.13)$$

which completes the solution of the problem.

In order for the solution to be real we must have

$$r_1^2 < R^2, \qquad r_1^2 < \frac{3}{8\pi\rho_{00}}, \qquad 2m < r_1,$$

$$(96.14)$$

which puts an upper limit on the possible size of a sphere of given density, and on the mass of a sphere of given radius. These limits are very generous, however, and have so far led to no conflict with astrophysical observation.

97. The energy of a sphere of perfect fluid

Before leaving the discussion of spheres of fluid, it will also be of interest to show the possibility of obtaining a very simple expression for their total energy when in a quasi-static state.†

† Tolman, *Phys. Rev.* **35**, 875 (1930).

To do this it will be simplest to consider the line element in the isotropic form

$$ds^2 = -e^\mu(dx^2 + dy^2 + dz^2) + e^\nu \, dt^2$$

$$\mu = \mu(r) \qquad \nu = \nu(r) \qquad r = \sqrt{(x^2 + y^2 + z^2)}. \qquad (97.1)$$

Since the coordinates x, y, z, t for this form of line element are evidently of the quasi-Galilean type, which become Galilean at great distances from the origin, we can then write in accordance with (92.4)

$$U = \iiint (\mathfrak{T}_4^4 - \mathfrak{T}_1^1 - \mathfrak{T}_2^2 - \mathfrak{T}_3^3) \, dxdydz$$

$$= \iiint (T_4^4 - T_1^1 - T_2^2 - T_3^3)e^{\frac{1}{2}(3\mu+\nu)} \, dxdydz, \qquad (97.2)$$

as an expression in these coordinates for the energy of a sphere of fluid having the above line element. We can substitute, moreover, for the component of the energy-momentum tensor the expressions in terms of density ρ_{00} and pressure p_0 given by (95.10), and rewrite the energy in the form

$$U = \iiint (\rho_{00} + 3p_0)e^{\frac{1}{2}(3\mu+\nu)} \, dxdydz. \qquad (97.3)$$

Or finally, noting that the proper spatial volume, corresponding to a coordinate range $dxdydz$, will be

$$dV_0 = e^{\frac{3}{2}\mu} \, dxdydz, \qquad (97.4)$$

we can re-express the energy for a static sphere of perfect fluid in the simple and physically understandable form

$$U = \int (\rho_{00} + 3p_0)e^{\frac{1}{2}\nu} \, dV_0, \qquad (97.5)$$

where the integration is to be taken over the whole proper volume of the sphere.

In the case of weak enough fields, i.e. small enough spheres so that the Newtonian theory of gravitation can be regarded as a satisfactory approximation, it is interesting to show that the above expression for energy would reduce to what might be expected on the basis of Newtonian ideas.

In weak fields, in accordance with (80.9), we can take the ordinary Newtonian potential ψ as given by the expression†

$$\psi = \tfrac{1}{2}(g_{44} - 1) = \tfrac{1}{2}(e^\nu - 1) \simeq \tfrac{1}{2}\nu,$$

† Where c in (80.9) has been set equal to one to correspond to our present units.

and hence can make the approximate substitution

$$e^{\frac{1}{2}\nu} = 1+\psi \tag{97.6}$$

in the formula for the energy of the sphere given by (97.5). This then gives us

$$U = \int (\rho_{00}+3p_0)(1+\psi) \, dV_0$$

$$= \int \rho_{00} \, dV_0 + \int \rho_{00}\psi \, dV_0 + 3 \int p_0 \, dV_0 + 3 \int p_0\psi \, dV_0. \tag{97.7}$$

This expression can be changed, however, into a more recognizable form. Since for weak fields ψ will be small compared with unity, and p_0 for ordinary matter small compared with ρ_{00}, we can neglect the last term in (97.7) in comparison with the other terms, and can drop the subscripts ($_0$) in the second and third terms which specify a proper system of coordinates for the measurement of quantities. We then have

$$U = \int \rho_{00} \, dV_0 + \int \rho\psi \, dV + 3 \int p \, dV. \tag{97.8}$$

On the basis of Newtonian theory, moreover, we can make a further substitution. Integrating over the total volume of the sphere contained within its radius r_1, we can write

$$3 \int p \, dV = 3 \int_0^{r_1} 4\pi r^2 p \, dr$$

$$= |4\pi r^3 p|_0^{r_1} - \int_0^{r_1} 4\pi r^3 \, dp$$

$$= - \int_0^{r_1} 4\pi r^3 \, dp,$$

since the pressure will fall to zero on the boundary of the sphere at r_1. And since $-4\pi r^2 \, dp$ is the total radial force acting outward on the spherical shell of material dM_r lying between the radii r and $r+dr$, we can equate this to the gravitational attraction acting on this shell and write

$$3 \int p \, dV = \int_0^r \frac{M_r}{r} \, dM_r.$$

Or finally, since the right-hand side of this expression is evidently the work which would be necessary to remove the total material of the shell to infinity, we can substitute the usual expression for potential energy and write

$$3 \int p \, dV = - \int \tfrac{1}{2}\rho\psi \, dV. \tag{97.9}$$

Substituting (97.9) in (97.8), we then have for the total energy of the sphere

$$U = \int \rho_{00} \, dV_0 + \int \tfrac{1}{2}\rho\psi \, dV. \tag{97.10}$$

We thus see, at the Newtonian level of approximation, that the relativistic formula for the total energy of a fluid sphere would reduce to the sum of the total proper energy and the usual Newtonian expression for potential gravitational energy. This satisfactory result can serve to increase our confidence in the practical advantages of Einstein's procedure in introducing the pseudo-tensor densities of potential gravitational energy and momentum $\mathfrak{t}_\mu^\nu$.

98. Non-static line elements with spherical symmetry

We must now turn to the more complicated case of *non-static* line elements with spherical symmetry. In accordance with (94.6), we can then assume the solution to be of the standard form

$$ds^2 = -e^\lambda \, dr^2 - r^2 \, d\theta^2 - r^2 \sin^2\theta \, d\phi^2 + e^\nu \, dt^2$$

$$\lambda = \lambda(r, t) \qquad \nu = \nu(r, t). \tag{98.1}$$

The Christoffel symbols corresponding to this form of line element are found to be

$$
\begin{array}{ll}
\{11,1\} = \tfrac{1}{2}\lambda' & \{21,2\} = 1/r \\
\{11,4\} = \tfrac{1}{2}e^{\lambda-\nu}\dot\lambda & \{22,1\} = -re^{-\lambda} \\
\{12,2\} = 1/r & \{23,3\} = \cot\theta \\
\{13,3\} = 1/r & \\
\{14,1\} = \tfrac{1}{2}\dot\lambda & \\
\{14,4\} = \tfrac{1}{2}\nu' &
\end{array}
$$

$$\tag{98.2}$$

$$
\begin{array}{ll}
\{31,3\} = 1/r & \{41,1\} = \tfrac{1}{2}\dot\lambda \\
\{32,3\} = \cot\theta & \{41,4\} = \tfrac{1}{2}\nu' \\
\{33,1\} = -r\sin^2\theta \, e^{-\lambda} & \{44,1\} = \tfrac{1}{2}e^{\nu-\lambda}\nu' \\
\{33,2\} = -\sin\theta\cos\theta & \{44,4\} = \tfrac{1}{2}\dot\nu,
\end{array}
$$

where the accents indicate differentiation with respect to r and the dots with respect to t, and all further three-index symbols vanish.

Using these values of the three-index symbols, the components of the energy-momentum tensor which do not vanish are then found to have the values†

$$8\pi T_1^1 = -e^{-\lambda}\left(\frac{\nu'}{r}+\frac{1}{r^2}\right)+\frac{1}{r^2}-\Lambda$$

$$8\pi T_2^2 = 8\pi T_3^3 = -e^{-\lambda}\left(\frac{\nu''}{2}-\frac{\lambda'\nu'}{4}+\frac{\nu'^2}{4}+\frac{\nu'-\lambda'}{2r}\right)+e^{-\nu}\left(\frac{\ddot\lambda}{2}+\frac{\dot\lambda^2}{4}-\frac{\dot\lambda\dot\nu}{4}\right)-\Lambda$$

$$8\pi T_4^4 = e^{-\lambda}\left(\frac{\lambda'}{r}-\frac{1}{r^2}\right)+\frac{1}{r^2}-\Lambda \qquad\qquad (98.3)$$

$$8\pi T_4^1 = -e^{-\lambda}\frac{\dot\lambda}{r}$$

$$8\pi T_1^4 = e^{-\nu}\frac{\dot\lambda}{r}.$$

It is interesting to compare these expressions for the components of the energy-momentum tensor with the corresponding ones (95.3) for the static case. As has been pointed out by Lemaître it will be noted that the difference lies only in the added term in the components of transverse stress T_2^2 and T_3^3 and the appearance of the new components T_4^1 and T_1^4. Roughly speaking, we can then say that the change from the static to the non-static case corresponds to the appearance of a transverse wave coupled with a radial flow of energy.

We could, of course, also use isotropic coordinates in the case of spherical symmetry, and assume the solution in accordance with (94.8) to be of the form

$$ds^2 = -e^{\mu}(dr^2+r^2\,d\theta^2+r^2\sin^2\theta\,d\phi^2)+e^{\nu}\,dt^2$$

$$\mu = \mu(r,t) \qquad\qquad \nu = \nu(r,t). \qquad\qquad (98.4)$$

The Christoffel symbols corresponding to this form of line element are found to be

$$\{11,1\} = \tfrac{1}{2}\mu' \qquad\qquad\qquad \{21,2\} = 1/r+\tfrac{1}{2}\mu'$$

$$\{11,4\} = \tfrac{1}{2}e^{\mu-\nu}\dot\mu \qquad\qquad \{22,1\} = -(r+\tfrac{1}{2}r^2\mu')$$

$$\{12,2\} = 1/r+\tfrac{1}{2}\mu' \qquad\qquad \{22,4\} = \tfrac{1}{2}r^2e^{\mu-\nu}\dot\mu$$

† The above values for the Christoffel three-index symbols and for the components of the energy-momentum tensor were calculated by Dr. Boris Podolsky and the present writer. The values of the T_μ^ν agree with those obtained for this same line element by Lemaître, *Monthly Notices*, **91**, 490 (1931).

$$\{13,3\} = 1/r + \tfrac{1}{2}\mu' \qquad\qquad \{23,3\} = \cot\theta$$

$$\{14,1\} = \tfrac{1}{2}\dot{\mu} \qquad\qquad\qquad \{24,2\} = \tfrac{1}{2}\dot{\mu}$$

$$\{14,4\} = \tfrac{1}{2}\nu' \tag{98.5}$$

$$\{31,3\} = 1/r + \tfrac{1}{2}\mu' \qquad\qquad \{41,1\} = \tfrac{1}{2}\dot{\mu}$$

$$\{32,3\} = \cot\theta \qquad\qquad\qquad \{41,4\} = \tfrac{1}{2}\nu'$$

$$\{33,1\} = -(r + \tfrac{1}{2}r^2\mu')\sin^2\theta \qquad \{42,2\} = \tfrac{1}{2}\dot{\mu}$$

$$\{33,2\} = -\sin\theta\cos\theta \qquad\qquad \{43,3\} = \tfrac{1}{2}\dot{\mu}$$

$$\{33,4\} = \tfrac{1}{2}r^2\sin^2\theta\, e^{\mu-\nu}\dot{\mu} \qquad \{44,1\} = \tfrac{1}{2}e^{\nu-\mu}\nu'$$

$$\{34,3\} = \tfrac{1}{2}\dot{\mu} \qquad\qquad\qquad \{44,4\} = \tfrac{1}{2}\dot{\nu},$$

where accents again denote differentiation with respect to r and dots with respect to t, and all further three-index symbols vanish.

The non-vanishing components of the energy-momentum tensor corresponding to this form of line element are[†]

$$8\pi T_1^1 = -e^{-\mu}\left(\frac{\mu'^2}{4} + \frac{\mu'\nu'}{2} + \frac{\mu'+\nu'}{r}\right) + e^{-\nu}\left(\ddot{\mu} + \tfrac{3}{4}\dot{\mu}^2 - \frac{\dot{\mu}\dot{\nu}}{2}\right) - \Lambda$$

$$8\pi T_2^2 = 8\pi T_3^3 = -e^{-\mu}\left(\frac{\mu''}{2} + \frac{\nu''}{2} + \frac{\nu'^2}{4} + \frac{\mu'+\nu'}{2r}\right) + e^{-\nu}\left(\ddot{\mu} + \tfrac{3}{4}\dot{\mu}^2 - \frac{\dot{\mu}\dot{\nu}}{2}\right) - \Lambda$$

$$8\pi T_4^4 = -e^{-\mu}\left(\mu'' + \frac{\mu'^2}{4} + \frac{2\mu'}{r}\right) + \tfrac{3}{4}e^{-\nu}\dot{\mu}^2 - \Lambda \tag{98.6}$$

$$8\pi T_4^1 = e^{-\mu}\left(\dot{\mu}' - \frac{\dot{\mu}\nu'}{2}\right)$$

$$8\pi T_1^4 = -e^{-\nu}\left(\dot{\mu}' - \frac{\dot{\mu}\nu'}{2}\right).$$

99. Birkhoff's theorem

The expressions (98.3) for the energy-momentum tensor corresponding to the standard form of line element (98.1) make it easy to derive an interesting theorem originally due to Birkhoff.[‡]

Consider a spherically symmetrical mass of material surrounded by empty space free from matter or radiation. Since all the components

---

[†] The above values for the Christoffel three-index symbols and for the components of the energy-momentum tensor were calculated by Dr. Boris Podolsky and have been checked by Dr. Dingle with the help of the more general results given in § 100.

[‡] Birkhoff, *Relativity and Modern Physics*, Harvard University Press, 1923. See p. 253, § 7.

of the energy-momentum tensor will have to be zero in this empty space, we shall have to have

$$\dot{\lambda} = 0 \tag{99.1}$$

in the space surrounding the sphere of material, as a result of the appearance of $\dot{\lambda}$ in the expressions for T_4^1 and T_1^4 in (98.3).

With $\dot{\lambda} = 0$, however, it will be seen that the expressions for the energy-momentum tensor (98.3) become identical in form with those for the static case given by (95.3), and hence in the empty space surrounding the sphere will again give Schwarzschild's exterior solution (see § 96)

$$e^{-\lambda} = e^{\nu} = 1 - \frac{2m}{r} - \frac{\Lambda r^2}{3}, \tag{99.2}$$

where m will again have to be a constant independent of the time to preserve the truth of (99.1).

Hence the condition of spherical symmetry alone is sufficient to secure Schwarzschild's static exterior solution for the empty space surrounding a sphere of material. And spherically symmetrical pulsations could take place in the sphere without any loss of mass or energy due to gravitational waves. For an actual loss of energy we should have to give up the requirement of empty space surrounding the sphere, and permit an actual flow of matter or radiation.

100. A more general line element

To conclude the present chapter, we may finally give the Christoffel symbols and components of the energy-momentum tensor corresponding to a very general form of line element, which have been computed by Dingle.†

For the line element we shall write

$$ds^2 = -A(dx^1)^2 - B(dx^2)^2 - C(dx^3)^2 + D(dx^4)^2, \tag{100.1}$$

where A, B, C, and D can be any functions of the coordinates, all four of them being regarded as essentially positive quantities so that x^1, x^2, x^3 will be space-like coordinates and x^4 time-like. This line element is more general than any of the previous ones which we have considered. It assumes the possibility of eliminating cross products, but does not require spherical symmetry.

† Dingle, *Proc. Nat. Acad.* **19**, 559 (1933).

The Christoffel symbols corresponding to this line element are

$$\{11,1\} = +\frac{1}{2A}\frac{\partial A}{\partial x^1} \quad \{21,1\} = +\frac{1}{2A}\frac{\partial A}{\partial x^2} \quad \{31,1\} = +\frac{1}{2A}\frac{\partial A}{\partial x^3} \quad \{41,1\} = +\frac{1}{2A}\frac{\partial A}{\partial x^4}$$

$$\{11,2\} = -\frac{1}{2B}\frac{\partial A}{\partial x^2} \quad \{21,2\} = +\frac{1}{2B}\frac{\partial B}{\partial x^1} \quad \{31,2\} = 0 \qquad\qquad \{41,2\} = 0$$

$$\{11,3\} = -\frac{1}{2C}\frac{\partial A}{\partial x^3} \quad \{21,3\} = 0 \qquad\qquad \{31,3\} = +\frac{1}{2C}\frac{\partial C}{\partial x^1} \quad \{41,3\} = 0$$

$$\{11,4\} = +\frac{1}{2D}\frac{\partial A}{\partial x^4} \quad \{21,4\} = 0 \qquad\qquad \{31,4\} = 0 \qquad\qquad \{41,4\} = +\frac{1}{2D}\frac{\partial D}{\partial x^1}$$

$$\{12,1\} = +\frac{1}{2A}\frac{\partial A}{\partial x^2} \quad \{22,1\} = -\frac{1}{2A}\frac{\partial B}{\partial x^1} \quad \{32,1\} = 0 \qquad\qquad \{42,1\} = 0$$

$$\{12,2\} = +\frac{1}{2B}\frac{\partial B}{\partial x^1} \quad \{22,2\} = +\frac{1}{2B}\frac{\partial B}{\partial x^2} \quad \{32,2\} = +\frac{1}{2B}\frac{\partial B}{\partial x^3} \quad \{42,2\} = +\frac{1}{2B}\frac{\partial B}{\partial x^4}$$

$$\{12,3\} = 0 \qquad\qquad \{22,3\} = -\frac{1}{2C}\frac{\partial B}{\partial x^3} \quad \{32,3\} = +\frac{1}{2C}\frac{\partial C}{\partial x^2} \quad \{42,3\} = 0$$

$$\{12,4\} = 0 \qquad\qquad \{22,4\} = +\frac{1}{2D}\frac{\partial B}{\partial x^4} \quad \{32,4\} = 0 \qquad\qquad \{42,4\} = +\frac{1}{2D}\frac{\partial D}{\partial x^2}$$

$$\{13,1\} = +\frac{1}{2A}\frac{\partial A}{\partial x^3} \quad \{23,1\} = 0 \qquad\qquad \{33,1\} = -\frac{1}{2A}\frac{\partial C}{\partial x^1} \quad \{43,1\} = 0$$

$$\{13,2\} = 0 \qquad\qquad \{23,2\} = +\frac{1}{2B}\frac{\partial B}{\partial x^3} \quad \{33,2\} = -\frac{1}{2B}\frac{\partial C}{\partial x^2} \quad \{43,2\} = 0$$

$$\{13,3\} = +\frac{1}{2C}\frac{\partial C}{\partial x^1} \quad \{23,3\} = +\frac{1}{2C}\frac{\partial C}{\partial x^2} \quad \{33,3\} = +\frac{1}{2C}\frac{\partial C}{\partial x^3} \quad \{43,3\} = +\frac{1}{2C}\frac{\partial C}{\partial x^4}$$

$$\{13,4\} = 0 \qquad\qquad \{23,4\} = 0 \qquad\qquad \{33,4\} = +\frac{1}{2D}\frac{\partial C}{\partial x^4} \quad \{43,4\} = +\frac{1}{2D}\frac{\partial D}{\partial x^3}$$

$$\{14,1\} = +\frac{1}{2A}\frac{\partial A}{\partial x^4} \quad \{24,1\} = 0 \qquad\qquad \{34,1\} = 0 \qquad\qquad \{44,1\} = +\frac{1}{2A}\frac{\partial D}{\partial x^1}$$

$$\{14,2\} = 0 \qquad\qquad \{24,2\} = +\frac{1}{2B}\frac{\partial B}{\partial x^4} \quad \{34,2\} = 0 \qquad\qquad \{44,2\} = +\frac{1}{2B}\frac{\partial D}{\partial x^2}$$

$$\{14,3\} = 0 \qquad\qquad \{24,3\} = 0 \qquad\qquad \{34,3\} = +\frac{1}{2C}\frac{\partial C}{\partial x^4} \quad \{44,3\} = +\frac{1}{2C}\frac{\partial D}{\partial x^3}$$

$$\{14,4\} = +\frac{1}{2D}\frac{\partial D}{\partial x^1} \quad \{24,4\} = +\frac{1}{2D}\frac{\partial D}{\partial x^2} \quad \{34,4\} = +\frac{1}{2D}\frac{\partial D}{\partial x^3} \quad \{44,4\} = +\frac{1}{2D}\frac{\partial D}{\partial x^4}$$

$$(100.2)$$

and the components of the energy-momentum tensor T^ν_μ are

$$-8\pi T^1_1 = \frac{1}{2}\left[\frac{1}{BC}\left(\frac{\partial^2 B}{\partial(x^3)^2}+\frac{\partial^2 C}{\partial(x^2)^2}\right)-\frac{1}{BD}\left(\frac{\partial^2 B}{\partial(x^4)^2}-\frac{\partial^2 D}{\partial(x^2)^2}\right)-\right.$$

$$\left.-\frac{1}{CD}\left(\frac{\partial^2 C}{\partial(x^4)^2}-\frac{\partial^2 D}{\partial(x^3)^2}\right)\right]-$$

$$-\frac{1}{4}\left[\frac{1}{BC^2}\left\{\frac{\partial B}{\partial x^3}\frac{\partial C}{\partial x^3}+\left(\frac{\partial C}{\partial x^2}\right)^2\right\}+\frac{1}{CB^2}\left\{\frac{\partial C}{\partial x^2}\frac{\partial B}{\partial x^2}+\left(\frac{\partial B}{\partial x^3}\right)^2\right\}-\right.$$

$$\left.-\frac{1}{BD^2}\left\{\frac{\partial B}{\partial x^4}\frac{\partial D}{\partial x^4}-\left(\frac{\partial D}{\partial x^2}\right)^2\right\}+\frac{1}{DB^2}\left\{\frac{\partial D}{\partial x^2}\frac{\partial B}{\partial x^2}-\left(\frac{\partial B}{\partial x^4}\right)^2\right\}-\right.$$

$$-\frac{1}{CD^2}\left\{\frac{\partial C}{\partial x^4}\frac{\partial D}{\partial x^4}-\left(\frac{\partial D}{\partial x^3}\right)^2\right\}+\frac{1}{DC^2}\left\{\frac{\partial D}{\partial x^3}\frac{\partial C}{\partial x^3}-\left(\frac{\partial C}{\partial x^4}\right)^2\right\}-$$

$$-\frac{1}{BCD}\left\{\frac{\partial C}{\partial x^2}\frac{\partial D}{\partial x^2}+\frac{\partial B}{\partial x^3}\frac{\partial D}{\partial x^3}-\frac{\partial B}{\partial x^4}\frac{\partial C}{\partial x^4}\right\}-\frac{1}{ABC}\frac{\partial B}{\partial x^1}\frac{\partial C}{\partial x^1}-$$

$$-\frac{1}{ABD}\frac{\partial B}{\partial x^1}\frac{\partial D}{\partial x^1}-\frac{1}{ACD}\frac{\partial C}{\partial x^1}\frac{\partial D}{\partial x^1}\Bigg]+\Lambda$$

$$-8\pi T_2^2=\frac{1}{2}\Bigg[\frac{1}{AC}\left(\frac{\partial^2 A}{\partial(x^3)^2}+\frac{\partial^2 C}{\partial(x^1)^2}\right)-\frac{1}{AD}\left(\frac{\partial^2 A}{\partial(x^4)^2}-\frac{\partial^2 D}{\partial(x^1)^2}\right)-$$

$$-\frac{1}{CD}\left(\frac{\partial^2 C}{\partial(x^4)^2}-\frac{\partial^2 D}{\partial(x^3)^2}\right)\Bigg]-$$

$$-\frac{1}{4}\Bigg[\frac{1}{AC^2}\left\{\frac{\partial A}{\partial x^3}\frac{\partial C}{\partial x^3}+\left(\frac{\partial C}{\partial x^1}\right)^2\right\}+\frac{1}{CA^2}\left\{\frac{\partial C}{\partial x^1}\frac{\partial A}{\partial x^1}+\left(\frac{\partial A}{\partial x^3}\right)^2\right\}-$$

$$-\frac{1}{AD^2}\left\{\frac{\partial A}{\partial x^4}\frac{\partial D}{\partial x^4}-\left(\frac{\partial D}{\partial x^1}\right)^2\right\}+\frac{1}{DA^2}\left\{\frac{\partial D}{\partial x^1}\frac{\partial A}{\partial x^1}-\left(\frac{\partial A}{\partial x^4}\right)^2\right\}-$$

$$-\frac{1}{CD^2}\left\{\frac{\partial C}{\partial x^4}\frac{\partial D}{\partial x^4}-\left(\frac{\partial D}{\partial x^3}\right)^2\right\}+\frac{1}{DC^2}\left\{\frac{\partial D}{\partial x^3}\frac{\partial C}{\partial x^3}-\left(\frac{\partial C}{\partial x^4}\right)^2\right\}-$$

$$-\frac{1}{ACD}\left\{\frac{\partial C}{\partial x^1}\frac{\partial D}{\partial x^1}+\frac{\partial A}{\partial x^3}\frac{\partial D}{\partial x^3}-\frac{\partial A}{\partial x^4}\frac{\partial C}{\partial x^4}\right\}-\frac{1}{ABC}\frac{\partial A}{\partial x^2}\frac{\partial C}{\partial x^2}-$$

$$-\frac{1}{ABD}\frac{\partial A}{\partial x^2}\frac{\partial D}{\partial x^2}-\frac{1}{BCD}\frac{\partial C}{\partial x^2}\frac{\partial D}{\partial x^2}\Bigg]+\Lambda$$

$$-8\pi T_3^3=\frac{1}{2}\Bigg[\frac{1}{AB}\left(\frac{\partial^2 A}{\partial(x^2)^2}+\frac{\partial^2 B}{\partial(x^1)^2}\right)-\frac{1}{AD}\left(\frac{\partial^2 A}{\partial(x^4)^2}-\frac{\partial^2 D}{\partial(x^1)^2}\right)-$$

$$-\frac{1}{BD}\left(\frac{\partial^2 B}{\partial(x^4)^2}-\frac{\partial^2 D}{\partial(x^2)^2}\right)\Bigg]-$$

$$-\frac{1}{4}\Bigg[\frac{1}{AB^2}\left\{\frac{\partial A}{\partial x^2}\frac{\partial B}{\partial x^2}+\left(\frac{\partial B}{\partial x^1}\right)^2\right\}+\frac{1}{BA^2}\left\{\frac{\partial B}{\partial x^1}\frac{\partial A}{\partial x^1}+\left(\frac{\partial A}{\partial x^2}\right)^2\right\}-$$

$$-\frac{1}{AD^2}\left\{\frac{\partial A}{\partial x^4}\frac{\partial D}{\partial x^4}-\left(\frac{\partial D}{\partial x^1}\right)^2\right\}+\frac{1}{DA^2}\left\{\frac{\partial D}{\partial x^1}\frac{\partial A}{\partial x^1}-\left(\frac{\partial A}{\partial x^4}\right)^2\right\}-$$

$$-\frac{1}{BD^2}\left\{\frac{\partial B}{\partial x^4}\frac{\partial D}{\partial x^4}-\left(\frac{\partial D}{\partial x^2}\right)^2\right\}+\frac{1}{DB^2}\left\{\frac{\partial D}{\partial x^2}\frac{\partial B}{\partial x^2}-\left(\frac{\partial B}{\partial x^4}\right)^2\right\}-$$

$$-\frac{1}{ABD}\left\{\frac{\partial B}{\partial x^1}\frac{\partial D}{\partial x^1}+\frac{\partial A}{\partial x^2}\frac{\partial D}{\partial x^2}-\frac{\partial A}{\partial x^4}\frac{\partial B}{\partial x^4}\right\}-\frac{1}{ABC}\frac{\partial A}{\partial x^3}\frac{\partial B}{\partial x^3}-$$

$$-\frac{1}{ACD}\frac{\partial A}{\partial x^3}\frac{\partial D}{\partial x^3}-\frac{1}{BCD}\frac{\partial B}{\partial x^3}\frac{\partial D}{\partial x^3}\Bigg]+\Lambda$$

$$-8\pi T_4^4 = \frac{1}{2}\left[\frac{1}{AB}\left(\frac{\partial^2 A}{\partial(x^2)^2}+\frac{\partial^2 B}{\partial(x^1)^2}\right)+\frac{1}{AC}\left(\frac{\partial^2 A}{\partial(x^3)^2}+\frac{\partial^2 C}{\partial(x^1)^2}\right)+\right.$$

$$\left.+\frac{1}{BC}\left(\frac{\partial^2 B}{\partial(x^3)^2}+\frac{\partial^2 C}{\partial(x^2)^2}\right)\right]-$$

$$-\frac{1}{4}\left[\frac{1}{AB^2}\left\{\frac{\partial A}{\partial x^2}\frac{\partial B}{\partial x^2}+\left(\frac{\partial B}{\partial x^1}\right)^2\right\}+\frac{1}{BA^2}\left\{\frac{\partial B}{\partial x^1}\frac{\partial A}{\partial x^1}+\left(\frac{\partial A}{\partial x^2}\right)^2\right\}+$$

$$+\frac{1}{AC^2}\left\{\frac{\partial A}{\partial x^3}\frac{\partial C}{\partial x^3}+\left(\frac{\partial C}{\partial x^1}\right)^2\right\}+\frac{1}{CA^2}\left\{\frac{\partial C}{\partial x^1}\frac{\partial A}{\partial x^1}+\left(\frac{\partial A}{\partial x^3}\right)^2\right\}+$$

$$+\frac{1}{BC^2}\left\{\frac{\partial B}{\partial x^3}\frac{\partial C}{\partial x^3}+\left(\frac{\partial C}{\partial x^2}\right)^2\right\}+\frac{1}{CB^2}\left\{\frac{\partial C}{\partial x^2}\frac{\partial B}{\partial x^2}+\left(\frac{\partial B}{\partial x^3}\right)^2\right\}+$$

$$-\frac{1}{ABC}\left(\frac{\partial B}{\partial x^1}\frac{\partial C}{\partial x^1}+\frac{\partial A}{\partial x^2}\frac{\partial C}{\partial x^2}+\frac{\partial A}{\partial x^3}\frac{\partial B}{\partial x^3}\right)+\frac{1}{ABD}\frac{\partial A}{\partial x^4}\frac{\partial B}{\partial x^4}+$$

$$\left.+\frac{1}{ACD}\frac{\partial A}{\partial x^4}\frac{\partial C}{\partial x^4}+\frac{1}{BCD}\frac{\partial B}{\partial x^4}\frac{\partial C}{\partial x^4}\right]+\Lambda$$

$$-8\pi A T_2^1 = -8\pi B T_1^2$$

$$= -\frac{1}{2}\left[\frac{1}{C}\frac{\partial^2 C}{\partial x^1\partial x^2}+\frac{1}{D}\frac{\partial^2 D}{\partial x^1\partial x^2}\right]+\frac{1}{4}\left[\frac{1}{C^2}\frac{\partial C}{\partial x^1}\frac{\partial C}{\partial x^2}+\frac{1}{D^2}\frac{\partial D}{\partial x^1}\frac{\partial D}{\partial x^2}+\right.$$

$$\left.+\frac{1}{AC}\frac{\partial A}{\partial x^2}\frac{\partial C}{\partial x^1}+\frac{1}{AD}\frac{\partial A}{\partial x^2}\frac{\partial D}{\partial x^1}+\frac{1}{BC}\frac{\partial B}{\partial x^1}\frac{\partial C}{\partial x^2}+\frac{1}{BD}\frac{\partial B}{\partial x^1}\frac{\partial D}{\partial x^2}\right]$$

$$-8\pi A T_3^1 = -8\pi C T_1^3$$

$$= -\frac{1}{2}\left[\frac{1}{B}\frac{\partial^2 B}{\partial x^1\partial x^3}+\frac{1}{D}\frac{\partial^2 D}{\partial x^1\partial x^3}\right]+\frac{1}{4}\left[\frac{1}{B^2}\frac{\partial B}{\partial x^1}\frac{\partial B}{\partial x^3}+\frac{1}{D^2}\frac{\partial D}{\partial x^1}\frac{\partial D}{\partial x^3}+\right.$$

$$\left.+\frac{1}{AB}\frac{\partial A}{\partial x^3}\frac{\partial B}{\partial x^1}+\frac{1}{AD}\frac{\partial A}{\partial x^3}\frac{\partial D}{\partial x^1}+\frac{1}{CB}\frac{\partial C}{\partial x^1}\frac{\partial B}{\partial x^3}+\frac{1}{CD}\frac{\partial C}{\partial x^1}\frac{\partial D}{\partial x^3}\right]$$

$$-8\pi B T_3^2 = -8\pi C T_2^3$$

$$= -\frac{1}{2}\left[\frac{1}{A}\frac{\partial^2 A}{\partial x^2\partial x^3}+\frac{1}{D}\frac{\partial^2 D}{\partial x^2\partial x^3}\right]+\frac{1}{4}\left[\frac{1}{A^2}\frac{\partial A}{\partial x^2}\frac{\partial A}{\partial x^3}+\frac{1}{D^2}\frac{\partial D}{\partial x^2}\frac{\partial D}{\partial x^3}+\right.$$

$$\left.+\frac{1}{AB}\frac{\partial A}{\partial x^2}\frac{\partial B}{\partial x^3}+\frac{1}{AC}\frac{\partial A}{\partial x^3}\frac{\partial C}{\partial x^2}+\frac{1}{DB}\frac{\partial D}{\partial x^2}\frac{\partial B}{\partial x^3}+\frac{1}{DC}\frac{\partial D}{\partial x^3}\frac{\partial C}{\partial x^2}\right]$$

$$-8\pi A T_4^1 = +8\pi D T_1^4$$

$$= -\frac{1}{2}\left[\frac{1}{B}\frac{\partial^2 B}{\partial x^1\partial x^4}+\frac{1}{C}\frac{\partial^2 C}{\partial x^1\partial x^4}\right]+\frac{1}{4}\left[\frac{1}{B^2}\frac{\partial B}{\partial x^1}\frac{\partial B}{\partial x^4}+\frac{1}{C^2}\frac{\partial C}{\partial x^1}\frac{\partial C}{\partial x^4}+\right.$$

$$\left.+\frac{1}{AB}\frac{\partial A}{\partial x^4}\frac{\partial B}{\partial x^1}+\frac{1}{AC}\frac{\partial A}{\partial x^4}\frac{\partial C}{\partial x^1}+\frac{1}{DB}\frac{\partial D}{\partial x^1}\frac{\partial B}{\partial x^4}+\frac{1}{DC}\frac{\partial D}{\partial x^1}\frac{\partial C}{\partial x^4}\right]$$

$$-8\pi BT_4^2 = +8\pi DT_2^4$$

$$= -\frac{1}{2}\left[\frac{1}{A}\frac{\partial^2 A}{\partial x^2 \partial x^4}+\frac{1}{C}\frac{\partial^2 C}{\partial x^2 \partial x^4}\right]+\frac{1}{4}\left[\frac{1}{A^2}\frac{\partial A}{\partial x^2}\frac{\partial A}{\partial x^4}+\frac{1}{C^2}\frac{\partial C}{\partial x^2}\frac{\partial C}{\partial x^4}+\right.$$

$$\left.+\frac{1}{AB}\frac{\partial A}{\partial x^2}\frac{\partial B}{\partial x^4}+\frac{1}{AD}\frac{\partial A}{\partial x^4}\frac{\partial D}{\partial x^2}+\frac{1}{CB}\frac{\partial C}{\partial x^2}\frac{\partial B}{\partial x^4}+\frac{1}{DC}\frac{\partial D}{\partial x^2}\frac{\partial C}{\partial x^4}\right]$$

$$-8\pi CT_4^3 = +8\pi DT_3^4$$

$$= -\frac{1}{2}\left[\frac{1}{A}\frac{\partial^2 A}{\partial x^3 \partial x^4}+\frac{1}{B}\frac{\partial^2 B}{\partial x^3 \partial x^4}\right]+\frac{1}{4}\left[\frac{1}{A^2}\frac{\partial A}{\partial x^3}\frac{\partial A}{\partial x^4}+\frac{1}{B^2}\frac{\partial B}{\partial x^3}\frac{\partial B}{\partial x^4}+\right.$$

$$\left.+\frac{1}{AC}\frac{\partial A}{\partial x^3}\frac{\partial C}{\partial x^4}+\frac{1}{AD}\frac{\partial A}{\partial x^4}\frac{\partial D}{\partial x^3}+\frac{1}{BC}\frac{\partial B}{\partial x^3}\frac{\partial C}{\partial x^4}+\frac{1}{BD}\frac{\partial B}{\partial x^4}\frac{\partial D}{\partial x^3}\right]$$

$$(100.3)$$

VIII

RELATIVISTIC ELECTRODYNAMICS

Part I. THE COVARIANT GENERALIZATION OF ELECTRICAL THEORY

101. Introduction

In the present chapter we shall give a brief account of the extension of electrodynamics to general relativity which is customarily made and which can be based on the electrodynamics of special relativity already considered in Chapter IV. We shall also consider some applications which are of interest for our further work.

We shall first consider the relativistic generalization of the Lorentz electron theory in spite of the difficulties, which we have previously emphasized, that arise from the fact that the Lorentz theory is developed from a microscopic point of view and yet ignores those restrictions on a correct microscopic treatment which must eventually be introduced in accordance with the more recent development of quantum theory. We shall then give some attention to the generalization of the macroscopic theory developed in the second part of Chapter IV.

102. The generalized Lorentz electron theory. The field equations

In the special theory of relativity it was found in § 46 that the Maxwell-Lorentz field equations could be expressed in Galilean coordinates with the help of two vectors, the *generalized potential* ϕ^μ, whose components are given in terms of the ordinary vector potential **A** and scalar potential ϕ, by the expression

$$\phi^\mu = (A_x, A_y, A_z, \phi), \tag{102.1}$$

and the *generalized current density* J^μ, whose components can be given in terms of proper charge density ρ_0 and coordinate velocity dx^μ/ds, or in terms of densities of charge ρ and current ρu referred to the coordinates being used, by the expressions

$$J^\mu = \rho_0 \frac{dx^\mu}{ds} = \left(\rho \frac{u_x}{c}, \quad \rho \frac{u_y}{c}, \quad \rho \frac{u_z}{c}, \quad \rho\right). \tag{102.2}$$

With the help of these vectors, the full content of the Maxwell-Lorentz field equations, using the Galilean coordinates permitted in the special theory of relativity, can then be expressed by the two equations

$$F_{\mu\nu} = \frac{\partial \phi_\mu}{\partial x^\nu} - \frac{\partial \phi_\nu}{\partial x^\mu} \tag{102.3}$$

and

$$\frac{\partial F^{\mu\nu}}{\partial x^\nu} = J^\mu, \tag{102.4}$$

where the first equation defines the antisymmetric electromagnetic *field tensor* $F_{\mu\nu}$ in terms of the potential, and the second equation relates the field tensor to the current vector.

The foregoing equations are to be taken as valid in the 'flat' space-time of special relativity and are expressed in the Galilean coordinates which may then be used. In accordance with the principle of equivalence, however, the analogous general relativity equations must also reduce to this same form when expressed in natural coordinates for the particular point of interest. Hence it is reasonable to assume that the completely relativistic field equations can be taken as being merely the covariant re-expression of the above equations of the special theory.

The covariant re-expression of the above equations is very simple. The equations of definition for the generalized potential and current (102.1) and (102.2) will need no modification, since by defining these vectors in one system of coordinates they have then been defined— with the help of the rules for tensor transformation—in all systems of coordinates. To obtain the covariant re-expression for the two remaining equations, we shall have only to substitute covariant differentiation for ordinary differentiation and write as the electro-magnetic field equations in general relativity

$$F_{\mu\nu} = (\phi_\mu)_\nu - (\phi_\nu)_\mu = \frac{\partial \phi_\mu}{\partial x^\nu} - \frac{\partial \phi_\nu}{\partial x^\mu} \tag{102.5}$$

and

$$(F^{\mu\nu})_\nu = J^\mu, \tag{102.6}$$

where the first of these equations is not even changed in form, owing to the mutual cancellation of the two terms containing Christoffel symbols which arise from the indicated covariant differentiation.

103. The motion of a charged particle

In addition to the field equations we shall also need to include in the theory an expression describing the motion of charged particles. This must of course be a covariant generalization of the fifth funda-mental equation of the Maxwell-Lorentz equation (41.4) for the force acting on a moving particle. The desired expression can be taken as

$$\frac{d^2x^\mu}{ds^2} + \{\alpha\beta, \mu\}\frac{dx^\alpha}{ds}\frac{dx^\beta}{ds} + \frac{e}{m_0}F^\mu_\alpha\frac{dx^\alpha}{ds} = 0, \tag{103.1}$$

where e/m_0 is the ratio of the charge of the particle to its rest mass and $F^\mu_\alpha = g^{\mu\epsilon}F_{\epsilon\alpha}$ is the electromagnetic field tensor already introduced. It will be seen that the equation gives expression to the combined action of the gravitational and electromagnetic fields on the particle.

To show that the above equation is a satisfactory generalization of the usual law of force for a charged particle, we must show in the first place that it is a covariant expression true in all coordinate systems if true in one, and in the second place that it reduces in natural coordinates to the usual equation for the force on a moving particle.

To show that it is a covariant expression, it is most convenient to note that it can evidently be rewritten in the form

$$\left[\left(\frac{dx^\mu}{ds}\right)_\alpha + \frac{e}{m_0}F^\mu_\alpha\right]\left(\frac{dx^\alpha}{ds}\right) = 0, \tag{103.2}$$

which is seen to be a tensor equation of rank one.

To show that it reduces in natural coordinates to the usual expression for electromagnetic force, we note that the Christoffel three-index symbols will then be zero, corresponding to the disappearance of gravitational effects with respect to freely falling axes. Making use of the expressions for the field tensor $F_{\mu\nu}$ in natural coordinates given by (46.9), and remembering in accordance with (20.5) that we can take ds/dt as the Lorentz factor of contraction $\sqrt{(1-u^2)}$ where u is the ordinary velocity of the particle in our present units, it will then be found that the four equations corresponding to (103.1) can be written in the familiar form

$$\frac{d}{dt}\left(\frac{m_0 u_r}{\sqrt{(1-u^2)}}\right) = eE_x + e(u_y H_z - u_z H_y)$$

$$\frac{d}{dt}\left(\frac{m_0 u_y}{\sqrt{(1-u^2)}}\right) = eE_y + e(u_z H_x - u_x H_z)$$

$$\frac{d}{dt}\left(\frac{m_0 u_z}{\sqrt{(1-u^2)}}\right) = eE_z + e(u_x H_y - u_y H_x) \tag{103.3}$$

$$\frac{d}{dt}\left(\frac{m_0}{\sqrt{(1-u^2)}}\right) = e(u_x E_x + u_y E_y + u_z E_z),$$

which—in our present units—are seen to be the usual equations for the action of the electromagnetic field in changing the momentum and energy of the particle.

104. The energy-momentum tensor

To complete the translation of the Lorentz electrodynamics into relativistic form we must also have a covariant expression for the electromagnetic energy-momentum tensor. This is found to be given in terms of the field tensor $F_{\mu\nu}$ by the formula

$$[T^{\mu\nu}]_{em} = -g^{\nu\beta}F^{\mu\alpha}F_{\beta\alpha} + \tfrac{1}{4}g^{\mu\nu}F^{\alpha\beta}F_{\alpha\beta}. \qquad (104.1)$$

This expression is easily seen to satisfy the necessary requirements. The expression is evidently covariant since it is a tensor equation of rank two. And substituting the values given by (46.9) for the components of the field tensor in natural coordinates, we find that the above expression does reduce in such coordinates to the special relativity expression for the electromagnetic energy-momentum, tensor as previously given by (46.20) and (46.21). Thus typical examples for the components of $[T^{\mu\nu}]_{em}$ are found, as would be expected in our present units which make $c = 1$, to be given by

$$\begin{aligned}
T^{11} &= -\tfrac{1}{2}(E_x^2 - E_y^2 - E_z^2 + H_x^2 - H_y^2 - H_z^2) \\
T^{12} &= -(E_x E_y + H_x H_y) \\
T^{14} &= (E_y H_z - E_z H_y) \\
T^{44} &= \tfrac{1}{2}(E_x^2 + E_y^2 + E_z^2 + H_x^2 + H_y^2 + H_z^2).
\end{aligned} \qquad (104.2)$$

Assuming the possibility already discussed in § 45 of combining analogous mechanical and electrical quantities, we could now state the energy-momentum principle for a combined mechanical and electrical system in the covariant form

$$(T^{\mu\nu})_\nu = ([T^{\mu\nu}]_{me} + [T^{\mu\nu}]_{em})_\nu = 0, \qquad (104.3)$$

corresponding to the previous special relativity form (46.22).

This completes all that is necessary for the covariant re-expression of the Lorentz electron theory in a form consonant with general relativity.

105. The generalized macroscopic theory

As already discussed and emphasized the Lorentz electron theory has a somewhat unsatisfactory status, owing to its microscopic character. It is hence interesting to find that the macroscopic theory, developed in the second part of Chapter IV, can also easily be re-expressed in a covariant form suitable for incorporation in general relativity.

The macroscopic theory in special relativity was based on two anti-symmetric *field tensors* $F^{\mu\nu}$ and $H^{\mu\nu}$ and on the *current vector* J^{μ}. These three tensors were defined in § 50 by giving their components in a system of Galilean coordinates so chosen that the electromagnetic medium under consideration would be macroscopically at rest. In these coordinates the field tensors have components which are directly given by the components of Maxwell's four familiar vectors of electric field strength **E**, electric displacement **D**, magnetic field strength **H**, and magnetic induction **B**, as they would be determined by an observer at rest in the medium. And the components of the current vector are given by the densities of current flow and of electric charge also as measured by such a special observer.

In building the macroscopic theory in general relativity, it is evident that we may at once take over these same tensors $F^{\mu\nu}$, $H^{\mu\nu}$, and J^{μ} into the general theory, since we can now define them in an entirely similar manner by reference to the measurements of a local observer using proper coordinates for the particular point of interest, and having defined the components in these proper coordinates we have then defined them by the rules of tensor transformation in all coordinates.

To proceed with the generalization of the macroscopic theory we must then make sure that our previous field equations given in § 50 are expressed in covariant form. This is already true for the first of the two equations

$$\frac{\partial F_{\mu\nu}}{\partial x^{\sigma}} + \frac{\partial F_{\nu\sigma}}{\partial x^{\mu}} + \frac{\partial F_{\sigma\mu}}{\partial x^{\nu}} = 0, \qquad (105.1)$$

owing to a mutual cancellation of three-index symbols, which arises when the corresponding covariant derivatives are taken on account of the antisymmetric character of the tensor $F_{\mu\nu}$. To make the second of the field equations covariant it is only necessary to replace ordinary differentiation by covariant differentiation and write

$$(H^{\mu\nu})_{\nu} = J^{\mu}. \qquad (105.2)$$

Finally as a possible set of equations to complete the macroscopic theory, we may take the constitutive equations in the covariant form in which they have already been written in § 51

$$H_{\alpha\beta}\frac{dx^\alpha}{ds} = \epsilon F_{\alpha\beta}\frac{dx^\alpha}{ds},$$

$$(g_{\alpha\beta}F_{\gamma\delta}+g_{\alpha\gamma}F_{\delta\beta}+g_{\alpha\delta}F_{\beta\gamma})\frac{dx^\alpha}{ds} = \mu(g_{\alpha\beta}H_{\gamma\delta}+g_{\alpha\gamma}H_{\delta\beta}+g_{\alpha\delta}H_{\beta\gamma})\frac{dx^\alpha}{ds},$$

$$J^\alpha-J_\beta\frac{dx^\beta}{ds}\frac{dx^\alpha}{ds} = \frac{\sigma}{c}g_{\beta\gamma}F^{\gamma\alpha}\frac{dx^\beta}{ds}, \tag{105.3}$$

where ϵ, μ, and σ are dielectric constant, magnetic permeability, and conductivity of the material as measured by a local observer, and dx^α/ds and dx^β/ds refer to the macroscopic velocity of the medium at the point of interest. These constitutive equations contain of course the usual approximations involved in assuming that the matter can be characterized at each point by the three scalars ϵ, μ, and σ.

The extension of the macroscopic theory to general relativity is thus straightforward. The equations obtained, however, are by no means simple and have as yet been little applied.

VIII

RELATIVISTIC ELECTRODYNAMICS (*contd.*)

Part II. SOME APPLICATIONS OF RELATIVISTIC ELECTRODYNAMICS

106. The conservation of electric charge

We may now turn to certain applications of relativistic electro-dynamics which will be of interest. The results obtained will suffer to some extent from the unsatisfactory character of the Lorentz electron theory which we have already emphasized.

We may first consider the relativistic analogue of the classical expression for the conservation of electric charge.

Introducing tensor densities, the second of our two field equations (102.6) can be written in the form (see Appendix III, equation 48),

$$\frac{\partial \mathfrak{F}^{\mu\nu}}{\partial x^{\nu}} = \mathfrak{J}^{\mu}, \tag{106.1}$$

and owing to the antisymmetry of $\mathfrak{F}^{\mu\nu}$ this leads to the result

$$\frac{\partial^2 \mathfrak{F}^{\mu\nu}}{\partial x^{\mu} \partial x^{\nu}} = \frac{\partial \mathfrak{J}^{\mu}}{\partial x^{\mu}} = 0. \tag{106.2}$$

In place of the current density, however, we may introduce the expression by which J^{μ} was defined (102.2), and rewrite this equation in the form

$$\frac{\partial}{\partial x^{\mu}} \left(\rho_0 \frac{dx^{\mu}}{ds} \sqrt{-g} \right) = 0, \tag{106.3}$$

where ρ_0 is the proper density of charge as it appears to a local observer and dx^{μ}/ds is the 'velocity' of the charge.

To show that this result implies the conservation of electricity, we may most conveniently examine its implications in a system of natural coordinates for the point of interest x, y, z, t. In agreement with the Galilean values which we shall then have for the $g_{\mu\nu}$ and with the disappearance of their first derivatives with respect to the coordinates, we can then substitute

$$\sqrt{-g} = 1 \qquad \frac{\partial}{\partial x^{\mu}} \sqrt{-g} = 0,$$

and rewrite (106.3) in the form

$$\frac{\partial}{\partial x} \left(\rho_0 \frac{dt}{ds} \frac{dx}{dt} \right) + \frac{\partial}{\partial y} \left(\rho_0 \frac{dt}{ds} \frac{dy}{dt} \right) + \frac{\partial}{\partial z} \left(\rho_0 \frac{dt}{ds} \frac{dz}{dt} \right) + \frac{\partial}{\partial t} \left(\rho_0 \frac{dt}{ds} \right) = 0;$$

or since dt/ds is the factor of Lorentz contraction this can again be rewritten in the form

$$\frac{\partial}{\partial x}(\rho u_x)+\frac{\partial}{\partial y}(\rho u_y)+\frac{\partial}{\partial z}(\rho u_z)+\frac{\partial\rho}{\partial t}=0, \tag{106.4}$$

where ρ is the density of charge and u_x, u_y, and u_z the ordinary components of velocity with respect to the present coordinates.

The result is, however, the usual equation of continuity for the 'substance' whose density is ρ, and the conservation of electricity has been demonstrated as desired.

107. The gravitational field of a charged particle

As a second application of relativistic electrodynamics, it will be interesting to consider the gravitational field of a charged particle.

Taking the particle as being at rest at the origin of our system of coordinates, we can evidently write the line element in the standard spherically symmetrical form

$$ds^2 = -e^\lambda\, dr^2 -r^2\, d\theta^2 -r^2\sin^2\theta\, d\phi^2 +e^\nu\, dt^2, \tag{107.1}$$

where λ and ν are functions of r alone which approach zero at very large values of r. To solve for λ and ν we must first consider the electric field surrounding the particle.

Taking the potentials ϕ_μ as functions of r alone, and substituting into the expression (102.5) which defines the field tensor $F_{\mu\nu}$ in terms of these potentials, we then see that the only components which could at the very most survive would be

$$F_{21} = -F_{12} \qquad F_{31} = -F_{13} \qquad F_{41} = -F_{14}.$$

It is easily shown, moreover, that the first two of these components would also vanish, since on substituting into the second of the two field equations in the form (106.1) we should have in the space surrounding the particle

$$\frac{\partial \mathfrak{F}^{2\nu}}{\partial x^\nu} = \frac{\partial}{\partial r}[g^{22}g^{11}F_{21}\sqrt{-g}] = \frac{\partial}{\partial r}(F_{21}\,e^{-\frac12(\lambda-\nu)}\sin\theta) = 0,$$

or
$$F_{21} = \text{const.}\,e^{\frac12(\lambda-\nu)},$$

together with a similar expression for F_{31}. At large distances from the particle, however, where λ and ν approach zero and the ordinary equations for the electromagnetic field become valid, we know that F_{21} would be zero from its relation to magnetic field strength and must hence conclude that the constant has the value zero, so that F_{21} and similarly F_{31} are zero throughout.

To obtain an expression for the sole remaining component of the field tensor, we have by again applying (106.1)

$$\frac{\partial \mathfrak{F}^{4\nu}}{\partial x^\nu} = \frac{\partial}{\partial r} [g^{44}g^{11}F_{41}\sqrt{-g}] = \frac{\partial}{\partial r}(-F_{41} r^2 e^{-\frac{1}{2}(\lambda+\nu)} \sin\theta) = 0,$$

or

$$-F_{41} = F_{14} = \frac{\epsilon}{r^2} e^{\frac{1}{2}(\lambda+\nu)}, \tag{107.2}$$

where ϵ is a constant of integration such that $4\pi\epsilon$ can be identified with the charge on the particle in our present (Heaviside, relativistic) units, owing to the known relation of F_{41} to the electric field strength at sufficient distances from the particle.

Having obtained this result for the surviving component of the field tensor, we can now substitute into the expression for the energy-momentum tensor (104.1) and readily obtain as the only components

$$T_1^1 = -T_2^2 = -T_3^3 = T_4^4 = \frac{1}{2}\frac{\epsilon^2}{r^4}. \tag{107.3}$$

These expressions for the energy-momentum tensor may now be equated to the expressions for this tensor in terms of λ and ν as furnished by (95.3) to give us the differential equations:

$$\frac{4\pi\epsilon^2}{r^4} = -e^{-\lambda}\left(\frac{\nu'}{r}+\frac{1}{r^2}\right)+\frac{1}{r^2},$$

$$\frac{4\pi\epsilon^2}{r^4} = e^{-\lambda}\left(\frac{\nu''}{2}-\frac{\lambda'\nu'}{4}+\frac{\nu'^2}{4}+\frac{\nu'-\lambda'}{2r}\right), \tag{107.4}$$

$$\frac{4\pi\epsilon^2}{r^4} = e^{-\lambda}\left(\frac{\lambda'}{r}-\frac{1}{r^2}\right)+\frac{1}{r^2},$$

where the cosmological constant Λ has been taken as equal to zero as not of present interest. And these equations are readily seen to have a solution corresponding to the line element

$$ds^2 = -\frac{dr^2}{1-\dfrac{2m}{r}+\dfrac{4\pi\epsilon^2}{r^2}} -r^2\,d\theta^2 -r^2\sin^2\theta\,d\phi^2 +\left(1-\frac{2m}{r}+\frac{4\pi\epsilon^2}{r^2}\right)dt^2. \tag{107.5}$$

This result is interesting in showing the contribution of the energy of the electric field surrounding the charge to the curvature of spacetime. For any actual charged particle the gravitational effect of the electrical energy would be negligible compared with that of the intrinsic mass m at reasonable distance from the particle. Thus, if we considered a particle with the mass and charge customarily assigned to the negative electron, we find, paying due regard to units, that the two terms contributing to the curvature would stand in the ratio

$$\frac{4\pi\epsilon^2/r^2}{2m/r} = \frac{2\pi\epsilon^2}{mr} = \frac{1\cdot5\times10^{-13}}{r},$$

where r is in centimetres. Hence we see that deviations from flat space-time due to the charge would be negligible compared with those due to the mass, except at exceedingly small distances from the particle.

108. The propagation of electromagnetic waves

We may next investigate the propagation of electromagnetic disturbances in space. In doing this we shall be interested in the propagation of the components of the field tensor $F_{\mu\nu}$ which have a fairly immediate physical significance, rather than in the propagation of the components of the potential ϕ_μ which are less directly interpretable.

Following a method due to Eddington,[†] we may write in accordance with the two field equations (102.5, 6), after differentiating with respect to ν,

$$J_{\mu\nu} = F^\alpha_{\mu\alpha\nu} = g^{\alpha\beta}F_{\mu\beta\alpha\nu}$$
$$= g^{\alpha\beta}(\phi_{\mu\beta\alpha\nu}-\phi_{\beta\mu\alpha\nu}),$$

and by a known theorem of the tensor analysis [see Appendix III, equation (43)] this can be re-expressed by introducing the Riemann-Christoffel tensor in the form

$$J_{\mu\nu} = g^{\alpha\beta}(\phi_{\mu\beta\nu\alpha}-\phi_{\beta\mu\nu\alpha})-g^{\alpha\beta}(R^\epsilon_{\mu\nu\alpha}\phi_{\epsilon\beta}+R^\epsilon_{\beta\nu\alpha}\phi_{\mu\epsilon}-R^\epsilon_{\beta\nu\alpha}\phi_{\epsilon\mu}-R^\epsilon_{\mu\nu\alpha}\phi_{\beta\epsilon})$$
$$= g^{\alpha\beta}(\phi_{\mu\beta\nu}-\phi_{\beta\mu\nu})_\alpha-g^{\alpha\beta}(R^\epsilon_{\mu\nu\alpha}F_{\epsilon\beta}-R^\epsilon_{\beta\nu\alpha}F_{\epsilon\mu})$$
$$= g^{\alpha\beta}(\phi_{\mu\beta\nu}-\phi_{\beta\mu\nu})_\alpha-R_{\mu\nu\alpha\epsilon}F^{\epsilon\alpha}-R^\epsilon_\nu F_{\epsilon\mu}.$$

Hence subtracting the analogous expression for $J_{\nu\mu}$ we obtain

$$J_{\mu\nu}-J_{\nu\mu} = g^{\alpha\beta}(\phi_{\mu\beta\nu}-\phi_{\nu\beta\mu}-\phi_{\beta\mu\nu}+\phi_{\beta\nu\mu})_\alpha-$$
$$-(R_{\mu\nu\alpha\epsilon}-R_{\nu\mu\alpha\epsilon})F^{\epsilon\alpha}-R^\epsilon_\nu F_{\epsilon\mu}+R^\epsilon_\mu F_{\epsilon\nu},$$

and making use of the symmetry properties of the tensors involved, and applying equation (42) in Appendix III, this then becomes

$$J_{\mu\nu}-J_{\nu\mu} = g^{\alpha\beta}(\phi_{\mu\nu\beta}-\phi_{\nu\mu\beta}+R^\epsilon_{\mu\beta\nu}\phi_\epsilon+R^\epsilon_{\nu\mu\beta}\phi_\epsilon+R^\epsilon_{\beta\nu\mu}\phi_\epsilon)_\alpha-$$
$$-2R_{\mu\nu\alpha\epsilon}F^{\epsilon\alpha}-R^\epsilon_\nu F_{\epsilon\mu}+R^\epsilon_\mu F_{\epsilon\nu},$$

† Eddington, *The Mathematical Theory of Relativity*, Cambridge, 1923, § 74.

which, in accordance with the cyclical property for the Riemann-Christoffel tensor $R^\epsilon_{\mu\beta\nu} + R^\epsilon_{\nu\mu\beta} + R^\epsilon_{\beta\nu\mu} = 0$,

gives us the desired result

$$g^{\alpha\beta}(F_{\mu\nu})_{\alpha\beta} = g^{\alpha\beta}(\phi_{\mu\nu} - \phi_{\nu\mu})_{\beta\alpha} = J_{\mu\nu} - J_{\nu\mu} + 2R_{\mu\nu\alpha\epsilon}F^{\epsilon\alpha} - R^\epsilon_\mu F_{\epsilon\nu} + R^\epsilon_\nu F_{\epsilon\mu}.$$
(108.1)

The operator $g^{\alpha\beta}(\quad)_{\alpha\beta}$ occurring on the left-hand side of this expression is a generalization of the dalembertian of the non-relativistic theory

$$-\frac{\partial^2}{\partial x^2} - \frac{\partial^2}{\partial y^2} - \frac{\partial^2}{\partial z^2} + \frac{\partial^2}{\partial t^2},$$

and the result (108.1) may be regarded as the analogue of the wave equation of the ordinary electromagnetic theory.

Adopting *natural* coordinates at the point of interest, and noting that in such coordinates the derivatives of the Christoffel three-index symbols will not vanish, while the symbols themselves become zero, it will be found that we can rewrite (108.1) in the form

$$g^{\alpha\beta}\left[\frac{\partial^2 F_{\mu\nu}}{\partial x^\alpha \partial x^\beta} - F_{\epsilon\nu}\frac{\partial}{\partial x^\beta}\{\mu\alpha, \epsilon\} - F_{\mu\epsilon}\frac{\partial}{\partial x^\beta}\{\nu\alpha, \epsilon\}\right]$$
$$= J_{\mu\nu} - J_{\nu\mu} + 2R_{\mu\nu\alpha\epsilon}F^{\epsilon\alpha} - R^\epsilon_\mu F_{\epsilon\nu} + R^\epsilon_\nu F_{\epsilon\mu}. \quad (108.2)$$

In natural coordinates, however, the components of $F_{\mu\nu}$ have the immediate interpretation in terms of field strength originally given by Tables (46.9, 10). Hence in the absence of current J_μ and at the limit of zero field strength, the wave equation reduces in natural coordinates to the familiar form

$$-\frac{\partial^2 F_{\mu\nu}}{\partial x^2} - \frac{\partial^2 F_{\mu\nu}}{\partial y^2} - \frac{\partial^2 F_{\mu\nu}}{\partial z^2} + \frac{\partial^2 F_{\mu\nu}}{\partial t^2} = 0, \quad (108.3)$$

which corresponds to the propagation of electromagnetic disturbances with unit velocity. It is interesting to note, however, that this result has been demonstrated only for vanishing field strengths and hence indeed for vanishing intensities of the electromagnetic disturbance itself.

With the help of (108.3), we may now give a new justification for our earlier procedure in taking the path of a ray of light as a space-time geodesic with $ds = 0$. As a solution of (108.3), corresponding to a plane wave, we find

$$F_{\mu\nu} = A\cos\frac{2\pi}{\lambda}(lx + my + nz - t) \quad (108.4)$$

provided we take the amplitude A, wave-length λ, and direction cosines l, m, and n as quantities whose first and second derivatives vanish. This wave corresponds in geometrical optics to a ray travelling with unit velocity

$$-dx^2-dy^2-dz^2+dt^2= 0$$

under the condition

$$\frac{d^2x}{dt^2} = \frac{d^2y}{dt^2} = \frac{d^2z}{dt^2} = 0,$$

when described in the natural coordinates, x, y, z, t being used. This result, however, can be re-expressed in a form valid for all systems of coordinates by stating that the path of the ray can be taken as a space-time geodesic with $ds = 0$. Thus our original principle receives the desired added justification.

109. The energy-momentum tensor for disordered radiation

We may next consider some problems connected with the energy-momentum tensor corresponding to different distributions of electromagnetic radiation.

Since a disordered distribution of electromagnetic radiation can be regarded as having the mechanical properties of a perfect fluid, we have already suggested in § 85 that we could assign to such radiation the usual expression for the energy-momentum tensor of a perfect fluid

$$T^{\mu\nu} = (\rho_{00}+p_0)\frac{dx^\mu}{ds}\frac{dx^\nu}{ds} - g^{\mu\nu}p_0, \qquad (109.1)$$

where the density and pressure of the radiation—as measured by a local observer who finds no net flow of energy—would be connected by the specially simple relation

$$\rho_{00} = 3p_0, \qquad (109.2)$$

and the quantities dx^μ/ds would be the components of velocity of such a local observer with respect to the coordinates actually being employed.

This method of deciding on the correct expression for the energy-momentum tensor of a disordered distribution of radiation, by treating it as a perfect fluid, is logically not unsatisfactory owing to the macroscopic character of the considerations involved. Nevertheless, it will be interesting to show† that we should also be led to the same

† Tolman and Ehrenfest, *Phys. Rev.* **36**, 1791 (1930).

result by taking the appropriate average of the microscopic expressions for the electromagnetic energy-momentum tensor which was considered in § 104.

To do this let us first take a system of proper coordinates in which there is on the average no net flow of energy at the point and time of interest. With respect to such a system of coordinates the components of the energy-momentum tensor looked at from a microscopic point of view will assume their classical values in terms of electric and magnetic field strengths as already given in § 104 by the typical examples

$$
\begin{aligned}
T^{11} &= -\tfrac{1}{2}(E_x^2 - E_y^2 - E_z^2 + H_x^2 - H_y^2 - H_z^2), \\
T^{12} &= -(E_x E_y + H_x H_y), \\
T^{14} &= (E_y H_z - E_z H_y), \\
T^{44} &= \tfrac{1}{2}(E_x^2 + E_y^2 + E_z^2 + H_x^2 + H_y^2 + H_z^2).
\end{aligned}
\tag{109.3}
$$

In using these expressions to obtain the corresponding macroscopic quantities, it is evident that we shall have the following relations holding *on the average*:

$$
\overline{E_x^2} = \overline{E_y^2} = \overline{E_z^2} \quad \text{and} \quad \overline{H_x^2} = \overline{H_y^2} = \overline{H_z^2}, \tag{109.4}
$$

since for disordered radiation the averaged field strengths will be independent of direction;

$$
\overline{E_x E_y} = \overline{E_y E_z} = \overline{E_z E_x} = 0 \quad \text{and} \quad \overline{H_x H_y} = \overline{H_y H_z} = \overline{H_z H_x} = 0, \tag{109.5}
$$

since for disordered radiation the lack of phase relations will make the instantaneous values of the above products positive or negative with equal probability; and

$$
\overline{E_y H_z - E_z H_y} = \overline{E_z H_x - E_x H_z} = \overline{E_x H_y - E_y H_x} = 0, \tag{109.6}
$$

since the coordinates now in use have been chosen so that there would be no net flow of energy.

Combining the foregoing results of the process of averaging with the expressions for the energy-momentum tensor (109.3), we then see that the only surviving components of the *macroscopic* averaged energy-momentum tensor can be written as

$$
T^{11} = T^{22} = T^{33} = p_0 \qquad T^{44} = \rho_{00} \tag{109.7}
$$

with

$$
\rho_{00} = 3p_0, \tag{109.8}
$$

where ρ_{00}, the proper macroscopic density of energy at the point of interest, is the average of the usual expression for the density of electromagnetic energy in the absence of matter, and the three surviving components of the Maxwell stresses are each equal to one-third of this amount, that is to the radiation pressure p_0.

Having obtained the expressions (109.7) for the components of the energy-momentum tensor in a particular system of coordinates, we can of course then obtain them in any system of coordinates by the rules of tensor transformation. And indeed, applying the same treatment as previously used, in § 85, to obtain in the case of an ordinary perfect fluid a general expression for the energy-momentum tensor from a knowledge of its components in proper coordinates, we are at once led to the expected expression

$$T^{\mu\nu} = (\rho_{00}+p_0)\frac{dx^\mu}{ds}\frac{dx^\nu}{ds} - g^{\mu\nu}p_0 \qquad (109.9)$$
$$\rho_{00} = 3p_0$$

for the energy-momentum tensor of a disordered distribution of radiation, where the 'velocities' dx^μ/ds are now to be interpreted as being those for a local observer who himself finds on the average no net flow of energy, and hence may be regarded as moving along with the radiation as a whole.

110. The gravitational mass of disordered radiation

Having shown that we are justified in taking the energy-momentum tensor for disordered radiation as having the same form as that for other perfect fluids, we may now draw an interesting conclusion as to the effectiveness of such radiation in producing a gravitational field.

If we take the line element for a static sphere of perfect fluid in the form

$$ds^2 = -e^\mu(dx^2+dy^2+dz^2)+e^\nu\,dt^2, \qquad (110.1)$$

we may set the mass of the sphere m equal to its energy U in accordance with § 91, and in accordance with § 97 express the latter in the form of an integral over the total volume of the fluid, giving us,

$$m = U = \int (\rho_{00}+3p_0)e^{\frac{1}{2}\nu}\,dV_0, \qquad (110.2)$$

where dV_0 is an element of proper volume of the fluid.

As a result of our considerations this expression should apply not only to spheres of fluid matter, but also to fluid mixtures of matter and radiation as well. Furthermore, the quantity m may be regarded as a measure of the field producing power of the sphere, since it was defined in § 90 so as to be the constant which would occur in the

Schwarzschild expression for the line element in the empty space surrounding the sphere. Hence, since the pressure p_0 of disordered radiation is necessarily equal to one-third its energy density ρ_{00}, and the pressure of matter is under ordinary circumstances only a minute fraction of its density, we are led to the interesting conclusion that disordered radiation in the interior of a fluid sphere contributes roughly speaking *twice* as much to the gravitational field of the sphere as the same amount of energy in the form of matter.

It is interesting to compare this conclusion with the fact, already mentioned at the end of § 83 (*b*), that the gravitational deflexion of light in passing an attracting mass is *twice* as much as would be calculated from a direct application of Newtonian theory for a particle moving with the velocity of light. In following sections we shall see additional examples of similar differences between the behaviour of matter and radiation.

111. The energy-momentum tensor corresponding to a directed flow of radiation

We may now turn from the consideration of disordered radiation to that of a directed flow of radiation. Using natural coordinates at the point of interest, we may then again take the components of the energy-momentum tensor, regarded from a microscopic point of view, as being given in terms of the electric and magnetic field strengths by the typical examples shown by (109.3). Lowering indices for later convenience, these can be written in the form

$$
\begin{aligned}
T_1^1 &= \tfrac{1}{2}(E_x^2 - E_y^2 - E_z^2 + H_x^2 - H_y^2 - H_z^2), \\
T_2^1 &= T_1^2 = (E_x E_y + H_x H_y), \\
T_4^1 &= -T_1^4 = (E_y H_z - E_z H_y), \\
T_4^4 &= \tfrac{1}{2}(E_x^2 + E_y^2 + E_z^2 + H_x^2 + H_y^2 + H_z^2).
\end{aligned}
\tag{111.1}
$$

Considering now for simplicity that the radiation is travelling in the x-direction and is plane polarized with its electric vector parallel to the y-direction, we shall have in accordance with the usual electromagnetic theory of light

$$
E_x = E_z = H_x = H_y = 0, \qquad E_y = H_z.
$$

And substituting in the above expressions shall obtain as the only surviving components of the energy-momentum tensor

$$
-T_1^1 = T_4^4 = T_4^1 = -T_1^4 = \frac{E_y^2 + H_z^2}{2},
\tag{111.2}
$$

all the components being thus numerically equal to the expression for the density of electromagnetic energy.

The result given by this expression has been obtained for plane polarized radiation and from a microscopic point of view, but should evidently also hold on the average for incoherent unpolarized radiation. We shall hence take as our general macroscopic expression for the energy-momentum tensor corresponding to a flow of radiation in the x-direction

$$-T_1^1 = T_4^4 = T_4^1 = -T_1^4 = \rho, \qquad (111.3)$$

where ρ is the density of radiant energy at the point of interest and these surviving components are expressed in *natural coordinates* for that point.

112. The gravitational field corresponding to a directed flow of radiation

With the help of this expression for the energy-momentum tensor for a directed flow of radiation, we may now determine the corresponding gravitational field, provided we take the field weak enough so that we can use Einstein's approximate solution of the field equations as developed in § 93.

We can then write the metrical tensor in the form

$$g_{\mu\nu} = \delta_{\mu\nu} + h_{\mu\nu}, \qquad (112.1)$$

where the $\delta_{\mu\nu}$ are the constant Galilean values of the $g_{\mu\nu}$, ± 1 and 0, and the $h_{\mu\nu}$ are small correction terms of the first order; and introducing the quantities

$$h_\mu^\lambda = \delta^{\lambda\alpha} h_{\mu\alpha} \qquad h = h_\alpha^\alpha = \delta^{\sigma\lambda} h_{\sigma\lambda}, \qquad (112.2)$$

where the $\delta^{\mu\nu}$ are the Galilean values of the $g^{\mu\nu}$, we can write the approximate solution of the field equations in the form

$$(h_\mu^\nu - \tfrac{1}{2}\delta_\mu^\nu h) = -4 \int \frac{[T_\mu^\nu]}{r} \, dxdydz, \qquad (112.3)$$

where the integration is to be carried out over the whole spatial volume, r is the distance from the point of interest, where the value of $(h_\mu^\nu - \tfrac{1}{2}\delta_\mu^\nu h)$ is desired, to the particular element of volume $dxdydz$ under consideration, and the square brackets indicate that we are to use the value of T_μ^ν at a time earlier than that of interest by the interval needed for a signal to pass with unit velocity from the element $dxdydz$ under consideration to the point of interest.

Furthermore, in applying this approximate solution to the case at hand, we can substitute for the T_μ^ν the values given above by (111.3) for natural coordinates, since the T_μ^ν are themselves quantities of the first order and the coordinates in actual use are approximately natural coordinates at each point under consideration. We thus obtain corresponding to a flow of radiation in the x-direction

$$h_1^1 - \tfrac{1}{2}h = 4 \int \frac{[\rho]\, dV}{r},$$

$$h_2^2 - \tfrac{1}{2}h = h_3^3 - \tfrac{1}{2}h = 0,$$

$$h_4^4 - \tfrac{1}{2}h = -4 \int \frac{[\rho]\, dV}{r}, \qquad (112.4)$$

$$h_4^1 = -h_1^4 = -4 \int \frac{[\rho]\, dV}{r},$$

and with the help of (112.2) can easily solve these equations in the form

$$-h_{11} = -h_{44} = h_{14} = h_{41} = 4 \int \frac{[\rho]\, dV}{r}, \qquad (112.5)$$

with all other components of $h_{\mu\nu}$ equal to zero.

This result then gives us a solution for the gravitational field corresponding to a flow of radiation in the x-direction, provided the field is weak enough to permit the use of Einstein's approximate solution. And this latter condition would presumably not be invalidated because of too large a contribution to the field from any ordinary beam or pulse of radiation that we might encounter in nature or the laboratory.

113. The gravitational action of a pencil of light

(a) The line element in the neighbourhood of a limited pencil of light. As an application of the foregoing expression for the gravitational field corresponding to a unidirectional flow of radiation, it would first be natural to try to consider the field in the neighbourhood of an infinite pencil of light, stretching in the x-direction from minus to plus infinity. This proves to be impossible, nevertheless, by the method adopted since the values of the $h_{\mu\nu}$ then come out infinite when the integration given in (112.5) is performed, which would invalidate the approximate solution of the field equations that has been employed.

This difficulty does not arise, however, if we consider a thin pencil of radiation of limited length l and constant linear density ρ, passing steadily along the x-axis between a source at $x = 0$ and an absorber at $x = l$. In accordance with (112.5), we can then write for the contribution of the radiation to the gravitational potentials at any point of interest x, y, z in the neighbourhood of the pencil

$$-h_{11} = -h_{44} = h_{14} = h_{41} = 4 \int \frac{[\rho]\, dV}{r}$$

$$= \int_{u=0}^{u=l} \frac{4\rho\, du}{[(x-u)^2+y^2+z^2]^{\frac{1}{2}}} \qquad (113.1)$$

$$= 4\rho \log \frac{[(l-x)^2+y^2+z^2]^{\frac{1}{2}}+(l-x)}{[x^2+y^2+z^2]^{\frac{1}{2}}-x},$$

and with a finite length of pencil l this can be made as small as desired by taking the density of radiation ρ small.

It should be noted that this expression has been derived on the assumption of a *steady* pencil of radiation, so that no explicit introduction of retarded potentials into the calculation was needed. The expression would hence of course not be applicable in the neighbourhood of times when the pencil was being started or stopped. It should also be noted that the treatment assumes a flow of radiation solely in the x-direction and thus neglects diffraction effects at the surface of the pencil. Finally, it should be pointed out that the expression gives only the contribution of the pencil of radiation to the field, and neglects the contribution of the bodies which act as source and absorber; and this includes a neglect of any effects resulting from changes in the motion or internal condition of these bodies which might themselves be connected with the flow of radiation.

With these restrictions, however, we may regard the gravitational field in the neighbourhood of this limited pencil of light as given by (113.1).

(*b*) **Velocity of a test ray of light in the neighbourhood of the pencil.** In order to appreciate the character of this gravitational field in the neighbourhood of a pencil of light, we may now consider, first the motion of test rays of light, and then the motion of test particles in the neighbourhood of the pencil.

To investigate the motion of the test rays, we may write the formula for interval in the neighbourhood of the pencil in accordance with (113.1) in the form

$$ds^2 = -(1-h_{11})\,dx^2 -dy^2-dz^2+(1+h_{11})\,dt^2 -2h_{11}\,dxdt,$$

and, setting this equal to zero in order to correspond to the track of the test ray of light, we obtain after dividing through by dt^2

$$(1-h_{11})\frac{dx^2}{dt^2}+\frac{dy^2}{dt^2}+\frac{dz^2}{dt^2}+2h_{11}\frac{dx}{dt} = 1+h_{11} \qquad (113.2)$$

as a general expression for the velocity of our test ray in the neighbourhood of the pencil.

Solving this general expression for velocity, first for the case of a test ray moving at the instant of interest parallel to the x-axis, and hence also to the pencil, we obtain the two cases

$$\frac{dx}{dt} = +1 \quad \text{and} \quad -\frac{1+h_{11}}{1-h_{11}}. \qquad (113.3)$$

On the other hand, solving for a test ray moving parallel to the y-axis, and hence in a plane perpendicular to the pencil, we obtain the two cases

$$\frac{dy}{dt} = \pm\sqrt{(1+h_{11})}. \qquad (113.4)$$

In accordance with the first of these expressions, we see that a test ray of light, moving parallel to the pencil and in the same direction as that for the light in the pencil, would have unit velocity at any point in the field. On the other hand, for test rays moving in other directions, we should have a variable velocity depending as might be expected on the position in the gravitational field of the pencil, since h_{11} will depend on position in the way given by (113.1).

We may also inquire into the acceleration which would be experienced by the rays. Differentiating (113.2), we obtain as a general expression for the accelerations

$$2\left(\frac{dx}{dt}-h_{11}\frac{dx}{dt}+h_{11}\right)\frac{d^2x}{dt^2}+2\frac{dy}{dt}\frac{d^2y}{dt^2}+2\frac{dz}{dt}\frac{d^2z}{dt^2}-$$
$$-\left(\frac{dx^2}{dt^2}-2\frac{dx}{dt}+1\right)\frac{dh_{11}}{dt} = 0, \qquad (113.5)$$

and for the special case of a ray which has at the instant of interest the components of velocity

$$\frac{dx}{dt} = 1 \qquad \frac{dy}{dt} = \frac{dz}{dt} = 0 \qquad (113.6)$$

this then leads to the result

$$\frac{d^2x}{dt^2} = 0. \tag{113.7}$$

Thus, such a test ray retains its unit velocity parallel to the pencil, and hence in accordance with the general expression for velocity (113.2) must also permanently retain zero components of velocity perpendicular to the pencil.

This result is of considerable interest, since it means, in the special case of parallel rays of light travelling in the same direction, that there will be no gravitational interaction between the rays. This conclusion is satisfactory from the point of view of the stability of our originally postulated pencil of light, and also from the point of view of interpreting the behaviour of parallel rays of light coming from distant astronomical objects.

(c) **Acceleration of a test particle in the neighbourhood of the pencil.** We may next consider the effect of the gravitational field of our pencil of radiation in accelerating stationary test particles placed in the neighbourhood.

In accordance with (74.13), the acceleration of such a test particle will be given by the geodesic equation

$$\frac{d^2x^\sigma}{ds^2} + \{\mu\nu, \sigma\} \frac{dx^\mu}{ds} \frac{dx^\nu}{ds} = 0,$$

and for stationary particles with

$$\frac{dx}{ds} = \frac{dy}{ds} = \frac{dz}{ds} = 0 \qquad \frac{dt}{ds} = 1,$$

this will reduce to

$$\frac{d^2x^\sigma}{ds^2} + \{44, \sigma\} = 0.$$

Working out the values of the Christoffel three-index symbols which correspond to the gravitational field of the pencil as given by (113.1) we easily find to the order of approximation of our solution of the field equations

$$\frac{d^2x}{ds^2} = -\frac{1}{2}\frac{\partial h_{44}}{\partial x} \qquad \frac{d^2y}{ds^2} = -\frac{1}{2}\frac{\partial h_{44}}{\partial y} \qquad \frac{d^2z}{ds^2} = -\frac{1}{2}\frac{\partial h_{44}}{\partial z}.$$

Furthermore, substituting the explicit value for h_{44} in the neighbourhood of the pencil provided by (113.1), we then obtain, with some rearrangement, for the acceleration of a *stationary* particle parallel to the pencil

$$\frac{d^2x}{dt^2} = 2\rho \left\{ \frac{1}{[x^2+y^2+z^2]^{\frac{1}{2}}} - \frac{1}{[(l-x)^2+y^2+z^2]^{\frac{1}{2}}} \right\}, \qquad (113.8)$$

and for its acceleration in a plane perpendicular to the pencil

$$\frac{d^2y}{dt^2} = -\frac{2\rho y}{y^2+z^2} \left\{ \frac{x}{[x^2+y^2+z^2]^{\frac{1}{2}}} + \frac{l-x}{[(l-x)^2+y^2+z^2]^{\frac{1}{2}}} \right\}, \qquad (113.9)$$

where x, y, z denotes the position of the test particle, and the track of the pencil lies as will be remembered along the x-axis from $x = 0$ to $x = l$.

For the case of a particle placed at a point equally distant from the two ends of the track, these general expressions reduce, for the case of acceleration parallel to the track to the simple result

$$\frac{d^2x}{dt^2} = 0, \qquad (113.10)$$

and, taking $z = 0$ for the acceleration towards the track, to the result

$$\frac{d^2y}{dt^2} = -\frac{2\rho l}{y[(\frac{1}{2}l)^2+y^2]^{\frac{1}{2}}}. \qquad (113.11)$$

These results for the acceleration of a test particle parallel and perpendicular to the track of the pencil are of considerable interest.

In the first place, it will be found that both of the general expressions (113.8) and (113.9) are just *twice* as great as would be calculated on the basis of Newtonian theory if we replaced the pencil of radiation by a material rod of the same density and length. This is another example (see § 110) of a case where radiation may be regarded as more effective in producing a gravitational field than a similar distribution of matter of the same density.

In the second place, it is of interest to emphasize in accordance with (113.8) and (113.10) that the acceleration parallel to the track of the pencil would be towards the longer segment of track for a particle placed at a point nearer one end of the track than the other, but would be zero for a particle placed equally distant from the two ends of the track. Hence for a particle, which is not actually situated in the path of the pencil, there is no preponderant gravitational action in the direction of motion of the light itself. This is to be contrasted with the effect of light pressure which would act on a particle actually placed in the pencil in the direction of radiation flow, and may also

be contrasted with the Compton effect on an electron placed in
the pencil which would also be preponderatingly in the forward
direction.

114. The gravitational action of a pulse of light

(a) **The line element in the neighbourhood of the limited track of
a pulse of light.** We may now turn to a consideration of the gravitational field in the neighbourhood of the track of a limited pulse of
radiation. This will be more complicated to treat than the case of
the steady pencil since the field will now be non-static, and we shall
have to make explicit use of the method of retarded potentials in
determining the way in which the gravitational effect spreads out
from the moving pulse.

Let us consider a pulse of radiation, of length λ, linear density ρ,
and negligible cross-section, travelling along the x-axis from $x = 0$
to $x = l$. These may be regarded as the points at which the pulse
emerges from the emitter and enters the absorber, or as giving an
arbitrary portion of the track selected for investigation, and we shall
neglect any effects coming from the pulse or parts of it that do not
lie within this range. We shall also neglect as before any gravitational effects due to the absorber and emitter or changes that may
take place within them. Some such restrictions appear to be necessary
in order to secure a determinable problem. In particular we shall
point out later that our method of attack has to be limited to a track
of finite length.

For simplicity we shall choose our time scale so as to make $t = 0$
when the front end of the pulse crosses the point $x = 0$. Then at
any later time the front end of the pulse will be located at $x = t$ and
the rear end at $x = t - \lambda$ since the pulse may be taken as travelling
with unit velocity.

Let us now take some point of interest x, y, z in the neighbourhood
of the track and calculate with the help of equation (112.5) the
gravitational field produced by the pulse at this point at the time t.
Since this equation for the gravitational field has to be applied in
accordance with the method of retarded potentials, let us take
$x = a$ as giving the position of the front end of the pulse and $x = b$
the position of the rear end of the pulse when they 'emit' the gravitational influence which is received at the point x, y, z at the time
t. In accordance with (112.5) we may then evidently write for the

gravitational potentials at x, y, z and time t

$$-h_{11} = -h_{44} = h_{14} = h_{41} = 4 \int \frac{[\rho]\,dV}{r}$$

$$= \int_{u=b}^{u=a} \frac{4\rho\,du}{[(x-u)^2+y^2+z^2]^{\frac{1}{2}}}$$

$$= 4\rho \log \frac{[(x-a)^2+y^2+z^2]^{\frac{1}{2}}-(x-a)}{[(x-b)^2+y^2+z^2]^{\frac{1}{2}}-(x-b)}. \qquad (114.1)$$

To evaluate this expression, however, we must determine a and b as functions of the time t. To do this we note with our choice of starting-point for time measurements that $a = x$ gives not only the position of the front end of the pulse when it emits the gravitational influence reaching the point of interest at time t, but also denotes the time at which this impulse is emitted. Hence $(t-a)$ is the time available for the gravitational influence to travel from the front end of the pulse to the point of interest and since this influence is propagated with unit velocity, we can write

$$(t-a)^2 = (x-a)^2+y^2+z^2,$$

and solving for a obtain

$$a = \frac{t^2-x^2-y^2-z^2}{2(t-x)}. \qquad (114.2)$$

Similarly, we obtain for b the expression

$$b = \frac{(t-\lambda)^2-x^2-y^2-z^2}{2(t-\lambda-x)}. \qquad (114.3)$$

These values for a and b apply of course to the positions of the front and rear end of the pulse only when they lie within the range of track from $x = 0$ to $x = l$ that we have selected for investigation. Since the pulse starts to enter this portion of track at $t = 0$, and $[x^2+y^2+z^2]^{\frac{1}{2}}$ is evidently the time needed for a gravitational influence to travel from the point of entrance to the point x, y, z of interest, we shall disregard the gravitational effect of the pulse completely until the time $t = [x^2+y^2+z^2]^{\frac{1}{2}}$ and then take

$$b = 0 \begin{cases} \text{from } t = [x^2+y^2+z^2]^{\frac{1}{2}} \\ \text{to} \quad t = [x^2+y^2+z^2]^{\frac{1}{2}}+\lambda. \end{cases} \qquad (114.4)$$

Similarly, since we are not interested in gravitational influences emitted from portions of the track lying beyond $x = l$, we shall take

$$a = l \begin{cases} \text{from } t = l+[(l-x)^2+y^2+z^2]^{\frac{1}{2}} \\ \text{to} \quad t = l+[(l-x)^2+y^2+z^2]^{\frac{1}{2}}+\lambda, \end{cases} \tag{114.5}$$

and disregard the gravitational effect completely after the latter of these times.

We may now substitute the foregoing expressions for a and b into (114.1) and obtain as explicit expressions, for the gravitational potentials at the point and time of interest x, y, z, t,

$$-h_{11} = -h_{44} = h_{14} = h_{41} \tag{114.6}$$

$$= 4\rho \log \frac{t-x}{[x^2+y^2+z^2]^{\frac{1}{2}}-x} \quad \begin{cases} \text{from } t = [x^2+y^2+z^2]^{\frac{1}{2}} \\ \text{to} \quad t = [x^2+y^2+z^2]^{\frac{1}{2}}+\lambda \end{cases}$$

$$= 4\rho \log \frac{t-x}{t-\lambda-x} \quad \begin{cases} \text{from } t = [x^2+y^2+z^2]^{\frac{1}{2}}+\lambda \\ \text{to} \quad t = l+[(l-x)^2+y^2+z^2]^{\frac{1}{2}} \end{cases}$$

$$= 4\rho \log \frac{[(l-x)^2+y^2+z^2]^{\frac{1}{2}}+(l-x)}{t-\lambda-x} \begin{cases} \text{from } t = l+[(l-x)^2+y^2+z^2]^{\frac{1}{2}} \\ \text{to} \quad t = l+[(l-x)^2+y^2+z^2]^{\frac{1}{2}}+\lambda. \end{cases}$$

With the help of these expressions, we can now appreciate why it is necessary to restrict our present kind of treatment to a finite portion of track in order to obtain a determinate problem. With an infinite track the second of the above expressions for the gravitational potentials would evidently be applicable at all times, and this would then become infinite at the time $t = x$ when the pulse comes abreast of the point of interest, which would invalidate the approximate method of solving the field equations that has been employed. Taking as we do, however, a limited portion of the track this difficulty is avoided since we consider only those gravitational effects which arise after the pulse starts to enter the track at $t = 0$, and this corresponds to a time of reception at the point of interest

$$t = [x^2+y^2+z^2]^{\frac{1}{2}}$$

so that the infinity no longer arises.

The foregoing expressions are derived on the assumption of a flow of radiation solely in the x-direction and hence neglect diffraction effects at the boundary of the pulse.

(b) **Velocity of a test ray of light in the neighbourhood of the pulse.** To appreciate the nature of the above gravitational field in the neighbourhood of the pulse, we may first consider the velocity of a test ray of light in the field of the pulse. The treatment will evidently be the same as that already given in § 113(b) for the case of the steady pencil and the results will evidently have exactly the same form as those obtained in that section, differing in content, nevertheless, since we should now have to substitute the expressions for h_{11} given by (114.6) instead of by (113.1) in order to get the specific dependence of velocity on position and time. For the case of a test ray of light moving in the same direction and parallel to the track of the pulse we shall, however, again evidently obtain the very satisfactory result of a velocity which remains permanently unity without being affected by gravitational interaction with the pulse.

(c) **Acceleration of a test particle in the neighbourhood of the pulse.** We may now investigate the gravitational acceleration which would be experienced by a neighbouring test particle as a result of the passage of the pulse of light. If we take the test particle as stationary the accelerations will again evidently be determined as in § 113(c) by the equation

$$\frac{d^2x^\sigma}{ds^2} + \{44, \sigma\} = 0.$$

The values of the Christoffel three-index symbols corresponding to our present non-static case are nevertheless now a little more complicated than before. We easily obtain, however, to the desired order of approximation

$$\frac{d^2x}{ds^2} = \frac{1}{2}\left(\frac{\partial h_{41}}{\partial t} + \frac{\partial h_{41}}{\partial t} - \frac{\partial h_{44}}{\partial x}\right),$$

$$\frac{d^2y}{ds^2} = -\frac{1}{2}\frac{\partial h_{44}}{\partial y},$$

$$\frac{d^2z}{ds^2} = -\frac{1}{2}\frac{\partial h_{44}}{\partial z}.$$

Substituting for the gravitational potentials the values given by (114.6), we then obtain for the acceleration—parallel to the track— of a *stationary* test particle placed for simplicity at a point where $z = 0$, the expressions

$$\frac{d^2x}{dt^2} = 2\rho\left(\frac{1}{t-x} + \frac{1}{[x^2+y^2]^{\frac{1}{2}}}\right) \qquad \begin{cases} \text{from } t = [x^2+y^2]^{\frac{1}{2}} \\ \text{to} \quad t = [x^2+y^2]^{\frac{1}{2}} + \lambda \end{cases}$$

$$\frac{d^2x}{dt^2} = 2\rho\left(\frac{1}{t-x} - \frac{1}{t-\lambda-x}\right) \qquad \begin{cases} \text{from } t = [x^2+y^2]^{\frac{1}{2}} + \lambda \\ \text{to} \quad t = l + [(l-x)^2+y^2]^{\frac{1}{2}} \end{cases}$$

$$\frac{d^2x}{dt^2} = -2\rho\left(\frac{1}{t-\lambda-x} + \frac{1}{[(l-x)^2+y^2]^{\frac{1}{2}}}\right) \qquad \begin{cases} \text{from} \\ \quad t = l+[(l-x)^2+y^2]^{\frac{1}{2}} \\ \text{to} \\ t = l+[(l-x)^2+y^2]^{\frac{1}{2}}+\lambda \end{cases}$$

$$(114.7)$$

and for the acceleration perpendicular to the track

$$\frac{d^2y}{dt^2} = \frac{-2\rho y}{[x^2+y^2]^{\frac{1}{2}}\{[x^2+y^2]^{\frac{1}{2}}-x\}} \qquad \begin{cases} \text{from } t = [x^2+y^2]^{\frac{1}{2}} \\ \text{to} \quad t = [x^2+y^2]^{\frac{1}{2}}+\lambda \end{cases}$$

$$\frac{d^2y}{dt^2} = 0 \qquad \begin{cases} \text{from } t = [x^2+y^2]^{\frac{1}{2}}+\lambda \\ \text{to} \\ \quad t = l+[(l-x)^2+y^2]^{\frac{1}{2}} \end{cases}$$

$$\frac{d^2y}{dt^2} = \frac{2\rho y}{[(l-x)^2+y^2]^{\frac{1}{2}}\{[(l-x)^2+y^2]^{\frac{1}{2}}+(l-x)\}} \qquad \begin{cases} \text{from} \\ \quad t = l+[(l-x)^2+y^2]^{\frac{1}{2}} \\ \text{to} \\ t = l+[(l-x)^2+y^2]^{\frac{1}{2}}+\lambda. \end{cases}$$

$$(114.8)$$

In accordance with these expressions, we see that parallel to the track, the calculated acceleration would first be in the same direction as the motion of the pulse, and then in the opposite direction. On the other hand, perpendicular to the motion of the pulse, the acceleration would first be towards the track and later away from it. These conclusions appear somewhat complicated, and a better idea of the actual nature of the gravitational interaction will be obtained if we consider the net integrated effect corresponding to the whole motion of the pulse over the selected portion of the track from $x = 0$ to $x = l$.

Making use of the expressions given by (114.7), integrating over the time intervals given, adding the results together and cancelling out a considerable number of balancing terms, it can easily be shown that we obtain for our *stationary* test particle the net acceleration parallel to the track

$$\int \frac{d^2x}{dt^2}\, dt = 2\rho\lambda\left\{\frac{1}{[x^2+y^2]^{\frac{1}{2}}} - \frac{1}{[(l-x)^2+y^2]^{\frac{1}{2}}}\right\}, \qquad (114.9)$$

and, making use of (114.8), for the net acceleration towards the track

$$\int \frac{d^2y}{dt^2}\, dt = -\frac{2\rho\lambda}{y}\left\{\frac{x}{[x^2+y^2]^{\frac{1}{2}}} + \frac{l-x}{[(l-x)^2+y^2]^{\frac{1}{2}}}\right\}; \qquad (114.10)$$

in both cases the result being that which corresponds of course only to the gravitational influences originating from the selected portion of the track $x = 0$ to $x = l$.

For the case of a particle placed at a point equally distant from the two ends of the track, these general results reduce to the simpler expressions analogous to (113.10) and (113.11) in the case of the steady pencil,

$$\int \frac{d^2x}{dt^2}\, dt = 0 \qquad (114.11)$$

and

$$\int \frac{d^2y}{dt^2}\, dt = -\frac{2\rho\lambda l}{y[(\tfrac{1}{2}l)^2 + y^2]^{\frac{1}{2}}}. \qquad (114.12)$$

These results for the net acceleration of stationary test particles parallel and perpendicular to the track of the pulse are of considerable interest.

In the first place, since these expressions give the total accelerative action produced by the pulse during the time l needed for it to traverse the selected portion of track, we may obtain expressions for the average accelerative action of the pulse ($\bar{a}_{\text{pulse}}$) by dividing through by l. Doing so and comparing with our previous expressions for the instantaneous accelerations (a_{pencil}) in the case of a steady pencil with the same length of track and the same location of the test particle, we obtain the general result

$$\frac{\bar{a}_{\text{pulse}}}{a_{\text{pencil}}} = \frac{\lambda\rho_{\text{pulse}}}{l\rho_{\text{pencil}}} = \frac{m_{\text{pulse}}}{m_{\text{pencil}}}. \qquad (114.13)$$

This result seems readily acceptable, moreover, on the basis of our usual physical notions. Since we have already pointed out that the gravitational action of the pencil on a test particle would be *twice* as great as would be calculated on the basis of Newtonian theory if we replaced the pencil of radiation by a material rod of the same density and length, the above result may be regarded as another example of a case where radiation is more effective in producing gravitational action than what might be regarded as the equivalent distribution of stationary matter.

In the second place, it is of interest to emphasize, in accordance with (114.9) and (114.11), that the net acceleration parallel to the track of the pulse would be towards the longer segment of the track for a particle placed at a point nearer one end of the track than the other, but would be zero for a particle placed equally distant from the two ends of the track. Again, as in the case of the pencil, a particle situated outside of the actual track of the pulse and hence not subject to such actions as light pressure or the Compton effect, would not appear liable to a preponderating effect in the direction of motion of the pulse itself, as a result of the gravitational influence coming from a definite selected portion of the track.

115. Discussion of the gravitational interaction of light rays and particles

The foregoing results as to the gravitational field of pencils and pulses of light have been presented in considerable detail† on account of the insight which they can give us into the gravitational inter-action of light rays and particles.

The most characteristic feature of these results lies in the discovery that the acceleration of a test particle by the gravitational influence proceeding from a selected portion of light track is—to the order of approximation employed—*twice* as great as would be calculated on a simple Newtonian basis if we regarded the track as filled during the time of interest with matter having the same average density as that provided by the passage of the light. This conclusion is very satis-factory since we have already seen in § 83 (*b*) that the gravitational bending of a ray of light in passing through the field of an attracting particle is also approximately *twice* as great as would be calculated on a simple Newtonian basis, if we regarded the ray of light as having the Newtonian acceleration of a particle moving through the field in question.

As a result of this double occurrence of the factor two, we are hence permitted in thinking about the mutual gravitational inter-action of particles and light rays to retain to a considerable extent in first approximation our usual ideas as to the conservation of momen-tum, without the necessity of resorting to the complete relativistic treatment in which the conservation laws are exactly preserved by introducing the pseudo tensor t_μ^ν of potential gravitational energy and momentum.

A simple example will serve to make this approximate conserva-tion of ordinary momentum clearer. In accordance with equation (114.12) in the last section, we are provided with an expression for the net acceleration—due to a selected portion l of the light track of a pulse—for a particle placed at the distance y from the track at a point equally distant from the two ends. Multiplying this expression by the

† Tolman, Ehrenfest, and Podolsky, *Phys. Rev.* **37**, 602 (1931).

mass of the particle M and denoting the mass of the pulse by $m = \rho\lambda$, we then obtain for the net momentum received by the particle in the y-direction

$$\int M \frac{d^2y}{dt^2}\, dt = -\frac{2mMl}{y[(\tfrac{1}{2}l)^2 + y^2]^{\frac{1}{2}}}. \tag{115.1}$$

On the other hand, in accordance with § 83 (b), we can take the momentum acquired by the pulse in the y-direction as being twice what would be calculated on a simple Newtonian basis, and hence, noting that velocity of light is unity, as given by

$$\int m \frac{d^2y}{dt^2}\, dt = \int_0^l \frac{2mMy}{[(\tfrac{1}{2}l-x)^2 + y^2]^{\frac{3}{2}}}\, dx = \frac{2mMl}{y[(\tfrac{1}{2}l)^2 + y^2]^{\frac{1}{2}}}. \tag{115.2}$$

Since these two expressions are equal in magnitude and opposite in sign, we thus have an illustration of the approximate conservation of momentum which obtains without taking into consideration any potential momentum of the gravitational field.

As a second important characteristic of the results obtained in the foregoing sections, we have the discovery that a stationary particle placed equally distant from the two ends of a selected portion of light track, but outside the actual track, receives therefrom no net acceleration in the direction of the motion of the light itself. From the point of view of the conservation of momentum, this discovery is the converse of the more familiar conclusion that light would receive no change in the total momentum in its direction of motion in passing through the field of a stationary particle between two points equally distant therefrom. This converse conclusion is sufficiently important from the point of view of astronomical observations to deserve separate consideration.

For the gravitational field surrounding a stationary particle, we can obviously write the line element in a spherically symmetrical static form

$$ds^2 = g_{11}(dr^2 + r^2\, d\theta^2 + r^2\sin^2\theta\, d\phi^2) + g_{44}\, dt^2. \tag{115.3}$$

See for example (82.12), where g_{11} and g_{44} are functions of r alone. If now we consider a ray of light passing through this field, it is evident from the static character of the line element that successive impulses travelling along the ray will take the same time Δt to travel from one given point r_1, θ_1, ϕ_1 to another r_2, θ_2, ϕ_2. Hence if two successive impulses are separated by an interval δt at the first point they will still be separated by the same interval δt at the second point.

Furthermore, if these two points are chosen at positions on the track at the same radius $r_1 = r_2$ from the particle, it is evident that this coordinate time interval δt will correspond to the same proper time interval $\delta t_0 = \sqrt{g_{44}}\, \delta t$ as measured by local observers located at the two points. Hence the period and frequency of light as measured by observers at rest in the field will not be altered by the passage of the light from one point in the field to another of the same gravitational potential g_{44}, a result which holds, moreover, in general for static fields. From the relation between frequency and total momentum in the direction of motion $g = h\nu/c$, we now reach, however, the con-clusion stated above as to the constancy of total momentum in the direction of motion.

This result has been obtained, of course, for the case of a static field, and we may legitimately ask whether the motion acquired by an originally stationary particle as a result of the passage of light would not in turn affect the field through which the light still has to pass, and thus by a second-order effect lead to a change in total momentum in the direction of motion. An exact analysis of such a second-order effect would appear to be complicated. Nevertheless, speaking roughly, since we may regard gravitational impulses and light as both travelling with the same fundamental velocity, it would seem difficult for any gravitational effect to emanate from an element of the oncoming radiation, travel down to the particle, and by changing the motion of the particle then produce an effect on the gravitational field through which that same element of radiation still has to pass. Hence we should in any case be inclined to expect that such a second-order effect on the total momentum and frequency of light travelling through the gravitational field of a particle would have to be exceedingly small compared with the first-order transverse effect on the direction of the momentum.

As a consequence it is usual to conclude that light giving sharply defined images of distant astronomical objects has not had its frequency appreciably affected by passing through the gravitational fields of particles lying along its path. This conclusion is important in interpreting the red-shift in the light from the extra-galactic nebulae, since it has been pointed out by Zwicky† that a gravitational effect on frequency, of the kind made improbable by the above considera-tions, might have offered another explanation for this red-shift

† Zwicky, *Proc. Nat. Acad.* **15**, 773 (1929).

rather than the now commonly accepted explanation based on the recessional motion of the nebulae.

116. The generalized Doppler effect

To conclude the present chapter, we may now give a schematic outline of the method of treating the general problem of the effect of gravitational fields and of the motions of source and observer on the measured wave-lengths of light.

To treat this problem we must have in the first place a knowledge of the line element in the region where the transmission of light from source to observer is taking place. This we may take in the general form

$$ds^2 = g_{11}\,dx^2 + 2g_{12}\,dxdy + \ldots + 2g_{34}\,dzdt + g_{44}\,dt^2. \quad (116.1)$$

In the second place we must know the positions of the source (x_1, y_1, z_1) and of the observer (x_2, y_2, z_2) as a function of the time t

$$(x_1, y_1, z_1) = f_1(t), \quad (116.2)$$

$$(x_2, y_2, z_2) = f_2(t). \quad (116.3)$$

Since the velocity of light is given by setting the expression for the line element (116.1) equal to zero, and since the position of source and observer are given by (116.2) and (116.3), we can then calculate the time of reception t_2 by the observer of a light impulse leaving the source at any desired time t_1 as a function of that time t_1, giving us an expression of the form

$$t_2 = f(t_1). \quad (116.4)$$

Furthermore, by differentiation we can also obtain an expression for the time interval δt_2 between the receipt of two successive wave crests in terms of t_1 and the time interval δt_1 between their emission, which will be of the form

$$\delta t_2 = \frac{df(t_1)}{dt_1}\,\delta t_1. \quad (116.5)$$

In accordance with the expression for the line element, however, we may write

$$\delta t_1^0 = \left[g_{11}\frac{dx^2}{dt^2} + 2g_{12}\frac{dx}{dt}\frac{dy}{dt} + \ldots + g_{44} \right]^{\frac{1}{2}}_{x_1 y_1 z_1 t_1} \delta t_1, \quad (116.6)$$

as an expression for the proper period of the emitting source as measured by a local observer moving with it, where dx/dt, dy/dt, and dz/dt are the components of velocity of the source at the time of emission. And similarly, we may write

$$\delta t_2^0 = \left[g_{11}\frac{dx^2}{dt^2} + 2g_{12}\frac{dx}{dt}\frac{dy}{dt} + \dots + g_{44} \right]_{x_2 y_2 z_2 t_2}^{\frac{1}{2}} \delta t_2 \qquad (116.7)$$

as the observed proper period of the oncoming signal as measured by the final observer, where dx/dt, etc., are now the velocities of this observer at the time of reception.

Substituting (116.6) and (116.7) into (116.5), and noting that the proper period δt_1^0 for the emitting source can be taken as proportional to the usually measured wave-length λ for the kind of luminous material involved, and that the observed proper period δt_2^0 for the oncoming signal can be taken as proportional to the observed wave-length $\lambda + \delta\lambda$ we obtain

$$\frac{\lambda + \delta\lambda}{\lambda} = \frac{\delta t_2^0}{\delta t_1^0} = \frac{df(t_1)}{dt_1} \frac{\left[g_{11}\frac{dx^2}{dt^2} + 2g_{12}\frac{dx}{dt}\frac{dy}{dt} + \dots + g_{44} \right]_{x_2 y_2 z_2 t_2}^{\frac{1}{2}}}{\left[g_{11}\frac{dx^2}{dt^2} + 2g_{12}\frac{dx}{dt}\frac{dy}{dt} + \dots + g_{44} \right]_{x_1 y_1 z_1 t_1}^{\frac{1}{2}}} \qquad (116.8)$$

as the desired expression for the shift in the observed wave-length of the light from a distant source.†

This general expression for shift in wave-length can, of course, be given specific content only when we have a given gravitational field, corresponding to a specific line element (116.1), and have a source and observer located and moving in a definite manner corresponding to specific forms for (116.2) and (116.3). It may be well to point out, however, that the expression implies in any case the same fractional shift in observed wave-length for all parts of the spectrum. It should also be noted that the fractional shift in wave-length is a definitely observable quantity, which will have the same calculated value, for any specified case, no matter what coordinates we may use in making the computation. Thus no changes will occur in the result, if we introduce changes in the form of the line element merely by substituting new variables as functions of the original ones before making the computation, a fact which is perhaps not always appreciated. Finally it may be remarked, that although the expression is formulated so as to give the value of the shift as a function of the time of

† This derivation has been obtained by taking $ds = 0$ as giving the velocity of light, and this expression has been justified for weak electromagnetic disturbances in § 108. For a treatment based directly on wave optics, see Laue, *Berl. Ber.* p. 3 (1931).

emission t_1, we can of course also obtain it as a function of the time of reception t_2 when we know the form of (116.4).

Our final expression may be regarded as that for a generalized Doppler effect, since its value depends not only on the direct effect of the motions of source and observer in changing the time for light to travel from one to the other, but also on the indirect effect of these motions in determining the relation between coordinate time intervals and proper time intervals, and furthermore on the effect of the gravitational potentials in determining the velocity with which the light travels. We shall have later use for this treatment of generalized Doppler effect in Chapter X.

IX

RELATIVISTIC THERMODYNAMICS

Part I. THE EXTENSION OF THERMODYNAMICS TO GENERAL
RELATIVITY

117. Introduction

In the development of the classical thermodynamics, two limitations were actually present. In the first place the thermodynamic
systems considered were tacitly taken as being at rest with respect
to the observer; and in the second place the systems considered were
either taken as unaffected by gravitation, or in any case as affected
by fields weak enough and small enough in extent, so that they could
be treated with the help of the Newtonian theory of gravitation and
the older ideas as to the nature of space and time. To remove the
first of these limitations and obtain a thermodynamic theory suitable
for moving systems, it is necessary to employ the principles for the
intercomparison of measurements made by observers in relative
motion to each other provided by the special theory of relativity. To
remove the second of the limitations, and obtain a theory suitable for
investigating the precise thermodynamic effects of gravity in fields
of any magnitude, and suitable for studying the thermodynamic
behaviour of systems large enough so that the curvature of space-
time cannot be neglected, it is necessary to make use of the more
precise theory of gravitation and more adequate ideas as to the
nature of space and time provided by the general theory of relativity.

In Chapter V we have already considered the extension of thermodynamics to special relativity, first carried out by Planck and by
Einstein. We obtained therefrom not only a thermodynamic theory for
moving systems, but also, with the help of the Lorentz transformation equations for heat, work, temperature, and entropy, we achieved
a deeper insight into the nature of thermodynamic quantities, and by
the introduction of our four-dimensional formulation of the second
law of thermodynamics we made a preliminary step in the direction
of covariant generalization.

In the present chapter we shall consider the extension of thermodynamics to general relativity, and the applications of the system of
relativistic thermodynamics which we thus obtain. To obtain this
extension it will merely be necessary to generalize our previous special

relativistic thermodynamic theory in what appears to be a straightforward and natural manner. Hence since this theory was itself obtained from the classical thermodynamics in a straightforward way, we shall have considerable confidence in the outcome. In addition we shall be able to confirm our confidence in this further extension with the help of examples which illustrate an agreement between the conclusions drawn from relativistic thermodynamics and those which can be drawn from relativistic mechanics alone.

Since the steps that must be taken to extend the classical thermodynamics first to special relativity and then to general relativity will appear to be almost self-evident and trivial, it might be supposed that the conclusions to be drawn from relativistic thermodynamics would necessarily have the same qualitative character as those familiar in the classical theory. As we shall see, nevertheless, on account of the great difference between classical and relativistic ideas as to the nature of space and time, conclusions of a qualitatively new kind can arise when we consider systems of sufficient extent so that gravitational curvature becomes important.

118. The relativistic analogue of the first law of thermodynamics

In the classical thermodynamics it was customary to express the requirements of the first law of thermodynamics with the help of the equation
$$\Delta E = Q - W. \tag{118.1}$$

This equation is to be regarded in the first place as expressing the principle of the conservation of energy by equating the total energy change in a system to that which is transferred across the boundary; and is to be regarded in the second place as introducing a distinction between the two methods of energy transfer—flow of heat and performance of work—which becomes especially important for the later application of the second law of thermodynamics.

In relativistic thermodynamics, in analogy with the classical first law, we shall in the first place have to satisfy the general principles of relativistic mechanics, which lead as we have seen in Chapter VII to the appropriate generalization of the classical laws for the conservation of energy and momentum.†

These principles of relativistic mechanics are all implicitly contained

† Tolman, *Proc. Nat. Acad.* **14**, 268 (1928).

in Einstein's field equations

$$-8\pi T^{\mu\nu} = R^{\mu\nu} - \tfrac{1}{2}Rg^{\mu\nu} + \Lambda g^{\mu\nu} \tag{118.2}$$

connecting the energy-momentum tensor $T^{\mu\nu}$ with the geometry of space-time. And since the tensor divergence of the right-hand side of this expression is found to vanish identically, these field equations lead at once to the familiar expressions for the equations of mechanics

$$(T^{\mu\nu})_\nu = 0 \tag{118.3}$$

and

$$\frac{\partial \mathfrak{T}^\nu_\mu}{\partial x^\nu} - \tfrac{1}{2}\mathfrak{T}^{\alpha\beta}\frac{\partial g_{\alpha\beta}}{\partial x^\mu} = 0, \tag{118.4}$$

or, by the introduction of the pseudo tensor density of potential energy and momentum $\mathfrak{t}^\nu_\mu$, to the expression

$$\frac{\partial(\mathfrak{T}^\nu_\mu + \mathfrak{t}^\nu_\mu)}{\partial x^\nu} = 0, \tag{118.5}$$

which is the form showing the closest resemblance to the classical energy-momentum principle. By requiring in relativistic thermodynamics that all thermodynamic processes should satisfy these principles of mechanics, we introduce the analogues of the classical requirements both for the conservation of energy and for the conservation of momentum, the introduction of the latter not having been explicitly necessary in the classical thermodynamics owing to the tacit restriction to stationary systems.

To complete the analogy with the classical first law of thermodynamics, we shall in the second place also have to introduce in relativistic thermodynamics a distinction between flow of heat and performance of work. This, however, must be postponed until the appropriate nature of the distinction has been made clear from our considerations of the relativistic extension of the second law of thermodynamics.

119. The relativistic analogue of the second law of thermodynamics

To assist us in obtaining the relativistic analogue of the usual second law of thermodynamics, we have in the first place the four-dimensional expression for the requirements of the second law in special relativity, as already given in § 71 *using Galilean coordinates* by the formula

$$\frac{\partial}{\partial x^\mu}\left(\phi_0 \frac{dx^\mu}{ds}\right) \delta x \delta y \delta z \delta t \geq \frac{\delta Q_0}{T_0}, \tag{119.1}$$

where ϕ_0 is the proper density of entropy at the point and time of interest as measured by a local observer at rest in the thermodynamic fluid or working substance, the quantities dx^μ/ds are the components of the macroscopic 'velocity' of the fluid at that point with respect to the coordinates in use, δQ_0 is the proper heat as measured by a local observer which flows at the proper temperature T_0 into the element of fluid and during the time denoted by $\delta x \delta y \delta z \delta t$, and the two signs of equality and inequality refer respectively to the cases of reversible and irreversible processes.

In addition, to guide us in obtaining the desired extension of the second law, we must also make use of the two fundamental ideas underlying general relativity which are expressed by the principle of covariance and by the principle of equivalence. In accordance with the principle of covariance, the axiom which we choose must be expressed in a general form which is the same for all coordinate systems, in order that we may avoid the introduction of unsuspected assumptions which might otherwise arise from the use of special coordinates. And in accordance with the principle of equivalence, the axiom must be chosen so as to agree with the requirements of the special theory of relativity, provided we use natural coordinates for the particular point of interest.

As a consequence of the foregoing considerations, we are then at once led to expect that the correct expression for the second law of thermodynamics in general relativity will be provided by taking the immediate covariant re-expression of the special relativity law (119.1) which can be written in the form†

$$\left(\phi_0 \frac{dx^\mu}{ds}\right)_\mu \sqrt{-g}\, \delta x^1 \delta x^2 \delta x^3 \delta x^4 \geqslant \frac{\delta Q_0}{T_0}, \qquad (119.2)$$

or defining the *entropy vector* by

$$S^\mu = \phi_0 \frac{dx^\mu}{ds} \qquad (119.3)$$

in the even more compact form

$$\mathfrak{S}^\mu_\mu\, \delta x^1 \delta x^2 \delta x^3 \delta x^4 \geqslant \frac{\delta Q_0}{T_0}. \qquad (119.4)$$

The above expression will evidently satisfy the principle of covariance owing to its character as a tensor expression of rank zero, $(\phi_0\, dx^\mu/ds)_\mu$ being a scalar since it is the contracted covariant

† Tolman, *Proc. Nat. Acad.* **14**, 701 (1928).

derivative of a vector, $\sqrt{-g}\,\delta x^1 \delta x^2 \delta x^3 \delta x^4$ being a scalar since it is the magnitude of a four-dimensional volume expressed in natural measure, and finally $\delta Q_0/T_0$ also being a scalar since it obviously does not depend on the particular coordinates in use. This expression will also satisfy the principle of equivalence since in natural coordinates for any point of interest, it will immediately reduce to the special relativity law (119.1), the contracted covariant derivative

$$(\phi_0\, dx^\mu/ds)_\mu$$

being replaced by the ordinary divergence, and the quantity $\sqrt{-g}$ assuming the value unity.

The principle given by (119.2) thus satisfies all the conditions that we now know how to impose and will be adopted in what follows as a statement of the relativistic second law of thermodynamics. It must be noted, however, that we have not been led absolutely uniquely to this law since other more complicated covariant expressions might be proposed which would also reduce in flat space-time to the special relativity law. Hence the proposed principle must be regarded as a postulate whose ultimate validity remains to be tested through comparison with the facts of observation.

It may be emphasized, nevertheless, that we have followed a sensible course of procedure in adopting the immediate covariant re-expression of the special relativity second law, since our previous satisfactory experience with the immediate covariant re-expressions of the special relativity formulae for space-time interval and geodesic trajectory have given us confidence that such procedure when feasible is likely to prove correct. It may also be emphasized that we shall find the theoretical consequences of the postulated relativistic second law to be coherent with the rest of relativity, and in particular shall find examples where the results of relativistic thermodynamics can be checked by using the methods of relativistic mechanics alone.

For purposes of practical computation it is often advantageous to re-express the statement of the relativistic second law given by (119.2) in the equivalent form

$$\frac{\partial}{\partial x^\mu}\left(\phi_0 \frac{dx^\mu}{ds}\sqrt{-g}\right)\delta x^1 \delta x^2 \delta x^3 \delta x^4 \geqslant \frac{\delta Q_0}{T_0}. \tag{119.5}$$

[See equation (46), Appendix III.]

120. On the interpretation of the relativistic second law of thermodynamics

Since we have taken the relativistic first law of thermodynamics as being merely a restatement of the principles of relativistic mechanics, a clear understanding of the relativistic second law is especially essential, as it is this latter principle which determines the whole character of relativistic thermodynamics. This understanding we shall now attempt to assure.

First of all it is to be remarked that the distinction between reversible and irreversible processes is still preserved in relativistic thermodynamics owing to the occurrence of the two signs of equality and inequality in the expression for the second law, the former being applicable to the case of reversible processes and the latter being applicable to irreversible processes.

The occurrence of the sign of inequality also implies a distinction for the case of irreversible processes between the forward and backward directions of time, similar to that in ordinary thermodynamics, since it is evident that the truth of our expression for the second law depends on the sign attached to the increment of coordinate time δx^4 occurring on the left-hand side of the inequality. Hence neglecting the occurrence of fluctuations, the entropy principle still indicates the unidirectional character of time in relativistic as in classical thermodynamics. This is of interest since the possibility of illuminating the principles of relativity by regarding time to be plotted as a fourth dimension perpendicular to space has sometimes had a tendency to obscure those reasons, whether fundamental or not, on which we customarily base our ideas as to the unidirectional quality of time.

Attention should also be drawn to the preservation of the essentially *macroscopic* and *phenomenological* character of thermodynamic considerations in the relativistic extension. Indeed, all the quantities occurring in the relativistic second law (119.2) are to be regarded as having significance only from a macroscopic point of view, and as being defined by perfectly definite empirical specifications which can be given for their determination.

Thus ϕ_0 is the entropy per unit volume of the fluid at the point and time of interest as measured by a local observer at rest with respect to the fluid. It is, of course, a macroscopic density owing to the nature of our concept of entropy, and is to be regarded as characterizing the fluid looked at from a large scale point of view without any microscopic analysis into atoms and radiation.

The quantities dx^μ/ds are the components of the macroscopic 'velocity' of the fluid at the position and instant of interest. Thus, for example, the value of dx^1/ds would be found by observing the motion of a macroscopically identifiable point of the fluid, and computing the rate of change of its x^1 coordinate with respect to the readings of a natural clock moving with the point. And dx^4/ds would be found by computing the rate of change in the value of the time-like coordinate x^4 for the point with respect to the readings of the same clock.

Similarly the quantity g is the determinant formed from the components of the metrical tensor $g_{\mu\nu}$ when these are determined macroscopically, an advantageous circumstance in view of our complete lack of knowledge even as to the significance of this tensor from an atomic point of view. Furthermore the coordinate range $\delta x^1 \delta x^2 \delta x^3 \delta x^4$ is to be regarded as denoting a macroscopically infinitesimal element of four-dimensional volume.

Turning finally, moreover, to the right-hand side of our expression, the quantity T_0 is to be taken as the absolute temperature of the fluid as measured in the usual manner by a local observer at rest in the fluid at the position and instant of interest. And δQ_0 is the heat as measured by this local observer which flows into an element of the fluid having the instantaneous proper volume δv_0 during the proper time δt_0, where these quantities are chosen to give the same magnitude of four-dimensional volume in natural measure as is included in the coordinate range $\delta x^1 \delta x^2 \delta x^3 \delta x^4$. These two final quantities δQ_0 and T_0 hence retain the macroscopic character of heat and temperature in the older thermodynamics.

It is important to emphasize the macroscopic and phenomenological character of relativistic thermodynamics, since it permits us to avoid all the complexities and uncertainties which attach to the atomic point of view, especially at the present time of partial incompleteness in the development of quantum theory, and allows us to proceed with the natural confidence belonging to an empirical approach.

121. On the interpretation of heat in relativistic thermodynamics

In the process of generalizing the second law of thermodynamics from the form which it assumes in special relativity to that taken in general relativity, it is not easy to follow the precise significance of the quantity δQ_0 appearing on the right-hand side of the expression and denoting a quantity of absorbed heat as measured by a local observer. Hence the specifications which we have given for the determination of δQ_0 in the preceding section may not be immediately evident, and the present section will be devoted to the interpretation of this quantity.†

To carry out this interpretation, let us first consider the expression for the relativistic second law (119.2), in the original form in which it was written

$$\left(\phi_0 \frac{dx^\mu}{ds} \right)_\mu \sqrt{-g} \, \delta x^1 \delta x^2 \delta x^3 \delta x^4 \geqslant \frac{\delta Q_0}{T_0}. \tag{121.1}$$

Assuming that we understand the significance of all the other quantities in this expression, we can then begin by showing with the help of the principle of covariance that δQ_0 is in any case a scalar, having a value proportional to the range $\delta x^1 \delta x^2 \delta x^3 \delta x^4$, but otherwise independent of the coordinate system.

To do this we note in the first place that the quantities $(\phi_0 \, dx^\mu/ds)_\mu$ and T_0 are necessarily scalars with numerical values entirely independent of the coordinate system—the first because it is the contracted covariant derivative of a vector and the second from the unique specifications given for its determination by a local observer. In the second place we note that the quantity $\sqrt{-g} \, \delta x^1 \delta x^2 \delta x^3 \delta x^4$ is also a scalar, having a numerical value proportional to the infinitesimal range $\delta x^1 \delta x^2 \delta x^3 \delta x^4$, but otherwise independent of the coordinate system, since it is an expression for the four-dimensional volume in natural measure specified by that range. Hence in accordance with the principle of covariance, it is evident that the remaining quantity appearing in the postulate δQ_0 must also be scalar with a numerical value proportional to the infinitesimal range $\delta x^1 \delta x^2 \delta x^3 \delta x^4$, but otherwise independent of the coordinate system, in order that the postulated law may agree with the requirements of covariance by having the same significance in all coordinate systems.

Having shown then that the quantity δQ_0 is necessarily a scalar with a value intrinsically independent of the coordinate system, we can now determine its value by making use of any specially convenient coordinate system. For this purpose we shall choose natural coordinates x, y, z, t for the particular point of interest. In accordance

† Tolman and Robertson, *Phys. Rev.* **43**, 564 (1933).

with the principle of equivalence, such natural coordinates can always be found, and their choice will make our previous principles of thermodynamics—as given for the special theory of relativity—valid in the immediate neighbourhood of the selected point.

Using these coordinates, covariant differentiation will reduce to ordinary differentiation and the quantity $\sqrt{-g}$ will assume the value unity, so that the left-hand side of our expression for the second law (121.1) will assume the form

$$\left(\phi_0 \frac{dx^\mu}{ds}\right)_\mu \sqrt{-g}\, \delta x^1 \delta x^2 \delta x^3 \delta x^4$$

$$= \left[\frac{\partial}{\partial x}\left(\phi_0 \frac{dx}{ds}\right) + \frac{\partial}{\partial y}\left(\phi_0 \frac{dy}{ds}\right) + \frac{\partial}{\partial z}\left(\phi_0 \frac{dz}{ds}\right) + \frac{\partial}{\partial t}\left(\phi_0 \frac{dt}{ds}\right)\right] \delta x \delta y \delta z \delta t,$$

$$(121.2)$$

and by substituting the evident expressions

$$\frac{dx}{ds} = u_x \frac{dt}{ds} \qquad \frac{dy}{ds} = u_y \frac{dt}{ds} \qquad \frac{dz}{ds} = u_z \frac{dt}{ds},$$

where u_x, u_y, and u_z are the components of the velocity of the fluid as ordinarily expressed, this can be rewritten as

$$\left[\frac{\partial}{\partial x}\left(\phi_0 \frac{dt}{ds} u_x\right) + \frac{\partial}{\partial y}\left(\phi_0 \frac{dt}{ds} u_y\right) + \frac{\partial}{\partial z}\left(\phi_0 \frac{dt}{ds} u_z\right) + \frac{\partial}{\partial t}\left(\phi_0 \frac{dt}{ds}\right)\right] \delta x \delta y \delta z \delta t.$$

In accordance with the special theory of relativity, however, entropy is an invariant for the Lorentz transformation, and hence entropy density will depend on the Lorentz contraction factor ds/dt in such a manner that we can substitute

$$\phi = \phi_0 \frac{dt}{ds},$$

where ϕ is the entropy density of the fluid referred to our present coordinates. Doing so, we then obtain in place of the above expression

$$\left[\frac{\partial}{\partial x}(\phi u_x) + \frac{\partial}{\partial y}(\phi u_y) + \frac{\partial}{\partial z}(\phi u_z) + \frac{\partial \phi}{\partial t}\right] \delta x \delta y \delta z \delta t,$$

which can be rewritten in the form

$$\left[\frac{\partial \phi}{\partial t} + \frac{\partial \phi}{\partial x} u_x + \frac{\partial \phi}{\partial y} u_y + \frac{\partial \phi}{\partial z} u_z + \phi\left(\frac{\partial u_x}{\partial x} + \frac{\partial u_y}{\partial y} + \frac{\partial u_z}{\partial z}\right)\right] \delta x \delta y \delta z \delta t$$

or

$$\left[\frac{d\phi}{dt} + \phi \operatorname{div} \mathbf{u}\right] \delta x \delta y \delta z \delta t,$$

where we now represent by the total derivative $d\phi/dt$ the rate of change in entropy density as we follow a point moving with the fluid.

And denoting by δv the volume of fluid instantaneously contained in the coordinate range $\delta x \delta y \delta z$ this becomes

$$\left[\frac{d\phi}{dt} \delta v + \phi \frac{d}{dt}(\delta v) \right] \delta t = \frac{d}{dt}(\phi \delta v)\, \delta t;$$

so that we can finally write

$$\left(\phi_0 \frac{dx^\mu}{ds} \right)_\mu \sqrt{-g}\ \delta x^1 \delta x^2 \delta x^3 \delta x^4 = \frac{d}{dt}(\phi \delta v)\, \delta t \qquad (121.3)$$

as an expression for the left-hand side of the relativistic second law (121.1), using natural coordinates for the point of interest.

Hence, using natural coordinates, it is evident that the left-hand side of the relativistic second law becomes the increase, in time δt, which takes place in the entropy of the small element of fluid instantaneously contained in the coordinate range $\delta x \delta y \delta z$. In accordance with the principle of equivalence, however, we can apply special relativistic thermodynamic theory to this small system and connect its increase in entropy with heat and temperature by the expression

$$\frac{d}{dt}(\phi \delta v) \delta t \geqslant \frac{\delta Q}{T}, \qquad (121.4)$$

where δQ is the heat absorbed by this element of fluid in time δt at temperature T, these quantities also being referred to our present system of coordinates. Furthermore, since the ratio of heat to temperature is an invariant for the Lorentz transformation we can also take

$$\frac{\delta Q}{T} = \frac{\delta Q_0}{T_0}, \qquad (121.5)$$

where δQ_0 and T_0 are the absorbed heat and temperature as measured in proper coordinates by a local observer moving with the element of fluid. Moreover, in accordance with the Lorentz contraction for volume elements and the Lorentz time dilation for time intervals we can write

$$\delta v \delta t = \delta v_0 \delta t_0, \qquad (121.6)$$

where δv_0 is the instantaneous volume of the element of fluid as measured in proper coordinates and δt_0 is the proper time during which the heat absorption takes place.

Hence, combining the information given by (121.3, 4, 5, 6), we have now obtained in natural coordinates an expression

$$\left(\phi_0 \frac{dx^\mu}{ds} \right)_\mu \sqrt{-g}\ \delta x^1 \delta x^2 \delta x^3 \delta x^4 \geqslant \frac{\delta Q_0}{T_0}, \qquad (121.7)$$

of the same form as postulated above for the relativistic second law, together with the desired specific interpretation of the quantity δQ_0 occurring on the right-hand side, as the heat—measured by a local observer at rest in the fluid at the point and time of interest—which flows into an element of the fluid having the instantaneous proper volume δv_0 during the proper time δt_0, these quantities being so chosen as to make

$$\delta v_0 \, \delta t_0 = \delta v \delta t = \sqrt{-g} \; \delta x^1 \delta x^2 \delta x^3 \delta x^4. \qquad (121.8)$$

This result has been obtained using natural coordinates. Nevertheless, in accordance with our earlier discussion of the scalar character of δQ_0, the interpretation is valid for any coordinate system. No specification of the shape of the element of fluid is given, since to the order of quantities considered, the heat absorbed depends only on the product of volume and time interval.

Since δQ_0 is the heat absorbed by a definite element of the fluid, it will be noted as in ordinary thermodynamics that heat flow is to be regarded as taking place relative to the material fluid or working substance of interest, rather than as relative to some system of spatial coordinates that happen to be in use.

It must finally be remarked, in order to remove uncertainties which could arise when we come to the integration of the second law expression, that in applying the second law each increment of heat entering a system of interest is to be taken as divided by the temperature at the location where it crosses the boundary separating the system from its surroundings, again as in the usual thermodynamics. Hence we can regard δQ_0 and T_0 as quantities which are determined by measurements made in the usual manner by observers located on the boundary of the element of fluid considered.

122. On the use of co-moving coordinates in thermodynamic considerations

The discussion of the preceding section has shown that heat flow is in general to be taken as relative to the material fluid of interest rather than relative to the particular spatial coordinates in use. This makes it especially convenient in thermodynamic considerations to select a coordinate system such that the fluid has permanently everywhere zero components of 'velocity'

$$\frac{dx^1}{ds} = \frac{dx^2}{ds} = \frac{dx^3}{ds} = 0 \qquad (122.1)$$

with respect to the spatial coordinates. Such coordinates may be called *co-moving* and are presumably always possible, since they can be obtained by taking the spatial coordinates as given by a network drawn so as to connect adjacent identifiable points of the fluid and then allowed to move therewith.

Referred to such coordinates, the relativistic second law gives specially simple and understandable results. Starting with the second form (119.5) in which we have expressed the law

$$\frac{\partial}{\partial x^\mu}\left(\phi_0\frac{dx^\mu}{ds}\sqrt{-g}\right)\delta x^1\delta x^2\delta x^3\delta x^4 \geqslant \frac{\delta Q_0}{T_0} \tag{122.2}$$

we at once obtain, on account of the permanent validity of the relations (122.1) at all points of the fluid, a reduction to the simple form

$$\frac{\partial}{\partial x^4}\left(\phi_0\frac{dx^4}{ds}\sqrt{-g}\right)\delta x^1\delta x^2\delta x^3\delta x^4 \geqslant \frac{\delta Q_0}{T_0}. \tag{122.3}$$

Furthermore, since the coordinates are mutually independent this can be rewritten in the form

$$\frac{\partial}{\partial x^4}\left(\phi_0\sqrt{-g}\,\delta x^1\delta x^2\delta x^3\frac{dx^4}{ds}\right)\delta x^4 \geqslant \frac{\delta Q_0}{T_0}. \tag{122.4}$$

In this form, however, the relation has very considerable advantages. In the first place, taking elements of four-dimensional volume at the point of interest which have equal volumes expressed in natural measure

$$\delta v_0\,dt_0 = \delta v_0\,ds = \sqrt{-g}\,\delta x^1\delta x^2\delta x^3dx^4, \tag{122.5}$$

it is evident that we can rewrite the above relation in the form

$$\frac{\partial}{\partial x^4}(\phi_0\,\delta v_0)\,\delta x^4 \geqslant \frac{\delta Q_0}{T_0}, \tag{122.6}$$

where δv_0 will be equal to the proper volume of the element of fluid permanently located in the coordinate range $\delta x^1\delta x^2\delta x^3$, as measured at any instant by a local observer moving with that element. In the second place, we can evidently re-express this latter relation in the form

$$\frac{\partial}{\partial t_0}(\phi_0\,\delta v_0)\,\delta t_0 \geqslant \frac{\delta Q_0}{T_0}, \tag{122.7}$$

where

$$\delta t_0 = \frac{dt_0}{dx^4}\,\delta x^4 \tag{122.8}$$

is the increment in proper time which corresponds at any instant to the increment in coordinate time δx^4.

The form for the second law given by (112.6) proves useful by containing an expression for the rate of change in the proper entropy of any given element of fluid with respect to the coordinate time x^4 which applies throughout all parts of the system under consideration. This is of considerable advantage in the treatment of finite systems.

The form for the law given by (122.7) is useful in again showing the validity of our previous interpretation of δQ_0. Since the left-hand side of (122.7) is the increase, as found by a local observer, which occurs in time δt_0 in the entropy of an element of fluid of volume δv_0, it is evident from the ordinary principles of thermodynamics—which must apply for such a local observer—that δQ_0 must be the heat which he finds to be absorbed by the element in that time. On the other hand, in accordance with (122.5) and (122.8) we have

$$\delta v_0 \, \delta t_0 = \sqrt{-g} \, \delta x^1 \delta x^2 \delta x^3 \delta x^4, \qquad (122.9)$$

which is our previous specification (121.8) for the volume of the element and time interval that are to be employed by the local observer in measuring the heat δQ_0.

The form for the second law given by (122.7) is also useful in emphasizing the principle that a local observer examining the thermodynamic behaviour of an element of fluid in his immediate neighbourhood must use the same methods of measuring entropy, heat, and temperature, and employ the same criteria of reversibility and irreversibility as have been made familiar by the classical thermodynamics. This principle serves to increase our confidence in the validity of relativistic thermodynamics, and to explain the fact that outstanding differences between the conclusions of classical and relativistic thermodynamics tend to appear only when large portions of the universe are under consideration.

RELATIVISTIC THERMODYNAMICS (*contd.*)

Part II. APPLICATIONS OF RELATIVISTIC THERMODYNAMICS

123. Application of the first law to changes in the static state of a system

We may now commence our investigation of the consequences of relativistic thermodynamics. Since our first interest will lie in determining the conditions for static thermodynamic equilibrium, we shall begin by examining the restrictions imposed by the principles of relativistic mechanics on the changes which might take place in a thermodynamic system from one static state to another without involving any changes in the surroundings that lie outside the selected region of interest.

Consider a system together with its surroundings which up to some initial 'time' x'^4 are in a given static state such that there are no changes taking place with respect to x^4, and then let a change take place inside the system without affecting the surroundings to some new static state at 'time' x''^4, after which there will again be no changes taking place with respect to the time-like coordinate x^4.

Since this change is to take place without involving any effects on the surroundings, it is evident (*a*) that there must be no transfer of energy or momentum between the system and surroundings, and (*b*) that the distribution of energy and momentum in the surroundings must remain unaltered. It is easy to show, however, that these conditions are met if the change inside the system involves no changes in the values of the gravitational potentials $g_{\mu\nu}$ and of their first and second derivatives $\partial g_{\mu\nu}/\partial x_\alpha$ and $\partial^2 g_{\mu\nu}/\partial x_\alpha \partial x_\beta$ at the boundary of the system and beyond.

To show that this restriction is sufficient to prevent the transfer of energy and momentum from the surroundings into the system, we may return to our previous expression (88.2) for the energy-momentum principle as applied to a finite system

$$\frac{d}{dx^4} \iiint (\mathfrak{T}_\mu^4 + \mathfrak{t}_\mu^4)\, dx^1 dx^2 dx^3 =$$

$$- \iint |\mathfrak{T}_\mu^1 + \mathfrak{t}_\mu^1|_{x^1}^{x'^1}\, dx^2 dx^3 - \iint |\mathfrak{T}_\mu^2 + \mathfrak{t}_\mu^2|_{x^2}^{x'^2}\, dx^1 dx^3 - \iint |\mathfrak{T}_\mu^3 + \mathfrak{t}_\mu^3|_{x^3}^{x'^3}\, dx^1 dx^2,$$

$$(123.1)$$

where the left-hand side of the equation gives the rate of change in the three components of momentum and in the energy of the system, according as we take $\mu = 1, 2, 3, 4$, and the right-hand side can be regarded as giving the flow of momentum or energy across the boundary from the surroundings into the system, provided we choose as usual coordinates such that the necessary limits of integration actually lie on the boundary separating the system from its surroundings.

Up until the initial time x'^4 when the change in state commences, the left-hand side of this equation will be zero since by hypothesis the system is then in some given static state. Hence the right-hand side of the equation will also be zero at time x'^4. The right-hand side, however, is a constant independent of x^4, since the quantities

$$-8\pi\mathfrak{T}_\mu^\nu = \mathfrak{R}_\mu^\nu - \tfrac{1}{2}\mathfrak{R}g_\mu^\nu + \Lambda g_\mu^\nu \sqrt{-g} \qquad (123.2)$$

and

$$16\pi\mathfrak{t}_\mu^\nu = -\mathfrak{g}_\mu^{\alpha\beta}\frac{\partial\mathfrak{L}}{\partial\mathfrak{g}_\nu^{\alpha\beta}} + g_\mu^\nu\mathfrak{L} + 2\Lambda g_\mu^\nu\sqrt{-g} \qquad (123.3)$$

are definitely determined by the $g_{\mu\nu}$ and their first and second derivatives, and by hypothesis these do not change at points on the boundary, corresponding to the limits of integration on the right-hand side of (123.1). Hence both sides of that equation remain zero and there is no transfer of energy or momentum between the system and its surroundings.

To show that the restriction is sufficient to prevent any change in the distribution of energy and momentum in the surroundings, we merely have to note again in accordance with (123.2) that the energy-momentum tensor is definitely determined by the gravitational potentials $g_{\mu\nu}$ and their first and second derivatives, and hence if these quantities remain constant for points outside the boundary the distribution of energy and momentum will also have to remain unchanged in the surroundings.

Hence, to sum up the results of this section, we may state, in accordance with the principles of relativistic mechanics or first law of relativistic thermodynamics, that a thermodynamic system can change from one static state to another without involving any changes in the surroundings, provided we subject the gravitational potentials and their first and second derivatives at points on the boundary and beyond to the restriction

$$\delta g_{\mu\nu} = \delta\left(\frac{\partial g_{\mu\nu}}{\partial x^\alpha}\right) = \delta\left(\frac{\partial^2 g_{\mu\nu}}{\partial x^\alpha\,\partial x^\beta}\right) = 0. \qquad (123.4)$$

124. Application of the second law to changes in the static state of a system

Having thus found restrictions—analogous to those imposed in classical thermodynamics by the ordinary first law—that are sufficient to guarantee changes in the static state of a system which could take place without involving any changes in the surroundings, we may now inquire into the restrictions which would be imposed by the second law on the possible changes in static state.

To investigate this it will be best to employ co-moving coordinates of the kind discussed in § 122. In any case the coordinates would be co-moving before and after the internal change takes place since the system is then by hypothesis in some static state, and by using coordinates which are also co-moving during the change we can then express the restrictions imposed by the second law on the nature of the change in the simple form (see 122.3)

$$\frac{\partial}{\partial x^4}\left(\phi_0 \frac{dx^4}{ds} \sqrt{-g}\right) \delta x^1 \delta x^2 \delta x^3 \delta x^4 \geqslant \frac{\delta Q_0}{T_0}. \tag{124.1}$$

In accordance with (122.6), we can regard the left-hand side of this expression as the increase which takes place in 'time' δx^4 in the proper entropy as measured by a local observer of the element of fluid permanently located in the 'spatial' range $\delta x^1 \delta x^2 \delta x^3$; and in accordance with (122.7) we can regard the right-hand side of this expression as given by the heat measured by a local observer which flows into this element of fluid in the increment of proper time δt_0 which corresponds to δx^4.

If we integrate this expression for a given element of the fluid over the total interval x'^4 to x''^4 during which change takes place, we shall obtain

$$\int_{x'^4}^{x''^4} \frac{\partial}{\partial x^4}\left(\phi_0 \frac{dx^4}{ds} \sqrt{-g}\right) \delta x^1 \delta x^2 \delta x^3 \delta x^4 \geqslant \int_{x'^4}^{x''^4} \frac{\delta Q_0}{T_0}, \tag{124.2}$$

where the left-hand side is the total change which the local observer finds in the entropy of the element, and the right-hand side—in accordance with the specifications given at the end of § 121—is the total result obtained by summing all the increments of heat that enter the element each divided by the temperature of the boundary at the time of passage, the measurements being made by observers on the boundary of the element.

If we now perform a second integration, this time over all the elements of fluid included in the system, it is evident that the right-hand side of (124.2) will lead to a null result giving

$$\int\limits_{x'^4}^{x''^4} \iiint \frac{\partial}{\partial x^4}\left(\phi_0 \frac{dx^4}{ds} \sqrt{-g}\right) \delta x^1 \delta x^2 \delta x^3 \delta x^4 \geqslant 0, \qquad (124.3)$$

since by hypothesis we shall be interested in changes which involve no flow of heat at the boundary of the system as a whole, and our method for the precise specification of δQ_0 and T_0 will lead to a cancellation between contiguous elements within the system.

Hence the conditions imposed by the relativistic second law of thermodynamics on changes in the interior of a system from one static state to another without affecting the surroundings can be expressed by the relation

$$\left[\iiint \phi_0 \frac{dx^4}{ds} \sqrt{-g}\, dx^1 dx^2 dx^3\right]_{x''^4} \geqslant \left[\iiint \phi_0 \frac{dx^4}{ds} \sqrt{-g}\, dx^1 dx^2 dx^3\right]_{x'^4}$$
$$(124.4)$$

where the subscripts x'^4 and x''^4 indicate that the values of the integrals are to be taken for the initial and final states of the system.

To emphasize the analogy with classical thermodynamics this result could also be stated as the requirement that the 'entropy' S can only remain constant or increase when the system changes from one static state to another, provided we define that quantity as the total integrated proper entropy of the elements of fluid in the system

$$S = \iiint \phi_0 \frac{dx^4}{ds} \sqrt{-g}\, dx^1 dx^2 dx^3. \qquad (124.5)$$

125. The conditions for static thermodynamic equilibrium

With the help of the two foregoing sections we may now express the conditions for static thermodynamic equilibrium in a finite system having no interaction with its surroundings in the form of the variational equation

$$\delta \iiint \phi_0 \frac{dx^4}{ds} \sqrt{-g}\, dx^1 dx^2 dx^3 = 0 \qquad (125.1)$$

under the subsidiary condition to be imposed at the boundary of the system

$$\delta g_{\mu\nu} = \delta\left(\frac{\partial g_{\mu\nu}}{\partial x^\alpha}\right) = \delta\left(\frac{\partial^2 g_{\mu\nu}}{\partial x^\alpha \partial x^\beta}\right) = 0. \qquad (125.2)$$

The first of these equations is the condition for a maximum value of the integral in question and is imposed by the second law which as shown by (124.4) will only permit increases if there is any change at all in this quantity when the system changes from one static state to another without interaction with the surroundings. And the second set of equations provides, as we have seen in § 123, a sufficient condition to prevent any interaction between the system and its surroundings when the internal change takes place.

126. Static equilibrium in the case of a spherical distribution of fluid

In the case of a *fluid* system held together by gravitational attraction, a state of static equilibrium will necessarily be one of spherical symmetry; and we may give special attention to the form assumed in that case by the above conditions for equilibrium. In accordance with (94.9) we may then write the line element to start with in the form

$$ds^2 = -e^{\mu}(dx^2 + dy^2 + dz^2) + e^{\nu} dt^2, \qquad (126.1)$$

where $\qquad \mu = \mu(r) \qquad \nu = \nu(r) \qquad r = \sqrt{(x^2 + y^2 + z^2)} \qquad (126.2)$

and the isotropic coordinates x, y, z, t are such that the limits of integration necessary to include any given region of interest will fall on the actual boundary separating that region from its surroundings, and hence are of the variety assumed in § 123 in obtaining the subsidiary conditions expressed by (125.2).

With the above form of line element we shall evidently have

$$\sqrt{-g} = e^{\frac{3}{2}\mu + \frac{1}{2}\nu} \quad \text{and} \quad \frac{dt}{ds} = e^{-\frac{1}{2}\nu}, \qquad (126.3)$$

the latter since the spatial components of fluid 'velocity' will be zero. Substituting in (125.1) and noting the implications of (125.2) and (126.2), we can then write the requirements for static thermodynamic equilibrium in the form

$$\delta \iiint \phi_0 e^{\frac{3}{2}\mu} \, dx\,dy\,dz = 0, \qquad (126.4)$$

under the subsidiary conditions at the boundary of the region of integration $\qquad \delta\mu = \delta\mu' = \delta\mu'' = \delta\nu = \delta\nu' = \delta\nu'' = 0, \qquad (126.5)$

where the accents denote differentiation with respect to

$$r = \sqrt{(x^2 + y^2 + z^2)}.$$

To obtain these conditions we have employed the coordinates

x, y, z, t since as remarked above they are of the kind used in § 123 in obtaining the relations (125.2). It will now be convenient for our later work, however, to transform to polar coordinates r, θ, ϕ, t.

We can then write the line element in the form

$$ds^2 = -e^\mu(dr^2 + r^2\,d\theta^2 + r^2\sin^2\theta\,d\phi^2) + e^\nu\,dt^2 \qquad (126.6)$$

$$\mu = \mu(r) \qquad \nu = \nu(r)$$

and, taking the region of integration as a spherical shell lying between r_1 and r_2, rewrite the requirements for static equilibrium for a spherical distribution of fluid in the form

$$\delta \int_{r_1}^{r_2} 4\pi\phi_0 e^{\frac{1}{2}\mu} r^2\,dr = 0 \qquad (126.7)$$

under the subsidiary conditions

$$\delta\mu = \delta\mu' = \delta\mu'' = \delta\nu = \delta\nu' = \delta\nu'' = 0 \quad \text{(at } r_1 \text{ and } r_2\text{)}. \quad (126.8)$$

In order to apply (126.7), however, we can introduce a more immediate and useful dependence on the form of the line element and on the composition of the fluid.

In the first place we recall in accordance with (95.15), that the expressions for the proper pressure p_0 and proper macroscopic density ρ_{00} of the fluid corresponding to the above form of line element are given by

$$8\pi p_0 = e^{-\mu}\left(\frac{\mu'^2}{4} + \frac{\mu'\nu'}{2} + \frac{\mu'+\nu'}{r}\right),$$

$$8\pi p_0 = e^{-\mu}\left(\frac{\mu''}{2} + \frac{\nu''}{2} + \frac{\nu'^2}{4} + \frac{\mu'+\nu'}{2r}\right),$$

$$8\pi\rho_{00} = -e^{-\mu}\left(\mu'' + \frac{\mu'^2}{4} + \frac{2\mu'}{r}\right), \qquad (126.9)$$

$$\frac{dp_0}{dr} = -\frac{\rho_{00}+p_0}{2}\nu'.$$

In the second place, since we can evidently take [see equation (51), Appendix III]

$$v_0 = 4\pi e^{\frac{1}{2}\mu} r^2\,dr \qquad (126.10)$$

as an expression for the infinitesimal proper spatial volume of the fluid lying between r and $r+dr$, and we can also take

$$S_0 = 4\pi\phi_0 e^{\frac{1}{2}\mu} r^2\,dr \qquad (126.11)$$

as an expression for the proper entropy of this spherical shell of fluid as it would be determined by local observers, both of these quantities being of course infinitesimals. The proper entropy of an element of fluid, however, will depend on its proper energy, volume, and composition in the same way as in the classical thermodynamics [see equation (60.4)]. Hence when we come to introduce the variation of the expression given by (126.11) into the condition for equilibrium (126.7) we can write

$$\delta S_0 = \frac{1}{T_0}\delta E_0 + \frac{p_0}{T_0}\delta v_0 + \left(\frac{\partial S_0}{\partial n_1}\right)_{E_0,v_0}\delta n_1 + \dots + \left(\frac{\partial S_0}{\partial n_n}\right)_{E_0,v_0}\delta n_n, \quad (126.12)$$

where T_0 is the proper temperature of the shell of fluid as measured by a local observer, E_0 is its proper energy and n_1, n_2, etc. are the number of mols of the different substances which determine its composition. Furthermore, in using this relation we can take in accordance with (126.9)

$$\delta E_0 = \delta(4\pi\rho_{00}e^{\frac{3}{2}\mu}r^2\,dr)$$

$$= -\left[\frac{e^{\frac{1}{2}\mu}}{2}\left(\delta\mu'' + \frac{\mu'}{2}\,\delta\mu' + \frac{2}{r}\delta\mu'\right) + \frac{e^{\frac{1}{2}\mu}}{4}\left(\mu'' + \frac{\mu'^2}{4} + \frac{2\mu'}{r}\right)\delta\mu\right]r^2\,dr$$

$$= \left[-\frac{e^{\frac{1}{2}\mu}}{2}\left(\delta\mu'' + \frac{\mu'}{2}\,\delta\mu' + \frac{2}{r}\,\delta\mu'\right) + 2\pi\rho_{00}e^{\frac{3}{2}\mu}\delta\mu\right]r^2\,dr, \quad (126.13)$$

and in accordance with (126.10)

$$\delta v_0 = \delta(4\pi e^{\frac{1}{2}\mu}r^2\,dr)$$

$$= 6\pi e^{\frac{1}{2}\mu}\delta\mu\,r^2\,dr, \quad (126.14)$$

and finally in accordance with (126.11)

$$\left(\frac{\partial S_0}{\partial n_i}\right)_{E_0,v_0}\delta n_i = 4\pi\left(\frac{\partial\phi_0}{\partial c_i^0}\right)_\mu\delta c_i^0\,e^{\frac{1}{2}\mu}r^2\,dr, \quad (126.15)$$

where $(\partial\phi_0/\partial c_i^0)_\mu$ is the partial derivative of proper entropy density with the concentration of the ith component, taken at constant energy density and constant specific volume, this latter being indicated by the subscript μ since energy density and specific volume are determined by this quantity and its derivatives.

Substituting these relations, we can now rewrite our earlier condition for equilibrium (126.7) in the form

$$\int_{r_1}^{r_2}\left[-\frac{e^{\frac{1}{2}\mu}}{2T_0}\left(\delta\mu'' + \frac{\mu'}{2}\,\delta\mu' + \frac{2}{r}\delta\mu'\right) + \frac{2\pi\rho_{00} + 6\pi p_0}{T_0}e^{\frac{1}{2}\mu}\,\delta\mu +\right.$$

$$\left. + 4\pi\sum_i\frac{\partial\phi_0}{\partial c_i^0}\delta c_i^0\,e^{\frac{1}{2}\mu}\right]r^2\,dr = 0.$$

This expression can be further simplified, however, in the usual manner by performing partial integrations and dropping terms which become zero on account of boundary conditions. Substituting

$$\delta\mu'' = \frac{d}{dr}(\delta\mu') \qquad \delta\mu' = \frac{d}{dr}(\delta\mu),$$

and, using from the boundary conditions (126.8), the relations

$$\delta\mu' = \delta\mu = 0 \qquad \text{(at } r_1 \text{ and } r_2\text{)}.$$

We thus readily obtain after some simplification

$$\int_{r_1}^{r_2} \left[-\frac{d}{dr}\left\{ e^{\frac{1}{2}\mu}r^2\frac{d}{dr}\left(\frac{1}{T_0}\right)\right\} + \frac{4\pi(\rho_{00}+3p_0)}{T_0}e^{\frac{1}{2}\mu}r^2 \right]\delta\mu \, dr \, +$$

$$+ \int_{r_1}^{r_2} 8\pi \sum_i \left(\frac{\partial\phi_0}{\partial c_i^0}\right)_\mu \delta c_i^0 \, e^{\frac{1}{2}\mu}r^2 \, dr = 0 \quad (126.16)$$

as our final expression for the condition of thermodynamic equilibrium in a fluid sphere.

In accordance with the method by which this expression was obtained, it will be seen that the variations in proper energy δE_0 and proper volume δv_0, originally occurring in (126.12), have both contributed to the variation in the metrical variable μ, while the variations δn_i in the number of mols of the different constituents in the shell of fluid between r and $r+dr$ have directly led to the variations δc_i^0 in the concentrations which determine the composition at each value of r. Since E_0 and v_0 originally entered our considerations as variables which were independent of those determining the composition n_i, it is evident that we can regard the variation indicated by $\delta\mu$ in equation (126.16) as independent of that indicated by the δc_i^0.

127. Chemical equilibrium in a gravitating sphere of fluid

We may now use the general condition for equilibrium obtained in the last section to investigate the chemical equilibrium between reacting substances in the interior of a gravitating sphere of fluid. Since the variations indicated by $\delta\mu$ and δc_i^0 in (126.16) are to be treated as independent, it is evident that we can take the second integral in this expression as itself equal to zero, and this can only be satisfied provided we have

$$\sum_i \left(\frac{\partial\phi_0}{\partial c_i^0}\right)_\mu \delta c_i^0 = 0, \qquad (127.1)$$

holding at each value of r, where the subscript μ indicates constancy of energy density and specific volume. Comparing this with (60.15), however, we see that this relation is the same as the classical condition for chemical equilibrium provided we use entropy densities and concentrations as measured by a local observer at rest in the fluid. Hence the chemical equilibria between reacting substances at any point in a gravitating sphere of fluid will be characterized by the same conditions—measured by a local observer—as would be calculated on a classical basis.

This is an example of the tendency already mentioned for relativistic considerations often to lead to the same conclusions, for the results of measurements by a local observer, as would be obtained from classical considerations. This tendency arises of course as a consequence of the original introduction of the principle of equivalence as a part of the axiomatic basis for the general theory of relativity, and a knowledge of this tendency can be used as a fairly safe intuitive guide in drawing conclusions when we feel sure that the phenomena under consideration do not depend on higher derivatives of the $g_{\mu\nu}$ than the first.

The definite demonstration of the principle that the conditions for chemical equilibrium are not directly affected by mere position in a gravitational field is very important, since this is tacitly assumed to be true in our usual consideration of stellar models. As a consequence of the principle, it should be noted that the results of our previous discussion of the equilibria between hydrogen and helium and that between matter and radiation would be applicable at any level in a star. Hence our previous difficulties as to their relative concentrations remain. In the last chapter we shall see that a similar principle holds for chemical equilibria in static cosmological models.

128. Thermal equilibrium in a gravitating sphere of fluid

We may also use the general condition for thermodynamic equilibrium obtained in § 126 to investigate the distribution of temperature in a fluid sphere which has come to thermal equilibrium. Again making use of the consideration that the variations indicated by $\delta\mu$ and δc_i^0 in (126.16) are to be treated as independent, we may this time conclude that the first integral in that expression is itself equal to zero. This, however, can evidently be true only if we have the relation

$$\frac{d}{dr}\left\{e^{\frac{1}{2}\mu}r^2\frac{d}{dr}\left(\frac{1}{T_0}\right)\right\} = \frac{4\pi(\rho_{00}+3p_0)}{T_0}e^{\frac{1}{2}\mu}r^2 \tag{128.1}$$

holding within the sphere at all values of r.

To put this equation in a form suitable for integration, we may re-express the right-hand side by substituting for $(\rho_{00}+3p_0)$ from (126.9), using the expression given there for ρ_{00}, plus the first expression for p_0, plus twice the second expression for p_0. Doing so we obtain

$$\frac{d}{dr}\left\{e^{\frac{1}{2}\mu}r^2\frac{d}{dr}\left(\frac{1}{T_0}\right)\right\} = \frac{e^{\frac{1}{2}\mu}}{T_0}\left(\frac{\nu''}{2}+\frac{\nu'^2}{4}+\frac{\mu'\nu'}{4}+\frac{\nu'}{r}\right)r^2$$

$$= \frac{e^{-\frac{1}{2}\nu}}{T_0}\frac{d}{dr}\left\{e^{\frac{1}{2}\mu}r^2\frac{d}{dr}(e^{\frac{1}{2}\nu})\right\}. \tag{128.2}$$

As a first integral of this equation we evidently have

$$e^{\frac{1}{2}\mu+\frac{1}{2}\nu}r^2\frac{d}{dr}\left(\frac{1}{T_0}\right) = \frac{e^{\frac{1}{2}\mu}r^2}{T_0}\frac{d}{dr}(e^{\frac{1}{2}\nu})+B,$$

where B is the constant of integration, and this may be rewritten in the form

$$\frac{d\log T_0}{dr} = -\frac{1}{2}\frac{d\nu}{dr}-\frac{Be^{-\frac{1}{2}\mu-\frac{1}{2}\nu}T_0}{r^2}. \tag{128.3}$$

By substituting from (126.9), however, this latter can be re-expressed as

$$\frac{d\log T_0}{dr} = \frac{1}{\rho_{00}+p_0}\frac{dp_0}{dr}-\frac{Be^{-\frac{1}{2}\mu-\frac{1}{2}\nu}T_0}{r^2}.$$

Hence if we assume on physical grounds that at the centre of the sphere, $r = 0$, we have dT_0/dr and dp_0/dr equal to zero, T_0 not equal to zero, and the other functions of r finite, it is evident that the constant B must be equal to zero. We may then write our final expression for the dependence of proper temperature on position in a static sphere of fluid at thermal equilibrium in the equivalent forms

$$\frac{d\log T_0}{dr} = -\frac{1}{2}\frac{d\nu}{dr}, \tag{128.4}$$

$$\frac{d\log T_0}{dr} = \frac{1}{\rho_{00}+p_0}\frac{dp_0}{dr}, \tag{128.5}$$

or by a second integration in the form

$$T_0e^{\frac{1}{2}\nu} = T_0\sqrt{(g_{44})} = C, \tag{128.6}$$

where C is a new constant of integration.

It would of course be interesting also to investigate the possibility for solutions of physical interest in which the constant of integration B was not taken equal to zero. Nevertheless for ordinary continuous distributions of fluid it seems clear that the simple equations just given must be regarded as the correct ones to use.

The first point to emphasize in connexion with the above results is the significant conclusion that the proper temperature of a fluid as measured by local observers using ordinary thermometric methods would not be constant throughout a fluid sphere which has come to thermal equilibrium, but would vary with gravitational potential, increasing with depth as we go toward the centre of the sphere. This conclusion is of course very different in character from the classical conclusion, as previously discussed in § 61, that uniform temperature throughout is a necessary condition for thermal equilibrium. Nevertheless, from the point of view of relativity, since all forms of energy must be expected to have weight as well as mass, the conclusion that a temperature gradient is necessary to prevent the flow of heat from regions of higher to those of lower gravitational potential seems a natural and appropriate result.†

A second important aspect of the new result which should be noted is the fact that the actual effect of gravitational potential on equilibrium temperature would be extremely small except in very strong fields. Thus in a field having the intensity of that at the earth's surface the change in temperature with radial position would have only the very small value

$$\frac{d \log T}{dr} \simeq -10^{-18}\,\text{cm.}^{-1} \tag{128.7}$$

This result is of course in agreement with the fact that we have as yet no observational evidence of any effect of gravitational field on thermal flow.

It is indeed questionable whether the new effect would even be large enough to have importance for theories of stellar structure, since as will be seen from the form of the principle given by (128.5) the percentage rate of increase in temperature as we proceed inward in a sphere which has come to thermal equilibrium would in any case be smaller than the percentage rate of increase in pressure, and indeed very much smaller for ordinary matter with ρ_{00} large compared with p_0. It is, however, conceivable that the new criteria for thermal flow might sometime be of interest in connexion with non-homogeneous cosmological models, having thermal flow from one portion of the model to another.

As a third point in connexion with the results of this section, it is interesting to note that the new relation between temperature and

† Tolman, *Phys. Rev.* **35**, 904 (1930).

gravitational potential is anyhow that which would be demanded in the case of a distribution of pure black-body radiation by the direct application of mechanical principles alone. Thus in the case of a spherical distribution of pure black-body radiation, such as might be thought of as surrounding a gravitating sphere of denser matter, we could conclude from the mechanical equations given by (126.9) that the pressure of the radiation would have to increase as we go inward, in order to support the weight of radiation above, at the rate

$$\frac{dp_0}{dr} = -\frac{\rho_{00}+p_0}{2}\frac{d\nu}{dr}. \tag{128.8}$$

For black-body radiation, however, we have the direct relations of Boltzmann and Stefan (see § 65) connecting the mechanical quantities, density and pressure, with the thermodynamic quantity, temperature

$$\rho_{00} = aT_0^4$$

and
$$p_0 = \frac{a}{3}T_0^4, \tag{128.9}$$

where a is Stefan's constant. And substituting these expressions above we at once obtain our previous relation between temperature and gravitational potential

$$\frac{d\log T_0}{dr} = -\frac{1}{2}\frac{d\nu}{dr}. \tag{128.10}$$

This direct verification, in a particular case, of a result previously obtained by taking the full apparatus of relativistic thermodynamics as a starting-point, can serve to increase our confidence in the validity of our extension of thermodynamics to relativity.

In concluding this section it should not be forgotten that the results here considered have been derived for the special case of a static distribution of fluid having spherical symmetry. It should also be noted that the quantity ν, appearing in the condition for thermal equilibrium (128.10), is to be taken as the ν which occurs in the special form (95.14) which can then be given to the line element, or could also be taken as the ν in the expression for the line element (95.12), since as shown in § 94 this quantity is not affected by the transformation between the two forms.

129. Thermal equilibrium in a general static field

We may now examine the conditions for thermal equilibrium in a more general static field, corresponding, for example, to a solid structure where spherical symmetry would not be a necessary characteristic of the final state of stability.† To investigate temperature equilibrium in such a case, we shall assume that the parts of the system whose temperatures are to be compared are in thermal contact with a small connecting tube containing black body radiation, or could be put into such contact without introducing any essential change in the nature of the system. Such a tube might be called a radiation thermometer, and by calculating the change in radiation pressure as we go from one portion of the tube to another we shall be able to determine the temperature distribution at equilibrium.

We shall take the line element for the system as having the very general static form

$$ds^2 = g_{ij}\,dx^i dx^j + g_{44}\,dt^2, \tag{129.1}$$

where we adopt the convention of using Latin indices to correspond to spatial coordinates and save Greek indices to indicate any coordinate. In accordance with the usual definition of a static system we take the potentials g_{14}, g_{24}, and g_{34} equal to zero, and take the other potentials g_{ij} and g_{44} as dependent in any arbitrary way desired on the spatial coordinates x^1, x^2, and x^3, but as independent of the time coordinate t. For the potential g_{44}, it will be noted from the form of the line element that we have the specially simple relation

$$g^{44} = \frac{1}{g_{44}}. \tag{129.2}$$

As an expression to use for the energy-momentum tensor of black-body radiation in the field defined by the above line element, we can take in accordance with § 109 the formula

$$T^{\mu\nu} = (\rho_{00} + p_0)\frac{dx^\mu}{ds}\frac{dx^\nu}{ds} - g^{\mu\nu}p_0, \tag{129.3}$$

with
$$\rho_{00} = 3p_0. \tag{129.4}$$

Furthermore, noting in the case of a static system that the overall macroscopic velocity of radiation flow would be equal to zero, we can write

$$\frac{dx^i}{ds} = \frac{dx^j}{ds} = 0 \quad (i,j = 1, 2, 3). \tag{129.5}$$

And, taking account of the form of the line element, can also write

† Tolman and Ehrenfest, *Phys. Rev.* **36**, 1791 (1930).

for the fourth component of 'velocity'

$$\frac{dx^4}{ds} = \frac{dt}{ds} = \frac{1}{\sqrt{(g_{44})}} = \sqrt{(g^{44})}. \tag{129.6}$$

Substituting these two expressions into (129.3), we then find that the energy-momentum tensor has as its only surviving components

$$T^{ij} = -g^{ij}p_0 \qquad T^{44} = g^{44}\rho_{00}. \tag{129.7}$$

And on lowering suffixes, we have

$$T^i_j = g_{j\alpha}T^{i\alpha} = -g_{j\alpha}g^{i\alpha}p_0 = -g^i_j p_0,$$
$$T^4_4 = g_{44}T^{44} = g_{44}g^{44}\rho_{00} = \rho_{00};$$

so that we obtain as the only surviving components of the energy-momentum tensor in its mixed form

$$T^1_1 = T^2_2 = T^3_3 = -p_0 \qquad T'^4_4 = \rho_{00}. \tag{129.8}$$

We are now ready to use the principles of relativistic mechanics in the familiar form

$$\frac{\partial \mathfrak{T}^\nu_\mu}{\partial x^\nu} - \tfrac{1}{2}\mathfrak{T}^{\alpha\beta}\frac{\partial g_{\alpha\beta}}{\partial x^\mu} = 0$$

to investigate the pressure in our postulated radiation thermometer. Taking the case $\mu = 1$ and substituting equations (129.7) and (129.8) we obtain

$$\frac{\partial}{\partial x^1}(-p_0\sqrt{-g}) - \tfrac{1}{2}(-g^{ij}p_0\sqrt{-g})\frac{\partial g_{ij}}{\partial x^1} - \tfrac{1}{2}(g^{44}\rho_{00}\sqrt{-g})\frac{\partial g_{44}}{\partial x^1} = 0,$$

and this can evidently be rewritten in the form

$$\sqrt{-g}\frac{\partial p_0}{\partial x^1} + p_0\frac{\partial\sqrt{-g}}{\partial x^1} - \tfrac{1}{2}p_0\sqrt{-g}\left(g^{ij}\frac{\partial g_{ij}}{\partial x^1} + g^{44}\frac{\partial g_{44}}{\partial x^1}\right) +$$
$$+ \tfrac{1}{2}(\rho_{00}+p_0)\sqrt{-g}\,g^{44}\frac{\partial g_{44}}{\partial x^1} = 0.$$

This can be easily simplified, however, by substituting, in accordance with Appendix III, equation (39),

$$g^{ij}\frac{\partial g_{ij}}{\partial x^1} + g^{44}\frac{\partial g_{44}}{\partial x^1} = g^{\alpha\beta}\frac{\partial g_{\alpha\beta}}{\partial x^1} = \frac{1}{g}\frac{\partial g}{\partial x^1},$$

which is seen to lead to a cancellation of the second and third terms above, giving us

$$\frac{\partial p_0}{\partial x^1} + \frac{\rho_{00}+p_0}{2}g^{44}\frac{\partial g_{44}}{\partial x^1} = 0.$$

And by making use of (129.2) and (129.4), this provides the final simple relation for the dependence of the pressure in the radiation thermometer on the coordinate x^1,

$$\frac{\partial \log p_0}{\partial x^1} + 2\frac{\partial \log g_{44}}{\partial x^1} = 0. \tag{129.9}$$

Since similar relations will hold for the dependence of pressure on the other spatial coordinates, we can now integrate and express the general dependence of radiation pressure on position in the remarkably compact form

$$p_0(g_{44})^2 = \text{const.}$$

and by substituting the relation between pressure and temperature given by the Boltzmann-Stefan relation

$$p_0 = \tfrac{1}{3}aT_0^4,$$

this leads at once to the desired expression for the dependence of proper temperature on position in a general static field

$$T_0\sqrt{(g_{44})} = C, \tag{129.10}$$

where C is the same in all parts of the system.

Several remarks may be made concerning this final simple result.

In the first place, comparing with (128.6) it will be noticed that the conditions for thermal equilibrium in a fluid sphere can now be regarded as a special case of this general result for any static field. Since the first result was obtained from the principles of relativistic thermodynamics and the second from mechanical principles alone, except for the final introduction of the Boltzmann-Stefan relation between pressure and temperature, we may regard the agreement as again furnishing justification for confidence in the new thermodynamics, similar to that pointed out at the end of the preceding section.

As a second point, it should be noted from the method of derivation that the constancy of $T_0\sqrt{(g_{44})}$ has been proved in the first instance solely for points within the radiation thermometer. Nevertheless, since T_0 and g_{44} are certainly continuous functions of position within the thermometer itself the result would also apply to the system itself where it comes in thermal contact with the thermometer.

As a further consideration, it should be noted that the derivation was carried out on the assumption that the system could be provided with a radiation thermometer, connecting the parts whose temperatures were to be compared, without disturbing the essential character of the system itself. Hence some discussion of the probable validity of this assumption would be desirable. For example, if we had a gravitating system containing solid parts it would be necessary to make a hole into the solid and insert a radiation thermometer if we wished to obtain information as to the temperature of the interior by the method we have suggested, and this procedure would certainly have some effect on the gravitational potentials $g_{\mu\nu}$ which are themselves directly related to the distribution of matter and energy in the system. Nevertheless, since the equation of connexion

$$-8\pi T_{\mu\nu} = R_{\mu\nu} - \tfrac{1}{2}Rg_{\mu\nu} + \Lambda g_{\mu\nu}$$

is a differential one giving the distribution of matter and energy in terms of the $g_{\mu\nu}$ *and* their first and second derivatives, it seems correct to assume that the insertion of a thermometer of small dimensions could be made without seriously altering the values of the $g_{\mu\nu}$ themselves. This question might be further investigated, however, since exceptional cases of interest might be found.

Finally, it is interesting to note that although the proper temperature T_0 varies from point to point in a gravitational system which has come to thermal equilibrium, nevertheless the constancy of the combined quantity $T_0\sqrt{(g_{44})}$ might provide some of the advantages of the classical principle of constant temperature as the criterion of thermal equilibrium. In this connexion it will be recalled that Einstein himself was led in his early speculations on the nature of gravitation to distinguish between a quantity, called 'wahre Temperatur', which would be constant throughout a system at thermal equilibrium and a second quantity, called at the suggestion of Ehrenfest 'Taschentemperatur', which would vary with gravitational potential. The considerations were of only a limited applicability since this was done at a time before the complete development of general relativity; the quantities mentioned, however, were respectively analogous to our present $T_0\sqrt{(g_{44})}$ and T_0. Nevertheless, since the proper temperature T_0 has an immediate physical significance from its direct relation to the measurements of a local observer, it will perhaps be best not to multiply the different kinds of temperature to which we might give names, and to regard T_0 as being fundamentally the thing that we mean by *the* temperature at a point.

130. On the increased possibility in relativistic thermodynamics for reversible processes at a finite rate

We must now consider the possibility in relativistic thermodynamics for certain kinds of thermodynamic processes to take place both reversibly and at a finite rate. This provides a second example of the differences between the conclusions of relativistic thermodynamics and those which seemed inevitable from the classical point of view.

We have already discussed in § 62 the general line of argument by which the classical thermodynamics was led to the conclusion that reversible thermodynamic processes would always have to be carried out at an infinitesimally slow rate in order to secure that maximum efficiency which would be necessary to permit a return both of the system and its surroundings to their original states. In the present section we shall use the expansion of a perfect monatomic gas as an example to illustrate the differences which can arise between classical and relativistic points of view as to reversibility and rate.

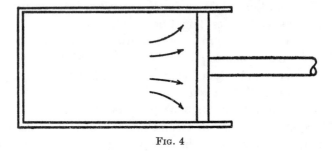

FIG. 4

Let us first consider the expansion of a sample of perfect monatomic gas by placing it in a cylinder provided with a movable piston as shown in Fig. 4. To begin with, it is evident in order to secure reversibility that we can allow no heat flow between the cylinder and its surroundings to take place at a finite rate, since this would involve a finite temperature gradient and hence the irreversible transfer of heat from regions of higher to those of lower temperature. Therefore the expansion would in any case have to be carried out adiabatically. Even then, however, it is evident that the expansion could not be carried out reversibly at a finite rate, in the first place because this would involve friction between the piston and the walls, and in the second place since the fluid in flowing in behind the moving piston would not be able to maintain as high a pressure at the expanding boundary as at an infinitesimal rate of expansion. Both of these effects would prevent the piston from doing as much work as would be necessary to recompress the gas.

Hence we may conclude that such an expansion of gas in an enclosed container with movable walls could not be carried out both reversibly and at a finite rate, since the expansion would not deliver sufficient work to the surroundings to secure reversibility. Furthermore, the argument would appear to be essentially the same and the conclusion valid whether we use classical thermodynamics or adopt a relativistic point of view.

Since this failure to secure reversibility in the case of an enclosed sample of gas, expanding at a finite rate, results from the inability of the expanding system to deliver the necessary work to its surroundings, let us next turn to the expansion of an unenclosed perfect gas without any other surroundings at all, the gas itself being the only thing in the universe that we consider. Here we find greater differences between classical and relativistic considerations, due principally to the fact that classical ideas did not provide a complete theory of gravitation.

Three remarks may be made which perhaps give a fair idea as to the classical point of view with regard to the expansion of an unenclosed gas. In the first place, classical thought was so strongly impressed by the frequent dependence of irreversibility on finite rate—as illustrated above by the expansion of gas in a cylinder—that the existence of reversible processes occurring at a finite rate was for the most part not even seriously entertained as a possibility for any process. In the second place, in accordance with the usual classical ideas of a three-dimensional Euclidean space having infinite extent, it seemed most natural to consider that the only important possibility for the expansion of an unenclosed gas would be its diffusion into the surrounding empty space. And this would be a process which would take place—to be sure at a finite rate—but with an irreversible increase in entropy owing in the last analysis to the entropy change

$$\Delta S = R \log \frac{p_1}{p_2} \qquad (130.1)$$

when a mol of gas drops from pressure p_1 to pressure p_2 without the performance of external work. Finally, the alternative possibility of a universe or cosmological model completely filled with an expanding gas was not considered from the classical point of view. It is this latter possibility that proves to be important for relativity. Classically, however, such models were not investigated, perhaps partly because it was known that the unmodified Newtonian theory of gravitation was incompatible with an infinite homogeneous distribution of gas in a static state,† and partly because the Newtonian theory—having provided no definite principles as to the velocity of propagation of gravitational action—was unable to carry through any unique treatment for such a non-static cosmological model.

Turning, however, to a relativistic consideration of the possibilities for the expansion of an unenclosed gas, we find that relativistic mechanics, with its definite theory as to the interrelated geometrical and gravitational aspect of the potentials $g_{\mu\nu}$, has been able to provide a perfectly unambiguous treatment for non-static cosmological models filled throughout their entire extent with any homogeneous distribution of expanding or contracting fluid. And it is these models —when filled with a sufficiently simple fluid such as a perfect monatomic gas—which furnish illustrations of the relativistic possibility for reversible processes to take place at a finite rate.

To obtain an understanding of this new possibility, we may anticipate the results to be derived in the next chapter by writing the line element for a non-static homogeneous model of the universe in the form

$$ds^2 = -\frac{e^{g(t)}}{[1+(r^2/4R_0^2)]^2}(dr^2+r^2\,d\theta^2+r^2\sin^2\theta\,d\phi^2)+dt^2, \quad (130.2)$$

where r, θ, and ϕ are spatial coordinates, t is the time-like coordinate, R_0 is a constant and the dependence of the line element on time is determined by the functional form of the exponent $g(t)$. This formula for interval can be shown to correspond to a cosmological model which is filled throughout its *entire* spatial extent with a homogeneous distribution of fluid. The coordinates used are of the co-moving variety discussed in § 122 so that an element of the fluid located in any given coordinate range $\delta r\delta\theta\delta\phi$ remains permanently therein. The proper volume of such an element of fluid

$$\delta v_0 = \frac{e^{\frac{3}{2}g(t)}}{[1+(r^2/4R_0^2)]^3}r^2\sin\theta\,\delta r\delta\theta\delta\phi, \quad (130.3)$$

will, however, in general be changing with the time owing to its dependence on $g(t)$. When g is increasing with t all the elements of the fluid in the whole model will be expanding at a rate which is proportionally the same in all parts of the model, and with g decreasing there will be a similar contraction throughout the model, and in

† Neumann, *Abh. d. Kgl. Sächs. Ges. d. Wiss. zu Leipzig, math.-nat. Kl.* **26**, 97 (1874); Seeliger, *Astr. Nachr.* **137**, 129 (1895).

general these changes in proper volume would take place at a finite rate.

We must now inquire whether such an expansion or contraction could take place reversibly as well as at a finite rate. Fig. 5 gives a symbolic two-dimensional representation of the space-like coordinates corresponding to the line element (130.2) which will assist in visualizing the differences between the expansion of gas in our present model and in the previous classical cylinder.

Applying the relativistic first law of thermodynamics—i.e. the principles of relativistic mechanics —to the model, it is easily found (see § 151) that the energy relations for each element of the fluid would be described by the familiar equation for an adiabatic expansion

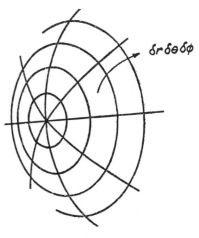

FIG. 5

$$\frac{d}{dt}(\rho_{00}\,\delta v_0) + p_0\frac{d}{dt}(\delta v_0) = 0, \quad (130.4)$$

the change in proper energy for each element of fluid being accounted for by the work which it does on its surroundings. Hence each element of the fluid would expand or contract adiabatically, without flow of heat from one portion of the model to another, as indeed would perhaps be intuitively evident from the homogeneity of conditions throughout the model.

Having ascertained the adiabatic quality of the process, we can then apply the relativistic second law (119.5) to our system after setting the right-hand side of the expression—which is proportional to heat absorbed—equal to zero. Doing so, the second law, for the model that we are considering and in the co-moving coordinates that we are using, then takes the simple form

$$\frac{d}{dt}\left\{\phi_0\frac{e^{\frac{3}{2}g(t)}}{[1+(r^2/4R_0^2)]^3}r^2\sin\theta\right\}\delta r\delta\theta\delta\phi\delta t \geqslant 0,$$

or, in accordance with (130.3),

$$\frac{d}{dt}(\phi_0\,\delta v_0) \geqslant 0. \qquad (130.5)$$

Hence the application of the relativistic second law of thermo-dynamics to such cosmological models shows that the proper entropy of each element of fluid as measured by a local observer can in any case only increase or remain constant. Since the equality sign applies to reversible processes, constant proper entropy for each element of fluid then becomes the necessary requirement for the reversible expansion of these models at a finite rate.

This requirement, however, can apparently be met provided the fluid filling the model is taken as sufficiently simple.

To show this, we first note that the nature of the model is such as to eliminate possibilities for entropy increase which might otherwise result from inefficient interaction between the elements of fluid and their surroundings. Thus there is no increase in entropy due to irreversible heat flow into any element of fluid owing to the entire absence of heat flow throughout the model; there is no entropy in-crease due to the friction of moving pistons or the like since no con-tainers for the elements of fluid are now involved; and there is no entropy increase due to an inability of the fluid to keep up its full pressure at the expanding boundary of an element owing to the uniform pressure throughout the whole model. There hence appears to be no irreversibility directly due to poor coupling of any element of fluid with its surroundings.

The remaining possibilities for entropy increase then lie in irre-versible changes taking place inside the elements of fluid in the actual material of which the fluid is composed. In the case of complicated fluids subjected to a finite rate of volume change such irreversible processes would certainly occur to an important extent. Thus if we took a bimolecular gas which tended to dissociate on expansion into its elements, it is evident that this chemical change could not com-pletely keep up with a finite rate of expansion and that the actual dissociation would take place under non-equilibrium conditions and hence be accompanied by increase in entropy. Indeed, with a finite rate of volume change, even the lag in such processes as the transfer of energy from the rotational to the translational degrees of freedom of a bimolecular gas would involve some irreversibility. In the case of simple enough fluids, however, such possibilities for inter-nal entropy increase would be almost or completely lacking. Thus if we took a perfect monatomic gas, as suggested for the fluid at the beginning of the section, there would be no possibility for internal irreversible processes, provided we neglect the small energy transfers that would take place between the gas and the slight amount of thermal radiation that would also actually be present. And in the case of a fluid composed of dust particles having negligible thermal pressure, or of one composed solely of black-body radiation itself, there would appear to be no possibilities for internal irreversibility at all.

The relativistic discovery of cosmological models, filled with material throughout their entire extent, thus provides possibilities for the expansion of an unenclosed fluid without its dissipation into empty space and without friction, irreversible heat flow, or pressure drop at the walls of any container, of a kind hitherto unknown. Analysed from the point of view of relativistic thermodynamics, this then leads to an increased possibility for processes to take place at a finite rate and yet also either with complete reversibility, or in any case with the elimination of sources of irreversibility that seemed classically inevitable.

Relativistic mechanics and relativistic thermodynamics have both contributed to the new result. Relativistic mechanics makes it possible to study the behaviour of cosmological models as a whole, and then from the relations of the fundamental tensor $g_{\mu\nu}$ to density, pressure, and proper volume to determine the behaviour of the individual elements of fluid in the model. With the help of the second law of relativistic thermodynamics we can then see if each of these elements of fluid behaves reversibly or irreversibly. Increased possibilities for reversible behaviour have thus been found for homogeneous systems, having uniform temperature and pressure throughout, and the investigation of non-homogeneous models from the same point of view would be interesting.

The main importance of the new result lies in its demonstration of the necessity of using relativistic rather than classical thermodynamics in any attempt to understand the behaviour of the universe as a whole. In the next chapter the result will be found applicable to an important class of cosmological models. It will there be shown in §§ 170 and 171, that the thermodynamic condition for reversibility, which we have obtained by taking the equality sign in the relativistic expression for the second law, actually agrees with the requirements for a real reversal in the motions of cosmological models; and it will be shown in § 173, that an observer in a reversibly expanding universe would be led to quite erroneous conclusions if he should try to interpret the behaviour of his surroundings by the use of classical rather than relativistic thermodynamics.

131. On the possibility for irreversible processes without reaching a final state of maximum entropy

It was shown in the last section that the theory of relativity as compared with classical theory provides an increased possibility for reversible thermodynamic processes. It was also evident from the discussion, however, that irreversible processes could in no way be eliminated from the considerations of relativistic thermodynamics, and indeed some degree of irreversibility would still appear to be the usual characteristic of the actual thermodynamic processes that take place in nature.

In the case of irreversible processes important differences between the conclusions of classical and relativistic thermodynamics can nevertheless arise. The classical thermodynamics, as shown in § 63, appeared to lead inevitably to the conclusion that the end result of irreversible processes would necessarily be a final state of maximum entropy where further thermodynamic change would be impossible. In the present section we shall discuss a relativistic possibility for irreversible processes to occur without ever reaching any unsurpassable maximum value of that quantity.

This new possibility for continuous irreversible change is also provided by the cosmological models, considered in the preceding section and discussed more completely in the next chapter. For our present purposes it is sufficient to note that there is an important class of these models, see § 163, such that expansion from any given finite proper volume would necessarily be followed by reversal in the direction of motion at some upper limit and return to smaller volumes. This behaviour can be deduced solely from the principles of relativistic mechanics alone, and does not depend on the nature or complexity of the fluid which we take as filling the model but only on its assumed homogeneity of distribution. Hence on purely mechanical grounds we are led to the consideration of a class of cosmological models which would undergo a continued succession of alternate expansions and contractions, without reference to the thermodynamic character of the processes taking place within the elements of fluid filling the model.

In the case of a fluid simple enough so that these internal processes would occur without increase in proper entropy, we have the conditions for perfect reversibility discussed in the last section, and we shall find in the next chapter that the model would continue to repeat a succession of identical expansions and contractions.

In the case of more complicated fluids, however, it is evident that entropy increases would occur within the elements of fluid composing the model as they expanded and contracted. Thus in the case of a diatomic gas capable of reacting to form its elements, dissociation would tend to take place during expansion and reassociation during compression, and with a finite rate of volume change there would be a lag so that these reactions would actually take place under non-equilibrium conditions and hence with increase in entropy. We must now inquire whether such irreversible behaviour would necessarily lead to a cessation in the succession of expansions and contractions.

From the classical point of view a gradual decay in the motions of expansion and contraction would have seemed inevitable, since the continued occurrence of irreversible processes in an isolated system would have ultimately led to a condition of maximum entropy where further change would be impossible. In the classical thermodynamics the entropy of a homogeneous fluid could be determined with the help of the familiar equation (60.4), previously developed in § 60,

$$dS = \frac{1}{T} dE + \frac{p}{T} dv + \frac{\partial S}{\partial n_1} dn_1 + \ldots + \frac{\partial S}{\partial n_n} dn_n, \qquad (131.1)$$

where the energy E, volume v, and number of mols $n_1, \ldots, n_n$ of the different chemical components present are the independent variables chosen to determine the state of the system. In applying this equation to an isolated system undergoing a succession of expansions and contractions, the energy change dE would have to be taken as zero owing to the classical principle of the conservation of energy, and the work pdv would have to be taken as zero owing to the isolation of the system. Hence in the long run the only possibility for increase in entropy in such a system would lie in the readjustment of composition, and this could not continue indefinitely since with a given value of energy and volume there is a maximum possible value for the entropy of a homogeneous fluid corresponding to the attainment of chemical equilibrium between its components. Thus the classical thermodynamics would have concluded that the irreversible increase in entropy could not permanently continue and that further change would cease at the condition of maximum entropy.

From the point of view of relativity theory, however, the foregoing reasoning has to be modified in an important manner owing to the changed status of the energy principle in relativistic mechanics. In accordance with relativistic thermodynamics, we can still apply an equation of the same form as (131.1)

$$d(\phi_0 \, \delta v_0) = \frac{1}{T_0} d(\rho_{00} \, \delta v_0) + \frac{p_0}{T_0} d(\delta v_0) + \frac{\partial(\phi_0 \, \delta v_0)}{\partial n_1^0} dn_1^0 + \dots + \frac{\partial(\phi_0 \, \delta v_0)}{\partial n_n^0} dn_n^0,$$
(131.2)

to determine the proper entropy $(\phi_0 \, \delta v_0)$ of each little element of fluid in the model in terms of its proper energy $(\rho_{00} \, \delta v_0)$, volume δv_0, and the number of mols $n_1^0, \dots, n_n^0$ of the different chemical components which it contains. In accordance with the principles of relativistic mechanics, nevertheless, we can no longer conclude that the total *proper* energy associated with the fluid would be a constant, owing to the well-known failure of the principle of energy conservation to hold in the theory of relativity, unless allowance is made for potential gravitational energy associated with the field as well as for the proper energy directly, associated with matter and radiation. Indeed, in accordance with the equation (130.4), already cited as applying to these cosmological models,

$$\frac{d}{dt}(\rho_{00} \, \delta v_0) + p_0 \frac{d}{dt}(\delta v_0) = 0,$$
(131.3)

it is evident that the proper energy of *every* element of fluid in the model will be decreasing when the model is expanding and increasing when it is contracting. Furthermore, there will be a general tendency for pressure to be too low to correspond to equilibrium on expansion and too high to correspond therewith on contraction. Thus instead of constant proper energy for each element of fluid in the model, we can expect in the long run a tendency for this energy to increase, thus removing the restrictions previously imposed by the classical principle of energy conservation on possible increases in entropy.

Hence in relativistic thermodynamics we can no longer assume that there would be an unsurpassable maximum value for the entropy of our system, which would limit the continuance of irreversible processes in the fluid and thus necessitate a final stagnant state. Indeed, as a result of the more detailed analysis of the next chapter we shall find in certain cases, instead of a decay in the amplitude of the successive irreversible expansions and contractions of these cosmological models, an actual tendency for gradual increase in the upper limit to which the model expands, always followed, however, by renewed contraction.

In order to appreciate the reasons for this new conclusion, it is well to emphasize as the most important step in the argument the removal of the classical requirement for a constant value of the energy *directly* associated with the fluid composing the system. Even in the classical thermodynamics, the removal of this restriction—by taking a system having interaction with its surroundings instead of an isolated one—is sufficient also to remove the restriction on possible entropy increase. Thus consider, for example, a sample of diatomic gas, capable of dissociating into its elements, and enclosed in an ordinary cylinder provided with non-conducting walls and a movable piston. On moving this piston in and out so as to secure an alternate expansion and compression, the gas will tend to dissociate on expansion and to recombine on compression. If this is carried out at a finite rate, however, equilibrium will not be maintained and the average pressure on expansion will be less than that necessary to secure recompression, so that a net amount of work will be necessary and the energy of the system will increase as the process is continued. As a further result of the failure to maintain equilibrium, moreover, the two reactions of dissociation and recombination will not take place reversibly, so that the entropy of the system will also increase as the process is continued. Thus as long as sufficient external energy is available to continue the succession of expansions and compressions, both the energy and entropy of the system will increase and there will be no unsurpassable maximum of the latter quantity. Hence from the classical point of view it would be the ultimate failure in the external energy supply, rather than the internal increase in entropy of the fluid in the cylinder, that would bring the proposed process to an end. From the relativistic point of view, on the other hand, since cosmological models can be constructed which have no limitation on total *proper* energy and hence also no limitation on total entropy, we must retain the possibility for irreversible processes which may continue without end.

The main importance of the new result again lies in its demonstration of the necessity of using relativistic rather than classical thermodynamics in any attempt to understand the behaviour of the universe as a whole.

132. Conclusion

In concluding the chapter, some apology should perhaps be offered for the apparently premature inclusion of the last two sections, on reversible processes at a finite rate, and irreversible processes having no final state of maximum entropy, since it is evident that these new possibilities can finally be satisfactorily understood only with the help of the cosmological models to be studied in the next chapter, and we shall take the matter up again in Part III of that chapter. The inclusion was made, nevertheless, in order to exhibit in a single connected form the differences between classical and relativistic thermodynamics. As indicated at the beginning of the chapter, it will be seen that conclusions of a qualitatively new kind are implied by the extension of thermodynamics to relativity, these differences being due, however, more to changes in concepts as to the nature of space and time than to fundamental modifications in the postulates of thermodynamics.

A number of developments of relativistic thermodynamics remain to be carried out.

Further study of the conditions governing thermal flow, together with an explicit expression for the energy-momentum tensor of a thermally conducting fluid would be desirable. The thermodynamic behaviour of *non*-homogeneous cosmological models, having thermal flow from one portion to another, should be investigated. The results might be of importance in interpreting the behaviour of the actual universe.

A study of thermodynamic fluctuations should also be made, especially as fluctuations may be very important at certain stages of cosmological development. Finally, the general interrelation between thermodynamics and statistical mechanics might well be treated from the point of view of general relativity. This would of course involve considerations that go beyond the macroscopic point of view adopted for the purposes of the present book.

of about 3×10^8 light years which can be reached with the 100-inch telescope, the investigations of Hubble have shown the presence of about 10^8 nebulae, which are individually roughly of the same character as our own galaxy of stars. These nebulae are to a certain extent gathered into clusters, but on the whole are distributed with a fair uniformity of about one nebula per 10^{18} cubic light years. From the red-shift in the spectra of these nebulae we can infer that they have a motion of mutual recession, and from their apparent diameters, luminosity, and colour we can get some limit as to the presence of intervening obscuring material. We thus have considerable knowledge as to the contents of the universe out to 3×10^8 *light years*, and indeed also some indication as to its probable behaviour within a past time span of 3×10^8 *years*.

There are, nevertheless, serious gaps in the information which we could desire. In the first place, although we can presumably make some extrapolation of the conditions observed in our immediate neighbourhood to greater distances, we have no real justification for assuming that the whole universe has the same properties as that portion which we have already seen. Hence, although we shall actually make great use of homogeneous models in our studies, we shall have to realize that we do this primarily in order to secure a definite and relatively simple mathematical problem, rather than to secure a correspondence with known reality. In the second place, although we have good information concerning the density of nebulae in our surroundings, we have very little information as to the density of other forms of matter or of radiation in the enormous extragalactic spaces lying between the observed nebulae. Indeed it seems possible from the work of Hubble that the density of matter in the form of extragalactic dust might be thousands of times as great as the averaged-out density of the nebulae, without giving rise to effects that would have so far been found. This is a very serious limitation on our knowledge, since it prevents any precise determination of gravitational curvature. As a result we do not know whether the actual universe is spatially closed or open, and can choose between universes which are finite and infinite in spatial extent only on the basis of dubious metaphysical predilections.

In view of these uncertainties in observational knowledge, much of our actual work must necessarily consist in a study of cosmological *models*, constructed in accordance with the theory of relativity, but

X

APPLICATIONS TO COSMOLOGY

Part I. STATIC COSMOLOGICAL MODELS

133. Introduction

In this final chapter, we must now investigate the applications of relativistic mechanics and relativistic thermodynamics to cosmology. This is an ambitious field of study characterized by danger as well as by interest.

The most fundamental although not the most pressing danger that threatens the validity of the study lies in the possibility that the relativistic theory of gravitation might not really be applicable to the universe as a whole, or even to that portion out to some hundred million light years, which can be observed with the help of the Mount Wilson 100-inch telescope. The three so-called crucial tests make us indeed confident that relativity provides a real advance over the Newtonian theory of gravitation, and that it furnishes an acceptable treatment for the field in the empty space surrounding a star out to distances of the order of the dimensions of the solar system. Nevertheless, the application of this same theory to the universe as a whole, filled with a distribution of matter and radiation rather than empty, involves an extension which cannot of course be made with certainty. To justify the extrapolation we can only depend in the first place on the remarkable rationality and inner logicality of the theory of relativity, which makes a wide range of applicability seem probable, and in the second place on the observed tendency for stars to cluster together in nebulae and for the nebulae themselves to occur with some clustering, which at least indicates for very great ranges of distance gravitational action of the general kind that would be predicted from relativistic theory. Furthermore, relativity certainly provides at the present time the only possible theory of gravitation that could be applied to the behaviour of large portions of the universe, and hence we are forced to make use of this theory if we are to carry out cosmological speculations at all.

Another source of difficulty for any kind of cosmological theory lies in the very real limitations in our observational knowledge as to the actual nature of the universe and its contents. Within the range not necessarily agreeing in all particulars with the real universe. Indeed we shall feel justified in studying some models, which are known to differ from the real universe in important ways, provided the results can illuminate our thinking by indicating the kind of phenomena that might actually occur without controverting established theory. With the help of such studies, however, we shall certainly make progress in understanding the behaviour of nature on the largest possible scale, and this presents a task as interesting as the human mind can set, and provides a goal as noble as the human spirit can conceive.

In Part I of the present chapter we shall consider static cosmological models. We shall first show that the only possibilities for a homogeneous static model are those provided by the original Einstein universe filled with a uniform distribution of material, by the de Sitter empty universe, and by the empty flat space-time of the special theory of relativity. We shall then give a brief discussion of these different possibilities, sufficient to show the reasons for abandoning them as providing satisfactory models for the actual universe. In Part II we shall then make use of the principles of relativistic mechanics to derive the line element for non-static homogeneous cosmological models, and to study their mechanical properties and behaviour. In Part III we shall apply the principles of relativistic thermodynamics to this behaviour. Finally, in Part IV, we shall compare the properties of such models with the phenomena of the actual universe.

134. The three possibilities for a homogeneous static universe

We now undertake the specific task of showing that the only possibilities for a static homogeneous universe are exhausted by the line elements of Einstein, and of de Sitter, and that corresponding to the special theory of relativity.

In obtaining any form of cosmological line element, we shall look at the universe from a large-scale point of view and neglect those local irregularities in gravitational field and in space-time curvature, which would occur in the immediate neighbourhood of individual stars or stellar systems. We can then treat the universe as filled with a continuous distribution of fluid of proper macroscopic density ρ_{00} and pressure p_0, and shall feel justified in making this simplification since our interest lies in obtaining a general framework for the behaviour of the universe as a whole, on which the details of local occurrences could later be superposed.

In the case of a static homogeneous universe it is evident that coordinates can certainly be chosen such that the line element will exhibit spherical symmetry around any desired origin, since all parts of the universe are intrinsically permanently alike. Hence we may evidently take the line element in the general spherically symmetrical static form

$$ds^2 = -e^\lambda \, dr^2 - r^2 \, d\theta^2 - r^2 \sin^2\theta \, d\phi^2 + e^\nu \, dt^2, \qquad (134.1)$$

with λ and ν functions of r alone as given by (95.12), and take the pressure and density in accordance with (95.13) as determined by the equations

$$8\pi p_0 = e^{-\lambda}\left(\frac{\nu'}{r} + \frac{1}{r^2}\right) - \frac{1}{r^2} + \Lambda, \qquad (134.2)$$

$$8\pi\rho_{00} = e^{-\lambda}\left(\frac{\lambda'}{r} - \frac{1}{r^2}\right) + \frac{1}{r^2} - \Lambda, \qquad (134.3)$$

$$\frac{dp_0}{dr} = -\frac{\rho_{00} + p_0}{2}\nu', \qquad (134.4)$$

where the accents signify differentiation with respect to r, and Λ is the cosmological constant.

From this simple starting-point we can now easily obtain the only possibilities for a static homogeneous model. To do so we have merely to investigate the consequence of imposing three necessary conditions on the foregoing equations. These are: *first* that the pressure p_0 as measured by a local observer shall everywhere be the same, owing to the assumed homogeneity of model; *secondly* that the proper macroscopic density ρ_{00} shall everywhere be the same, again owing to the homogeneity of the model; and *thirdly* that the line element shall reduce for small values of r to the special relativity form for flat spacetime with $\lambda = \nu = 0$, owing to the known validity of the special theory of relativity for a limited space-time region, when we neglect local gravitational fields as postulated above.

In accordance with the first of these conditions, it is evident—since p_0 is to have the same value in all parts of the model—that equation (134.4) can only be satisfied by taking

$$\frac{\rho_{00} + p_0}{2}\nu' = 0, \qquad (134.5)$$

and this can itself in turn only be satisfied by the three possibilities of setting ν', or $(\rho_{00} + p_0)$, or both equal to zero.

These three possibilities
$$\nu' = 0, \tag{134.6}$$
or
$$\rho_{00} + p_0 = 0, \tag{134.7}$$
or both
$$\nu' = 0 \quad \text{and} \quad \rho_{00} + p_0 = 0, \tag{134.8}$$

lead respectively to the Einstein, to the de Sitter, and to the special relativity line elements for the universe as we may now show in detail.

135. The Einstein line element

We may first consider the Einstein line element which arises from the first of the above three possibilities

$$\nu' = 0. \tag{135.1}$$

Integrating this equation, and remembering that the line element is to reduce to the special relativity form, with $\nu = 0$ for small values of r, we at once obtain
$$\nu = \text{const.} = 0, \tag{135.2}$$

as the only possible solution.

On the other hand, substituting this result in the expression for the pressure given by (134.2) and solving, we obtain

$$e^{-\lambda} = 1 - (\Lambda - 8\pi p_0) r^2. \tag{135.3}$$

Hence, defining for convenience a new constant R by the equation

$$\Lambda - 8\pi p_0 = \frac{1}{R^2}, \tag{135.4}$$

we can then write as an expression for the resulting line element

$$ds^2 = -\frac{dr^2}{1 - r^2/R^2} - r^2\, d\theta^2 - r^2 \sin^2\theta\, d\phi^2 + dt^2. \tag{135.5}$$

This is one of the well-known forms for the original Einstein line element for a static universe,† and we shall return later to a discussion of some of its properties.

136. The de Sitter line element

We may next consider the de Sitter line element which arises from the second of the possibilities given above

$$\rho_{00} + p_0 = 0. \tag{136.1}$$

---

† Einstein, *Berl. Ber.* 1917, p. 142.

Adding the individual expressions for ρ_{00} and p_0 given by (134.2) and (134.3) we must then set

$$8\pi(\rho_{00}+p_0) = e^{-\lambda}\left(\frac{\lambda'}{r}+\frac{\nu'}{r}\right) = 0,$$

or
$$\lambda' = -\nu',$$

and since λ and ν must both become zero at $r = 0$, in order for the line element to reduce to the special relativity form at the origin, this can only be satisfied by
$$\lambda = -\nu. \tag{136.2}$$

On the other hand, since ρ_{00} is to be a constant independent of position we can immediately integrate (134.3), and obtain as a solution, which may be readily verified by redifferentiation,

$$e^{-\lambda} = 1-\frac{\Lambda+8\pi\rho_{00}}{3}r^2+\frac{A}{r},$$

where A is the constant of integration. And, again making use of the reduction of the line element to the special relativity form with $\lambda = \nu = 0$ at $r = 0$, we see that we must take this constant A equal to zero. Hence, noting (136.2), we at once have

$$e^{-\lambda} = e^{\nu} = 1-\frac{\Lambda+8\pi\rho_{00}}{3}r^2 \tag{136.3}$$

as a complete solution for the form of the line element. Hence, now defining for convenience a new constant R by the equation

$$\frac{\Lambda+8\pi\rho_{00}}{3} = \frac{1}{R^2}, \tag{136.4}$$

we can write as the complete expression for the line element

$$ds^2 = -\frac{dr^2}{1-r^2/R^2}-r^2\,d\theta^2-r^2\sin^2\theta\,d\phi^2+\left(1-\frac{r^2}{R^2}\right)dt^2. \tag{136.5}$$

This, however, is one of the well-known forms for the original de Sitter line element,† and we shall later return to a discussion of some of its properties.

137. The special relativity line element

Finally, we may turn to the third possibility for a static homogeneous universe in which we require in accordance with (134.8) both

$$\nu' = 0 \quad \text{and} \quad \rho_{00}+p_0 = 0. \tag{137.1}$$

† de Sitter, *Proc. Akad. Wetensch. Amsterdam,* **19**, 1217 (1917).

Under these circumstances, however, we can take both the equations (135.2) from the Einstein case and (136.2) from the de Sitter case as valid, and hence can write as a complete solution the simple result,

$$\lambda = \nu = 0, \qquad (137.2)$$

corresponding to the special relativity form of line element

$$ds^2 = -dr^2 - r^2\,d\theta^2 - r^2\sin^2\theta\,d\phi^2 + dt^2, \qquad (137.3)$$

which applies to the perfectly empty and 'flat' space-time of the special theory of relativity.

In accordance with the discussion of § 134, this now exhausts all the possibilities for a static homogeneous universe;† and hence when we find that none of these three possibilities gives a satisfactory representation of the actual universe, we shall then have to turn to the consideration of some less restricted class of models.

We may now undertake a brief survey of some of the more important properties of the Einstein and de Sitter line elements, both of which include the special relativity line element as a particular case when the constant R assumes the value infinity. The survey will be of interest not only for historical reasons, but for the insight which it can give into the more adequate models which we shall later study.

138. The geometry of the Einstein universe

By the transformation of coordinates the Einstein line element

$$ds^2 = -\frac{dr^2}{1 - r^2/R^2} - r^2\,d\theta^2 - r^2\sin^2\theta\,d\phi^2 + dt^2 \qquad (138.1)$$

can be written in several different forms which are sometimes convenient or can be of assistance in understanding the implied geometry.

By the substitution

$$r = \frac{\rho}{1 + \rho^2/4R^2} \qquad (138.2)$$

we obtain an isotropic form

$$ds^2 = -\frac{1}{[1 + \rho^2/4R^2]^2}(d\rho^2 + \rho^2\,d\theta^2 + \rho^2\sin^2\theta\,d\phi^2) + dt^2, \qquad (138.3)$$

---

† This proof that the Einstein, de Sitter, and special relativity line elements exhaust the possibilities for a *static solution* follows the treatment of Tolman, *Proc. Nat. Acad.* **15**, 297 (1929). For an earlier proof, see Friedmann, *Zeits. f. Physik*, **10**, 377 (1922); and for a proof that there are no additional *stationary solutions*, in the sense of § 142, see Robertson, *Proc. Nat. Acad.* **15**, 822 (1929).

which by an obvious further transformation can also be written as

$$ds^2 = -\frac{1}{[1+\rho^2/4R^2]^2}(dx^2+dy^2+dz^2)+dt^2. \qquad (138.4)$$

By the substitution $r = R \sin \chi$ (138.5)

we obtain

$$ds^2 = -R^2(d\chi^2+\sin^2\chi\, d\theta^2 +\sin^2\chi \sin^2\theta\, d\phi^2)+dt^2. \qquad (138.6)$$

Finally, by introducing a larger number of variables, with the help of the equations

$$z_1 = R\sqrt{\left(1-\frac{r^2}{R^2}\right)} \qquad z_2 = r \sin \theta \cos \phi$$

$$z_3 = r \sin \theta \sin \phi \qquad z_4 = r \cos \theta, \qquad (138.7)$$

where $z_1^2+z_2^2+z_3^2+z_4^2 = R^2,$ (138.8)

the line element assumes the form

$$ds^2 = -dz_1^2-dz_2^2-dz_3^2-dz_4^2+dt^2, \qquad (138.9)$$

which permits us to regard our original space-time as embedded in a Euclidean space of higher dimensions.

The kind of geometry corresponding to these different expressions for the line element is not completely determined, since different hypotheses as to connectivity and as to the identification of points can in general be made for a given differential formula for interval. It will be simplest, however, as suggested by the last form in which we have written the line element, to regard the spatial extent of the Einstein universe as being the whole three-dimensional spherical surface $z_1^2+z_2^2+z_3^2+z_4^2 = R^2$ embedded in the four-dimensional Euclidean space (z_1, z_2, z_3, z_4). The geometry corresponding to the space-like variables in the Einstein line element would then be that for so-called *spherical* space of radius R. By the identification of antipodal points of the sphere and the introduction of suitable connectivity the spatial geometry could also be taken as of the so-called *elliptical* kind.

Taking the spatial geometry as spherical, the total proper spatial volume of the Einstein universe would be given in accordance with (138.6) by

$$v_0 = \int_0^{2\pi} \int_0^{\pi} \int_0^{\pi} R^3\sin^2\chi \sin \theta\, d\chi d\theta d\phi = 2\pi^2R^3, \qquad (138.10)$$

and the total proper distance around the universe would be

$$l_0 = 2\pi R. \qquad (138.11)$$

Taking the geometry as elliptical, the corresponding quantities would be one-half as great, and this difference could provide in principle a method for distinguishing between the two possibilities of spherical and elliptical space.

Introducing the time-like as well as the space-like variables, the complete space-time geometry corresponding to the Einstein universe could be regarded as that for a four-dimensional cylindrical surface embedded in five-dimensional space.

Perhaps the chief importance of this investigation into the nature of the geometry implied by the Einstein line element lies in the assistance thereby provided to our intuitional appreciation of the homogeneity of the model. In accordance with the symmetrical form given to the line element by (138.9) it is immediately evident that on transforming back to r, θ, ϕ, and t the origin of spatial coordinates and the zero point for the time coordinate could be taken at will, in agreement of course with our original assumptions as to the static and spatially homogeneous character of the model. It may, nevertheless, be emphasized in conclusion that for most problems of immediate interest there is no necessity to go beyond the results which can be obtained by usual analytical methods directly from the differential formula for interval, and no necessity to attempt to visualize the geometry as a whole.

139. Density and pressure of material in Einstein universe

We may now turn to more physical aspects of the Einstein universe, by investigating the relations which would govern the density and pressure of the material in the model.

Returning to the general form (134.1) for the line element

$$ds^2 = -e^\lambda \, dr^2 - r^2 \, d\theta^2 - r^2\sin^2\theta \, d\phi^2 + e^\nu \, dt^2 \qquad (139.1)$$

and introducing the values

$$e^{-\lambda} = 1 - \frac{r^2}{R^2} \quad \text{and} \quad \nu = 0, \qquad (139.2)$$

found in § 135, into the expressions for pressure and density (95.13) which correspond to this general form of line element, we easily obtain

$$8\pi p_0 = -\frac{1}{R^2} + \Lambda, \qquad (139.3)$$

and

$$8\pi \rho_{00} = \frac{3}{R^2} - \Lambda \qquad (139.4)$$

as expressions for the proper pressure and proper density of the material filling the model in terms of the two constants Λ and R.

Alternatively, these equations may be solved for the two constants in the form

$$\Lambda = 4\pi(\rho_{00}+3p_0) \tag{139.5}$$

and

$$\frac{1}{R^2} = 4\pi(\rho_{00}+p_0). \tag{139.6}$$

Hence, since the density ρ_{00} of the fluid taken as filling the model could on physical grounds only be a positive quantity, and the pressure p_0 could only be positive or—assuming the possibility of reasonable cohesive forces—could only be negative to a very limited extent, we may conclude at once that Λ and R^2 would both be essentially positive quantities.

If we regard Λ and R^2 as in the nature of adjustable parameters the model could be taken as filled with a fluid having any desired values for its pressure and density.

Thus if we assumed the fluid to be composed of incoherent matter exerting no pressure, for example free particles (stars) having negligible relative motions, as originally considered by Einstein, we should have from the above

$$\Lambda = \frac{1}{R^2} \tag{139.7}$$

and

$$4\pi\rho_{00} = \frac{1}{R^2}, \tag{139.8}$$

and in accordance with (138.10) the total mass of the universe would be

$$M = \rho_{00}v_0 = \tfrac{1}{2}\pi R. \tag{139.9}$$

On the other hand, if we took the model as filled solely with radiation, which has the highest known ratio of pressure to density for any possible fluid,

$$\rho_{00} = 3p_0,$$

we should have

$$\Lambda = \frac{3}{2R^2} \tag{139.10}$$

and

$$4\pi p_0 = \frac{1}{4R^2} \qquad 4\pi\rho_{00} = \frac{3}{4R^2}. \tag{139.11}$$

Comparing (139.8) with (139.11), we again see an example of the tendency first mentioned in § 110 for radiation to produce greater gravitational curvature than the same density of ordinary matter.

Finally, if we took the model as completely empty with ρ_{00} and p_0, both equal to zero, we should have

$$\Lambda = \frac{1}{R^2} = 0 \qquad (139.12)$$

and the Einstein universe would degenerate into the flat space-time of the special theory of relativity.

Several important conclusions may be drawn from the results of this section. In the first place, it is to be noted that the discussion does demonstrate in accordance with the principles of relativity at least the conceptual possibility for cosmological models which would agree to some extent with the actual universe by containing a uniform distribution of material of finite concentration. In the second place, since we have seen above that R^2 would be positive and finite except in the degenerate case of a completely empty universe, it is to be noted that the radius R of the Einstein model would have to be real corresponding to an unbounded but nevertheless closed universe with a finite spatial volume. Finally, it may be emphasized, as seen above, that the cosmological constant Λ would have to be a positive quantity greater than zero if the model is to contain any matter at all. This necessity was perhaps the strongest argument which led to Einstein's addition of the logically possible but otherwise surprising cosmological term to his original field equations. If we later find models which could contain a finite concentration of matter without the Λ-term, we can look with favour on the possibility of taking Λ equal to zero.

140. Behaviour of particles and light rays in the Einstein universe

We may now turn to a further discussion of the physical properties of the original Einstein universe by investigating the behaviour of particles and light rays in such a model.

In accordance with (74.13), the motion of a free particle in the gravitational field corresponding to the Einstein line element

$$ds^2 = -\frac{dr^2}{1-r^2/R^2} - r^2\,d\theta^2 - r^2\sin^2\theta\,d\phi^2 + dt^2 \qquad (140.1)$$

would be given by the equations for a geodesic

$$\frac{d^2x^\sigma}{ds^2} + \{\mu\nu, \sigma\}\frac{dx^\mu}{ds}\frac{dx^\nu}{ds} = 0. \qquad (140.2)$$

We shall be specially interested in the case of particles which are at least temporarily at rest with respect to the spatial coordinates. This geodesic equation would then reduce to

$$\frac{d^2x^\sigma}{ds^2} + \{44, \sigma\}\left(\frac{dt}{ds}\right)^2 = 0, \tag{140.3}$$

since the spatial components of the 'velocity' of the particle dr/ds, $d\theta/ds$, and $d\phi/ds$ would be zero. Comparing the expression for the Einstein line element (140.1), however, with the expressions for the Christoffel three-index symbols given by (95.2) for this general form of line element, we see that all symbols of the form $\{44, \sigma\}$ would vanish, and we are thus led at once to the conclusion that particles at rest with respect to the spatial coordinates would also have zero acceleration

$$\frac{d^2r}{ds^2} = \frac{d^2\theta}{ds^2} = \frac{d^2\phi}{ds^2} = 0 \tag{140.4}$$

and hence would remain permanently at rest.

This conclusion is of importance, since the Einstein model could not be expected to persist at all in the assumed static state, if the free particles contained in it could not remain at rest. The result is, nevertheless, not a sufficient criterion for complete stability as we shall see later.

The velocity of light in the Einstein universe can be obtained by setting the expression for the interval (140.1) equal to zero. Doing so and focusing attention on the case of light travelling in the radial direction we obtain for the velocity of light from or towards the origin

$$\frac{dr}{dt} = \pm\sqrt{\left(1 - \frac{r^2}{R^2}\right)}, \tag{140.5}$$

where it is to be specially noted, as a result of the form of the line element (104.1), that the time-like variable t agrees with proper time as it would be measured by a local observer at rest in the model with respect to the spatial coordinates.

In accordance with this result, the time necessary for light to travel from the origin around the universe and back would be

$$t = 4\int_0^R \frac{dr}{\sqrt{(1 - r^2/R^2)}} = 2\pi R, \tag{140.6}$$

if we assumed spherical space, or one-half this amount if we assumed elliptical space. The amusing theoretical possibility thus provided, for the light issuing from a star to travel around the universe and by refocusing lead to the appearance of a 'ghost' star in the neighbourhood should not be taken very seriously in view of the idealization and inadequacy of the original Einstein model as a representation of the actual universe.

The most important application of the expression for the velocity of light given by (140.5) lies in its use in showing that we could expect no systematic shift in the wave-length of light from distant objects in the static Einstein universe. Consider an observer for convenience at the origin of coordinates $r = 0$ and a luminous source (nebula) at $r = r$, both being taken as permanently at rest with respect to the spatial coordinates in agreement with the zero acceleration for stationary particles demonstrated above, and in agreement with the static character ascribed to the model. In accordance with (140.5) the 'time' t_2 for the reception by the observer of light leaving the source at 'time' t_1 would be

$$t_2 = t_1 + \int_0^r \frac{dr}{\sqrt{(1-r^2/R^2)}} = t_1 + R\sin^{-1}\frac{r}{R}.$$

Hence, since r is a constant, the interval δt_2 between the receipt of two successive wave crests would be equal to the interval δt_1 between their emission

$$\delta t_2 = \delta t_1 \qquad (140.7)$$

On the other hand, however, in accordance with the form of the line element (140.1) the time-like variable t agrees with the proper time as measured by local observers at rest with respect to the spatial coordinates. Hence the equality (140.7) also implies an equality between the proper periods of the emitted and received light as they would be determined by observers at rest with respect to the original source and at rest at the origin. As a result, the light on reception would be observed to have the same period and wave-length as is found for the particular luminous material involved when it is used in the laboratory to provide a stationary source of light for a spectroscope.

The method of obtaining this result gives a particularly simple illustration of the general method for treating the generalized Doppler effect, schematically outlined at the end of Chapter VIII. In accordance with the result we can conclude in the case of the original Einstein model of the universe that there would be no systematic

connexion between observed wave-length and distance from observer
to luminous sources such as the nebulae. There could of course be
small Doppler effects due to the individual motions of the nebulae,
but as a result of the general static character of the model we should
expect these effects to be positive and negative with equal frequency
and with no great spread from a mean of zero.

141. Comparison of Einstein model with actual universe

To conclude our brief consideration of the properties of the Einstein
universe we must now make some comparison with the properties of
the actual universe.

The most satisfactory feature of the Einstein model is its correspondence as shown in § 139 with a universe which could actually
contain a finite concentration of uniformly distributed matter. In
this respect it gives us a cosmology which is superior to that provided
by the de Sitter model which as we shall see in § 143 would have to
be regarded as empty. It may again be emphasized, nevertheless,
that this advantage is gained only at the expense of introducing the
extra cosmological term $\Lambda g_{\mu\nu}$ into Einstein's original field equations,
which is a device similar to the modification in Poisson's equation
proposed in the past† in order to permit a uniform static distribution
of matter in the flat space of the Newtonian theory.

In accordance with the estimate of Hubble (see § 177) the density
of matter in the actual universe in the form of visible nebulae would
have a value of about

$$\rho = (1\cdot3 \text{ to } 1\cdot6)\times 10^{-30} \text{ gm./cm.}^3 \qquad (141.1)$$

if averaged out over the whole of intergalactic space, as of course must
be done in replacing the actual universe by a model filled with a continuous distribution of fluid. On the other hand, in accordance with
§ 139, we have found that the density ρ_{00} in an Einstein universe
filled with incoherent matter exerting negligible pressure would be
related to radius R and cosmological constant Λ by the equation

$$4\pi\rho_{00} = \frac{1}{R^2} = \Lambda. \qquad (141.2)$$

Hence, neglecting the density of unseen matter and neglecting the
pressure and density of the radiation in intergalactic space, and

† See Neumann, *Allgemeine Untersuchungen über das Newtonsche Prinzip der
Fernwirkungen*, Leipzig, 1896.

introducing the factor for the conversion of grammes to gravitational units given by (81.7) we obtain

$$\Lambda = \frac{1}{R^2} = \frac{4\pi \times 10^{-30}}{1 \cdot 349 \times 10^{28}} \text{cm.}^{-2},$$

or $\Lambda \simeq 9 \cdot 3 \times 10^{-58} \text{cm.}^{-2},$ (141.3)

$R \simeq 3 \cdot 3 \times 10^{28} \text{cm} \simeq 3 \cdot 5 \times 10^{10}$ light years.   (141.4)

In obtaining these values of Λ and R we have taken the density of matter as 10^{-30} gm./cm.³, and this is presumably a lower limit for that quantity since it neglects dust and gas in the enormous reaches of internebular space. Hence the value for Λ must be regarded as a lower limit and that for R as an upper limit.

The value for Λ is small enough to be compatible with known planetary motions in the solar system, since if we write the Schwarzschild line element in the complete form

$$ds^2 = -\frac{dr^2}{1 - \frac{2m}{r} - \frac{\Lambda}{3}r^2} - r^2\,d\theta^2 - r^2\sin^2\theta\,d\phi^2 + \left(1 - \frac{2m}{r} - \frac{\Lambda}{3}r^2\right) dt^2,$$

the ratio of the previously neglected term $\Lambda r^2/3$ to the main term $2m/r$ at the distance of Neptune's orbit would only be

$$\frac{\Lambda r^2/3}{2m/r} = \frac{\Lambda r^3}{6m} \simeq \frac{9 \cdot 3 \times 10^{-58} \times (4 \cdot 5 \times 10^{14})^3}{6 \times 1 \cdot 5 \times 10^5} \simeq 10^{-19}, \quad (141.5)$$

where we have taken the gravitational mass of the sun as $1 \cdot 5 \times 10^5$.

With regard to the value obtained for the radius R ($3 \cdot 5 \times 10^{10}$ light years), there is also no trouble since as yet our telescopes have only penetrated to about 3×10^8 light years.

The most unsatisfactory feature of the Einstein model as a basis for the cosmology of the actual universe is the finding discussed at the end of the last section, that it provides no reason to expect any systematic shift in the wave-length of light from distant objects. In the actual universe, however, the work of Hubble and Humason shows a definite red-shift in the light from the nebulae which increases at least very closely in linear proportion to the distance. This is of course the main consideration which will lead us to prefer non-static to static models of the universe as a basis for actual cosmology.

Closely connected with this unsatisfactory feature of the static Einstein model will be our later finding that the Einstein universe would not be stable. To be sure as we have seen in connexion with

(140.4), free particles at rest in the model would not be subject to acceleration. Nevertheless, we shall later find, for example, that a static Einstein universe would start contracting as a whole if the matter in it should commence to be transformed into radiation, or vice versa start expanding if the radiation in it should commence a condensation into matter. And we shall find in general the possibility for a wide variety of models that could expand or contract, as compared with very severe restrictions necessary for the permanence of a static model.

142. The geometry of the de Sitter universe

Having found the original Einstein universe, although very important for an understanding of relativistic cosmology, not entirely satisfactory as a model for the actual universe, we may now turn to a consideration of the other static possibility provided by the de Sitter universe.

By the transformation of coordinates we can change the original form in which we obtained the de Sitter line element

$$ds^2 = -\frac{dr^2}{1-r^2/R^2} - r^2\,d\theta^2 - r^2\sin^2\theta\,d\phi^2 + \left(1-\frac{r^2}{R^2}\right)dt^2 \quad (142.1)$$

into other forms which are sometimes convenient or geometrically illuminating.

By the substitution
$$r = R\sin\chi \tag{142.2}$$
we obtain

$$ds^2 = -R^2\,d\chi^2 - R^2\sin^2\chi\,d\theta^2 - R^2\sin^2\chi\sin^2\theta\,d\phi^2 + \cos^2\chi\,dt^2, \quad (142.3)$$

which is a form that has often been employed in discussing the de Sitter universe.

A more interesting result may be obtained, however, by introducing five variables and transforming in accordance with the equations

$$\alpha = r\sin\theta\cos\phi \qquad \beta = r\sin\theta\sin\phi \qquad \gamma = r\cos\theta$$
$$\delta+\epsilon = Re^{t/R}\sqrt{(1-r^2/R^2)} \qquad \delta-\epsilon = Re^{-t/R}\sqrt{(1-r^2/R^2)}. \quad (142.4)$$

This leads to the form
$$ds^2 = -d\alpha^2 - d\beta^2 - d\gamma^2 - d\delta^2 + d\epsilon^2.$$

And by the further transformation
$$z_1 = i\alpha \qquad z_2 = i\beta \qquad z_3 = i\gamma \qquad z_4 = i\delta \qquad z_5 = \epsilon \quad (142.5)$$
we obtain the result

$$ds^2 = dz_1^2 + dz_2^2 + dz_3^2 + dz_4^2 + dz_5^2, \tag{142.6}$$

where as a consequence of (142.4) we have the relation

$$z_1^2+z_2^2+z_3^2+z_4^2+z_5^2 = (iR)^2 \qquad (142.7)$$

as the equation which determines that four-dimensional surface in the five-dimensional manifold that corresponds to space-time. In accordance with this result we can regard the geometry of the de Sitter universe as that holding on the surface of a sphere embedded in five-dimensional Euclidean space. And, as in the case of the Einstein universe, we gain an added intuitional appreciation of the homogeneity of the de Sitter model. It may be emphasized, nevertheless, that the formal simplicity in the expression for the line element given by (142.6) is achieved at the expense of losing track of the physical distinction between space-like intervals which are to be measured in principle by the use of metre sticks and time-like intervals which are measurable with the help of clocks.

Finally, we may examine an interesting and important transformation of coordinates discovered independently by Lemaître[†] and by Robertson,[‡] and specially employed by the latter. The transformation is obtainable by introducing the new variables

$$\bar{r} = \frac{r}{\sqrt{(1-r^2/R^2)}} e^{-t/R} \qquad \bar{t} = t+\tfrac{1}{2}R\log\left(1-\frac{r^2}{R^2}\right). \qquad (142.8)$$

This leads to the expression

$$ds^2 = -e^{2\bar{t}/R}(d\bar{r}^2 +\bar{r}^2 \, d\theta^2 +\bar{r}^2\sin^2\theta \, d\phi^2)+d\bar{t}^2,$$

which by dropping the bars over $\bar{r}$ and $\bar{t}$, and also introducing for simplicity

$$k = \frac{1}{R}, \qquad (142.9)$$

can be written in the form

$$ds^2 = -e^{2kt}(dr^2 +r^2 \, d\theta^2 +r^2\sin^2\theta \, d\phi^2)+dt^2, \qquad (142.10)$$

or by an obvious further substitution in the form

$$ds^2 = -e^{2kt}(dx^2+dy^2+dz^2)+dt^2. \qquad (142.11)$$

In this form for the line element the gravitational potentials $g_{\mu\nu}$ are no longer independent of the time-like coordinate t, which is now being employed. This, however, need occasion no surprise since it is obvious that any static form of line element can be changed into a non-static form by a suitable substitution of new coordinates which are functionally dependent on the original coordinates of both space-

† Lemaître, *J. Math. and Phys.* (M.I.T.), **4**, 188 (1925).
‡ Robertson, *Phil. Mag.* **5**, 835 (1928).

like and time-like character.† Moreover, as shown by Robertson, in the present case the properties of the manifold defined by (142.10) may be regarded to a certain extent as intrinsically independent of the new t, since a transformation to the variables

$$\bar{r} = re^{kt_0} \qquad \bar{t} = t - t_0,$$

which may be considered as a change in spatial scale combined with change to a new zero point for the time-like variable, leaves the form of the line element unaltered.

The line elements (142.10) and (142.11) may be designated in the language proposed by Robertson as *stationary* rather than *static*. This designation must not be confused, however, with another usage‡ in which the term stationary is used to denote line elements in which the potentials $g_{\mu\nu}$ are all independent of x^4 with some of the components g_{14}, g_{24}, and g_{34} present, and the term static is reserved for line elements in which these cross terms are missing. We shall find later use for the Robertson form of expression for the de Sitter line element [see §§ 144 (d), 183, and 184].

143. Absence of matter and radiation from de Sitter universe

We must now turn to more physical aspects of the de Sitter universe by investigating the possibility for matter and radiation in the model. In accordance with the general treatment of the requirements for a static homogeneous model discussed in § 134, the de Sitter line element was obtained in § 136 by assuming that the necessary conditions for such a model were to be met by taking the proper density and pressure in the model as connected by the relation

$$\rho_{00} + p_0 = 0. \tag{143.1}$$

The proper density of material ρ_{00} is, nevertheless, from its physical nature a quantity which could only be zero or positive. Furthermore, even if we permitted the idealized fluid filling the model to exhibit cohesive forces, it is evident that a negative pressure equal to the density in our present units could not be even remotely approached by any known material. Hence the above condition is evidently to be met only by taking the density and pressure each individually equal to zero

$$\rho_{00} = 0 \quad \text{and} \quad p_0 = 0 \tag{143.2}$$

---

† For another non-static form for the de Sitter line element, see Lanczos, *Physik. Zeits.* **23**, 539 (1922).

‡ See Weyl, *Raum, Zeit, Materie*, Berlin (1921), p. 244.

corresponding to a completely empty universe, containing no appreciable amount either of matter or radiation.

As a consequence of taking $\rho_{00} = 0$, we obtain a simplification in our previous equation (136.4) connecting cosmological constant and radius of the universe, so that the de Sitter line element can now be written as

$$ds^2 = -\frac{dr^2}{1-r^2/R^2} - r^2\,d\theta^2 - r^2\sin^2\theta\,d\phi^2 + \left(1 - \frac{r^2}{R^2}\right)dt^2, \quad (143.3)$$

together with the simple expression for the constant R in terms of the cosmological constant

$$\frac{1}{R^2} = \frac{\Lambda}{3}. \quad (143.4)$$

In accordance with this result, the de Sitter model can be regarded as spatially closed if the cosmological constant is positive, as degenerating into the open 'flat' space-time of the special theory of relativity if the cosmological constant is equal to zero, and as spatially open but 'curved' if the cosmological constant should be a negative quantity. In what follows we shall regard Λ as positive and R as real corresponding to a closed model.

It is also interesting to note in accordance with (143.4) that Schwarzschild's exterior solution (96.3)

$$ds^2 = -\frac{dr^2}{1-(2m/r)-(\Lambda r^2/3)} - r^2\,d\theta^2 - r^2\sin^2\theta\,d\phi^2 + \left(1 - \frac{2m}{r} - \frac{\Lambda}{3}r^2\right)dt^2$$

$$(143.5)$$

for the static field in the empty space surrounding a spherical mass of matter, goes over into the de Sitter line element for a completely empty universe if we let the mass m of the sphere of matter at the origin go to zero. The expression given by (143.5) is interesting as being an actual if not very important example of a cosmological line element corresponding to a non-homogeneous model.

144. Behaviour of test particles and light rays in the de Sitter universe

(a) **The geodesic equations.** Since the de Sitter line element corresponds to a model which must strictly be taken as completely empty, the presence of matter and radiation in the actual universe would necessarily produce some distortion away from the de Sitter model, a question to which we shall later return in § 183. The introduction of test particles and test light rays into the model to study the gravitational field therein can, however, of course be considered.

The motion of test particles and light rays will be governed by the equations for a geodesic

$$\frac{d^2x^\sigma}{ds^2} + \{\mu\nu, \sigma\}\frac{dx^\mu}{ds}\frac{dx^\nu}{ds} = 0. \tag{144.1}$$

Taking the de Sitter line element in the general form

$$ds^2 = -e^\lambda\, dr^2 - r^2\, d\theta^2 - r^2\sin^2\theta\, d\phi^2 + e^\nu\, dt^2 \tag{144.2}$$

where

$$e^{-\lambda} = e^\nu = 1 - \frac{r^2}{R^2}, \tag{144.3}$$

and substituting into this geodesic equation the values for $\{\mu\nu, \sigma\}$ given by (95.2), we obtain the four following cases for $\sigma = 1, 2, 3, 4$.

$$\frac{d^2r}{ds^2} + \frac{1}{2}\frac{d\lambda}{dr}\left(\frac{dr}{ds}\right)^2 - re^{-\lambda}\left(\frac{d\theta}{ds}\right)^2 - r\sin^2\theta\, e^{-\lambda}\left(\frac{d\phi}{ds}\right)^2 + \tfrac{1}{2}e^{\nu-\lambda}\frac{d\nu}{dr}\left(\frac{dt}{ds}\right)^2 = 0,$$

$$\frac{d^2\theta}{ds^2} + \frac{2}{r}\frac{dr}{ds}\frac{d\theta}{ds} - \sin\theta\cos\theta\left(\frac{d\phi}{ds}\right)^2 = 0,$$

$$\frac{d^2\phi}{ds^2} + \frac{2}{r}\frac{dr}{ds}\frac{d\phi}{ds} + 2\cot\theta\frac{d\theta}{ds}\frac{d\phi}{ds} = 0,$$

$$\frac{d^2t}{ds^2} + \frac{d\nu}{dr}\frac{dr}{ds}\frac{dt}{ds} = 0.$$

Without loss of generality, however, these equations can be readily simplified by choosing coordinates such that the motion of interest is initially in the plane $\theta = \tfrac{1}{2}\pi$. In accordance with the second of the above equations the motion will then remain permanently in that plane and the equations will reduce to

$$\frac{d^2r}{ds^2} + \frac{1}{2}\frac{d\lambda}{dr}\left(\frac{dr}{ds}\right)^2 - re^{-\lambda}\left(\frac{d\phi}{ds}\right)^2 + \tfrac{1}{2}e^{\nu-\lambda}\frac{d\nu}{dr}\left(\frac{dt}{ds}\right)^2 = 0,$$

$$\frac{d^2\phi}{ds^2} + \frac{2}{r}\frac{dr}{ds}\frac{d\phi}{ds} = 0,$$

$$\frac{d^2t}{ds^2} + \frac{d\nu}{ds}\frac{dt}{ds} = 0.$$

The first integrals corresponding to these equations can be easily obtained, since the form of the line element (144.2) itself provides one integral and the second and third equations can be readily integrated by inspection. We thus obtain

$$e^{\lambda}\left(\frac{dr}{ds}\right)^2 + r^2\left(\frac{d\phi}{ds}\right)^2 - e^{\nu}\left(\frac{dt}{ds}\right)^2 + 1 = 0,$$

$$\frac{d\phi}{ds} = \frac{h}{r^2},$$

$$\frac{dt}{ds} = ke^{-\nu},$$

where h and k are constants of integration. Finally, substituting the last two of these equations into the first and introducing the values for λ and ν given by (144.3) we obtain the equations of motion in the form

$$\frac{dr}{ds} = \pm\sqrt{\left(k^2-1+\frac{r^2}{R^2}-\frac{h^2}{r^2}+\frac{h^2}{R^2}\right)},$$

$$\frac{d\phi}{ds} = \frac{h}{r^2}, \tag{144.4}$$

$$\frac{dt}{ds} = \frac{k}{1-r^2/R^2}.$$

In accordance with these equations it will be noted that h is a parameter which can assume either positive or negative values depending on the direction of motion. It should be noted, however, that for all values of $r < R$ the parameter k must be a positive quantity, since we shall take increases in coordinate time t as directly correlated with increases in proper time s. In the case of light rays the parameters h and k will assume infinite values, owing to the relation $ds = 0$ then obtaining.

(b) **Orbits of particles.** We may now use the foregoing integrals of the geodesic equations to secure information as to the motion of particles in a de Sitter universe.

We may first investigate the *shape* of orbit. Combining the first of the above equations with the second and rearranging we easily obtain

$$d\phi = \frac{h\, dr}{r^2\sqrt{\left\{\dfrac{r^2}{R^2}+\left(k^2-1+\dfrac{h^2}{R^2}\right)-\dfrac{h^2}{r^2}\right\}}}. \tag{144.5}$$

This equation can be readily integrated to give an analytical expression for the shape of the orbits taken by particles in the de Sitter universe. An immediate intuitive appreciation of these shapes can be obtained, however, by noting that (144.5) is well known in

Newtonian mechanics† as applying to the shape of orbit taken by a particle with a central repulsive force proportional to the radius r. Hence in the de Sitter model, the orbits of free particles, plotted in the present coordinates r, θ, ϕ, would be in general curved away from the origin as though the particles were repelled by it.

We may next investigate the *velocity* of motion in the orbit. This will, of course, not be the same as in the Newtonian analogue mentioned above. In terms of increments of proper time ds for the particle itself, the two components of orbital velocity are already given by the foregoing first integrals of the geodesic equations. It will be noted, however, in accordance with the form (144.2, 3) in which we have taken the line element, that the coordinate time t is the proper time as it would be measured by an observer at rest at the origin. Hence, since it will be convenient in making comparisons with the actual universe to regard ourselves as located at the origin of coordinates, it will be advantageous to express the velocities for different particles in terms of t. To do this we have merely to eliminate ds from equations (144.4), which gives us

$$\frac{dr}{dt} = \pm \frac{(1-r^2/R^2)}{k} \bigg/ \sqrt{\left(k^2-1+\frac{r^2}{R^2}-\frac{h^2}{r^2}+\frac{h^2}{R^2}\right)} \qquad (144.6)$$

and
$$\frac{d\phi}{dt} = \frac{h(1-r^2/R^2)}{kr^2} \qquad (144.7)$$

for the two components of orbital velocity in terms of ordinary time as measured at the origin.

As a result of these equations the radial velocity of the particle would be zero when

$$k^2-1+\frac{r^2}{R^2}-\frac{h^2}{r^2}+\frac{h^2}{R^2} = 0, \qquad (144.8)$$

and both components of velocity would be zero at

$$r = R. \qquad (144.9)$$

The first of these equations determines the value of r at perihelion when the particle most closely approaches the origin. And in accordance with the second equation all particle motion ceases at the radius R, which we shall later designate as the apparent horizon of the universe.

For the particular case of purely radial motion with $h = 0$, the

---

† See, e.g., Boltzmann, *Vorlesungen über die Principe der Mechanik*, Teil I, 20, equation (40).

condition for closest approach (144.8) reduces to

$$r = R\sqrt{(1-k^2)} \qquad (144.10)$$

perihelion only occurring when the parameter k is less than unity, the particle passing through the origin for larger values of k.

Differentiating (144.6) and (144.7), we can also obtain expressions for the *acceleration* of a particle in its orbit. With some rearrangement of terms these become

$$\frac{d^2r}{dt^2} = -\frac{2r/R^2}{1-r^2/R^2}\left(\frac{dr}{dt}\right)^2 + \left(\frac{1-r^2/R^2}{k}\right)^2\left(\frac{r}{R}+\frac{h^2}{r^3}\right) \qquad (144.11)$$

and

$$\frac{d^2\phi}{dt^2} = -\frac{2h}{kr^3}\frac{dr}{dt}. \qquad (144.12)$$

In accordance with (144.11) we see that the radial acceleration of a particle which has zero radial velocity is necessarily positive at any point between $r = 0$ and $r = R$. Hence a free particle which once reaches perihelion and starts to move away from the origin would never again return. It will also be noted that for a particle at rest at the origin, with $r = 0$ and $h = 0$, the acceleration would vanish. Hence such a particle would remain permanently located at the origin, thus removing any conflict with our previous statement as to the convenience of regarding ourselves as located at the origin of the coordinates which we are using.

(c) **Behaviour of light rays in the de Sitter universe.** We may now turn to the behaviour of light rays in the model. In accordance with the remarks made in connexion with the integrals of the geodesic equations (144.4), the parameters h and k would have to be infinite to correspond to the path of light in our present coordinates. Introducing this condition into the equation for the shape of orbit (144.5) we then obtain

$$d\phi = \frac{dr}{r^2\sqrt{\left(\dfrac{k^2-1}{h^2}+\dfrac{1}{R^2}-\dfrac{1}{r^2}\right)}} \qquad (144.13)$$

for the path of light in the de Sitter model. This will be recognized as corresponding in Newtonian mechanics† to the orbit of a particle in the limiting case where the central force becomes zero. Furthermore, the equation can be integrated in the form $r\cos\phi+ar\sin\phi = b$, where a and b are constants. Hence the trajectories of light rays in the coordinates chosen would correspond to straight lines. This

† See, e.g., Boltzmann, loc. cit.

provides an advantage for these coordinates in interpreting astronomical measurements of distance.

To determine the velocity of light in the model we may return to the expression for the line element itself (144.2, 3), and set $ds^2 = 0$. Doing so we obtain the general result

$$\left(\frac{dr}{dt}\right)^2 + \left(1-\frac{r^2}{R^2}\right)\left\{r^2\left(\frac{d\theta}{dt}\right)^2 + r^2\sin^2\theta\left(\frac{d\phi}{dt}\right)^2\right\} = \left(1-\frac{r^2}{R^2}\right)^2, \quad (144.14)$$

and for the case of purely radial motion this reduces to

$$\frac{dr}{dt} = \pm\left(1-\frac{r^2}{R^2}\right). \quad (144.15)$$

Integrating this result from $r = 0$ to $r = R$, it is found that an infinite length of time as measured by an observer at the origin would be necessary for light to travel between the origin and $r = R$. Hence an observer at the origin could never have any information of events happening at R or beyond and could speak of a *horizon* to the universe at this distance. It should be remarked, however, that another observer located at a different origin would locate his horizon differently, and hence the spatially closed character of the model, mentioned in connexion with (143.4), is to be regarded as applying to the findings of a particular observer.

(*d*) **Doppler effect in the de Sitter universe.** With the help of our knowledge as to the behaviour of particles and light rays in a de Sitter universe we can now investigate the wave-length—as measured at the origin—of light coming from freely moving particles in the model.

In accordance with the expression for the velocity of light given by (144.15), light leaving a particle located at the radius r at 'time' t_1 would arrive at the origin at the later 'time' t_2 given by

$$t_2 = t_1 + \int_0^r \frac{dr}{1-r^2/R^2}.$$

And hence by differentiation the 'time' interval δt_2 between the reception of two successive wave crests would be related to the 'time' interval δt_1, between their emission by the equation

$$\delta t_2 = \left(1+\frac{1}{1-r^2/R^2}\frac{dr}{dt}\right)\delta t_1, \quad (144.16)$$

where dr/dt is the radial velocity of the particle at the time of emission.

On the other hand, the proper time interval δt_1^0 for an observer on the moving particle corresponding to the interval δt_1, assuming motion in the plane $\theta = \frac{1}{2}\pi$, would evidently be

$$\delta t_1^0 = \frac{1-r^2/R^2}{k}\delta t_1 \tag{144.17}$$

in accordance with the third of equations (144.4), while the proper time interval between crests as measured at the origin would be

$$\delta t_2^0 = \delta t_2. \tag{144.18}$$

Hence combining the three foregoing equations we obtain—by the method of § 116—for the shift $\delta\lambda$ in wave-length measured at the origin

$$\frac{\lambda+\delta\lambda}{\lambda} = \frac{\delta t_2^0}{\delta t_1^0} = \frac{1+\dfrac{1}{1-r^2/R^2}\dfrac{dr}{dt}}{\dfrac{1-r^2/R^2}{k}}$$

or

$$\frac{\lambda+\delta\lambda}{\lambda} = \frac{k}{1-r^2/R^2} + \frac{k}{[1-r^2/R^2]^2}\frac{dr}{dt}, \tag{144.19}$$

where the first term depends on the parameter k for the orbit and the radial position r of the particle at the time of emission, while the second term depends also on the radial velocity of the particle at the time of emission.

Since the parameter k as mentioned above would necessarily be a positive quantity, we see that the shift can be either towards the red or the violet according to the sign and magnitude of the velocity of the particle dr/dt at the time of emission. When this velocity is positive the shift is necessarily in the direction of longer wave-lengths, but when it is negative the shift will be in the opposite direction only if the second term is great enough to overweigh the first. For example, for the case of a particle at perihelion with no component of radial motion at all we find, by introducing the condition for perihelion given by (144.8), a red-shift of the amount

$$\frac{\lambda+\delta\lambda}{\lambda} = \frac{k}{1-r^2/R^2} = +\frac{\sqrt{\left(1-\dfrac{r^2}{R^2}+\dfrac{h^2}{r^2}-\dfrac{h^2}{R^2}\right)}}{1-r^2/R^2}, \tag{144.20}$$

where r is now the radius at which perihelion occurs.

In the de Sitter universe, we thus find the possibility both for red- or violet-shifts in the light coming from distant particles, but nevertheless some tendency to favour the occurrence of red- over violet-shifts. This leads to the suggestion that the de Sitter model might account for the great preponderance of red-shifts over violet-shifts in the case of the nearer spiral nebulae discovered by Slipher, and the linear relation between red-shift and distance as we go to the more distant nebulae discovered by the extensive work of Hubble and Humason.

To examine this suggestion it is evident that we cannot proceed solely on the basis of the expression for the generalized Doppler effect given by (144.19). This formula tells us, to be sure, what the observed wave-length of light from a given particle would be, provided we know its orbit and its position therein at the time of emission. But this information would have to be supplemented by some hypothesis as to the orbits and positions for the particles actually present, in order to make predictions as to phenomena in the real universe.

At first sight, the most natural hypothesis to introduce in this connexion might appear to be one which would maintain conditions in our immediate neighbourhood permanently in an approximately steady state. To secure this result we should have to assume an approximate equality between the number of particles (nebulae) which are entering our range of vision at any given time and the number which are leaving after having passed perihelion within that range.

This hypothesis of continuous entry has been examined in some detail, however, by the present writer[†] and found to show little promise as furnishing an account of the actual universe. In accordance with (144.19) and (144.20) there would indeed be some excess of red-shifts over violet-shifts in the observed light from the moving particles, since the red-shift would commence prior to the passage of perihelion and continue permanently thereafter. Nevertheless, it would be hard to account for the complete absence of violet-shifts actually found for all but a very few of the nearest nebulae, or to account for the fairly precise, observational, linear relation between red-shift and distance on the proposed basis.

An alternative hypothesis suggested by Weyl and investigated by himself[‡] and by Robertson[§] has shown more immediate promise of possibly furnishing an account of the observed relation between red-shift and distance. In accordance with this hypothesis the

† Tolman, *Astrophys. Jour.* **69**, 245 (1929).
‡ Weyl, *Phys. Zeits.* **24**, 230 (1923); *Phil. Mag.* **9**, 936 (1930).
§ Robertson, *Phil. Mag.* **5**, 835 (1928).

nebulae in the actual universe are to be regarded as lying on a coherent pencil of geodesics which diverge from a common point in the past.

To investigate the detailed nature of the Weyl hypothesis it is most convenient to use the coordinates of Robertson, which were found in § 142 to lead to the very simple expression for the line element,
$$ds^2 = -e^{2kt}(dr^2 + r^2\,d\theta^2 + r^2\sin^2\theta\,d\phi^2) + dt^2. \qquad (144.21)$$

Using these coordinates and applying the geodesic equation
$$\frac{d^2x^\sigma}{ds^2} + \{\mu\nu, \sigma\}\frac{dx^\mu}{ds}\frac{dx^\nu}{ds} = 0,$$

to the case of particles having no spatial components of 'velocity' $(dr/ds = d\theta/ds = d\phi/ds = 0)$, it is at once seen from the expressions for the Christoffel symbols $\{\mu\nu,\sigma\}$ provided by (98.5), that the accelerations would vanish and that such particles would remain permanently at rest with respect to r, θ, and ϕ.

The Weyl hypothesis then consists in assuming that the nebulae in the actual universe are to be treated as a uniformly distributed set of free particles, which—except for small peculiar motions— remain at rest with respect to the spatial coordinates now being employed. It will be noted from the form of the line element that the present coordinate t is now the proper time not only for a particle at rest at the origin but also for any of these other particles which are at rest with respect to r, θ, and ϕ. It will also be seen that any one of these particles could be taken as at the origin of coordinates without change in form of the line element. Hence all these particles may be regarded as *equivalent*, in the sense that observers thereon would all find approximately the same phenomena occurring in the universe.

Although these particles are chosen so as to remain at rest with respect to our present spatial coordinates, it is evident that the proper distance between them as measured by rigid scales laid end to end would be changing with the time t, owing to the occurrence of this quantity in the components of $g_{\mu\nu}$. Hence we should expect a *Doppler shift* in the light passing from one such particle to another.

In accordance with the form of the line element the radial velocity of light in terms of our present coordinates would be
$$\frac{dr}{dt} = \pm e^{-kt}.$$

Hence, considering a particle permanently located at the radius r, the times t_1 and t_2 for the emission of radiation from the particle and its reception at the origin would be connected by the equation

$$\int_{t_1}^{t_2} e^{-kt}\, dt = \int_0^r dr = r = \text{const.},$$

which on differentiation gives

$$\delta t_2 = e^{k(t_2 - t_1)}\, \delta t_1$$

as the relation connecting the time interval δt_2 between the reception of two successive wave crests with the time interval δt_1 between their emission. Since t, however, is the proper time for observers both at the particle and origin, this now gives as a general expression for the Doppler shift observed at the origin

$$\frac{\lambda + \delta\lambda}{\lambda} = \frac{\delta t_2}{\delta t_1} = e^{k(t_2 - t_1)} \qquad (144.22)$$

or as an approximation for values of r which are not too great

$$\frac{\delta\lambda}{\lambda} \simeq k(t_2 - t_1) \simeq kr. \qquad (144.23)$$

It is evident, moreover, as will be discussed for the general case of non-static homogeneous models in detail in § 179, that the coordinate distance r would in first approximation be proportional to astronomical determinations of distance. Hence, with the help of the Weyl hypothesis, we have obtained a distribution of nebulae in the de Sitter model which would exhibit an approximately linear relation between red-shift and distance as is found in the actual universe.

It should perhaps be emphasized, nevertheless, that this result is due fully as much to the assumption we have made concerning the distribution of the nebulae in space-time as to the inherent properties of the de Sitter model. It may also be pointed out that a reversal in the signs of the terms $-t/R$ and $(R/2)\log\sqrt{\{1 - (r^2/R^2)\}}$ in the transformation equations (142.8), by which we obtained the Robertson expression for the de Sitter line element, would give us a set of coordinates equally appropriate for discussing the reverse case of a system of approaching particles which would exhibit a Doppler shift towards the violet instead of towards the red. It should also be emphasized, however, that the Weyl hypothesis has the very attractive feature of putting all the particles (nebulae) in the model on the same footing, so that there would be nothing unique about the phenomena observed from any particular nebula.

145. Comparison of de Sitter model with actual universe

The most satisfactory feature of the de Sitter model is the possibility which we have just discussed for it to contain a distribution of moving particles so chosen as to imitate the linear relation between red-shift and distance discovered by Hubble and Humason for the light from the nebulae in the actual universe.

In accordance with (144.23) we have as the expression for this relation

$$\frac{\delta\lambda}{\lambda} = kr, \tag{145.1}$$

and as a result of the astronomical measurements, see § 177 (d), we may give to k the approximate numerical value

$$k \simeq 6 \cdot 0 \times 10^{-28} \text{ cm.}^{-1} \simeq 5 \cdot 7 \times 10^{-10} \text{ (light years)}^{-1}. \tag{145.2}$$

On the other hand, in accordance with (142.9) and (143.4) we may write for k in the case of the de Sitter model the theoretical expressions

$$k = \frac{1}{R} = \sqrt{\frac{\Lambda}{3}}. \tag{145.3}$$

And this gives us

$$R \simeq 1 \cdot 66 \times 10^{27} \text{ cm.} \simeq 1 \cdot 75 \times 10^9 \text{ light years} \tag{145.4}$$

and $$\Lambda = 1 \cdot 08 \times 10^{-54} \text{ cm.}^{-2}. \tag{145.5}$$

These results may be compared with the previous case of the Einstein universe as given by (141.3) and (141.4). It will be noted that the cosmological constant Λ comes out very considerably greater in the case of the de Sitter universe than in that of the Einstein universe. Nevertheless, comparing with the result given by (141.5), it will be seen that Λ is not large enough to affect known planetary orbits. It will also be noted that the distance R to the horizon in the de Sitter universe comes out appreciably less than the radius R of the Einstein universe, and perhaps dangerously close to the distances of the order 3×10^8 light years which have already been penetrated by the telescope.

The most unsatisfactory feature of the de Sitter model, as a basis for the cosmology of the actual universe, is the finding discussed in § 143 that the line element when strictly taken corresponds to a completely empty universe containing neither matter nor radiation.

Hence the actual presence of matter and radiation in the real universe must be regarded as producing a distortion away from the proposed line element. And we shall later be able to show in § 183 that this distortion might be serious.

It is interesting to note the contrast in the successful and unsuccessful features of the two original static models. The Einstein model permits a finite concentration of matter in the universe, but does not allow for any red-shift in the observed light coming from the nebulae. The de Sitter model permits, with the introduction of the Weyl hypothesis, a red-shift in the light from distant particles, but does not allow for the observed finite concentration of matter in the actual universe. The non-static models, to which we now turn in Part II of this Chapter, will be found to permit the successful features of both the older models.

APPLICATIONS TO COSMOLOGY (*contd.*)

Part II. THE APPLICATION OF RELATIVISTIC MECHANICS TO NON-STATIC HOMOGENEOUS COSMOLOGICAL MODELS

146. Reasons for changing to non-static models

The original static universes of Einstein and of de Sitter are certainly very important in furnishing examples of the kind of cosmological model that can be constructed within the theoretical framework of general relativity. Moreover, as we shall see later, it is possible, although not necessarily probable, that these models might really correspond to a considerable extent with the initial and final states of the actual universe. Nevertheless, it is evident that neither of these models gives a satisfactory description of the present state of the actual universe, the one because it permits no shift in the wavelength of light from the nebulae, and the other because it permits no matter or radiation to be present in space.

We must hence turn to some less restricted class of models in our attempts to describe the behaviour of the actual universe, and may begin by investigating the effects of dropping our previous requirement that the line element for the universe should be expressible in a static form independent of the time-like coordinate x^4.

There are several reasons which make it natural to abandon this assumption that our cosmological models should necessarily be static in character. In the first place, it is of course evident that any increase in generality which can be brought about by the removal of previous restrictions will be of advantage in increasing the range of possible applicability. The non-static models which we shall now study are, to be sure, mathematically more complicated than our previous static ones; nevertheless, the history of human endeavours to understand the universe would certainly indicate no *a priori* right to demand mathematical simplicity of nature. In the second place, although there was some observational evidence for ascribing a reasonably stationary character to our surroundings at a time when our knowledge of the universe was practically limited to the stars in our own galaxy, this evidence must now be regarded as completely replaced by the observed red-shift in the light from the extra-galactic nebulae which at least leads to the presumption that these objects are not static but are moving away from each other. In the third place, even if some successful alternative hypothesis should be proposed for explaining this red-shift, it should be emphasized that processes are certainly observed in the universe, such as the emission of radiation from the stars at the presumable expense of their mass, which—unless compensated in some unknown and ingenious manner—certainly lead to changes in gravitational field with the time and hence necessarily to a non-static universe.† Finally, as we shall see later, we shall find that an originally static Einstein universe would in any case not be stable but would start to expand or contract as a result of disturbances.‡

By dropping the previous restriction to static models, we are at once led to the study of a considerable group of non-static homogeneous models,§ which were first theoretically investigated by Friedmann,‖ and first considered in connexion with the phenomena of the actual universe by Lemaître.††

147. Assumption employed in deriving non-static line element

We shall commence our investigation by considering the derivation of the form of line element which applies to the proposed models. The first completely satisfactory derivation of this line element was given by Robertson,‡‡ who based his deduction on two simple geometrical assumptions—first, that space-time from a large-scale point of view should be separable into space and a 'cosmic' time orthogonal thereto in such a way that the line element could be written at the start in the form $ds^2 = g_{ij}\,dx^i dx^j + dt^2$ $(i, j = 1, 2, 3)$, and secondly, that space-time should be spatially homogeneous and isotropic when looked at from a large-scale point of view. This was followed by a derivation by the present writer§§ based on a set of assumptions, selected on grounds of their immediate physical character, but not chosen as simply and critically as is possible. The somewhat similar derivation to be given below will be based essentially on a single

---

† Tolman, *Proc. Nat. Acad.* **16**, 320 (1930).
‡ Eddington, *Monthly Notices*, **90**, 668 (1930).
§ For an excellent summary of the work on static and non-static models up until the end of 1932, see Robertson, *Reviews of Modern Physics*, **5**, 62 (1933).
‖ Friedmann, *Zeits. f. Physik*, **10**, 377 (1922).
†† Lemaître, *Ann. Soc. Sci. Bruxelles*, **47 A**, 49 (1927).
‡‡ Robertson, *Proc. Nat. Acad.* **15**, 822 (1929). The earlier deduction of Friedmann was not entirely satisfactory.
§§ See loc. cit., § 146.

assumption as to spatial isotropy, having an immediate observational significance which will be evident from the beginning.

In accordance with the results of Hubble, the large-scale properties of the universe do not appear to depend in any significant way on the direction of observation as far out as the 100-inch Mount Wilson telescope is able to penetrate. Thus with respect to our own location, on a particular one of the galaxies or nebulae which constitute the observed portion of the universe, we actually find the universe to be spatially isotropic. Generalizing this observed fact, we shall then take as our only essential hypothesis—necessary in addition to the principles of relativistic mechanics for deriving the desired line element—the assumption that an observer, located anywhere in the universe and at rest with respect to the mean motion of the matter in his neighbourhood, would also obtain observations showing a similar large-scale independence of direction. In other words, we shall assume spatial isotropy for the physical findings of any such observer.

This assumption is a natural one to introduce, since it avoids the anthropocentric assignment of a unique importance to our own location in the universe, and proceeds as best we may by regarding the observations that we obtain as fairly representing the character of those which would be obtained from similar locations in other portions of the universe. Before investigating the consequences of this assumption, nevertheless, several critical remarks may be made concerning it.

In the first place, it should be emphasized that the assumption is in any case meant to be only a rough principle applying on the average to regions large enough to contain many nebulae. In the second place, it should be noted that the requirement of spatial isotropy is to apply, of course, only as stated, to the findings of observers who are at rest with respect to the matter in their part of the universe, since observers moving through this matter would certainly obtain findings which were dependent on the direction of the relative motion.

Most important of all, however, it is to be emphasized that the assumption is to be regarded merely as a working hypothesis, suggested by the present state of observational knowledge, but necessarily subject to some modification if we desire to allow for the finer details of the observed irregularities in nebular distribution, and perhaps subject to far-reaching modification if more powerful telescopes should reveal a systematic lack of uniformity in different parts of the universe. The assumption is not in the least intended to be taken as a fundamental law of nature, on the same footing as the principle of relativity, but should be regarded more nearly as a mere statement defining the kind of cosmological model we shall next discuss.† Furthermore, it is specially important to realize the possibility that this assumption of spatial isotropy might not agree with the facts in the actual universe, since even if the model we obtain does prove successful in correlating a certain number of cosmological phenomena, we must always keep an open mind as to changes and improvements which could make a better or more extended theory possible. To this we shall return later.

148. Derivation of line element from assumption of spatial isotropy

We must now turn to the details of deriving the general form of line element for the class of models that we are to discuss. As a result of our assumption of spatial isotropy, it is evident that we may at the start require our coordinate system to be such that the line element will explicitly exhibit spherical symmetry around the origin of coordinates, which can be taken at any desired point in the model which remains at rest with respect to the matter in its neighbourhood. Furthermore, it is evident that we can at the same time employ—as will prove most convenient—a co-moving coordinate system obtained by taking the spatial components as determined by a network of meshes which is drawn so as to connect adjacent material particles (nebulae) in the model and is allowed to move therewith. Hence as a starting-point, we shall assert the possibility of expressing the line element in co-moving coordinates in the most general possible form exhibiting spatial spherical symmetry.

$$ds^2 = -e^\lambda\, dr^2 - e^\mu (r^2\, d\theta^2 + r^2\sin^2\theta\, d\phi^2) + e^\nu\, dt^2 + 2a\, drdt. \quad (148.1)$$

We take this most general form as a starting-point, rather than either of the simpler forms exhibiting spherical symmetry previously discussed in §§ 94 and 98, on account of the assertion that we are to use co-moving coordinates, which necessitates a special investigation

† The procedure is very different from that of Milne, *Zeits. f. Astrophys.*, **6**, 1 (1933), who would regard the homogeneity of the universe as a fundamental principle from which even the laws of gravitation might be deduced.

to see if simplifications can be introduced without disturbing this desired character of the coordinate system. It is easy to demonstrate, however, that a reduction to the second of the two previous simplified forms can be made, still maintaining the co-moving character of the coordinates.

In order to obtain simplifications, we may obviously consider any transformations of coordinates which do not upset the relations of the type

$$\frac{dr}{ds} = \frac{d\theta}{ds} = \frac{d\phi}{ds} = 0, \qquad (148.2)$$

which must hold for the spatial components of the 'velocity' of particles in the model, if our coordinates are to be co-moving as desired.

Without disturbing these relations we can evidently substitute a new time-like variable t' defined by the equation

$$dt' = \eta(a\, dr + e^\nu\, dt), \qquad (148.3)$$

where η is an integrating factor which makes the right-hand side of (148.3) a perfect differential. In accordance with (148.3) we shall have

$$e^\nu\, dt^2 + 2a\, dr dt = \frac{dt'^2}{\eta^2 e^\nu} - \frac{a^2}{e^\nu}\, dr^2. \qquad (148.4)$$

So that on substitution into (148.1), and dropping primes the line element can be written in the simpler form

$$ds^2 = -e^\lambda\, dr^2 - e^\mu(r^2\, d\theta^2 + r^2\sin^2\theta\, d\phi^2) + e^\nu\, dt^2, \qquad (148.5)$$

where λ, μ, and ν are now functions of r and the present t, and the relations (148.2) have not been upset since r, θ, and ϕ are still the same variables as before. We have now reduced the line element to the general form studied by Dingle as discussed in § 100.

To proceed farther in the simplification, we may next consider the components of gravitational acceleration for a free test particle in the model. These would be determined by the equations for a geodesic (74.13), and for the case of a particle at rest with respect to r, θ, and ϕ this would give us

$$\frac{d^2r}{ds^2} = -\{44, 1\}\left(\frac{dt}{ds}\right)^2 \qquad \frac{d^2\theta}{ds^2} = -\{44, 2\}\left(\frac{dt}{ds}\right)^2 \qquad \frac{d^2\phi}{ds^2} = -\{44, 3\}\left(\frac{dt}{ds}\right)^2.$$

$$(148.6)$$

Since this test particle is spatially at rest with respect to our present system of co-moving coordinates, it is also at rest with respect to a local observer moving with the matter in the neighbourhood. In accordance with our assumption of spatial isotropy, however, such a local observer must obtain physical results which are independent of direction. Hence these accelerations can only have the value zero,† and we are led to the conclusion that the three Christoffel symbols appearing in (148.6) must themselves be zero. And from Dingle's values for these quantities as given by (100.2) we then obtain

$$\frac{\partial \nu}{\partial r} = \frac{\partial \nu}{\partial \theta} = \frac{\partial \nu}{\partial \phi} = 0$$

as a condition on the quantity ν occurring in the expression for the line element (148.5). This shows that ν is a function of t alone, and permits us to introduce a new time variable defined by the expression

$$t' = \int e^{\frac{1}{2}\nu} \, dt \tag{148.7}$$

without disturbing the co-moving character of the coordinates. Doing so and dropping primes, we then obtain the further reduction to the form
$$ds^2 = -e^{\lambda} \, dr^2 - e^{\mu}(r^2 \, d\theta^2 + r^2 \sin^2\theta \, d\phi^2) + dt^2. \tag{148.8}$$
We thus obtain a separation of space-time into space and a universal time t orthogonal thereto, without the necessity of introducing any further hypothesis.

In accordance with this form of line element
$$t = t_0$$
would now be the proper time as it would be measured by a local observer at rest with respect to the matter in his neighbourhood, and

$$\delta l_1 = e^{\frac{1}{2}\lambda} \, \delta r \qquad \delta l_2 = e^{\frac{1}{2}\mu} r \, \delta\theta \qquad \delta l_3 = e^{\frac{1}{2}\mu} r \sin\theta \, \delta\phi$$

would be the proper distances as measured by this observer between particles belonging in the model which would permanently retain the above differences in coordinate position. For the fractional rate of change in these proper distances with proper time, we then obtain

$$\frac{\partial}{\partial t_0} \log \delta l_1 = \frac{1}{2} \frac{\partial \lambda}{\partial t} \qquad \frac{\partial}{\partial t_0} \log \delta l_2 = \frac{\partial}{\partial t_0} \log \delta l_3 = \frac{1}{2} \frac{\partial \mu}{\partial t},$$

and, by our hypothesis of spatial isotropy for the findings of the local observer, are led to the useful relation

$$\frac{\partial \lambda}{\partial t} = \frac{\partial \mu}{\partial t}. \tag{148.9}$$

† We cannot in general use the co-moving character of coordinates as necessary justification for the requirement that the acceleration of such a particle must be zero, since gravitational action might be balanced by a pressure gradient.

This result now shows the possibility of a further simplification in the line element by the substitution

$$\frac{dr'}{r'} = e^{\frac{1}{2}(\lambda-\mu)}\frac{dr}{r} \quad \text{or} \quad \log r' = \int e^{\frac{1}{2}(\lambda-\mu)}\frac{dr}{r}. \qquad (148.10)$$

This substitution will not disturb the co-moving character of the coordinates, since we can evidently write for the radial velocity dr'/ds of a particle in our new coordinates

$$\frac{1}{r'}\frac{dr'}{ds} = \frac{\partial}{\partial r}\left[\int e^{\frac{1}{2}(\lambda-\mu)}\frac{dr}{r}\right]\frac{dr}{ds} + \frac{\partial}{\partial t}\left[\int e^{\frac{1}{2}(\lambda-\mu)}\frac{dr}{r}\right]\frac{dt}{ds}$$

$$= \frac{\partial}{\partial r}\left[\int e^{\frac{1}{2}(\lambda-\mu)}\frac{dr}{r}\right]\frac{dr}{ds} + \frac{1}{2}\left[\int e^{\frac{1}{2}(\lambda-\mu)}\left(\frac{\partial\lambda}{\partial t}-\frac{\partial\mu}{\partial t}\right)\frac{dr}{r}\right]\frac{dt}{ds},$$

and in accordance with (148.9) this will be zero for any particle which is at rest, with dr/ds equal to zero, in our original coordinates. Introducing (148.10) in (148.8) and dropping primes, we shall then be able to write the line element in the second of the forms considered in § 98,

$$ds^2 = -e^{\mu}(dr^2 + r^2\,d\theta^2 + r^2\sin^2\theta\,d\phi^2) + dt^2, \qquad (148.11)$$

where μ is now a function of the present r and t.

To continue with the derivation, we may again consider the proper distance

$$\delta l_0 = e^{\frac{1}{2}\mu}\,\delta r$$

between neighbouring particles belonging to the model which are permanently separated by the coordinate distance δr. For the fractional rate of change of such a measured distance with the time we can write

$$\frac{\partial \log \delta l_0}{\partial t_0} = \frac{1}{2}\frac{\partial\mu}{\partial t},$$

and from our assumption of spatial isotropy it is evident that this quantity could not be found by the local observer either to increase or decrease with r. Hence we are led to the conclusion

$$\frac{\partial}{\partial r}\frac{\partial \log \delta l_0}{\partial t_0} = \frac{1}{2}\frac{\partial^2\mu}{\partial r\partial t} = 0, \qquad (148.12)$$

and must take μ as the sum of a function of r and t

$$\mu(r,t) = f(r) + g(t). \qquad (148.13)$$

Introducing (148.13) into (148.11), we may now write the line element in the still more explicit form

$$ds^2 = -e^{f(r)+g(t)}(dr^2 + r^2\,d\theta^2 + r^2\sin^2\theta\,d\phi^2) + dt^2. \qquad (148.14)$$

To proceed from this point, we could now make use of a known principle of Riemannian geometry (Schur's theorem) in accordance with which the spatial isotropy at every point of the sub-space $(r,\ \theta,\ \phi)$ with t constant would necessitate its spatial homogeneity, and thus permit us to write for $e^{f(r)}$ a known form of solution. In accordance, nevertheless, with our desire to emphasize the physical character of our considerations, we shall actually proceed in a different manner.

Comparing the form for the line element (148.14) with the expressions for the energy-momentum tensor given for this general form by (98.6), we can now write as expressions for the only surviving components of the energy-momentum tensor

$$8\pi T_1^1 = -e^{-\mu}\left(\frac{f'^2}{4} + \frac{f'}{r}\right) + \ddot{g} + \tfrac{3}{4}\dot{g}^2 - \Lambda,$$

$$8\pi T_2^2 = 8\pi T_3^3 = -e^{-\mu}\left(\frac{f''}{2} + \frac{f'}{2r}\right) + \ddot{g} + \tfrac{3}{4}\dot{g}^2 - \Lambda, \qquad (148.15)$$

$$8\pi T_4^4 = -e^{-\mu}\left(f'' + \frac{f'^2}{4} + \frac{2f'}{r}\right) + \tfrac{3}{4}\dot{g}^2 - \Lambda,$$

where
$$\mu(r,t) = f(r) + g(t),$$

and accents denote differentiation with respect to r and dots with respect to t.

These expressions give, of course, the components of the energy-momentum tensor referred to our present system of coordinates (r, θ, ϕ, t). At any point of interest, however, we may evidently introduce proper coordinates (x_0, y_0, z_0, t_0) for a local observer at rest with respect to r, θ, and ϕ, in such a way that we shall have the relations

$$dx_0 = e^{\frac{1}{2}\mu}\,dr \qquad dy_0 = e^{\frac{1}{2}\mu}r\,d\theta \qquad dz_0 = e^{\frac{1}{2}\mu}r\sin\theta\,d\phi \qquad dt_0 = dt$$

holding in the neighbourhood of that point. And in accordance with the general rules for the transformation of tensors, it is then seen that the above expressions would also give the analogous components of the energy-momentum tensor referred to these proper coordinates.

Thus, for example, we should have

$$T^1_{01} = \frac{\partial x^1_0}{\partial x^\alpha}\frac{\partial x^\beta}{\partial x^1_0}T^\alpha_\beta = e^{\frac12\mu}e^{-\frac12\mu}T^1_1 = T^1_1.$$

Hence the above expressions (148.15) may also be taken as giving the components of the energy-momentum tensor referred to proper coordinates as used by a local observer at rest with respect to the matter in the model. And from the assumed spatial isotropy in the findings of such an observer, we can then conclude that his measurements of stress will have to lead to a symmetry between the x, y, and z directions, such that we shall have

$$T^1_1 = T^2_2 = T^3_3. \tag{148.16}$$

Making use of this result, we then see from (148.15), that we obtain the relation

$$\frac{f'^2}{4}+\frac{f'}{r} = \frac{f''}{2}+\frac{f'}{2r},$$

or

$$\frac{d^2f}{dr^2}-\frac12\left(\frac{df}{dr}\right)^2-\frac1r\frac{df}{dr} = 0, \tag{148.17}$$

as an equation for determining the form of $f(r)$. As a first integral of this equation we have

$$\frac{df}{dr} = c_1 r e^{\frac12 f},$$

where c_1 is the constant of integration. And as the second integral we then obtain

$$e^{f(r)} = \frac{1/c_2^2}{[1-c_1 r^2/4c_2]^2}, \tag{148.18}$$

where c_2 is the second constant of integration.

This now completes the derivation. Returning to our previous expression for the line element (148.14), absorbing the constant factor $1/c_2^2$ in $e^{g(t)}$, and to agree with familiar forms of expression putting

$$-\frac{c_1}{c_2} = \frac{1}{R_0^2}, \tag{148.19}$$

where R_0^2 is a *constant* which can be positive, negative, or infinite, we can then write the line element in the final form

$$ds^2 = -\frac{e^{g(t)}}{[1+r^2/4R_0^2]^2}(dr^2+r^2\,d\theta^2 +r^2\sin^2\theta\,d\phi^2)+dt^2, \tag{148.20}$$

where $g(t)$ is still an undetermined function of the time t.

We have been interested in presenting this long derivation in order to show, by a line of reasoning each step of which has an immediate physical interpretation, that the assumption of spatial isotropy for the large-scale physical findings obtained by observers at rest with respect to the matter in their neighbourhood, combined with the principles of relativistic mechanics, does inevitably lead to the proposed line element. Hence, if we should later be dissatisfied on observational or philosophical grounds with the results to be obtained from the proposed model, we must modify either the principles of relativistic mechanics, or the assumption that all observers in the universe must be expected to obtain large-scale results which are independent of the direction of observation.

149. General properties of the line element

(a) **Different forms of expression for the line element.** By the transformation of coordinates the line element for our present non-static models

$$ds^2 = -\frac{e^{g(t)}}{[1+r^2/4R_0^2]^2}(dr^2+r^2\,d\theta^2+r^2\sin^2\theta\,d\phi^2)+dt^2 \quad (149.1)$$

can be written in several different forms which are sometimes convenient or can be of assistance in understanding the implied geometry.

By the obvious substitutions

$$x = r\sin\theta\cos\phi \qquad y = r\sin\theta\sin\phi \qquad z = r\cos\theta \quad (149.2)$$

we obtain the form

$$ds^2 = -\frac{e^{g(t)}}{[1+r^2/4R_0^2]^2}(dx^2+dy^2+dz^2)+dt^2, \quad (149.3)$$

with
$$r = \sqrt{(x^2+y^2+z^2)},$$

which makes the spatial isotropy at any point perhaps more obvious.

By the substitution
$$\bar{r} = \frac{r}{1+r^2/4R_0^2} \quad (149.4)$$

the line element assumes the form

$$ds^2 = -e^{g(t)}\left(\frac{d\bar{r}^2}{1-\bar{r}^2/R_0^2}+\bar{r}^2\,d\theta^2+\bar{r}^2\sin^2\theta\,d\phi^2\right)+dt^2, \quad (149.5)$$

which has the advantage of showing the relation of this non-static line element to one of the most familiar forms for the static Einstein line element.†

† Comparing the transformation equation (149.4) with the previous transformation equation (138.2) used in connexion with the Einstein universe, it is to be noted that

By the further substitution

$$\bar{r} = R_0 \sin \chi \qquad (149.6)$$

we can now write the line element in the form

$$ds^2 = -R_0^2 e^{g(t)}(d\chi^2 + \sin^2\chi \, d\theta^2 + \sin^2\chi \sin^2\theta \, d\phi^2) + dt^2. \qquad (149.7)$$

Finally, by introducing a larger number of dimensions, with the help of the equations

$$z_1 = R_0 \sqrt{(1-\bar{r}^2/R_0^2)}, \qquad z_2 = \bar{r}\sin\theta\cos\phi,$$

$$z_3 = \bar{r}\sin\theta\sin\phi, \qquad z_4 = \bar{r}\cos\theta, \qquad (149.8)$$

where

$$z_1^2 + z_2^2 + z_3^2 + z_4^2 = R_0^2, \qquad (144.9)$$

the line element assumes the form

$$ds^2 = -e^{g(t)}(dz_1^2 + dz_2^2 + dz_3^2 + dz_4^2) + dt^2, \qquad (149.10)$$

which at any given time t permits us to regard our original space as embedded in a Euclidean space of a larger number of dimensions.

(b) **Geometry corresponding to line element.** As in the case of the static Einstein universe, the kind of geometry corresponding to these different expressions for the line element is not completely determined since different hypotheses as to connectivity and as to the identification of points could be made.

It will be simplest, however, as suggested by the last form in which we have written the line element, to regard the spatial extent of this non-static universe at any given time t as the whole three-dimensional spherical surface defined by

$$z_1^2 + z_2^2 + z_3^2 + z_4^2 = R_0^2, \qquad (149.11)$$

embedded in the four-dimensional Euclidean space (z_1, z_2, z_3, z_4). Since the proper distance at time t corresponding to the coordinate interval dz_1, would from the form of the line element (149.10) evidently be

$$dl_0 = e^{\frac{1}{2}g(t)} dz_1, \qquad (149.12)$$

with similar expressions for the other spatial coordinates, it is evident that the radius of this spherical surface would be

$$R = R_0 e^{\frac{1}{2}g(t)}. \qquad (149.13)$$

Hence this quantity is often spoken of as the radius of the non-static universe, and the geometry is spoken of as being that for the surface of a sphere in four dimensions whose radius is a function of the time.

---

our present r is analogous to the previous ρ, and our present $\bar{r}$ is analogous to the previous r.

It should be noted, however, in accordance with the equation (148.19) by which R_0 was introduced, that this radius could be real, imaginary, or infinite.

If we assume the radius real, the total integrated proper spatial volume of the model at any selected time t would be given in accordance with (149.7) by

$$v_0 = \int_0^{2\pi} \int_0^{\pi} \int_0^{\pi} R_0^3 \, e^{\frac{3}{2}g(t)} \sin^2\chi \sin\theta \, d\chi d\theta d\phi = 2\pi^2 R_0^3 \, e^{\frac{3}{2}g(t)}, \quad (149.14)$$

and the total integrated proper distance around the universe would be

$$l_0 = 2\pi R_0 \, e^{\frac{1}{2}g(t)}. \quad (149.15)$$

Taking the spatial geometry as elliptical rather than spherical, the corresponding quantities would be half as great.

If we assume the radius infinite or imaginary, the model would be spatially open rather than closed and the total proper volume could be most conveniently calculated, in accordance with the form (149.5) for the line element, from the expression

$$v_0 = \int_0^{2\pi} \int_0^{\pi} \int_0^{\infty} \frac{e^{\frac{3}{2}g(t)}}{(1+\bar{r}^2/A^2)^{\frac{3}{2}}} \bar{r}^2 \sin\theta \, d\bar{r} d\theta d\phi = \infty,$$

where A^2 is a positive quantity which can assume the value infinity, and the upper limit for $\bar{r}$ can be taken as infinity without disturbing the possibility for a physical interpretation of the line element by changing its signature. Evaluating the integral we then obtain an infinite total proper volume for open models.

The symmetrical form (149.10), which we have been able to give to the line element by the device of considering a larger number of dimensions, is valuable in clearly showing the spatial homogeneity of the model already mentioned in connexion with (148.14). It is an interesting extension of Schur's theorem of Riemannian geometry, that the spatial isotropy which we have assumed for observers at all points in space-time should lead to an orthogonal separation into space and time and to homogeneity for the sub-manifold of space. It is in accordance with this result that our present models of the universe have been designated as non-static *homogeneous* cosmological models.

(c) **Result of transfer of origin of coordinates.** The spatial homogeneity of the model makes it evident that the origin for the spatial coordinates can be selected at will at any desired point in the model without affecting the forms in which the line element can be expressed. Furthermore, on transferring the origin of coordinates from one point in the model to another, it can be shown not only that the line element can be written in an unaltered form, but also that the coordinates of the new origin in the old system of coordinates will be related in the expected way to the coordinates of the old origin in the new system of coordinates. We may now demonstrate this principle.†

To agree with our later use of the result, we shall employ a system of coordinates $(\bar{r}, \theta, \phi, t)$ corresponding to the third form (149.5) in which we have written the line element

$$ds^2 = -e^{g(t)}\left(\frac{d\bar{r}^2}{1-\bar{r}^2/R_0^2}+\bar{r}^2\,d\theta^2+\bar{r}^2\sin^2\theta\,d\phi^2\right)+dt^2, \quad (149.16)$$

and shall consider a *nebula* as being at rest at the origin of these coordinates, and an *observer* permanently located at $\bar{r}=a$ as providing the new origin of the coordinates to which we wish to transform.

In this original system of coordinates S we may then tabulate the spatial coordinates for the nebula and observer in the form

System S	$\bar{r}$	θ	ϕ
Nebula	0	..	..
Observer	a	0	0

$$(149.17)$$

where the angular coordinates for the nebula at the origin are of course indeterminate, and for simplicity we have given the observer the polar and equatorial angles $\theta = \phi = 0$, since the starting-points for measuring these angles can evidently be chosen in any arbitrary way that proves convenient.

We now desire to find the result of transforming to a new system of coordinates S' of the same type as S but with the origin of coordinates located at the observer. To carry out this transformation it will be simplest to introduce intermediate steps in which we employ coordinates of the type given by the expression for the line element (149.10), corresponding to a treatment of our original space as embedded in a Euclidean space of one more dimension. Making use of the transformation equations (149.8), we shall then first transform to a new system of coordinates S_z, in which the spatial coordinates for

† Tolman, *Proc. Nat. Acad.* **16**, 511 (1930).

nebula and observer will be seen to have the values

System S_z	z_1	z_2	z_3	z_4
Nebula	R_0	0	0	0
Observer	$R_0 \sqrt{(1-a^2/R_0^2)}$	0	0	a

(149.18)

We may now consider a further change of coordinates to a new system S_z', which may be regarded as a rotation in the $z_1 z_4$ plane, and which we define by the transformation equations

$$z_1' = z_1 \cos\alpha + z_4 \sin\alpha, \qquad z_2' = z_2,$$
$$z_4' = -z_1 \sin\alpha + z_4 \cos\alpha, \qquad z_3' = z_3, \qquad (149.19)$$

where we take

$$\sin\alpha = \frac{a}{R_0} \qquad \cos\alpha = \sqrt{(1-a^2/R_0^2)}. \qquad (149.20)$$

Applying these transformation equations we then easily obtain as our new coordinates for the nebula and observer

System S_z'	z_1'	z_2'	z_3'	z_4'
Nebula	$R_0 \sqrt{(1-a^2/R_0^2)}$	0	0	$-a$
Observer	R_0	0	0	0

(149.21)

This last transformation of coordinates, however, will be seen to have been such as to leave the line element in the form (149.10) and to preserve the form of the relation (149.9) which determines the three-dimensional surface in the four-dimensional manifold which corresponds to physical space. Hence we may now employ transformation equations of the form (149.8) to go back to a coordinate system S' in which the line element will again have the form (149.16) with which we started. Doing so we then easily obtain for the coordinates of nebula and observer,

System S'	$\bar{r}'$	θ'	ϕ'
Nebula	a	π	..
Observer	0	..	..

(149.22)

where the values for all but one of the angular coordinates are undetermined.

Comparing the tables (149.17) and (149.22), we now see that we have actually carried out a transformation from an original system of coordinates with the nebula at the origin $\bar{r} = 0$ and the observer at $\bar{r} = a$, to a new system of coordinates having the same form of line element (149.16) but with the observer at the origin of coordinates

$\bar{r}' = 0$ and the nebula at $\bar{r}' = a$. That this simple relation should hold for such a transformation of coordinates was indeed to be expected. Nevertheless, the result will be of sufficient importance for our later considerations to justify the explicit proof which we have given here.

(d) **Physical interpretation of line element.** In accordance with our general principles for the physical interpretation of formulae for interval, we can of course relate any of the foregoing expressions for the line element, in a non-static homogeneous universe, with the results that would be obtained by suitable measurements made in the ordinary manner with metre sticks or clocks. Thus if we write the line element in its original form

$$ds^2 = -\frac{e^{g(t)}}{[1+r^2/4R_0^2]^2}(dr^2 + r^2\,d\theta^2 + r^2\sin^2\theta\,d\phi^2) + dt^2, \quad (149.23)$$

we see that measurements of proper distance dl_0 made by a local observer at rest with respect to $r, \theta,$ and ϕ, would be connected with coordinate differences by the equation

$$dl_0 = \frac{e^{\frac{1}{2}g(t)}}{[1+r^2/4R_0^2]}\sqrt{(dr^2 + r^2\,d\theta^2 + r^2\sin^2\theta\,d\phi^2)}. \quad (149.24)$$

Furthermore, we see that his measurements of proper time dt_0 made with his local clock would be connected with differences in coordinate time by the very simple relation

$$dt_0 = dt. \quad (149.25)$$

Similarly, of course, in the case of observers who are not at rest with respect to r, θ, and ϕ we could find the somewhat more complicated relations between measurements of proper distance or time and coordinate differences.

As a result of the simple relation (149.25), we see that the coordinate time t, in all of the expressions which we have given for our non-static line element, would agree with proper time as measured on his own clock by any local observer at rest with respect to the mean motion of matter in his part of the universe. It is important to emphasize this result, since it means that we can identify the coordinate t with our own measurements and estimates of past and future time. Hence any estimates of the time scale needed for astronomical changes are appropriately expressed in terms of the coordinate t, and no real changes in time scale are brought about by the mere substitution of a new time-like coordinate in place of t.

150. Density and pressure in non-static universe

Up to the present stage in our consideration of these homogeneous non-static models, we have made no hypothesis as to the nature of the material filling the model, beyond the assumption that we could neglect local irregularities from the large-scale point of view employed in cosmology, and the assumption that the material could then be taken as obeying Einstein's field equations

$$-8\pi T_{\mu\nu} = R_{\mu\nu} - \tfrac{1}{2} R g_{\mu\nu} + \Lambda g_{\mu\nu},$$

where, in accordance with the large-scale point of view, the components of the energy-momentum tensor would have to be assigned values which could be regarded as appropriate averages for the position and instant of interest.

We may now, however, introduce a more specific hypothesis by assuming that the material filling the model can be treated as a perfect fluid. It was advantageous to delay the introduction of this assumption until the spatial homogeneity of the model had been demonstrated. With a non-homogeneous distribution we should expect to encounter phenomena, such as the net outward flow of radiation from a region containing a greater concentration of luminous matter than its surroundings, which could not be appropriately represented by replacing the actual material by a perfect fluid, owing to the circumstance noted in § 86 that the expression for the energy-momentum tensor of a perfect fluid restricts the behaviour of the fluid to adiabatic processes without flow of heat, and hence provides no analogy for a transfer of energy by radiation flow from one portion of matter to another. Having found homogeneity, however, for the class of models under consideration we may now regard the radiation derived from the nebulae in any given large region as suffering no net increase or decrease by exchange with the surroundings, and introduce the definite hypothesis that the material in the actual universe, consisting of nebulae together with dispersed intergalactic matter and radiation, can be treated for the purposes of the model as a perfect fluid.

As a consequence of this hypothesis, we can now apply, to the material filling the model, the specific expression obtained in § 85 for the energy-momentum tensor of a perfect fluid

$$T^{\mu\nu} = (\rho_{00} + p_0) \frac{dx^\mu}{ds} \frac{dx^\nu}{ds} - g^{\mu\nu} p_0, \tag{150.1}$$

where ρ_{00} and p_0 are the proper macroscopic density and pressure as they would be measured by a local observer at rest in the fluid, and the quantities dx^μ/ds are the components of the macroscopic 'velocity' of this fluid with respect to the coordinates in use.

Employing our original coordinate system (r, θ, ϕ, t) in which the line element (149.1) assumes the form

$$ds^2 = -\frac{e^{g(t)}}{[1+r^2/4R_0^2]^2}(dr^2 + r^2\,d\theta^2 + r^2\sin^2\theta\,d\phi^2)+dt^2 \quad (150.2)$$

the spatial components of the 'velocity' of the fluid would be zero

$$\frac{dr}{ds} = \frac{d\theta}{ds} = \frac{d\phi}{ds} = 0, \quad (150.3)$$

owing to the fact that these coordinates have been chosen so as to be co-moving. And the temporal component would be

$$\frac{dt}{ds} = 1, \quad (150.4)$$

owing to the form of the line element. This introduces considerable simplification when combined with (150.1), and we then find as the only surviving components of the energy-momentum tensor

$$T^{11} = -g^{11}p_0 \quad T^{22} = -g^{22}p_0 \quad T^{33} = -g^{33}p_0 \quad T^{44} = \rho_{00}, \quad (150.5)$$

or, on lowering indices,

$$T_1^1 = T_2^2 = T_3^3 = -p_0 \quad T_4^4 = \rho_{00}. \quad (150.6)$$

The line element (150.2) is written, however, in a standard form, and expressions for the components of the energy-momentum tensor corresponding to this form have already been given by (98.6). Applying these expressions to the case of the present line element and introducing the pressure and density as given by (150.6), we then readily obtain

$$8\pi p_0 = -\frac{1}{R_0^2}e^{-g(t)}-\ddot{g}-\tfrac{3}{4}\dot{g}^2+\Lambda \quad (150.7)$$

and

$$8\pi\rho_{00} = \frac{3}{R_0^2}e^{-g(t)}+\tfrac{3}{4}\dot{g}^2-\Lambda, \quad (150.8)$$

as simple expressions for the local pressure and density of the fluid in the model, where the dots indicate differentiation with respect to the time.

Several remarks may be made concerning these expressions. In the first place, it will be noticed that the pressure and density are functions of the time t alone, and at a given value of t would be independent of position in the universe, in agreement with the spatial homogeneity of the model which we have already discussed. In the second place, it will be seen that by taking $g(t)$ as a constant independent of t the expressions for pressure and density would reduce to those given for the static Einstein universe by (139.3) and (139.4) with $R = R_0 e^{\frac{1}{2}g}$ as the constant value of the radius. Perhaps most important of all, however, in contrast to the case of the original static Einstein universe, it will be seen—when $g(t)$ does vary with t— that it is no longer essential for the constant R_0 to be real and the cosmological constant Λ to be positive in order to obtain a positive density of energy ρ_{00} in the model and a pressure p_0 which is not negative. This is especially significant since it removes the older *a priori* arguments for a necessarily closed universe and for the necessary introduction of a cosmological term, and leaves these questions still open for observational decision.

In interpreting the expressions for density ρ_{00} and pressure p_0 given by (150.7) and (150.8), it must be remembered that these quantities apply to the idealized fluid in the model, which we have substituted in place of the matter and radiation actually present in the real universe. In making this substitution, it would appear reasonable to take ρ_{00} as the averaged-out density of energy, corresponding to the nebulae and the internebular matter and internebular radiation present in a sufficiently large region of the universe to be representative, including of course as an important item, and at the moment as the best known item, the energy mc^2 corresponding to the mass of the nebulae. For the pressure p_0 of the fluid, it would appear reasonable to take the sum of the partial pressures, corresponding firstly to the random motions of the nebulae themselves, secondly to the random motions of dust or other particles of matter present in internebular space, and thirdly to the density of internebular radiation.

With the help of this picture of the factors responsible for the values of total energy density ρ_{00} and pressure p_0, we can also obtain a rough expression for that part of the energy density ρ_m which directly corresponds to the mass of the nebulae and other particles of matter present in the universe. In the case of the *nebulae* the pressure corresponding to the random motion of these enormous particles would be equal to two-thirds of their kinetic energy per unit volume

$$p = \tfrac{2}{3}\rho_k$$

from ordinary kinetic theory considerations. This of course would be very small. In the case of *particles* of dust or other matter present in internebular space, the pressure would vary from two-thirds the density of their kinetic energy

$$p = \tfrac{2}{3}\rho_k$$

for slow random motions, down to one-third this quantity

$$p = \tfrac{1}{3}\rho_k$$

for particles with velocities approaching that of light. Finally, for the case of *radiation* the pressure would in general be one-third the energy density

$$p = \tfrac{1}{3}\rho_k.$$

For the nebulae and slow moving particles, however, it is evident that the density of kinetic energy would be negligible compared with the density corresponding directly to the mass of the particles. Hence we may roughly take

$$\rho_m = \rho_{00} - 3p_0 \tag{150.9}$$

as that part of the total energy density which corresponds directly to the mass of the nebulae and whatever internebular matter may be present; and this expression becomes exact when the pressure due to matter can be completely neglected.

Combining with equations (150.7) and (150.8) we can then write for the density of matter in the universe the approximate expression

$$8\pi\rho_m = \frac{6}{R_0^2}e^{-g(t)} + 3\ddot{g} + 3\dot{g}^2 - 4\Lambda. \tag{150.10}$$

151. Change in energy with time

By substituting the values for the components of the energy-momentum tensor (150.5) and (150.6) into the general equation of relativistic mechanics

$$\frac{\partial \mathfrak{T}_\mu^\nu}{\partial x^\nu} - \tfrac{1}{2}\mathfrak{T}^{\alpha\beta}\frac{\partial g_{\alpha\beta}}{\partial x^\mu} = 0, \tag{151.1}$$

for the case $\mu = 4$, we at once obtain

$$\frac{\partial}{\partial t}(\rho_{00}\sqrt{-g}) + \tfrac{1}{2}p_0\sqrt{-g}\left(g^{11}\frac{\partial g_{11}}{\partial t} + g^{22}\frac{\partial g_{22}}{\partial t} + g^{33}\frac{\partial g_{33}}{\partial t}\right) = 0,$$

since g_{44} has the constant value unity. And introducing the values for the components of the metrical tensor corresponding to the line element (150.2) which we are employing

$$ds^2 = -\frac{e^{g(t)}}{[1 + r^2/4R_0^2]^2}(dr^2 + r^2\,d\theta^2 + r^2\sin^2\theta\,d\phi^2) + dt^2 \tag{151.2}$$

we easily find that this reduces to

$$\frac{\partial}{\partial t}\left(\frac{\rho_{00}\,r^2\sin\theta\,e^{\frac{3}{2}g(t)}}{[1+r^2/4R_0^2]^3}\right)+p_0\frac{\partial}{\partial t}\left(\frac{r^2\sin\theta\,e^{\frac{3}{2}g(t)}}{[1+r^2/4R_0^2]^3}\right)=0. \qquad (151.3)$$

This result, however, can be given an immediate physical interpretation. In accordance with the form of the line element (151.2), it will be seen that the proper volume as measured by a local observer corresponding to a small coordinate range $\delta r\delta\theta\delta\phi$ would be given at any time t by the expression

$$\delta v_0 = \frac{r^2\sin\theta\,e^{\frac{3}{2}g(t)}}{[1+r^2/4R_0^2]^3}\,\delta r\delta\theta\delta\phi. \qquad (151.4)$$

Moreover, since the coordinates in use are co-moving, this is the volume as it would appear to a local observer of an element of the fluid which would remain permanently in that range. Hence combining (151.3) and (151.4), we can now write

$$\frac{d}{dt}(\rho_{00}\,\delta v_0)+p_0\frac{d}{dt}(\delta v_0)=0, \qquad (151.5)$$

and interpret this as relating the changes, which a local observer would find in the energy $(\rho_{00}\,\delta v_0)$ of any element of the fluid, with the work done on the surroundings in the way to be expected for adiabatic changes in volume.

As a consequence of (151.4) and (151.5), we now see that the volume of every element of fluid in the model would be increasing with the time if $g(t)$ is increasing with t, and decreasing when $g(t)$ is decreasing, and furthermore, if the pressure p_0 is a positive quantity greater than zero, that the proper energy of *every* element of fluid in the model would be decreasing when $g(t)$ is increasing, and increasing when $g(t)$ is decreasing. Hence, except for the special case of zero pressure, the total proper energy of the fluid will not in general be a constant; and the principle of energy conservation can only be made to apply by introducing a quantity to represent the potential energy of the gravitational field in the way already discussed in § 87.

For some purposes it will be more convenient to write (151.3) in the form

$$\frac{d}{dt}(\rho_{00}\,e^{\frac{3}{2}g(t)})+p_0\frac{d}{dt}(e^{\frac{3}{2}g(t)})=0, \qquad (151.6)$$

which can evidently be done owing to the mutual independence of the coordinates r,θ,ϕ, and t.

This latter form of expression can be readily verified with the help of the explicit expressions for density ρ_{00} and pressure p_0 given by (150.7) and (150.8), a necessary result since the fundamental equation for the components of the energy-momentum tensor

$$-8\pi T^{\mu\nu} = R^{\mu\nu} - \tfrac{1}{2}Rg^{\mu\nu} + \Lambda g^{\mu\nu}$$

must provide—as we have seen—all the information which can be obtained from the equation of mechanics,

$$\frac{\partial \mathfrak{T}^\nu_\mu}{\partial x^\nu} - \tfrac{1}{2}\mathfrak{T}^{\alpha\beta}\frac{\partial g_{\alpha\beta}}{\partial x^\mu} = 0,$$

which can be derived from it.

In what follows we shall often find it convenient to take (151.6) together with (150.8) as being the two equations which relate the pressure and density of the fluid to the line element for the model, thus replacing the second-order equation (150.7) by the first-order equation (151.6).

152. Change in matter with time

With the help of our rough expression (150.9) for that part of the total energy density
$$\rho_m = \rho_{00} - 3p_0, \tag{152.1}$$
which corresponds directly to the mass of the nebulae and whatever intergalactic matter may be present, we can also investigate the dependence of the matter in the model on the time.† Combining (152.1) with (151.5) we can put

$$\frac{d}{dt}(\rho_m\,\delta v_0) + 3\frac{d}{dt}(p_0\,\delta v_0) + p_0\frac{d}{dt}(\delta v_0) = 0, \tag{152.2}$$

and regarding
$$M = \rho_m\,\delta v_0 \tag{152.3}$$
as the total proper mass of the nebulae and other particles of matter in a given coordinate range, we can rewrite this with the help of (151.4) in a form

$$-\frac{1}{M}\frac{dM}{dt} = \frac{6p_0}{\rho_m}\frac{dg}{dt} + \frac{3}{\rho_m}\frac{dp_0}{dt}, \tag{152.4}$$

which gives the fractional rate of change in the proper mass of the matter present in the model.

For the special case in which we take the pressure as permanently equal to zero, this rate would become equal to zero and we should have conservation of mass, as well as the conservation of total proper energy already noted for this special case in connexion with equation (151.5).   † Tolman, *Proc. Nat. Acad.* **16**, 409 (1930).

Also for the special case determined by the condition

$$6p_0\frac{dg}{dt}+3\frac{dp_0}{dt}=0,\qquad(152.5)$$

which can also be expressed in the form

$$3\frac{d}{dt}(p_0\,\delta v_0)+p_0\frac{d}{dt}(\delta v_0)=0,\qquad(152.6)$$

we should have conservation of mass. And this will be seen to be the condition which would apply to the case of a model containing a constant amount of matter exerting negligible pressure, and containing radiation which exerts the pressure $p_r=\rho_r/3$, which will later be treated in § 160.

In general, however, we should desire to allow some change in the proper mass associated with the matter in the model, since changes of this kind are presumably occurring in the actual universe. Thus, in accordance with the Einstein relation between mass and energy, the emission of radiation from the nebulae would be accompanied by a decrease in their mass, irrespective of the possibilities that the ultimate source of this radiation might lie in destructive processes such as the mutual annihilation of electrons and protons, or might lie in synthetic processes such as the formation of helium from hydrogen with an accompanying decrease in mass.† Similarly, if the source of the cosmic rays should lie in the annihilation of internebular particles of matter or in the synthesis of more complicated atoms from hydrogen, there would also be a decrease in the mass of the matter in the universe.

For some purposes, see §§ 165 and 184, it is more useful to re-express equation (152.4) in a form which shows the direct dependence of the rate of loss of mass on the rate of change in $g(t)$. Making use of our previous expressions for ρ_{00} and p_0 (150.7) and (150·8), we readily obtain after some rearrangement

$$-\frac{1}{M}\frac{dM}{dt}=\frac{3}{2}\left[\frac{\rho_{00}+\frac{5}{8}p_0}{\rho_m}-\frac{1}{4\pi\rho_m}\left(\ddot{g}+\frac{\ddot{g}}{g}\right)\right]\dot{g},\qquad(152.7)$$

where the dots indicate differentiation with respect to time. This expression for the fractional rate of decrease in the mass of matter in the model shows that the annihilation of matter would in any case

---

† This conclusion would have to be modified if the radiation from the stars should prove to be due to a failure of the principle of energy conservation in their interiors, as has been suggested by Bohr.

necessarily lead to a non-static model with $g(t)$ not a constant. This result provides the specific justification for one of the general reasons given in § 146 for changing to non-static models.

153. Behaviour of particles in the model

We may next consider the behaviour of free particles in the non-static model corresponding to our line element

$$ds^2 = -\frac{e^{g(t)}}{[1+r^2/4R_0^2]^2}(dr^2+r^2\,d\theta^2 +r^2\sin^2\theta\,d\phi^2)+dt^2. \quad (153.1)$$

In accordance with the principles of relativistic mechanics, the motion of free particles in the model would be determined by the equations for a geodesic

$$\frac{d^2x^\sigma}{ds^2}+\{\mu\nu,\sigma\}\frac{dx^\mu}{ds}\frac{dx^\nu}{ds} = 0. \quad (153.2)$$

And since the line element (153.1) is written in a standard form, we can employ our previous expressions for the Christoffel symbols $\{\mu\nu,\sigma\}$ as given by (98.5), in using these equations.

We may first investigate the case of a particle which is at rest with respect to the spatial coordinates r, θ, and ϕ which will give us

$$\frac{dr}{ds} = \frac{d\theta}{ds} = \frac{d\phi}{ds} = 0, \qquad \frac{dt}{ds} = 1. \quad (153.3)$$

The equations for a geodesic then reduce to

$$\frac{d^2x^\sigma}{ds^2}+\{44,\sigma\} = 0,$$

and since all values of $\{44,\sigma\}$ are seen from (98.5) to be zero for our present line element, we find that all components of acceleration for such a particle would vanish

$$\frac{d^2r}{ds^2} = \frac{d^2\theta}{ds^2} = \frac{d^2\phi}{ds^2} = \frac{d^2t}{ds^2} = 0, \quad (153.4)$$

and the particle would remain permanently at rest with respect to the spatial coordinates, and increments ds in proper time as measured by a local observer on the particle would permanently agree with increments dt in the coordinate time which we are using.

The conclusion that particles at rest with respect to our spatial coordinates would experience no gravitational acceleration, tending to set them in motion, is of course in agreement with the fact that we have chosen co-moving coordinates such that the fluid filling the model remains permanently at rest with respect to r, θ, and ϕ. The result applies only to the gravitational acceleration, but owing to the homogeneity of the model, it is evident that other kinds of acceleration arising from collisions or radiation pressure would also be zero on the average for particles at rest relative to our spatial coordinates. The conclusion that particles of matter in the model would remain permanently at rest with respect to the coordinates r, θ, and ϕ must not be confused with the fact that the proper distance between two such particles as determined by fitting metre sticks from one to the other would be changing with the time if $g(t)$ is so changing.

To investigate the more general case of particles having any arbitrary initial velocity with respect to the coordinates, it will prove most expeditious to start with the equation for a geodesic (153.2) for the case $\sigma = 4$, which will give us

$$\frac{d^2t}{ds^2} + \{\mu\nu, 4\}\frac{dx^\mu}{ds}\frac{dx^\nu}{ds} = 0. \tag{153.5}$$

Substituting from (98.5) the values of $\{\mu\nu, 4\}$ which correspond to our line element (153.1), we then obtain

$$\frac{d^2t}{ds^2} + \tfrac{1}{2}e^\mu\dot\mu\left(\frac{dr}{ds}\right)^2 + \tfrac{1}{2}e^\mu\dot\mu r^2\left(\frac{d\theta}{ds}\right)^2 + \tfrac{1}{2}e^\mu\dot\mu r^2\sin^2\theta\left(\frac{d\phi}{ds}\right)^2 = 0,$$

where
$$e^\mu = \frac{e^{g(t)}}{[1+r^2/4R_0^2]^2},$$

and from the form of the line element itself, it is evident that the above result can be rewritten as

$$\frac{d^2t}{ds^2} + \frac{1}{2}\frac{dg}{dt}\left[\frac{dt^2}{ds^2} - 1\right] = 0,$$

or
$$\frac{2\dfrac{dt}{ds}\dfrac{d}{dt}\left(\dfrac{dt}{ds}\right)}{\dfrac{dt^2}{ds^2} - 1} = -\frac{dg}{dt},$$

which can be integrated to give us

$$\frac{dt^2}{ds^2} - 1 = Ae^{-g(t)}, \tag{153.6}$$

where A is the constant of integration.

To interpret this result, we may now again return to the line element (153.1) and note that this can be written in the form

$$\frac{ds^2}{dt^2} = 1 - \frac{e^{g(t)}}{[1+r^2/4R_0^2]^2}\left(\frac{dr^2}{dt^2} + r^2\frac{d\theta^2}{dt^2} + r^2\sin^2\theta\frac{d\phi^2}{dt^2}\right)$$

which can then be applied to the motion of particles in the form

$$\frac{ds^2}{dt^2} = 1 - \frac{u^2}{c^2}, \tag{153.7}$$

where c is the velocity of light, and u is the velocity of the particle as measured in the ordinary manner by an observer in its neighbourhood who is at rest with respect to r, θ, and ϕ, and who uses his own determinations of increments in proper time and proper distance

$$dt_0 = dt, \qquad dl_0 = \frac{e^{\frac{1}{2}g(t)}}{[1 + r^2/4R_0^2]}\, dr, \text{ etc.} \tag{153.8}$$

Substituting (153.7) in (153.6) we then obtain

$$\frac{u^2/c^2}{1 - u^2/c^2} = A e^{-g(t)}, \tag{153.9}$$

as an expression for the time dependence of the velocity u with which a free particle would be found to be moving by local observers along its path, who themselves remain at rest with respect to the average motion of matter in their neighbourhood.

In accordance with (153.9), if $g(t)$ is increasing with time, and the proper volumes of elements of the fluid in the model hence expanding, the velocities of such free particles will be decreasing with time, and vice versa if the model is contracting these velocities will be increasing. If we apply this result to particles which are themselves regarded as constituents of the fluid in the model, and correlate the random velocities of such particles with the contributions they make to the energy density and pressure of this fluid, it can readily be shown that the dependence of velocity on time given by (153.9) is in entire agreement with the relation between energy density and pressure for the fluid previously given by (151.6).

With the help of (153.9) we can also discuss the energy of free particles in the model as a function of time. This can be of interest in connexion with the energies of the cosmic rays in the actual universe, since at least a portion of these rays may be due to fast-moving particles. By solving (153.9) we can easily obtain the result

$$E = \frac{E_0}{\sqrt{(1 - u^2/c^2)}} = E_0\sqrt{(1 + Ae^{-g(t)})}, \tag{153.10}$$

where E is the expression given by the special theory of relativity for the total energy of a particle, including its proper energy $E_0 = m_0 c^2$ corresponding to its mass. In accordance with this result, we see that the energy of such particles as measured by local observers at rest with respect to the mean motion of matter in their neighbourhood would be decreasing with the time if $g(t)$ is increasing and the model expanding.

To procure a better idea as to the rate at which the energy of such particles would be changing with time, we can obtain by the differentiation and rearrangement of (153.10)

$$-\frac{1}{(E-E_0)}\frac{d}{dt}(E-E_0) = \frac{1}{2}\left(1+\frac{E_0}{E}\right)\frac{dg}{dt} \qquad (153.11)$$

as an expression for the fractional rate of decrease in the kinetic energy of the particles $(E-E_0)$, which in the case of the cosmic rays would be that portion of the energy available for producing ionization.

This formula has the advantage of expressing the rate of energy change in terms of the quantity $\dot{g} = dg/dt$, which as we shall later find is closely related to the red-shift in the light from the nebulae that would correspond to our model. In accordance with (153.11), the fractional rate of decrease in the kinetic energy of free particles in the model would vary from $\dot{g}$ for slow moving particles with $E \simeq E_0$, down to $\dot{g}/2$ for particles having velocities approaching that of light with $E \gg E_0$. As will be shown in § 156, the limiting case of particles having zero rest mass and moving with the precise velocity of light, would correspond to the behaviour of light quanta or photons.

With the help of the equations for a geodesic, we can also investigate the form of the trajectories for free particles as well as their velocities. For our later purposes it will be sufficient to consider a particle which is originally moving in the radial direction with

$$\frac{d\theta}{ds} = \frac{d\phi}{ds} = 0. \qquad (153.12)$$

In accordance with the geodesic equations (153.2), we should then have

$$\frac{d^2\theta}{ds^2}+\{11,2\}\left(\frac{dr}{ds}\right)^2+2\{14,2\}\frac{dr}{ds}\frac{dt}{ds}+\{44,2\}\left(\frac{dt}{ds}\right)^2 = 0,$$

and

$$\frac{d^2\phi}{ds^2}+\{11,3\}\left(\frac{dr}{ds}\right)^2+2\{14,3\}\frac{dr}{ds}\frac{dt}{ds}+\{44,3\}\left(\frac{dt}{ds}\right)^2 = 0,$$

and since the Christoffel symbols are all six found from (98.5) to be zero, we obtain the result

$$\frac{d^2\theta}{ds^2} = \frac{d^2\phi}{ds^2} = 0. \qquad (153.13)$$

We may hence conclude that free particles having their original motion directed to or from the origin of coordinates will continue to move in a radial direction. It is evident that this result is a direct consequence of the spatial isotropy of the model, and that it would also hold on transforming the line element to the form (149.5) since the coordinates θ and ϕ are not affected by the transformation.

154. Behaviour of light rays in the model

We may now turn to the behaviour of light rays in our present model, still making use of the line element in the form

$$ds^2 = -\frac{e^{g(t)}}{[1+r^2/4R_0^2]^2}(dr^2+r^2\,d\theta^2 +r^2\sin^2\theta\,d\phi^2)+dt^2. \quad (154.1)$$

In accordance with the principles of relativistic mechanics, the equations for a geodesic (153.2) would apply to the motion of light rays as well as particles provided we consider the limiting case with $ds = 0$.

Setting $ds = 0$ in the expression for the line element, we can at once write as a general expression for the velocity of light in our model, except when it is actually passing through matter,

$$\frac{e^{g(t)}}{[1+r^2/4R_0^2]^2}\left(\frac{dr^2}{dt^2}+r^2\frac{d\theta^2}{dt^2}+r^2\sin^2\theta\frac{d\phi^2}{dt^2}\right) = 1. \quad (154.2)$$

And noting our previous expressions for increments in proper time and proper distance (153.8), we see that the velocity of light in empty space at any point in the model will be found to have the normal value

$$u = c, \quad (154.3)$$

when measured in the ordinary manner by any observer at rest with respect to the mean motion of matter in his neighbourhood.

For the special case of a ray of light moving in the radial direction, we have in accordance with (154.2) the coordinate velocity

$$\frac{dr}{dt} = \pm e^{-\frac{1}{2}g(t)}\left[1+\frac{r^2}{4R_0^2}\right]. \quad (154.4)$$

Furthermore, in agreement with the treatment given to the radial motion of a particle, see (153.13), it is evident that a ray, travelling to or from the origin of coordinates, would permanently maintain its motion on a radial path.

Integrating (154.4) over the time interval t_1 to t_2 needed for a ray of light to travel between the origin and any desired coordinate position, we obtain the expression

$$\int_0^r \frac{dr}{1+r^2/4R_0^2} = \int_{t_1}^{t_2} e^{-\frac{1}{2}g(t)}\, dt \tag{154.5}$$

or

$$2R_0 \tan^{-1} \frac{r}{2R_0} = \int_{t_1}^{t_2} e^{-\frac{1}{2}g(t)}\, dt, \tag{154.6}$$

where the integral on the right-hand side can be evaluated only on the basis of some specific information or assumption as to the form of $g(t)$.

If we assume that $g(t)$ can be taken as linear with respect to t

$$g = 2kt, \tag{154.7}$$

over the time interval of interest, we can then easily compute the right-hand side of (154.6) and obtain

$$r = 2R_0 \tan \frac{e^{-kt_1} - e^{-kt_2}}{2kR_0}. \tag{154.8}$$

This formula for the time t_1 to t_2 necessary for light to travel in either direction between the origin $r = 0$ and the coordinate distance $r = r$ can be applied, when the time interval is short enough so that the effect of the derivatives of g with respect to t higher than the first can be neglected. The result can find a possible application in interpreting the reception of cosmic rays from intergalactic space.

In the case of a closed ever-expanding model of the universe, the relation (154.6) between r and t_1 to t_2 can lead to interesting restrictions on the coordinate distance which light could travel in a finite time. Let us assume—merely for purposes of illustration—a model having the exact linear dependence of g on t (154.7) for all times from minus to plus infinity, and having the real radius $R = R_0 e^{kt}$, which would increase from zero to infinity between $t = -\infty$ and $t = +\infty$. In the first place, it is then evident from (154.8) that light could be received at the origin at any given finite time t_2 coming from any desired coordinate distance r, provided one chooses the time of starting t_1—which can go to minus infinity—early enough. On the other hand, for light which leaves the origin at time t_1, it is evident that there would be a maximum coordinate distance

$$r = 2R_0 \tan \frac{e^{-kt_1}}{2kR_0}, \tag{154.9}$$

which could be reached even at $t = \infty$. According to the values of k and R_0, there would hence be a specific starting-time after which light could no longer travel completely around the model. Thus, under the assumptions taken, an observer at rest in the fluid filling the model could theoretically obtain information concerning sufficiently early states of all parts of the universe, but even by waiting an infinite length of time could not obtain information as to their behaviour later than a certain epoch. The discussion applies of course only to a particular assumed model, but is perhaps valuable in widening our views as to conceptual possibilities.

155. The Doppler effect in the model

We may next examine the Doppler effect on the observed wavelength of light, coming from distant objects in the model which corresponds to our line element

$$ds^2 = -\frac{e^{g(t)}}{[1+r^2/4R_0^2]^2}(dr^2+r^2\,d\theta^2+r^2\sin^2\theta\,d\phi^2)+dt^2. \quad (155.1)$$

Since we shall be interested in comparing the wave-lengths of light from different objects as observed at a single location, it will be simplest to take the observer as *permanently* located at the origin of coordinates, and the luminous source as at any desired coordinate distance r which, however, may be varying with the time. We can then readily obtain an expression for the generalized Doppler effect following the schematic method outlined in § 116.

In accordance with our expression for the radial velocity of light in the model, see (154.4) and (154.5), we can write

$$\int_{t_1}^{t_2} e^{-\frac{1}{2}g(t)}\,dt = \int_0^r \frac{dr}{1+r^2/4R_0^2} \quad (155.2)$$

as an equation, connecting the 'time' t_1 at which light leaves a source located at r, with the 'time' t_2 at which it arrives at the origin. Hence, differentiating this expression with respect to the time of departure t_1, we can obtain

$$e^{-\frac{1}{2}g_2}\,\delta t_2 - e^{-\frac{1}{2}g_1}\,\delta t_1 = \frac{1}{1+r^2/4R_0^2}\left(\frac{dr}{dt}\right)\delta t_1,$$

as an equation connecting the 'time' interval δt_1 between the departure of two wave crests from the source with the 'time' interval δt_2 between their arrival at the origin, where g_1 and g_2 denote the values of $g(t)$ at t_1 and t_2, and (dr/dt) is the radial component of the 'coordinate velocity' of the source at the time of emission. And, noting the expressions for proper distances and times which would correspond to the form of the line element, we can evidently rewrite this in the form

$$e^{-\frac{1}{2}g_2}\delta t_2 = e^{-\frac{1}{2}g_1}\delta t_1 + e^{-\frac{1}{2}g_1}\frac{u_r}{c}\delta t_1, \tag{155.3}$$

where c is the velocity of light and u_r is now the radial component of the velocity of the source, as it would be measured in the ordinary manner by an observer at rest with respect to r, θ, and ϕ.

In accordance with the form of the line element, however, the proper time interval δt_1^0 between the emission of these wave crests, as measured by a local observer moving with the source, would be related to the coordinate interval δt_1 by the expression

$$\delta t_1^0 = \left\{ -\frac{e^{g_1}}{[1+r^2/4R_0^2]^2}\left(\frac{dr^2}{dt^2}+r^2\frac{d\theta^2}{dt^2}+r^2\sin^2\theta\frac{d\phi^2}{dt^2}\right)+1 \right\}^{\frac{1}{2}} \delta t_1.$$

And noting again the implications of the line element, this can evidently be rewritten as

$$\delta t_1^0 = \sqrt{(1-u^2/c^2)}\,\delta t_1, \tag{155.4}$$

where u is the total velocity of the source at the time of emission as measured in the ordinary manner by a local observer who is at rest with respect to r, θ, and ϕ. Furthermore, for the proper time interval δt_2^0 between the reception of the wave crests by an observer at rest at the origin we shall evidently have

$$\delta t_2^0 = \delta t_2. \tag{155.5}$$

Substituting (155.4) and (155.5) in (155.3), and equating the ratio of the proper periods of the emitted and received light to the ratio of the corresponding wave-lengths, we then finally obtain as the complete expression for the generalized Doppler effect

$$\frac{\lambda+\delta\lambda}{\lambda} = \frac{\delta t_2^0}{\delta t_1^0} = \frac{e^{\frac{1}{2}(g_2-g_1)}}{\sqrt{(1-u^2/c^2)}}\left(1+\frac{u_r}{c}\right), \tag{155.6}$$

where $\lambda+\delta\lambda$ can evidently be taken as the wave-length of the light as ultimately observed at the origin, while λ is the wave-length of the same light as measured by an observer who is located at the source and moves along therewith.

The most important term in this expression for the generalized Doppler effect is $e^{\frac{1}{2}(g_2-g_1)}$, which is due to the general motion of particles (nebulae) in the model connected with changes in the value of $g(t)$. The next most important term is $(1+u_r/c)$ which is due to any peculiar radial velocity which the source in question may have relative to the mean motion of the matter in its neighbourhood. The least important term is $\sqrt{(1-u^2/c^2)}$, which may be regarded as due to the effect of velocity on the rate of a moving clock. (Transverse Doppler effect.)

In studying with the help of the model the red-shift in the light from the extra-galactic nebulae, it is usually sufficient to regard the nebulae as having the mean motion of matter appropriate to their neighbourhood and hence as at rest with respect to r, θ, and ϕ. We then have to consider only the most important term, connected with the general expansion of the model, and can write

$$\frac{\lambda+\delta\lambda}{\lambda} = e^{\frac{1}{2}(g_2-g_1)}, \tag{155.7}$$

or for the fractional change in wave-length,

$$\frac{\delta\lambda}{\lambda} = e^{\frac{1}{2}(g_2-g_1)}-1. \tag{155.8}$$

This result can also be written in other forms which prove illuminating. Introducing the radius of the model

$$R = R_0 e^{\frac{1}{2}g}, \tag{155.9}$$

the result takes the form

$$\frac{\delta\lambda}{\lambda} = \frac{R_2-R_1}{R_1}, \tag{155.10}$$

where R_1 is the radius of the model at the time the light leaves the source and R_2 is the radius at the time it arrives at the observer. This makes it clear that a red-shift in the light from distant objects would be correlated on the basis of this model with a general expansion of the model, and the consequent recession of the source from the observer.

This dependence of red-shift on recession can be made even clearer if we introduce, in accordance with the form of the line element, the total proper distances from observer to source

$$l_1 = e^{\frac{1}{2}g_1} \int_0^r \frac{dr}{1+r^2/4R_0^2} \quad \text{and} \quad l_2 = e^{\frac{1}{2}g_2} \int_0^r \frac{dr}{1+r^2/4R_0^2} \tag{155.11}$$

as they would be determined at times t_1 and t_2 by noting the number of metre sticks necessary to reach from the origin to the coordinate distance r. Introducing these expressions into (155.7), we can then write

$$\frac{\lambda+\delta\lambda}{\lambda} = \frac{l_2}{l_1} = 1+\frac{l_2-l_1}{l_1}, \qquad (155.12)$$

where (l_2-l_1) is the increase in proper distance from source to observer that takes place during the time taken by the light to travel from the one to the other. Since this time of travel in first approximation will equal in relativistic units the proper distance l_1, we can also write the last result in the approximate form

$$\frac{\lambda+\delta\lambda}{\lambda} \simeq 1+\frac{\delta l}{\delta t} \simeq 1+\frac{u}{c}, \qquad (155.13)$$

where u may be roughly regarded as the velocity of recession of the source.

In accordance with these different expressions, it will be seen on the basis of the present model that the red-shift in the light from the extragalactic nebulae is to be interpreted as due to a real motion of recession, and is to be assigned approximately the amount which would be calculated from the usual expression for the ordinary Doppler effect. It is to be emphasized, however, as a consequence of the homogeneity of the model, that there is nothing unique about the recession of the nebulae away from any particular (our own) location, and that similar red-shifts would be obtained by observers at rest with respect to the matter in other portions of the model.

156. Change in Doppler effect with distance

To investigate the change in the Doppler effect as we go to more distant sources (nebulae), we may differentiate our previous expression

$$\frac{\delta\lambda}{\lambda} = e^{\frac{1}{2}(g_2-g_1)}-1, \qquad (156.1)$$

with respect to the coordinate distance to the source r. In doing so we may regard g_2 as a constant, since this is the value of $g(t)$ at the time light is received at the origin, and we are actually interested in comparing the Doppler effects for different sources, which are all seen at the origin at the same given time t_2 which we can take as the present. On the other hand, g_1 must be regarded as a variable, since by going to greater coordinate distances r we shall have to go to earlier times of emission t_1 in order for the light to reach the origin at t_2. Hence on the differentiation of (156.1), we must write

$$\frac{d}{dr}\left(\frac{\delta\lambda}{\lambda}\right) = -\tfrac{1}{2}e^{\frac{1}{2}(g_2-g_1)}\frac{dg_1}{dt}\frac{dt}{dr}, \qquad (156.2)$$

where dt will be the change in time of emission corresponding to the change in position dr.

Noting, however, the expression for the radial velocity of light that would correspond to the line element

$$ds^2 = -\frac{e^{g(t)}}{[1+r^2/4R_0^2]^2}(dr^2 +r^2\,d\theta^2 +r^2\sin^2\theta\,d\phi^2)+dt^2, \quad (156.3)$$

which we are using, it is evident that dt and dr will be so connected that we can rewrite our expression (156.2) for the change in Doppler effect with distance in the form

$$\frac{d}{dr}\left(\frac{\delta\lambda}{\lambda}\right) = \frac{e^{\frac{1}{2}g_2}}{1+r^2/4R_0^2}\frac{\dot g_1}{2}, \qquad (156.4)$$

where $\dot g_1$ is the rate of change in $g(t)$ at the time the light is emitted.

For our later purposes, it will also be convenient to have this result in the form which it assumes when we use the alternative expression (149.5) for the line element

$$ds^2 = -e^{g(t)}\left(\frac{d\bar r^2}{1-\bar r^2/R_0^2}+\bar r^2\,d\theta^2 +\bar r^2\sin^2\theta\,d\phi^2\right)+dt^2, \quad (156.5)$$

obtained in § 149 by transforming from r to $\bar r$ with the help of (149.4). With this form for the line element it is evident that the change in Doppler effect with coordinate distance will be given by

$$\frac{d}{d\bar r}\left(\frac{\delta\lambda}{\lambda}\right) = \frac{e^{\frac{1}{2}g_2}}{\surd(1-\bar r^2/R_0^2)}\frac{\dot g_1}{2}. \qquad (156.6)$$

Since we shall later find both r and $\bar r$ in first approximation to be proportional to astronomically measured distances, these formulae indicate an approximately linear relation between red-shift and distance, until we go out to distances where the change in $\dot g_1$ becomes important. A more complete discussion of the change in Doppler effect with distance will be made possible in Part IV of the present chapter where we shall express $g(t)$ as a series in t.

To conclude this somewhat lengthy consideration of the Doppler effect in expanding or contracting cosmological models, it will also be useful for some purposes to show that the wave-length or frequency associated with any individual light quantum or photon as measured by observers lying along its path would be changing in a definite manner with the time, provided these observers are at rest with respect to our coordinates r, θ, and ϕ. To see this we may return to our exact formula (155.6) for the wave-length of light $(\lambda+\delta\lambda)$ leaving a source at time t_1 having any arbitrary position and motion, as finally measured at time t_2 by an observer at rest in the coordinate system. If now we consider different such observers lying along the path of the photon, it is evident that the only quantity which will be changed in this formula as later and later observers examine the photon will be the quantity g_2, which is the value of $g(t)$ at the time of observation. Hence taking a logarithmic differentiation of (155.6) with respect to the time we can write for such observers

$$\frac{d\log(\lambda+\delta\lambda)}{dt} = \frac{1}{2}\frac{dg_2}{dt}, \qquad (156.7)$$

which by changing to frequencies can be expressed in the more convenient form

$$-\frac{1}{\nu}\frac{d\nu}{dt} = \frac{1}{2}\frac{dg}{dt}, \qquad (156.8)$$

where ν is the frequency of any photon as measured by observers at rest with respect to the coordinate system r, θ, and ϕ, and g is the value of $g(t)$ for the model as a whole at the time of interest.

157. General discussion of dependence on time for closed models

Our derivation of the non-static line element

$$ds^2 = -\frac{e^{g(t)}}{[1+r^2/4R_0^2]^2}(dr^2+r^2\,d\theta^2+r^2\sin^2\theta\,d\phi^2)+dt^2$$

for homogeneous cosmological models placed no immediate restrictions on the behaviour of these models as a function of the time, and we must now turn to the discussion of the form of the hitherto undetermined function $g(t)$. This we shall undertake by several different lines of attack.

In accordance with equations (150.7) and (150.8), the pressure and density of the fluid taken as filling the model are definite functions of $g(t)$ and its derivatives. Hence the time behaviour of the model can be regarded as determined by the properties of this fluid. In the present section on closed models of the universe with R_0 real, and in the next section on open models with R_0 infinite or imaginary, we shall give a general discussion of the different possible types of time behaviour that could occur if we impose only very general restrictions on the properties of the fluid, such as the requirement that its pressure and density could never assume negative values. In the following sections of this part of the present chapter, we shall then discuss the time behaviour, making more specific assumptions as to the nature of the model or the fluid filling it. In Part III of the present chapter, we shall turn to the thermodynamic aspects of the changes that could take place in cosmological models with time. And finally in Part IV, in connexion with the correlation of actual observational data, we shall have occasion to treat the time dependence by the more phenomenological method of expressing $g(t)$ as a power series in t, with coefficients which are to be determined as far as possible from actual knowledge as to red-shift and as to pressure and density in the universe.

(a) **General features of time dependence, R real, $\rho_{00} \geqslant 0$, $p_0 \geqslant 0$.** We shall commence our discussion of time dependence† by assuming a closed model with R_0 real, and by assuming a fluid filling the model which cannot withstand tension, so that the density ρ_{00} and pressure p_0 can on physical grounds only be zero or positive.

As the two equations which relate the density and pressure of this fluid to $g(t)$, it will be most convenient to take (151.6) and (150.8)

$$\frac{d}{dt}(\rho_{00}\,e^{\frac{3}{2}g(t)}) + p_0 \frac{d}{dt}(e^{\frac{3}{2}g(t)}) = 0,$$

and
$$8\pi\rho_{00} = \frac{3}{R_0^2}e^{-g(t)} + \frac{3}{4}\left(\frac{dg}{dt}\right)^2 - \Lambda, \tag{157.1}$$

these being equivalent to the information originally given as to p_0 and ρ_{00} by (150.7, 8). Also for simplicity of expression, it will be convenient if we re-express these equations by introducing the radius of the model

$$R = R_0\,e^{\frac{1}{2}g(t)}. \tag{157.2}$$

Doing so, the first of these two expressions—the energy equation—can be rewritten in the form

$$\frac{d}{dt}(\rho_{00}\,R^3) + p_0 \frac{d(R^3)}{dt} = 0, \tag{157.3}$$

which gives
$$\frac{d(\rho_{00}\,R^3)}{dR} = -3p_0\,R^2 \tag{157.4}$$

and
$$\frac{d\rho_{00}}{dR} = -\frac{3(\rho_{00}+p_0)}{R}. \tag{157.5}$$

† The treatment given in this and the next section closely follows that of Robertson, *Reviews of Modern Physics*, **5**, 62 (1933).

Thus in accordance with our assumptions as to R, ρ_{00}, and p_0, the quantities $(\rho_{00} R^3)$ and ρ_{00} could only decrease or remain constant as R increases. Furthermore, in accordance with (157.5) it is evident that the density of the fluid would go to zero if the radius goes to infinity, and hence all ever-expanding models would finally have the properties of the original de Sitter model.

Introducing (157.2) into the second of our original equations (157.1), this reduces to a form

$$\left(\frac{dR}{dt}\right)^2 = \frac{8\pi\rho_{00} R^2}{3} + \frac{\Lambda R^2}{3} - 1,$$

or

$$\frac{dR}{dt} = \pm\sqrt{\left(\frac{8\pi\rho_{00} R^2}{3} + \frac{\Lambda R^2}{3} - 1\right)}, \qquad (157.6)$$

which conveniently expresses the rate of change of the radius of the model with time.

Since the quantity under the radical sign must necessarily be positive or zero, we are then led for any given value of the cosmological constant Λ, to the expression

$$\frac{3}{R^2} - 8\pi\rho_{00} \leqslant \Lambda \qquad (157.7)$$

as a necessary restriction on R if the behaviour of the model is to be real, and to

$$\frac{3}{R^2} - 8\pi\rho_{00} = \Lambda \qquad (157.8)$$

as the condition that the change in the radius R with time shall cease or reverse its direction.

(b) **Curve for the critical function of R.** In order to examine the behaviour of the critical quantity

$$Q = \left(\frac{3}{R^2} - 8\pi\rho_{00}\right) = \frac{1}{R^2}\left(3 - \frac{8\pi\rho_{00} R^3}{R}\right) \qquad (157.9)$$

as a function of the radius R, as this latter changes with the time in the case of any given model, it will be convenient to try to construct as nearly as possible a rough plot of $Q(R)$ against R.

Differentiating Q with respect to R and setting the result equal to zero we obtain

$$\frac{dQ}{dR} = -\frac{6}{R^3} - 8\pi\frac{d\rho_{00}}{dR} = 0,$$

or introducing (157.5)

$$\frac{dQ}{dR} = -\frac{6}{R^3} + \frac{24\pi(\rho_{00}+p_0)}{R} = 0, \qquad (157.10)$$

as a necessary condition for a maximum, minimum, or point of inflexion on the curve. And combining with (157.9), this gives

$$Q = \frac{1}{R^2} + 8\pi p_0 \quad (> 0) \tag{157.11}$$

as an equation for the value of Q itself when such a change in the curve takes place.

Differentiating a second time, we then obtain

$$\frac{d^2Q}{dR^2} = \frac{18}{R^4} - \frac{24\pi(\rho_{00}+p_0)}{R^2} + \frac{24\pi}{R}\frac{d\rho_{00}}{dR} + \frac{24\pi}{R}\frac{dp_0}{dr} \left\{ \begin{array}{l} < 0 \\ = 0, \\ > 0 \end{array} \right.$$

or introducing (157.10) and (157.5)

$$\frac{dp_0}{dR} \left\{ \begin{array}{l} < \\ = \\ > \end{array} \right\} \frac{\rho_{00}+p_0}{R}, \tag{157.12}$$

with the respective signs $(<)$, $(=)$, and $(>)$, as the further conditions, sufficient to distinguish the three cases of a maximum, point of inflexion, or minimum. This result shows that the curve can have no points of inflexion or minima unless we are willing to assume that the pressure of the fluid could be increasing during expansion.

With the help of these results, we can now make a rough plot as shown in Fig. 6, for $Q(R)$ as a function of R as this increases with the time.

The features of this plot, concerning which we have sufficient information to be sure, are shown by the full lines at A, B, and C. They can be justified as follows. (A) In accordance with (157.4) the quantity $(\rho_{00} R^3)$ can only decrease or remain constant as R increases. Hence, omitting the case of a completely empty model as not of present interest, it is evident from (157.9) that Q rises asymptotically from minus infinity at $R = 0$, and continues to increase as long as R increases without reaching any maximum or passing through any point of inflexion until after crossing the axis $Q = 0$, since by (157.11) such points can only occur for positive values of that quantity. (B) If R continues to increase, the curve must ultimately exhibit at least one maximum, since by (157.9) the quantity Q would ultimately have to decrease with R. (C) Finally, if R still continues to increase, the curve must ultimately approach $Q = 0$, asymptotically as $3/R^2$, as R goes to infinity.

The features of the plot, concerning which we do not have sufficient information to be sure, are shown by the dotted lines at a, b, and c. They consist in the possibility for points of inflexion as shown at a, and minima followed by later maxima as shown at b and c. In accordance with (157.11) such features could exist only in the range between

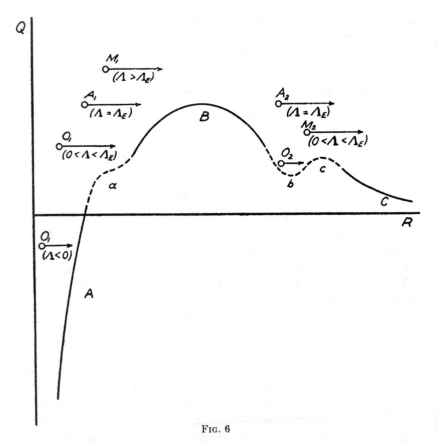

FIG. 6

$Q = 0$ and $Q = Q_{max}$, where Q_{max} is the highest maximum on the curve as shown at B. And in accordance with (157.12), they could exist then, only if the pressure of the fluid should be able at certain points to rise during expansion.

With the help of this plot we can now make some predictions as to the possible types of time behaviour for models, assuming different values for the cosmological constant Λ. During the course of an expansion, Λ, which has the same dimensions as the critical quantity Q, will stay constant as indicated by the horizontal lines in the figure.

Furthermore, in accordance with (157.7), the critical quantity

$$Q = (3/R^2 - 8\pi\rho_{00})$$

must be smaller than Λ during expansion, and the motion must cease or reverse when Q becomes equal to Λ. Hence the different types of motion, to the discussion of which we now turn, will correspond to horizontal lines $\Lambda = $ const. lying above the critical curve $Q(R)$, and the nature of the motion will be determined by the points where these horizontals intersect the critical curve.

(c) **Monotonic universes of type M_1 for $\Lambda > \Lambda_E$.** We shall denote the value of Q at the highest maximum on the curve as shown at B by Λ_E, since as we shall see later this would be the value of Λ for a conceivable static Einstein universe.

If the actual value of Λ is greater than Λ_E, the time behaviour can be qualitatively described without ambiguity. Since the line $\Lambda = $ const. then makes no intersections with the critical curve, the model would be of an ever-expanding type which proceeds from some singular state at $R_s \geqslant 0$ to the final state of an empty de Sitter universe as $R \to \infty$. If we consider the behaviour of the model at times earlier than that of the singular state, the motion would consist in a contraction from larger radii down to $R = R_s$. The present equations are not sufficient to describe the mechanism of passage through the singular state, and this may be regarded as occurring at a point at or in the neighbourhood of $R = 0$, where the idealizations involved in setting up the model are not suited for the treatment of that mechanism.

In accordance with the expression for rate of expansion given by (157.6), and the condition on $(\rho_{00} R^3)$ given by (157.4), it will be seen that such a model would leave a singular state at $R = 0$ with an infinite velocity. And by considering the integration of the above expression for rate of expansion, it will be seen that any finite value of R would then be reached in a finite time, but that an infinite time would elapse during the passage from R finite to R infinite.

Such a model, which expands without reversal from a singular state in the past to infinity in the future, we designate as a monotonic universe of the first kind, type M_1. As a model for the actual universe, it has the disadvantage of spending only an infinitesimal fraction of its total existence in a condition which differs appreciably from that of a completely empty de Sitter universe. Hence, if we are willing to take our observations on the actual universe as giving a fair sample of the kind of conditions that would be found anywhere and at any time, we should then rule out this model.

(d) **Asymptotic universe of types A_1 and A_2, for $\Lambda = \Lambda_E$.** We next turn to cases in which the cosmological constant is just equal to the value of Q at the maximum point of the curve as shown at B in Fig. 6. This value we have denoted as Λ_E. In accordance with (157.11), we then have

$$8\pi p_E = \Lambda_E - \frac{1}{R_E^2},\qquad(157.13)$$

where p_E and R_E are the pressure and radius at this point. And in accordance with (157.8), if we consider a static universe $(dR/dt = 0)$ with the above radius and cosmological constant, we should also have

$$8\pi \rho_E = \frac{3}{R_E^2} - \Lambda_E.\qquad(157.14)$$

These, however, are the conditions for pressure and density for a static Einstein universe of radius R_E and cosmological constant Λ_E, as given by (139.3) and (139.4). Hence a static Einstein universe could exist under the conditions corresponding to the maximum point of the curve. It would, nevertheless, be unstable as we shall later see.

With $\Lambda = \Lambda_E$, two types of behaviour for a non-static model would be possible.

The first type would be given by a model which starts expanding from a singular state at $R_s < R_E$, and asymptotically approaches the condition of a static Einstein universe at $R = R_E$, where in accordance with (157.13, 14) combined with (150.7, 8) the quantities dR/dt and d^2R/dt^2 would both become zero. Considering the behaviour of the model at times earlier than that of the singular state the behaviour would consist in a contraction from larger radii down to $R = R_s$. Such a model which expands from a singular state to a final static Einstein state, we designate as an asymptotic universe of the first kind, type A_1.

The second type of behaviour with $\Lambda = \Lambda_E$, would be given by a model which can be regarded as having asymptotically started from the static Einstein state at $R = R_E$ at an infinite time in the past and as expanding permanently in the future into the condition of an empty de Sitter universe. Such a model we designate as an asymptotic universe of the second kind, type A_2.

As models for the actual universe, both of these types have the same disadvantage as type M_1 of presumably spending only an inappreciable fraction of their total existence in a condition comparable to that which we find in the actual universe. Type A_2, however, has the advantage of apparently originating from a nonsingular state of finite volume at an infinite time in the past. This will be discussed more fully in § 159 on the stability of the static Einstein universe, and in § 161 specially devoted to these models.

(e) **Monotonic universes of type M_2 and oscillating universes of types O_1 and O_2, for $0 < \Lambda < \Lambda_E$.** We next consider cases in which Λ lies between zero and Λ_E. Here two different types of behaviour are definitely possible, and further types possible if the critical curve does have more than one maximum as indicated at b and c in Fig. 6.

As the first type of behaviour, we have those models which expand continuously into the future from some point on the critical curve at $R_1 > R_E$ past the maximum, where a reversal in the direction of motion from a preceding contracting phase takes place. Such a model, which has a true minimum finite radius and then expands without reversal to the state of an empty de Sitter universe, we designate as a monotonic universe of the second kind, type M_2. As a model for the actual universe, it again has the disadvantage of spending all but an infinitesimal fraction of its total existence in a condition unlike that which we observe.

As the second type of behaviour, we have models which expand from a singular state at $R_s < R_E$ to a maximum radius which lies on the critical curve where the direction of motion will reverse. The contraction thus initiated then continues, until expansion would again start at a singular state, which from physical considerations must at least be located at a radius which is not less than $R = 0$. Such a model we designate as an oscillating universe of the first kind, type O_1. As a model for the actual universe, it has the advantage of spending all its life in a condition where there is a finite density of matter, provided irreversible processes do not take place which alter the conditions for successive maxima (see § 175). It has, of course, the disadvantage of a singular state at the lower limit of contraction, through which the mechanism of passage is not described by the present equations.

In case we allow a second maximum on the critical curve lower than the highest maximum at R_E with an intervening minimum as shown in Fig. 6 at b and c, a very interesting new type of behaviour would be conceivable. This would arise with a value of Λ between the values of the critical quantity Q at this maximum and minimum, which would evidently permit an oscillation between a true minimum and maximum radius, where R assumes the critical values given by the curve for Q as a function of R. In the case of reversible behaviour this would give a strictly periodic motion without singular states. Such a model we designate as an oscillating universe of the second kind, type O_2. As a model for the actual universe it might at first sight seem to have great advantages, but as pointed out in connexion with (157.12), the necessary minimum on the critical curve could only occur if the pressure in the model could increase during expansion. We shall return later to the discussion of this matter in § 172, and shall have to conclude that such models would not be of great importance.

If a second maximum of the critical curve should exist, we should also evidently have possibilities for asymptotic universes of types A_1 and A_2, in the range $0 < \Lambda < \Lambda_E$, but these would be similar to those already discussed above in § 157 (d).

(f) **Oscillating universe of type O_1, for $\Lambda \leqslant 0$.** Finally, for the case of closed homogeneous models with the radius R real, we must consider the possibilities if the cosmological constant should lie in the range $\Lambda \leqslant 0$. Here it is immediately evident from Fig. 6, that the only possible kind of behaviour would be an oscillation of type O_1 back and forth between singular states at the lower limit which the radius reaches, and maxima of the radius which lie on the critical curve. As a model for the actual universe, this behaviour would have the advantages and disadvantages already mentioned above for type O_1.

In conclusion it should be specially emphasized that such an oscillatory behaviour of type O_1, is the only possibility for a closed homogeneous model with the cosmological constant Λ equal to zero. This is important, since $\Lambda = 0$ certainly seems the most reasonable assumption to make at the present time. In the first place the original argument, as discussed in § 139, for Einstein's addition of the logically permissible but otherwise surprising cosmological term to his original field equations in order to obtain a universe with a finite density of matter, now no longer exists in view of the wider possibilities presented by non-static models. In the second place, we have at the present time no accepted theory for any value at all for the cosmological constant, although interesting considerations concerning this matter have been presented by Eddington.† And in the third place, from the observational point of view we can at least say that the value of Λ must be small in order not to upset the application of relativistic theory to the orbits of the planets. Hence in what follows we shall lay special stress on the behaviour of models with the cosmological term omitted.

158. General discussion of dependence on time for open models

To complete our discussion we must also consider the behaviour in time for open models of the universe with R_0 imaginary or infinite. Here the possibilities for different kinds of behaviour are quite restricted.

We may again start with our previous equations for the dependence of density and pressure on the time (151.6) and (150.8)

$$\frac{d}{dt}(\rho_{00}\,e^{\frac{3}{2}g(t)})+p_0\frac{d}{dt}(e^{\frac{3}{2}g(t)})= 0, \tag{158.1}$$

and

$$8\pi\rho_{00} = \frac{3}{R_0^2}e^{-g(t)}+\frac{3}{4}\left(\frac{dg}{dt}\right)^2-\Lambda, \tag{158.2}$$

but now since the radius $R = R_0\,e^{\frac{1}{2}g(t)}$ would be an infinite or imaginary quantity without direct appeal to our physical intuition, there will be no advantage in introducing the radius of the model.

For our further purposes, the first of these equations may be re-expressed in the forms

$$\frac{d(\rho_{00}\,e^{\frac{3}{2}g})}{dg} = -\tfrac{3}{2}p_0\,e^{\frac{3}{2}g} \tag{158.3}$$

and

$$\frac{d\rho_{00}}{dg} = -\tfrac{3}{2}(\rho_{00}+p_0), \tag{158.4}$$

which show that $\rho_{00}\,e^{\frac{3}{2}g}$ and ρ_{00} are both quantities which could only decrease or remain constant as g increases, if we again introduce the assumption that the fluid in the model cannot withstand tension.

The second of our original equations (158.2) can be written in the form

$$\left(\frac{de^{\frac{1}{2}g}}{dt}\right)^2 = \frac{8\pi\rho_{00}}{3}e^g+\frac{\Lambda}{3}e^g-\frac{1}{R_0^2},$$

and since by the hypothesis of an open model R_0 is either infinite or

† Eddington, *The Expanding Universe*, Cambridge, 1933.

imaginary this can be re-expressed as

$$\frac{de^{\frac{1}{2}g}}{dt} = \pm \sqrt{\left(\frac{8\pi\rho_{00}\,e^g}{3} + \frac{\Lambda}{3}e^g + A^2\right)}, \qquad (158.5)$$

where A is a real quantity which would assume the value zero if R_0 is infinite.

Since the quantity under the radical sign must necessarily be positive or zero, we are then led to the expression

$$-3A^2e^{-g} - 8\pi\rho_{00} \leqslant \Lambda, \qquad (158.6)$$

as a necessary restriction on g if the behaviour of the model is to be real, and to

$$-3A^2e^{-g} - 8\pi\rho_{00} = 0, \qquad (158.7)$$

as the condition for a reversal in the direction of the rate of change of g with t.

We can also easily construct a plot of the critical quantity

$$Q = -3A^2e^{-g} - 8\pi\rho_{00} = -\frac{1}{e^g}\left[3A^2 + \frac{8\pi\rho_{00}\,e^{\frac{1}{2}g}}{e^{\frac{1}{2}g}}\right] \qquad (158.8)$$

as a function of $e^{\frac{1}{2}g}$ as shown in Fig. 7. In accordance with (158.8) Q is always negative, asymptotically approaching the value $Q = -\infty$ as $e^{\frac{1}{2}g}$ goes to zero as a result of (158.3), and asymptotically approaching the value $Q = 0$ as $e^{\frac{1}{2}g}$ goes to infinity, without any maxima, minima, or points of inflexion.

With the help of this plot of the critical curve, we then readily see that only two kinds of behaviour would now be possible. The first would occur with $\Lambda \geqslant 0$, and would consist in the monotonic increase of $e^{\frac{1}{2}g}$ from a singular state to infinity, giving us a universe of the type previously labelled M_1 which ultimately goes over into an empty de Sitter world, including the possibility of a Euclidean space with $\Lambda = 0$. The second type of behaviour would occur with $\Lambda < 0$, and would consist in the oscillation of $e^{\frac{1}{2}g}$ from a singular state to a maximum and return, giving us a universe of the type previously labelled O_1.

In treating the previous case of closed universes, it simplified the form of statement to describe the behaviour of the radius of the universe $R = R_0\,e^{\frac{1}{2}g(t)}$. In the present case of open universes, however, it seemed simpler to speak of the behaviour of $e^{\frac{1}{2}g(t)}$ itself, since R_0 would be infinite or imaginary. In both cases, nevertheless, it should be noted that the proper volume of any given element of fluid in the homogeneous model—as measured by a local observer—would always be proportional to $e^{\frac{1}{2}g(t)}$. Hence the changes, which we found to take place in the above quantities with time, can be immediately interpreted in terms of the expansion and contraction of the fluid filling the model, both for the case of closed models having a finite total

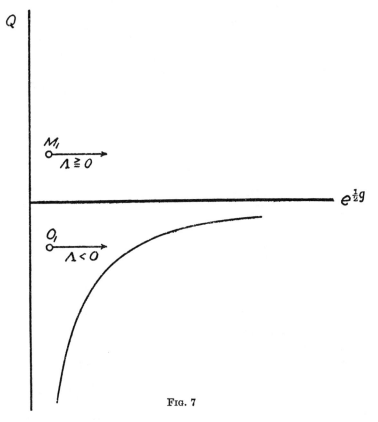

FIG. 7

proper volume and for the case of open models having an infinite total proper volume.

159. On the instability of the Einstein static universe

We may now turn to a number of specific treatments which have been given to the behaviour of homogeneous cosmological models with time. It will first be of interest to investigate the stability of the original Einstein static model with the help of our present knowledge of the behaviour of non-static models.

We may first look at the stability of the Einstein universe from the point of view of the plot which we have given in Fig. 6 for the critical quantity $Q(R)$ as a function of the radius R. In accordance with the consideration which we have given to this critical curve, it is evident that the conditions for a static Einstein universe would correspond to a maximum, minimum, or point of inflexion on this curve, the radius of the universe being equal to the value of R at that point and the cosmological constant being equal to the value of $Q(R)$ at the point. This is immediately seen from equations (157.11) for the value of Q at such a point, and (157.8) for the condition that the radius shall not be changing with time, which give us our previous conditions

$$8\pi p_0 = -\frac{1}{R^2} + \Lambda$$

and

$$8\pi \rho_{00} = \frac{3}{R^2} - \Lambda$$

for the pressure and density in a static Einstein universe.

With the help of Fig. 6, we then immediately see that a static Einstein universe corresponding to a maximum point on the critical curve would be unstable, since the radius would continue to change in the same direction if the model once started to expand or contract. Also for a model corresponding to a point of inflexion, we should have instability, since there would be one direction in which the radius could change without crossing the critical curve.

On the other hand, for a static Einstein universe corresponding to a minimum on the critical curve, we should evidently have stability, since the radius could not change at all without crossing the critical curve. This latter possibility could not be realized, however, as shown by (157.12) unless the pressure of the fluid in the model should increase with expansion. And on physical grounds we should not expect to find this for any actual fluid in an equilibrium condition. Hence we may conclude in general that a static Einstein universe would be in unstable equilibrium against changes in radius, and if it once started to expand or contract it would continue in such motion.

We may also inquire into what change might occur in a static Einstein model that would initiate an expansion or contraction away from the state of rest. In the case of a fluid filling the model whose pressure would be decreased by expansion and increased by contraction, this can easily be found with the help of our general expression for the pressure in a homogeneous non-static model,

$$8\pi p_0 = -\frac{1}{R_0^2}e^{-g}-\ddot{g}-\tfrac{3}{4}\dot{g}^2+\Lambda. \qquad (159.1)$$

Assuming the model originally in the state of a static Einstein universe, this expression must reduce at that time to the usual equation for the pressure in such a model

$$8\pi p_0 = -\frac{1}{R_0^2}e^{-g}+\Lambda, \qquad (159.2)$$

with $\qquad \ddot{g} = 0 \qquad$ and $\qquad \dot{g} = 0 \qquad$ (159.3)

and $R_0 e^{\frac{1}{2}g}$ equal to the prescribed radius for the static Einstein model. Hence if we should now suppose some process to take place in this momentarily non-expanding and non-contracting model which led to a change in pressure with the time, we could write in accordance with (159.1) and (159.3),

$$-8\pi\frac{dp_0}{dt} = \frac{d^3g}{dt^3} \qquad (159.4)$$

as an expression for the effect of the changing pressure on $g(t)$.

As a consequence of this equation, we now see that the initiation of any process in such a model involving decrease in pressure would also initiate an expansion which would then continue, since by hypothesis the expansion itself would lead to still further decreases in pressure. Vice versa, an increase in pressure would initiate a contraction.

Hence, if we had a static Einstein universe, and free radiation in it should commence to condense into matter, or freely moving particles in it should be captured by condensation, the model would start to expand.† Or, on the other hand, if the matter in it should commence to transform into radiation the model would start to contract. We may thus conclude, not only that a static Einstein universe would be in unstable equilibrium, but that processes are easily conceivable which would initiate a change away from the equilibrium value of its radius.

160. Models in which the amount of matter is constant

We may now consider in some detail the time behaviour of certain specific models which will be selected so as to illustrate different possibilities.

---

† The nature of such processes has been specially investigated by Lemaître, *Monthly Notices*, **91**, 490 (1931).

In the case of closed models, containing a mixture composed of a constant amount of incoherent matter (nebulae, dust) exerting negligible pressure, together with radiation exerting the pressure that corresponds to its density, a general expression for the radius as a function of time was first obtained by Lemaître.†

Since the pressure exerted by radiation is one-third its energy density, we can evidently write the energy equation (157.3) for these models in the form

$$\frac{d}{dt}[(\rho_m+3p_0)R^3]+p_0\frac{d}{dt}(R^3) = 0, \qquad (160.1)$$

where ρ_m is the density of matter. Since matter, however, is itself to be conserved this gives us

$$\rho_m R^3 = \text{const.} \quad \text{and} \quad p_0 R^4 = \text{const.},$$

and employing the symbols used by Lemaître we can then write

$$8\pi\rho_m = \frac{\alpha}{R^3}$$

$$8\pi p_0 = \frac{\beta}{R^4} \qquad (160.2)$$

$$8\pi\rho_{00} = 8\pi(\rho_m+3p_0) = \frac{\alpha}{R^3}+\frac{3\beta}{R^4},$$

where α and β are constants.

Introducing these expressions into our general equation (157.6), we then obtain

$$\frac{dR}{dt} = \pm\sqrt{\left(\frac{\Lambda R^2}{3}-1+\frac{\alpha}{3R}+\frac{\beta}{R^2}\right)} \qquad (160.3)$$

as an explicit expression for the radius as a function of time. This result applies to models in which matter exerts negligible pressure and is conserved in amount. Putting $\beta = 0$, we obtain the special case originally investigated by Friedmann, in which the total pressure is zero and energy as well as matter is conserved.

The integration or quadrature of the above expression has been specially studied by de Sitter.‡ And a slightly more general expression, in which the pressure of matter is not neglected, and in which explicit allowance is made for the case of open as well as closed models has been studied by Heckmann.§

† Lemaître, *Ann. Soc. Sci. Bruxelles*, **47 A**, 49 (1927).
‡ de Sitter, *Bull. Astron. Inst. Netherlands*, 5, 211 (1930); ibid., **6**, 141 (1931).
§ Heckmann, *Nachr. Ges. Wiss. Göttingen*, 1932, p. 97.

161. Models which expand from an original static state

In the case of models which can be regarded as expanding from the original state of a static Einstein universe, a direct integration of the foregoing expression for the change in radius with time can be obtained.

Combining the expressions for pressure and density given by (160.2) with the original expressions for pressure and density (139.3, 4) as found for a static Einstein universe, we can write for the special case of such models

$$8\pi p_E = \frac{\beta}{R_E^4} = -\frac{1}{R_E^2} + \Lambda \qquad (161.1)$$

and

$$8\pi\rho_E = \frac{\alpha}{R_E^3} + \frac{3\beta}{R_E^4} = \frac{3}{R_E^2} - \Lambda, \qquad (161.2)$$

where R_E is the radius in the original static state. Substituting these expressions into (160.3), we can obtain after considerable rearrangement the simple form

$$\frac{dR}{dt} = \frac{R - R_E}{\sqrt{3}R_E R} \Bigg/ \left\{ \left(1 + \frac{\beta}{R_E^2}\right)(R^2 + 2R_E R) + 3\beta \right\}, \qquad (161.3)$$

where R_E and β are the only parameters.

To prepare this expression for integration it is simplest to express R in terms of its increase over the original value R_E by substituting

$$R = R_E(1+x) \qquad x = \frac{R - R_E}{R_E}. \qquad (161.4)$$

Doing so we can then rewrite (161.3) in a form suitable for evaluation

$$\int dt = \frac{\sqrt{3}R_E}{\sqrt{(1 + \beta/R_E^2)}} \int \frac{(x+1)\,dx}{x\sqrt{X}}, \qquad (161.5)$$

where we shall use as abbreviations

$$\sqrt{X} = \sqrt{(x^2 + 4x + C^2)} \qquad (161.6)$$

and

$$C^2 = 3\left(1 + \frac{\beta/R_E^2}{1 + \beta/R_E^2}\right). \qquad (161.7)$$

Integrating (161.5), we then obtain

$$t = \frac{\sqrt{3}R_E}{\sqrt{(1 + \beta/R_E^2)}} \left[\log(x + \sqrt{X} + 2) + \frac{1}{C} \log\frac{x + \sqrt{X} - C}{x + \sqrt{X} + C} \right] + \text{const.}, \qquad (161.8)$$

as a definite expression for x and hence also for the radius $R = R_0 e^{ig(t)}$ as a function of the time.

Since the second term in this expression becomes minus infinity when $x = 0$, and the first term becomes plus infinity when $x = \infty$, it will be seen that the model would expand from the original static Einstein state with $R = R_E$ at $t = -\infty$ to the final empty de Sitter state at $t = +\infty$, both of these states being approached asymptotically.

The effect of the pressure of radiation on the rate of expansion is given by the appearance of β/R_E^2 in the above expressions. It can readily be seen, however, that the effect of pressure must in any case be small, since we can write in accordance with (161.1) and (161.2),

$$\frac{\beta}{R_E^2} = \frac{2p_E}{\rho_E + p_E}, \tag{161.9}$$

where p_E and ρ_E are the pressure and density in the original static Einstein state. Hence β/R_E^2 can in any case only vary between 0 for a model containing matter without radiation to $\frac{1}{2}$ for a model containing nothing but radiation. And we may conclude from the way in which this term enters the above expressions, that the course of the expansion will be primarily determined only by the radius R_E of the original static state.

This matter was specially investigated by de Sitter (loc. cit) who compared the time behaviour of the two models given by

$$\beta = 0 \quad \text{and} \quad \frac{\beta}{R_E^2} = \frac{1}{2} \tag{161.10}$$

for the respective cases of no radiation and no matter present. For the first of these cases equation (161.8) reduces to

$$t = \sqrt{3}R_E\left[\log(x+\sqrt{X}+2)+\frac{1}{\sqrt{3}}\log\frac{x+\sqrt{X}-\sqrt{3}}{x+\sqrt{X}+\sqrt{3}}\right]+\text{const.,} \tag{161.11}$$

and for the second to the much simpler form

$$t = \frac{R_E}{\sqrt{2}}\log 4(x^2+2x)+\text{const.} = \frac{R_E}{\sqrt{2}}\log(R^2-R_E^2)+\text{const.} \tag{161.12}$$

For a given value of R_E, however, the two expressions give very similar histories of expansion as shown by curves I and VII in Fig. 8, taken from de Sitter's article. We may hence conclude, that the case with zero pressure specially studied by Lemaître in 1927 as a model for the actual universe is sufficiently representative of the class.

Universes which expand from an original static Einstein state have

sometimes been favoured by cosmologists, since the equations then
lead to no singular states and the models appear to offer an infinite
time scale for past cosmological processes. More recently, however,

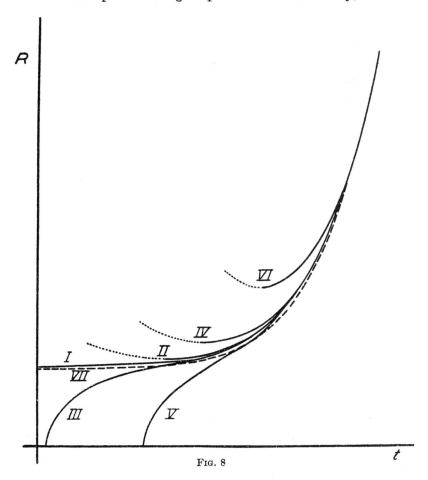

FIG. 8

as emphasized by Eddington† and others‡ it has been felt that the
logarithmic infinity for past time provided by these models was
likely to have no real physical significance, in view of the unstable
character of the static Einstein state, which we have investigated
in § 159.

† Eddington, *Monthly Notices*, **90**, 668 (1930).
‡ McCrea and McVittie, *Monthly Notices*, **91**, 128 (1930); ibid. **92**, 7 (1931).
Lemaître, *Monthly Notices*, **91**, 490 (1931).

162. Ever-expanding models which do not start from a static state

Turning next to the case of ever-expanding models which do not start from an original static state, it is not in general possible to obtain a simple integral for the relation between radius and time, given by (160.3) for the case in which matter is conserved. By numerical quadrature, however, de Sitter has calculated the dependence of radius on time for a number of such models as shown in Fig. 8.

The curves in this figure may be divided into four groups. Curves I and VII are the cases already mentioned for a model containing matter without radiation and a model containing radiation without matter, which start from an original static state. Curves II and IV are for models of type M_2 discussed in § 157 (e), which first contract to a true minimum radius having a value lying on the critical curve, and then expand monotonically to the empty de Sitter state. Curves III and V are for models of type M_1 discussed in § 157 (c), which expand monotonically from a singular state, here taken as located at $R = 0$. Finally curve VI is calculated for the limiting case of an entirely empty universe.

From the point of view of a representation for the actual universe, it will be noted that monotonic universes of the first kind, which expand from a singular state, might offer some advantages in providing a reasonably long time scale subsequent to the singular state. Lemaître† has more recently advocated such models and has picturesquely described the original singular state as that of a giant atom.

163. Oscillating models ($\Lambda = 0$)

In view of the rationality already emphasized of taking the value zero for the unknown cosmological constant Λ, and entirely omitting the cosmological term from Einstein's field equations, we must pay special attention to oscillating models, which then become the only possibility for a closed universe. For two such models a simple treatment of the time behaviour can be given.

The first of these models was originally considered by Friedmann‡ as early as 1922, and has since been advocated by Einstein.§ The

† Lemaître, *Revue des questions scientifiques*, 1931, p. 391.
‡ Friedmann, *Zeits. f. Physik*, **10**, 377 (1922).
§ Einstein, *Berl. Ber.* 1931, p. 235.

fluid in the model is taken as being incoherent matter which is conserved in amount and exerts negligible pressure. Referring to equations (160.2), we immediately see that the radius for such models will be given as a function of the time by setting Λ and β equal to zero in (160.3), and writing

$$\frac{dR}{dt} = \pm \sqrt{\left(\frac{\alpha}{3R} - 1\right)}, \qquad (163.1)$$

where α is a constant connected with the density of matter and radius by the equation

$$8\pi\rho_m R^3 = \alpha = \text{const.} \qquad (163.2)$$

The integral of this equation is readily seen to be a cycloid in the Rt-plane given by

$$R = \frac{\alpha}{6}(1 - \cos\psi) \qquad t = \frac{\alpha}{6}(\psi - \sin\psi), \qquad (163.3)$$

in accordance with which the radius oscillates between a singular state with $R = 0$ at $t = 0$ and a maximum of $R = \alpha/3$ at $t = \pi\alpha/6$.

A second closed model with $\Lambda = 0$, for which the time behaviour has been calculated† is obtained by taking the fluid as consisting solely of black-body radiation. Referring again to equations (160.2) and (160.3), we see that the radius for such models will be given as a function of time by

$$\frac{dR}{dt} = \pm \sqrt{\left(\frac{\beta}{R^2} - 1\right)}, \qquad (163.4)$$

where β is a constant connected with the radius and pressure of the radiation by the equation

$$8\pi p_0 R^4 = \beta = \text{const.} \qquad (163.5)$$

The integral of this equation is readily seen to be

$$R = \sqrt{(\beta - t^2)}, \qquad (163.6)$$

with the maximum of R falling at $t = 0$.

As shown in § 157, oscillating models can also be obtained with values of Λ other than zero, and a number of plots for the time behaviour of such models are given in Fig. 9, also taken from de Sitter (loc. cit). Curve IX represents the cycloid for the case $\Lambda = 0$ with the pressure zero. Curve VII is the limiting case $\Lambda = \Lambda_E$ of type A_1 which expands asymptotically to a static Einstein state, and separates the oscillating models of type O_1 from the ever expanding models of type M_1. The diagram also gives the time behaviour for

† Tolman, *Phys. Rev.* **38**, 1758 (1931).

several ever-expanding models plotted, however, with somewhat different units.

From the point of view of representing the actual universe, the oscillating models with $\Lambda = 0$ tend to have a short time scale from

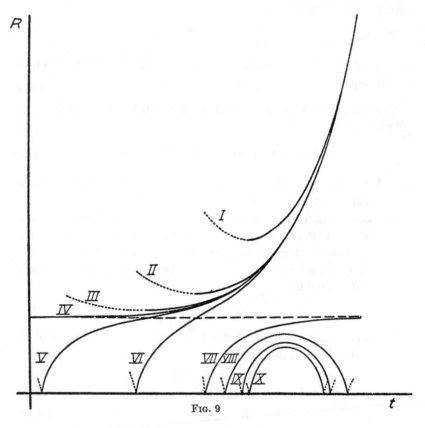

Fig. 9

the singular state, and further investigations would be necessary to describe the mechanism of passage through that state.

To investigate the time scale from the singular state, it is most convenient to start with the equation for pressure in its original form (150.7)

$$8\pi p_0 = -\frac{1}{R_0^2}e^{-g} - \ddot{g} - \tfrac{3}{4}\dot{g}^2 + \Lambda.$$

Setting $\Lambda = 0$ this can then be rewritten as

$$-\frac{\ddot{g}}{\dot{g}^2} = \frac{3}{4} + \frac{1}{R_0^2}\frac{e^{-g}}{\dot{g}^2} + \frac{8\pi p_0}{\dot{g}^2}, \tag{163.7}$$

so that we can in any case put

$$\frac{d}{dt}\left(\frac{1}{\dot g}\right) \geqslant \frac{3}{4}.$$

Integrating over the course of expansion from the singular state $(t_s, \dot g_s)$ to the present state $(t, \dot g)$, we obtain

$$t - t_s \leqslant \frac{4}{3}\left(\frac{1}{\dot g} - \frac{1}{\dot g_s}\right).$$

Hence for oscillating models with $\Lambda = 0$, we can in any case write

$$\Delta t \leqslant \frac{4}{3\dot g}, \tag{163.8}$$

where Δt is the elapsed time since the singular state, and $\dot g$ is the present value of that quantity, as could be determined for the actual universe from observations on the red-shift.

164. The open model of Einstein and de Sitter ($\Lambda = 0$, $R_0 = \infty$)

Mathematically the simplest of all models can be obtained by taking the cosmological constant Λ equal to zero, and setting the constant R_0 in the fundamental expression for the line element (149.1) equal to infinity

$$\Lambda = 0 \qquad R_0 = \infty, \tag{164.1}$$

as has been proposed by Einstein and de Sitter.†

The line element can then be written in the form

$$ds^2 = -e^{g(t)}(dr^2 + r^2\,d\theta^2 + r^2\sin^2\theta\,d\phi^2) + dt^2, \tag{164.2}$$

or also as $\qquad ds^2 = -e^{g(t)}(dx^2 + dy^2 + dz^2) + dt^2$

and space-time becomes spatially flat and spatially infinite in extent.

Furthermore, substituting (164.1) into the general expressions for pressure and density (150.7) and (150.8), these then reduce to the very simple form

$$8\pi p_0 = -\ddot g - \tfrac{3}{4}\dot g^2, \tag{164.3}$$

and $\qquad\qquad\qquad 8\pi \rho_{00} = \tfrac{3}{4}\dot g^2. \tag{164.4}$

The first of these equations requires that the acceleration $\ddot g$ always be negative to prevent negative pressures. The second of the equations provides an immediate relation between density and Doppler effect. For the two limiting cases of a fluid consisting solely of matter

---

† Einstein and de Sitter, *Proc. Nat. Acad.* **18**, 213 (1932).

exerting a negligible pressure, or a fluid consisting solely of radiation exerting the pressure corresponding to its density, these equations can be immediately integrated.

For the case of matter having zero pressure we can write from (164.3)

$$\frac{d^2g}{dt^2} + \frac{3}{4}\left(\frac{dg}{dt}\right)^2 = 0, \tag{164.5}$$

which has the integral

$$e^{\frac{3}{2}g} = at + b, \tag{164.6}$$

where a and b are constants, and by combining with (164.4) we obtain

$$a = \sqrt{(6\pi\rho_{00} e^{\frac{3}{2}g})}, \tag{164.7}$$

as the value of the first of these constants.

For the case of radiation, having a pressure equal to one-third its energy density, we obtain by combining (164.3) and (164.4) the equation

$$\frac{d^2g}{dt^2} + \left(\frac{dg}{dt}\right)^2 = 0, \tag{164.8}$$

which has the integral

$$e^g = at + b, \tag{164.9}$$

where the constant a this time has the value

$$a = \sqrt{\left(\frac{32\pi}{3}\rho_{00} e^{2g}\right)}. \tag{164.10}$$

By the method employed at the end of the last section, the elapsed time from the singular state is also found for models of the Einstein-de Sitter type to be in any case as small as

$$\Delta t \leqslant \frac{4}{3\dot{g}}. \tag{164.11}$$

165. Discussion of factors which were neglected in studying special models

The special models, which we have used in the foregoing sections to illustrate the different possibilities of temporal behaviour, were purposely limited, for the sake of simplicity, by the assumptions that the pressure due to the particles of matter in the fluid could be neglected, and that the total mass of this matter should remain constant. It is evident at least in theory, however, that these assumptions were not necessary for a solution of the problem of behaviour in time, provided we have sufficient information as to the properties of the fluid filling the model.

In general, for the treatment of temporal behaviour, the expressions for pressure and density obtained in § 150, written most conveniently in the form

$$\frac{d}{dt}(\rho_{00}\,e^{\frac{3}{2}g}) + p_0\frac{d}{dt}(e^{\frac{3}{2}g}) = 0 \qquad (165.1)$$

and

$$\frac{d}{dt}(e^{\frac{1}{2}g}) = \pm\sqrt{\left(\frac{8\pi\rho_{00}\,e^g}{3} + \frac{\Lambda e^g}{3} - \frac{1}{R_0^2}\right)}, \qquad (165.2)$$

will give two of the necessary equations for determining ρ_{00}, p_0, and g as functions of the time, and the third equation will be provided by the nature of the fluid.

In case the fluid behaves reversibly, so that its pressure is a definite function of energy and volume, this third relation will be of the nature of an equation of state of the general form

$$p_0 = p_0(E, v) = p_0(\rho_{00}e^{\frac{3}{2}g},\, e^{\frac{3}{2}g}), \qquad (165.3)$$

where $e^{\frac{3}{2}g}$, as we have seen (151.4), is a quantity which is proportional to the proper volume of any given element of the fluid. In case the fluid behaves irreversibly, the third equation would have to contain time derivatives of $g(t)$ as well as $g(t)$ itself. In any case, however, with the help of the three equations and assumptions as to initial conditions and as to the values of the constants Λ and R_0, the problem should be theoretically soluble.

The fact that we have neglected in our illustrative examples any contribution to pressure due to the random motion of particles of matter in the model can hardly be regarded as immediately serious for the purposes of cosmology. In the case of the *present* state of the actual universe, we should certainly regard the random motions of the nebulae themselves as properly correlated with a negligible pressure for the idealized fluid filling our model, and should presumably also regard this as legitimate for the random motion of dust or other particles which may be present in intergalactic space.

The fact, however, that we have taken the total mass of the matter in the models as constant, deserves a little more attention, since this means when applied to the real universe that we are neglecting the actual flow of radiation from the nebulae into internebular space. In accordance with equation (152.7) we can write

$$-\frac{1}{M}\frac{dM}{dt} = \frac{3}{2}\left[\frac{\rho_{00}+\frac{5}{8}p_0}{\rho_m} - \frac{1}{4\pi\rho_m}\left(\ddot{g}+\frac{\dddot{g}}{\dot{g}}\right)\right]\dot{g} \qquad (165.4)$$

as an expression for the fractional rate at which the mass of the particles would be changing with time. And writing for simplicity

$$\gamma = 3\left[\frac{\rho_{00}+\frac{5}{8}p_0}{\rho_m} - \frac{1}{4\pi\rho_m}\left(\ddot{g}+\frac{\dddot{g}}{\dot{g}}\right)\right] \tag{165.5}$$

we can re-express the above equation in the form

$$-\frac{1}{M}\frac{dM}{dt} = -\frac{1}{(\rho_m e^{\frac{3}{2}g})}\frac{d}{dt}(\rho_m e^{\frac{3}{2}g}) = \gamma\frac{\dot{g}}{2}, \tag{165.6}$$

where ρ_m is the density of matter in the model.

Concerning the present value of γ in the actual universe, we do not have complete information since the cosmic rays have an unknown origin which may very likely involve a decrease in the mass of matter. If, however, we assume for all the matter in the universe a rate of fractional decrease the same as that for the sun we should obtain as small a value as

$$\gamma \simeq 10^{-4}, \tag{165.7}$$

and the true value may well be smaller yet. (de Sitter estimates $\gamma \simeq 2\times 10^{-7}$.)

Also concerning the rate of change of γ with time we have no knowledge. Nevertheless, over a reasonable time interval we may take γ as a constant. Doing so, we can then integrate equation (165.6) and obtain

$$\rho_m = \rho_1 e^{-\frac{1}{2}(3+\gamma)g} \tag{165.8}$$

as an expression for the density of matter in our model, where ρ_1 is a constant. Furthermore, for the density of radiation in the model we can write

$$\rho_r = 3p_0, \tag{165.9}$$

if we neglect the contribution of random particle motion to the total pressure p_0.

Substituting these two expressions into the energy equation (165.1) we shall then have

$$\frac{d}{dt}(\rho_1 e^{-\frac{1}{2}\gamma g}+3p_0 e^{\frac{3}{2}g})+p_0\frac{d}{dt}(e^{\frac{3}{2}g}) = 0,$$

and by performing the indicated differentiations, resubstituting from (165.8) and (165.9), and rearranging, we can obtain therefrom

$$\frac{1}{p_0}\frac{dp_0}{dt} = \frac{1}{\rho_r}\frac{d\rho_r}{dt} = -\left(2-\frac{\gamma}{2}\frac{\rho_m}{\rho_r}\right)\dot{g}, \tag{165.10}$$

as an expression for the logarithmic rate of change in the pressure and in the density of radiation.

This result is of interest in showing, for the case of an expanding model with $\ddot{g}$ positive, that the pressure and density of radiation would be decreasing with time except for large values of γ or ρ_m/ρ_r. Assigning to γ the value given by (165.7), the pressure of radiation would cease to build up as soon as the density of radiation compared with that of matter reached the value

$$\rho_r = 2{\cdot}5 \times 10^{-5}\rho_m. \tag{165.11}$$

Hence as first pointed out by de Sitter,[†] the theory of an expanding universe is capable of accounting for the apparent disappearance of the radiation which pours from the stars into internebular space. It will also be noted, when the pressure does decrease monotonically with expansion, that there can be no minima (see 157.12) on the critical curve, Fig. 6, and hence no chance for an oscillating behaviour of the second kind, type O_2.

As a final remark concerning the simplifications which were made in obtaining specific illustrations of the different varieties of temporal behaviour, it is to be noticed that the fluids filling the models were so chosen that their changes in density could be assumed to take place reversibly as the models themselves expanded or contracted, and the possibilities for irreversible behaviour were neglected. In Part III of the present chapter, to which we now turn, the differences between thermodynamically reversible and irreversible expansions and contractions of cosmological models will be specially considered.

[†] de Sitter, *Bull. Astron. Inst. Netherlands*, **5**, 211 (1930).

APPLICATIONS TO COSMOLOGY (*contd.*)

166. Application of the relativistic first law

We must now undertake a brief consideration of the thermo-dynamic behaviour of the homogeneous cosmological models which correspond to the line element

$$ds^2 = -\frac{e^{g(t)}}{[1+r^2/4R_0^2]^2}(dr^2+r^2\,d\theta^2+r^2\sin^2\theta\,d\phi^2)+dt^2. \quad (166.1)$$

In accordance with the principles of relativistic thermodynamics as developed in Chapter IX, we may regard the relativistic analogue of the first law of ordinary thermodynamics as provided by the principles of relativistic mechanics as expressed in the form of the energy-momentum equation

$$\frac{\partial \mathfrak{T}^\nu_\mu}{\partial x^\nu}-\tfrac{1}{2}\mathfrak{T}^{\alpha\beta}\frac{\partial g_{\alpha\beta}}{\partial x^\mu}=0. \quad (166.2)$$

And in applying this expression to the case of the above line element we may take the only surviving components of the energy-momentum tensor, in accordance with (150.5) and (150.6), as given by

$$T^{11}=-g^{11}p_0 \qquad T^{22}=-g^{22}p_0 \qquad T^{33}=-g^{33}p_0 \qquad T^{44}=\rho_{00},$$

or, on lowering indices, $\quad(166.3)$

$$T^1_1=T^2_2=T^3_3=-p_0 \qquad T^4_4=\rho_{00},$$

where ρ_{00} and p_0 are the proper macroscopic density and pressure of the fluid as they would be measured by a local observer at rest therein at the position and instant of interest.

Substituting (166.3) into (166.2), we obtain, for the case $\mu=1$,

$$-\frac{\partial}{\partial r}(p_0\sqrt{-g})+\tfrac{1}{2}p_0\sqrt{-g}\left(g^{11}\frac{\partial g_{11}}{\partial r}+g^{22}\frac{\partial g_{22}}{\partial r}+g^{33}\frac{\partial g_{33}}{\partial r}+g^{44}\frac{\partial g_{44}}{\partial r}\right)=0,$$

where the last term in the parenthesis can be added on account of the constancy of g_{44}. As a result of equation (39) in Appendix III, however, this immediately reduces to

$$-\sqrt{-g}\frac{\partial p_0}{\partial r}-p_0\frac{\partial\sqrt{-g}}{\partial r}+p_0\frac{\partial\sqrt{-g}}{\partial r}=0.$$

And since similar expressions result from taking μ as 2 or 3, the only information that we obtain, by applying the energy-momentum equation to our present line element for the cases $\mu = 1, 2, 3$ is the independence of pressure on position

$$\frac{\partial p_0}{\partial r} = \frac{\partial p_0}{\partial \theta} = \frac{\partial p_0}{\partial \phi} = 0, \tag{166.4}$$

which is already evident from the known spatial homogeneity of the model.

Substituting (166.3) in (166.2) for the case $\mu = 4$, however, we obtain as already seen in § 151, the equation

$$\frac{\partial}{\partial t}(\rho_{00} \sqrt{-g}) + \tfrac{1}{2} p_0 \sqrt{-g} \left(g^{11} \frac{\partial g_{11}}{\partial t} + g^{22} \frac{\partial g_{22}}{\partial t} + g^{33} \frac{\partial g_{33}}{\partial t} \right) = 0,$$

and by inserting the expressions for the $g_{\mu\nu}$ given by the line element, this reduces to the important result

$$\frac{\partial}{\partial t}\left(\rho_{00} \frac{r^2 \sin \theta \, e^{\frac{3}{2}g(t)}}{[1 + r^2/4R_0^2]^3} \right) + p_0 \frac{\partial}{\partial t}\left(\frac{r^2 \sin \theta \, e^{\frac{3}{2}g(t)}}{[1 + r^2/4R_0^2]^3} \right) = 0. \tag{166.5}$$

Noting that the proper volume of any element of fluid, permanently located in the case of the present co-moving coordinates in any desired range $\delta r \delta \theta \delta \phi$, would be given by

$$\delta v_0 = \frac{r^2 \sin \theta \, e^{\frac{3}{2}g(t)}}{[1 + r^2/4R_0^2]^3} \, \delta r \delta \theta \delta \phi, \tag{166.6}$$

we can also rewrite (166.5) in the form

$$\frac{d}{dt}(\rho_{00} \, \delta v_0) + p_0 \frac{d}{dt}(\delta v_0) = 0. \tag{166.7}$$

This equation shows—as previously remarked—that the proper energy of each element of the fluid in the model as measured by a local observer would change with the proper volume of the element, in accordance with the ordinary equation for the adiabatic expansion or compression of the fluid.

The result is thermodynamically important since it shows that there will be no heat flow into or out of the elements of fluid composing the model. This conclusion may also be regarded as a consequence of the spatial homogeneity of the model.

167. Application of the relativistic second law

In accordance with the principles of relativistic thermodynamics, the analogue of the ordinary second law of thermodynamics as discussed in § 119 can be taken as given by the expression

$$\frac{\partial}{\partial x^\mu}\left(\phi_0 \frac{dx^\mu}{ds}\sqrt{-g}\right)\delta x^1 \delta x^2 \delta x^3 \delta x^4 \geqslant \frac{\delta Q_0}{T_0}, \qquad (167.1)$$

where ϕ_0 is the proper entropy density of the fluid at the position and instant of interest, the quantities dx^μ/ds are the components of the macroscopic 'velocity' of this fluid referred to the coordinates in use, T_0 is the proper temperature, and δQ_0 is the heat flowing into the element of fluid and during the time denoted by $\delta x^1 \delta x^2 \delta x^3 \delta x^4$ as measured by the local observer. The sign of equality in this expression refers to reversible processes and the sign of inequality to irreversible processes.

In applying this expression to the models under consideration using coordinates corresponding to the line element in the form (166.1), we can take

$$\frac{dr}{ds}=\frac{d\theta}{ds}=\frac{d\phi}{ds}=0 \qquad \frac{dt}{ds}=1, \qquad (167.2)$$

owing to the co-moving character of the coordinates, and can set

$$\delta Q_0 = 0, \qquad (167.3)$$

owing to the adiabatic character of the changes demonstrated in the preceding section. Substituting in (167.1), we can then write the relativistic second law for these models in the form

$$\frac{d}{dt}\left(\phi_0 \frac{r^2\sin\theta\, e^{\frac{3}{2}g(t)}}{[1+r^2/4R_0^2]^3}\,\delta r\delta\theta\delta\phi\right)\geqslant 0, \qquad (167.4)$$

and on substituting the expression for proper volume given by (166.6), this can be written in the form

$$\frac{d}{dt}(\phi_0\,\delta v_0)\geqslant 0, \qquad (167.5)$$

which shows that the proper entropy for each element of fluid in the model can only increase or at best remain constant as time proceeds.

With the help of the two relations (166.7) and (167.5), we thus obtain the very satisfactory result that a local observer who examines an element of fluid in his immediate neighbourhood would find therefor the same behaviour as would be predicted from the classical principles of thermodynamics for an element of fluid undergoing an adiabatic expansion or contraction.

168. The conditions for thermodynamic equilibrium in a static Einstein universe

Since we have seen that the static Einstein universe can be regarded as a special case given by our non-static models when $g(t)$ becomes constant, we can now use the foregoing information as to the thermodynamic behaviour of non-static models to investigate the conditions for equilibrium in the original Einstein model.

We can investigate the conditions for a state of thermodynamic equilibrium in the usual manner, by considering the possibilities for change to a neighbouring state of the model, by varying the radius of the model

$$R = R_0 e^{\frac{1}{2}g}, \tag{168.1}$$

and the number of mols

$$n_1^0, n_2^0, ..., n_n^0, \tag{168.2}$$

of the different chemical constituents which would give the composition of any selected element of fluid and hence of the model as a whole.

During the progress of such a variation, the model could be regarded as temporarily non-static with the energy and entropy of each element of the fluid subject to our previous relations (166.7) and (167.5). Hence since by (167.5) the entropy of each element of the fluid can only remain constant or increase with time, we can take

$$\delta(\phi_0 v_0) = 0, \tag{168.3}$$

under the subsidiary condition

$$\delta(\rho_{00} v_0) + p_0 \delta v_0 = 0, \tag{168.4}$$

as the necessary requirement for thermodynamic equilibrium, where we have now written

$$v_0 = \frac{r^2 \sin \theta \, e^{\frac{1}{2}g}}{[1 + r^2/4R_0^2]^3} \, \delta r \delta \theta \delta \phi, \tag{168.5}$$

as the proper volume of the particular element of the fluid considered.

To use the above conditions for equilibrium, we can evidently write in accordance with the classical thermodynamics

$$\delta(\phi_0 v_0) = \delta S_0 = \frac{1}{T_0} \, \delta E_0 + \frac{p_0}{T_0} \, \delta v_0 + \left(\frac{\partial S_0}{\partial n_1^0}\right)_{E_0, v_0} \delta n_1^0 + ... + \left(\frac{\partial S_0}{\partial n_n^0}\right)_{E_0, v_0} \delta n_n^0, \tag{168.6}$$

since the proper entropy $S_0 = \phi_0 v_0$ as measured by a local observer will evidently depend in the classical manner on proper energy, volume, and composition. Hence since we have $E_0 = \rho_{00} v_0$ for the proper energy, we obtain by combining (168.4) and (168.6)

$$\sum_i \left(\frac{\partial S_0}{\partial n_i^0}\right) \delta n_i^0 = 0, \qquad (168.7)$$

as a necessary condition for thermodynamic equilibrium in a static Einstein universe.

This result is of interest since by comparsion with (60.12), we see that it gives the classical condition for chemical equilibrium between different substances in the fluid. Hence the relative proportions between different materials which might be able to change into each other, for example hydrogen and helium, or indeed matter and radiation, would have the same values at thermodynamic equilibrium in a static Einstein universe as we should calculate for flat space-time. This is important since any effect of the gravitational curvature in the models on such ratios could have been very important for cosmology.†

Although the pair of relations (168.3) and (168.4), or the equivalent pair (168.4) and (168.7), can be taken as necessary conditions for thermodynamic equilibrium, it is of course evident that further investigation is necessary to determine whether they are sufficient conditions for the equilibrium to be stable. And the investigations of § 159 have actually shown that the equilibrium state for an Einstein universe would in general be unstable towards small variations of the radius, unless indeed we could have a fluid whose pressure would increase on expansion.

169. The conditions for reversible and irreversible changes in non-static models

With the help of our expression (167.5) for the second law as applied to homogeneous cosmological models.

$$\frac{d}{dt}(\phi_0 \delta v_0) \geqslant 0, \qquad (169.1)$$

we can readily distinguish between the characteristics of reversible and irreversible changes in such models.

For the case of reversible processes, we shall have to use the equality sign in this expression, and can thus take constant proper entropy

---

† Such an effect was originally supposed to be present by Lenz, *Phys. Zeits.* **27**, 642 (1926). See, however, Tolman, *Proc. Nat. Acad.* **14**, 353 (1928) and ibid. **17**, 153 (1931).

for each element of fluid in the model as the criterion of reversibility. Hence to investigate the possibility for reversible changes in the model we must examine the causes which could lead to an increase in the entropy of an element of the fluid.

In doing so, we note—as already pointed out in § 130—in the first place that no entropy increases could occur as a result of irreversible heat flow, since we have seen, from our application of the first law and also from the homogeneity of the model, that there is no heat flow in these models from one portion of the fluid to another. In the second place, we note that no entropy increase could occur from the friction of moving members against the walls of any container for the fluid, as in familiar examples of adiabatic changes in volume, since now no such parts or container are involved. In the third place, we note that no entropy increases could result from an inability of the fluid to maintain the same pressure in the interior and at the boundary of any element of fluid, as in ordinary cases of expansion or compression in a cylinder where a pressure gradient is set up by the motion of the piston, since as a result of the homogeneity of the model the pressure (see 166.4) is uniform throughout.

We thus see that the familiar sources for entropy increase, connected in ordinary engineering practice with heat flow at a finite rate and imperfect interaction of the working fluid with its surroundings, would be eliminated in the case of the elements of fluid in our cosmological models. We can hence conclude that the changes in the model will be reversible, provided the internal physical-chemical processes which occur in the fluid itself as the model expands or contracts involve no entropy increase.

The actual attainment of reversible behaviour for our non-static cosmological models will then depend on the possibility of selecting fluids of a simple enough constitution so that no internal irreversible processes, which would change the proper entropy of any given element of the fluid filling the model, can occur. We have already pointed out in § 130 of the chapter on relativistic thermodynamics that two such fluids would be provided by a distribution of particles of incoherent matter (dust) exerting zero pressure, and by a distribution of black-body radiation. And in the next two sections we shall give special attention to the reversible behaviour of models filled with these two fluids.

In the case of more complicated fluids, however, it is evident that internal processes would in general accompany a finite rate of change in the volume of an element of the fluid which would lead to increases in its entropy. This would then lead to the sign of inequality in the second law expression (169.1), and hence also to the conditions for a thermodynamically irreversible behaviour of the model. As a simple example of such a fluid, we have already pointed out in § 131 the case of a diatomic gas, which with a finite rate of expansion or compression would dissociate into its elements or recombine under non-equilibrium conditions and hence with increase in entropy. In later sections of this Part of Chapter X we shall give special attention to the irreversible expansion and contraction of cosmological models.

170. Model filled with incoherent matter exerting no pressure as an example of reversible behaviour

We may now give a little detailed consideration to a model filled with a distribution of incoherent matter or dust particles exerting negligible pressure, as furnishing an example of thermodynamically reversible behaviour at a finite rate. In such a model, the proper entropy associated with any element of the fluid would always be merely the sum total of the entropies of its constituent unchanging particles. Thus the entropy would have to remain constant, even with a finite rate of expansion or contraction of the model, and we should have the conditions for reversibility given by the equality sign in (169.1).

Hence we should expect the expansion or contraction of such models to take place reversibly, with nothing to prevent the return of the model to an earlier state provided the conditions are such that a reversal in the direction of motion does take place. Indeed, if we set the cosmological constant Λ equal to zero, and thus obtain the conditions for closed models with an oscillating behaviour of the first kind, type O_1, we have already seen in § 163 that the radius would symmetrically increase and decrease with the time in a manner which can be described as a cycloid in the Rt-plane by the equations

$$R = \frac{\alpha}{6}(1-\cos\psi) \qquad t = \frac{\alpha}{6}(\psi - \sin\psi), \qquad (170.1)$$

where α is a constant.

Thus the behaviour of such models would not only be thermodynamically reversible, but within a finite time would be subject to actual reversal as well if we set Λ equal to zero. Furthermore, even if we are uncertain as to the mechanism of passage through the singular state at $R = 0$, we can at least conclude that the model would return again from its maximum expansion to states having the same radius R as before, and with exactly the same rate of change (dR/dt) as before but in the reverse direction.

171. Model filled with black-body radiation as an example of reversible behaviour

As a second example of reversible behaviour with a finite rate of change, we may take a model filled solely with black-body radiation. Here too it is perhaps immediately evident that the entropy associated with the contents of any element having the coordinate range $\delta r \delta \theta \delta \phi$ would be constant, since the absence of irreversibility due to pressure gradients or friction of moving parts, combined with the absence of any other material present which could interact irreversibly with the radiation, means that changes in the proper volume of such an element even at a finite rate could be regarded as the reversible adiabatic expansion or compression of black-body radiation, which from the point of view of classical thermodynamics leads to no change in entropy.

Nevertheless, the situation is sufficiently complicated so that it may be desirable to give a more detailed analysis. We shall first show that an expansion or contraction of the model would lead to a new black-body distribution of radiation corresponding to a new temperature; and show that the change in proper volume and temperature for any element $\delta r \delta \theta \delta \phi$ would then be such as to leave the entropy unchanged.

As the definition of a black-body distribution of radiation, i.e. a distribution which is in thermodynamic equilibrium, we have the Planck distribution law (65.6), which at any desired initial time t_1 would give us

$$dE_1 = \frac{8\pi h \nu_1^3}{c^3} \frac{1}{e^{h\nu_1/kT_1} - 1} \, d\nu_1 \, dv_1 \qquad (171.1)$$

for the radiational energy dE_1 which a local observer at rest in the coordinates r, θ, ϕ would find in the frequency range ν_1 to $\nu_1 + d\nu_1$, and in the volume dv_1, at the temperature T_1.

At any later time t_2 when the quantity $g(t)$ which determines the temporal behaviour of the model has changed from g_1 to g_2, the frequency as measured by a local observer of the photons originally responsible for the above energy will have become in accordance with (156.8)

$$\nu_2 = e^{\frac{1}{2}(g_1-g_2)}\nu_1, \qquad (171.2)$$

and hence, owing to the proportionality with frequency, their energy will have become

$$dE_2 = e^{\frac{1}{2}(g_1-g_2)} dE_1. \qquad (171.3)$$

Furthermore, in accordance with the dependence of proper volume on time given by (151.4), the volume now containing these photons, or rather their equivalent, will have become

$$dv_2 = e^{\frac{3}{2}(g_2-g_1)} dv_1. \qquad (171.4)$$

Substituting these three equations into (171.1), we then obtain for the distribution of radiation at time t_2 as measured again by a local observer

$$dE_2 = \frac{8\pi h \nu_2^3}{c^3} \frac{1}{e^{h\nu_2/kT_2}-1} dv_2 dv_2, \qquad (171.5)$$

provided we take T_2 as given by

$$T_2 = T_1 e^{\frac{1}{2}(g_1-g_2)}. \qquad (171.6)$$

This thus gives the desired demonstration that the expansion or contraction of the model leads to a new distribution of black-body radiation with the new temperature determined by (171.6).

Hence since the entropy of black-body radiation is given in accordance with (65.5) in terms of its temperature and volume by the well-known formula

$$S = \frac{4}{3}aT^3v \qquad (171.7)$$

we see that the entropy associated with any given element $\delta r \delta \theta \delta \phi$ would remain constant, since we can write therefor in accordance with the foregoing equations

$$S = \frac{4}{3}aT_1^3 \delta v_1 = \frac{4}{3}aT_2^3 \delta v_2 = \text{const.} \qquad (171.8)$$

Thus also in the case of a model filled solely with black-body radiation we should have constant entropy for each element $\delta r \delta \theta \delta \phi$, and hence the condition for expansions or contractions at a finite rate reversibly. Moreover, here too as in the preceding case, by taking the cosmological constant Λ as equal to zero, we could obtain closed models in which the motion would not only be reversible but actually reversed as well, the relation between radius and time being given in accordance with (163.6) by

$$R = \sqrt{(\beta-t^2)}, \qquad (171.9)$$

where β is a constant.

This case of reversible behaviour in a model containing black-body radiation is perhaps more interesting than the previous one of a model composed of dust particles exerting no pressure, since now the processes of expansion and contraction with the accompanying changes in temperature seem definitely thermodynamic in character, as compared with the previous expansions and contractions which seemed purely mechanical in character and hence perhaps quite naturally reversible.

Also in the case of models containing a mixture of dust particles and radiation having no appreciable interaction, it is evident that we should expect reversible behaviour. But as soon as we go to particles small enough so that their thermal motion cannot be neglected it is evident that we must expect some slight irreversibility, since with a finite rate of volume change there would be a delay in the transfer of energy between the particles and the accompanying radiation, which would lead to a lag behind the conditions for true equilibrium.

In connexion with the foregoing discussion, however, it is perhaps unnecessary to stress the precise—in any case hypothetical—conditions under which completely reversible volume changes could take place at a finite rate. It is more important to emphasize the absence from our present cosmological models of the factors of irreversible heat flow, friction, and pressure gradients which are such common sources of irreversibility in ordinary thermodynamic processes taking place at a finite rate, that we may not realize that it is their presence rather than the mere finiteness of the rate itself which is leading to irreversibility.†

172. Discussion of failure to obtain periodic motions without singular states

The foregoing examples of a reversible oscillation in the radius of a closed model, between values corresponding to a lower singular state and an upper maximum, suggest an investigation of the possibilities for a strictly periodic behaviour in which the volume of any element of the fluid would pass continuously back and forth between a true minimum and maximum. This would be a periodic oscillation of the second kind, type O_2, already mentioned as conceivable in § 157 (e). We may now show, nevertheless, assuming reasonable

---

† This example of reversible behaviour at a finite rate, together with a more complicated one, will be found in Tolman, *Phys. Rev.* **37**, 1639 (1931); ibid. **38**, 797 (1931).

properties for the fluid in the model, that no such strictly periodic oscillations would be possible,† and that even non-periodic oscillations of this kind would not appear important.

Remembering that the dependence of the line element (166.1) on time for the models under discussion is given by the quantity $g(t)$, we shall write the conditions for an oscillation of the model between a true minimum and maximum in the form

$$g_1 < g_2 \quad \dot{g}_1 = \dot{g}_2 = 0 \quad \ddot{g}_1 \geqslant 0 \quad \ddot{g}_2 \leqslant 0, \quad (172.1)$$

where the dots indicate differentiation with respect to time, and the subscripts 1 and 2 indicate the value of the given quantity at the minimum and maximum respectively. By combining these expressions with our expressions for the proper density and pressure of the fluid in the model

$$8\pi\rho_{00} = \frac{3}{R_0^2}e^{-g} + \tfrac{3}{4}\dot{g}^2 - \Lambda$$

$$(172.2)$$

and $$8\pi p_0 = -\frac{1}{R_0^2}e^{-g} - \ddot{g} - \tfrac{3}{4}\dot{g}^2 + \Lambda,$$

we can then find what properties the fluid would have to show in order to permit the postulated minimum and maximum. Since we shall wish to consider both closed and open models, we may distinguish the three separate cases $R_0^2 > 0$, $R_0^2 = \infty$, and $R_0^2 < 0$, corresponding respectively to closed, open flat, and open curved models.

For the case $R_0^2 > 0$, we can readily obtain from the foregoing

$$\rho_1 > \rho_2$$

$$p_1 < p_2 \quad (172.3)$$

as relations which must hold for the densities and pressures of the fluid at the minimum and maximum, in order for an oscillation of the type in question to occur. If such behaviour takes place, the density of the fluid would then *decrease* as the volume of each element of fluid in the model increases in the ratio $e^{\frac{3}{2}g_1}$ to $e^{\frac{3}{2}g_2}$ in passing from minimum to maximum, but the pressure would have to *increase* in passing from minimum to maximum, in agreement with a necessary condition for oscillations of type O_2 already found for the special case of closed models with positive pressure in § 157 (e).

For strictly periodic oscillations, nevertheless, between a definite minimum and maximum it is evident that the behaviour of each

† Tolman, *Phys. Rev.* 38, 1758 (1931).

element of the fluid would have to be thermodynamically reversible since otherwise the same state could not be returned to over and over again. Hence, in connexion with the above, we can rule out such strictly periodic oscillations, unless we are willing to assume a fluid filling the model which has the unusual properties of an *increase* in pressure accompanying *reversible* adiabatic *expansion*.

For oscillations which are not strictly periodic but which might occur once or more between minima and maxima which do not have to remain fixed, the requirement of thermodynamic reversibility could be dropped. Hence the above conditions would be compatible with such an oscillation if, for example, there should be an irreversible rush in the formation of radiation during expansion so that the pressure would be sufficiently high at maximum to bring about reversal. It could hardly be expected, nevertheless, that the pressure could then decrease again on contraction so as to permit a second minimum.

For the case $R_0^2 = \infty$, to which we now turn, we obtain from the combination of (172.1) with (172.2)

$$\rho_1 = \rho_2; \tag{172.4}$$

and for the case $R_0^2 < 0$, we obtain

$$\rho_1 < \rho_2 \tag{172.5}$$

as necessary conditions for oscillatory motion of the type under consideration.

In accordance with the energy equation (151.6), nevertheless, we can write
$$d\rho_{00} = -\tfrac{3}{2}(\rho_{00}+p_0)\,dg, \tag{172.6}$$

for the change in density with g. Hence the above conditions could be met only if we assumed an unknown kind of fluid which can support a negative pressure at least equal to its energy density.

As a result of the above discussion, it is evident, at least at the present stage of the theory, that we may neglect homogeneous models in which the elements of fluid would undergo either a strictly periodic expansion and contraction or any kind of successive oscillations in volume between a true minimum and maximum. This finding, nevertheless, affects of course in no way the possibilities for oscillation between a lower singular state and a true upper maximum which we have previously studied.

173. Interpretation of reversible expansions by an ordinary observer

In earlier sections we have studied the possibility for expansions or contractions to take place in our cosmological models at a finite rate, and yet thermodynamically either completely reversibly, or at least with the elimination of sources of irreversibility which commonly accompany a finite rate of change in small-size systems. This leads to the possibility that cosmological processes, which might be interpreted by an ordinary unsophisticated observer as irreversible merely on account of their finite rate, could actually be taking place reversibly. Such a possible confusion must be avoided in order to obtain clear notions as to cosmological phenomena.

To investigate the matter, we shall take to start with the extremely simple model of § 171, filled solely with black-body radiation, and at the time of interest undergoing a reversible expansion with the quantity $g(t)$ in the formula for the line element

$$ds^2 = -\frac{e^{g(t)}}{[1+r^2/4R_0^2]^2}(dr^2 + r^2 d\theta^2 + r^2\sin^2\theta\, d\phi^2) + dt^2, \quad (173.1)$$

increasing with t. In considering the model we shall carefully distinguish between the *results*, which would be obtained by a *local observer* at rest with respect to r, θ, and ϕ and hence at rest with respect to the mean flow of energy in the model, and the *interpretation*, which he would place on these results from the point of view of classical thermodynamics, if he were an *ordinary observer* unfamiliar with relativistic thermodynamics and uninformed as to the general expansion taking place in the model.

In determining the results which this ordinary observer would desire for his interpretation, we shall consider him for convenience as located at the origin of coordinates and let him examine the contents of the universe in a small region in his immediate neighbourhood. In doing this, in view of his ignorance as to the general expansion taking place, we shall assume that he marks this region off, not so as to contain a given element of the fluid in the model, but by laying measuring rods end to end from the origin so as to obtain a sphere of constant proper radius

$$l_0 = \text{const.} \quad (173.2)$$

around the origin. Taking this sphere as small enough so that terms of the order $r^2/4R_0^2$ can be neglected in comparison with unity, the coordinate r at its boundary will have the value

$$r = l_0 e^{-\frac{1}{2}g} \qquad (173.3)$$

which will be varying with the time at the rate

$$\frac{dr}{dt} = -\tfrac{1}{2}l_0 e^{-\frac{1}{2}g}\frac{dg}{dt}. \qquad (173.4)$$

And the proper volume of the sphere will have the constant value

$$v_0 = \tfrac{4}{3}\pi l_0^3. \qquad (173.5)$$

Furthermore, in order to ascertain the results obtained by our observer from the measurements made in this region, we can use the energy equation (166.7) which can be conveniently written in the form

$$\frac{d}{dt}(\rho_{00}e^{\frac{3}{2}g})+p_0\frac{d}{dt}(e^{\frac{3}{2}g}) = 0, \qquad (173.6)$$

together with the relations connecting proper density, pressure, and temperature

$$\rho_{00} = aT_0^4 \quad \text{and} \quad p_0 = \frac{a}{3}T_0^4, \qquad (173.7)$$

which hold in case the fluid is black-body radiation as assumed.

With the help of the foregoing, it is then easily seen, that our observer will find

$$\frac{1}{\rho_{00}}\frac{d\rho_{00}}{dt} = \frac{1}{p_0}\frac{dp_0}{dt} = -2\frac{dg}{dt}, \qquad (173.8)$$

for the rate at which energy density and pressure are decreasing in his neighbourhood;

$$\frac{1}{T_0}\frac{dT_0}{dt} = -\frac{1}{2}\frac{dg}{dt}, \qquad (173.9)$$

for the rate at which the temperature is dropping in his neighbourhood; and

$$\frac{1}{n_0}\frac{dn_0}{dt} = -\frac{3}{2}\frac{dg}{dt}, \qquad (173.10)$$

for the rate at which the number of photons n_0 inside his sphere of constant volume is decreasing with time owing to net flow across the boundary.

Moreover, it is evident, if our observer stations an assistant on the boundary of his sphere and directs him to compare the frequency of photons escaping with those that are entering from outside, that he will report an average shift towards the red for the entering photons, since in accordance with (173.4) this assistant would be moving with respect to a local observer, chosen so as to remain at rest in the coordinate system r, θ, and ϕ, and hence also so as to obtain isotropic findings for the frequency of radiation.

Thus our ordinary observer would have at his disposal a continually dropping temperature in his own neighbourhood, and a flow of energy away therefrom towards regions of apparently lower temperature in the depths of space beyond, which he would be inclined to interpret from the classical point of view as evidences for a general process of energy degradation. He could hence be led to the erroneous conclusion that the universe was behaving irreversibly, in spite of the fact that the more legitimate considerations of relativistic thermodynamics have shown that such a model would actually be behaving reversibly, and indeed with a suitable value for the cosmological constant Λ would pass through a maximum expansion and return again to its original volume with reversed velocities.

The above model is of course highly idealized, containing as it does nothing but black-body radiation. By neglecting the interaction between radiation and matter, however, reversible behaviour could also be obtained with a model containing a mixture of black-body radiation and incoherent matter; and the same results would be found as to the flow of radiation away from any given location during expansion. Computations have also been made† with a model containing a mixture of black-body radiation and a perfect monatomic gas, assuming the possibility of transforming radiation into matter and vice versa, and assuming—contrary to the presumable possibilities—that the interaction between matter and radiation could take place rapidly enough to maintain equilibrium conditions. In such a model, in addition to the outward flow of radiation, it is found except for extraordinarily high temperatures that a reversible expansion would be accompanied by the annihilation of matter.

The main point to be stressed in connexion with the foregoing is the feasibility of mimicking with the help of reversibly expanding models —at least to some extent—the kind of behaviour, which in the case of the real universe would naturally be interpreted from older points of view as irreversible. This of course does not mean that actually irreversible processes are not taking place in the real universe, but it does emphasize the necessity of using relativistic rather than classical thermodynamics in the study of cosmology.

† Tolman, *Phys. Rev.* **38**, 797 (1931).

174. Analytical treatment of succession of expansions and contractions for a closed model with $\Lambda = 0$

Since processes which are actually thermodynamically irreversible appear to take place in the real universe, we may now turn to a consideration of the irreversible behaviour of cosmological models. To prepare for this we shall devote the present section to an analytical treatment of the behaviour—whether reversible or irreversible—of closed models with the cosmological constant Λ set equal to zero. We make this selection partly because this assigns to Λ what seems to be—as already emphasized—the most natural value to take, and partly because closed models with this value of Λ provide a good illustration of the new relativistic features of irreversible processes which we shall wish to study.

We have already seen in § 157 (f), that models of the above kind could only undergo an expansion from a lower singular state to an upper maximum followed by return to smaller volumes. We shall now investigate this behaviour in more detail.†

For the models under consideration we may take the line element in the form

$$ds^2 = -\frac{e^{g(t)}}{[1+r^2/4R_0^2]^2}(dr^2+r^2\,d\theta^2\ +r^2\sin^2\theta\,d\phi^2)+dt^2 \quad (174.1)$$

and base the treatment on the expressions for proper pressure and density given by (150.7) and (150.8):

$$8\pi p_0 = -\frac{1}{R_0^2}e^{-g}-\ddot{g}-\tfrac{3}{4}\dot{g}^2 \quad (174.2)$$

and

$$8\pi\rho_{00} = \frac{3}{R_0^2}e^{-g}+\tfrac{3}{4}\dot{g}^2, \quad (174.3)$$

where Λ has been set equal to zero, and where in agreement with our assumption of a closed model, we must take

$$R_0^2 > 0. \quad (174.4)$$

Furthermore, in agreement with physical possibilities we must take

$$\rho_{00} \geqslant 0, \quad (174.5)$$

since the density of material in the model could not be zero. We shall also take

$$p_0 \geqslant 0, \quad (174.6)$$

since we shall regard the model as filled with a mixture of matter and

† Tolman and Ward, *Phys. Rev.* **39**, 835 (1932).

radiation capable of exerting positive pressure, but incapable of withstanding tension.

(a) **The upper boundary of expansion.** Assuming that at some initial time $t = 0$, the model has a finite volume and a finite rate of expansion corresponding to

$$g = g_0 \quad \text{and} \quad \dot{g} = \dot{g}_0 \tag{174.7}$$

we may first show that there will be a finite upper boundary beyond which $g(t)$ cannot increase, without reference to the reversibility or irreversibility of behaviour.

Combining equation (174.2) with the inequality (174.6), we can write in general

$$\ddot{g} + \tfrac{3}{4}\dot{g}^2 + \frac{1}{R_0^2} e^{-g} \leqslant 0, \tag{174.8}$$

and, since $\dot{g}$ will be positive as long as expansion continues, we can multiply this by the positive quantity $2e^{\frac{3}{2}g}\dot{g}$ and write

$$2e^{\frac{3}{2}g}\dot{g}\ddot{g} + \tfrac{3}{2}e^{\frac{3}{2}g}\dot{g}^3 + \frac{2}{R_0^2}e^{\frac{3}{2}g}\dot{g} \leqslant 0,$$

or
$$\frac{d}{dt}(e^{\frac{3}{2}g}\dot{g}^2) + \frac{4}{R_0^2}\frac{d}{dt}(e^{\frac{3}{2}g}) \leqslant 0, \tag{174.9}$$

as an expression which will hold as long as g continues to increase.

Integrating (174.9) between $t = 0$ and any later time of interest $t = t$, and substituting the initial values of g and $\dot{g}$ as given by (174.7), we then obtain

$$e^{\frac{3}{2}g}\dot{g}^2 + \frac{4}{R_0^2}e^{\frac{3}{2}g} \leqslant e^{\frac{3}{2}g_0}\dot{g}_0^2 + \frac{4}{R_0^2}e^{\frac{3}{2}g_0}, \tag{174.10}$$

or noting in accordance with (174.4) that R_0^2 is positive

$$e^{\frac{3}{2}g} \leqslant e^{\frac{3}{2}g_0} + \frac{R_0^2}{4}e^{\frac{3}{2}g_0}\dot{g}_0^2 - \frac{R_0^2}{4}e^{\frac{3}{2}g}\dot{g}^2 \tag{174.11}$$

as an expression which will hold as long as g continues to increase. Hence, since g_0 and $\dot{g}_0$ are by hypothesis finite, there will be a finite upper boundary which g cannot surpass. This result may be expressed in the form

$$g \leqslant \gamma, \tag{174.12}$$

where γ is a finite quantity.

(b) **Time necessary to reach maximum.** With the help of the above, we can now show further that g will reach its maximum value and start to decrease within a finite time.

Combining the two inequalities (174.8) and (174.12), we can evidently write

$$\ddot{g} \leqslant -\frac{1}{R_0^2} e^{-\gamma} - \tfrac{3}{4}\dot{g}^2$$

or

$$\frac{d\dot{g}}{dt} \leqslant -\frac{1}{R_0^2} e^{-\gamma}, \qquad (174.13)$$

and integrating this between $t = 0$ and any later time of interest, we obtain

$$\dot{g} \leqslant \dot{g}_0 - \frac{1}{R_0^2} e^{-\gamma} t,$$

where $\dot{g}_0$ is the initial value of dg/dt. In accordance with this expression, however, we see that at a finite time

$$t \leqslant R_0^2 e^{\gamma} \dot{g}_0, \qquad (174.14)$$

$g(t)$ will reach its maximum and start to decrease.

(c) **Time necessary to complete contraction.** It will also be of interest to consider the behaviour of the model after passing through the maximum and starting to contract. As $\dot{g}$ will then evidently be negative, we may now multiply (174.8) by the negative quantity $2e^{\frac{1}{2}g}\dot{g}$, and integrating as was done before in order to obtain (174.10), write as the result for the present case

$$e^{\frac{1}{2}g}\dot{g}^2 + \frac{4}{R_0^2} e^{\frac{1}{2}g} \geqslant e^{\frac{1}{2}g_m}\dot{g}_m^2 + \frac{4}{R_0^2} e^{\frac{1}{2}g_m},$$

where g_m and $\dot{g}_m$ are the values of the quantities indicated, on passing through the maximum at time $t = t_m$. Moreover, since the velocity will be zero at this maximum, we shall actually have $\dot{g}_m = 0$, and may rewrite the above result in the form

$$e^{\frac{1}{2}g}\dot{g}^2 \geqslant \frac{4}{R_0^2}(e^{\frac{1}{2}g_m} - e^{\frac{1}{2}g}).$$

Furthermore, with $\dot{g}$ negative and R_0 real and positive corresponding to a closed model, this is equivalent to

$$e^{\frac{1}{2}g}\frac{dg}{dt} \leqslant -\frac{2}{R_0}\sqrt{(e^{\frac{1}{2}g_m} - e^{\frac{1}{2}g})}. \qquad (174.15)$$

This expression, however, can readily be integrated between the time t_m at which the maximum was passed and any later time of interest t, to give

$$(t - t_m) \leqslant R_0\{e^{\frac{1}{2}g}\sqrt{(e^{\frac{1}{2}g_m} - e^{\frac{1}{2}g})} - e^{\frac{1}{2}g_m}\sin^{-1}\frac{e^{\frac{1}{2}g}}{e^{\frac{1}{2}g_m}} + \tfrac{1}{2}\pi e^{\frac{1}{2}g_m}\}. \qquad (174.16)$$

In accordance with this expression, we then see that within a finite time
$$(t-t_m) \leqslant \tfrac{1}{2}\pi R_0 e^{\frac{3}{2}g_m} \qquad (174.17)$$

after passing through its maximum, the value of g would decrease to minus infinity, provided the singular state at the lower limit of the motion did not occur earlier.

(d) **Behaviour at lower limit of contraction.** To summarize the foregoing conclusions, we see that the model starting at a selected initial time with any finite value of g_0 and finite rate of expansion $\dot{g}_0$, would then reach a maximum value of g and start contracting within a finite time later. And furthermore this contraction would proceed at a sufficient rate so that g could decrease to minus infinity again within a finite time. We must now inquire as to the behaviour of the model on reaching the lower limit of contraction.

In the first place, since the proper volume of any element of fluid in the model would always be proportional to $e^{\frac{3}{2}g}$, we realize on physical grounds alone that $e^{\frac{3}{2}g} = 0$, $g = -\infty$, would in any case set a lower limit for possible contraction. In the second place, nevertheless, in accordance with (174.15) when $e^{\frac{3}{2}g}$ reaches the value zero we should have
$$\dot{g} = -\infty, \qquad (174.18)$$
and hence also in accordance with (174.8)
$$\ddot{g} = -\infty, \qquad (174.19)$$

at this point. Thus the conditions for an analytical minimum are completely unsatisfied, and the analysis would fail to describe the passage of the model through the point.

Hence, since on physical grounds the contraction cannot proceed further than the point $e^{\frac{3}{2}g} = 0$, it is evident on mathematical grounds that we can maintain the validity of the fundamental differential equations (174.2) and (174.3) which control the behaviour of the model, only by introducing a renewed expansion which starts from some singular state at the lower limit of contraction. This singular state may of course lie near rather than exactly at the point $e^{\frac{3}{2}g} = 0$.

It is, to be sure, unfortunate that our differential equations for the motion of the model are not sufficient to describe the mechanism of passage through the lower limit of contraction, the existence of which is physically inevitably necessary. As suggested by Einstein,† it is

† Einstein, *Berl. Ber.* 1931, p. 235.

possible that the idealizations—such for example as the complete homogeneity of the model—on which the analysis has been founded are to be regarded in the case of an actual physical system as failing in the neighbourhood of this lower limit of contraction. The situation is perhaps similar to that which would be furnished by an attempt to describe the behaviour of an elastic ball, bouncing up and down from the floor, solely with the help of the usual equation for gravitational acceleration

$$\frac{d^2h}{dt^2} = -g. \qquad (174.20)$$

This equation would be sufficient to describe the motion of the ball as it rose to its maximum height and fell from that point. It would fail, however, to give a description of the mechanism of reversal when the ball reached the floor, and further considerations involving the size and elastic properties of the ball would be necessary to describe the passage through that point.

As the end result of this section, we may then conclude, for the case $\Lambda = 0$, that the only possible behaviour, for a closed homogeneous model of the universe filled with a fluid unable to withstand tension, would be a continued succession of expansions and contractions, such that $g(t)$ would increase from a singular state at the lower limit of the previous contraction up to a true maximum, and then return again to a singular state where renewed expansion would again set in. Furthermore, if at any given initial time the value of g and its rate of increase $\dot{g}$ were finite, the upper limit reached by g would be finite and only a finite time would be necessary to complete the cycle of expansion and contraction. Finally, it is to be emphasized that these conditions have been obtained without any reference to the reversibility or irreversibility of the behaviour of the model, and would be equally valid for the succession of *identical* expansions and contractions which would correspond to reversible behaviour and for the succession of *changing* expansions and contractions which would be obtained with irreversible behaviour.

175. Application of thermodynamics to a succession of irreversible expansions and contractions

As already pointed out in Chapter IX, a continued succession of irreversible expansions and contractions, as found for the models considered in the preceding section, would seem very strange from the point of view of classical thermodynamics, which would predict an ultimate state of maximum entropy and rest as the result of continued irreversible processes in an isolated system. Hence we must now examine the bearing of relativistic thermodynamics on this finding.†

In accordance with our general discussion in § 169 of the conditions for reversibility and irreversibility in the behaviour of homogeneous models, the succession of irreversible expansions and contractions, which we are now considering, would be characterized by a continued increase in the proper entropy of any selected element of fluid in the model, as given by the sign of inequality in the expression

$$\frac{d}{dt}(\phi_0 \, \delta v_0) > 0. \tag{175.1}$$

Thus, although the model might pass through states in the course of an expansion or contraction in which the conditions momentarily correspond to those for physical-chemical equilibrium, it is evident that the entropy of any element of the fluid would ultimately have to increase without limit as the irreversible expansions and contractions continued. Hence we must now show that this can be possible, since the classical thermodynamics has accustomed us to the idea of a maximum upper value for the possible entropy of any isolated system.

To investigate this point it is evident that we may take the proper entropy, measured for any small element of the fluid by a local observer, as depending on the state in accordance with the classical equation

$$d(\phi_0 \, \delta v_0) = \frac{1}{T_0} d(\rho_{00} \, \delta v_0) + \frac{p_0}{T_0} d(\delta v_0) + \frac{\partial(\phi_0 \, \delta v_0)}{\partial n_1^0} dn_1^0 + \dots + \frac{\partial(\phi_0 \, \delta v_0)}{\partial n_n^0} dn_n^0, \tag{175.2}$$

where the proper energy of the element $(\rho_{00} \, \delta v_0)$, its proper volume δv_0, and the number of mols $n_1^0, \dots, n_n^0$ of its different chemical constituents are taken as the independent variables which determine its state.

In applying this equation to the continued increase in the entropy of the element, which must take place if the irreversible expansions and contractions continue, we note in accordance with the result obtained from the first law in § 166

$$\frac{d}{dt}(\rho_{00} \, \delta v_0) + p_0 \frac{d}{dt}(\delta v_0) = 0, \tag{175.3}$$

† Tolman, *Phys. Rev.* **39**, 320 (1932).

that the immediate cause which leads to entropy increase cannot be due to the presence of the first two terms on the right-hand side of (175.2), since their sum will always be equal to zero. Hence the internal mechanism by which the entropy increase is actually occurring at any time must be due to the presence of the remaining terms on the right-hand side, which correspond to the irreversible adjustment of composition in the direction of equilibrium.

At first sight it might seem that such an adjustment of concentrations could provide only a limited increase in entropy, since the classical thermodynamics has made us familiar with the existence of a maximum possible entropy for a system having a given energy and volume; the present case differs, however, from the classical case of an isolated system, since the proper energy of any selected element of fluid in the model does not have to remain constant. Indeed, in accordance with (175.3), the proper energy of every element of fluid in the model would be decreasing with time during expansion and increasing with time during contraction. Hence if the pressure tends to be greater during a compression than during the previous expansion, as would be expected with a lag behind equilibrium conditions, an element of the fluid can return to its original volume with increased energy and hence also with increased entropy. Thus, although the internal mechanism of entropy increase would always be due at any instant to the adjustment of concentrations, for example in the direction of dissociation during the later stages of expansion and in the direction of recombination during the later stages of contraction, the possibility for continued entropy increase would have to be due in the long run to an increase in the proper energy of the elements of fluid in the model.

As shown in § 131 of the last chapter, the situation is analogous to the continued increase in entropy and energy which would occur in the classical case of a continued succession of irreversible adiabatic expansions and compressions for a dissociating gas in a cylinder with non-conducting walls and a movable piston, so long as external energy was available to complete the desired compressions; and in the relativistic case this external energy can be regarded as coming from the potential energy of the gravitational field associated with Einstein's pseudo-tensor density t_μ^ν. Similar considerations could also be given to the irreversible expansion and contraction of a mixture of matter and radiation, assuming a delay in their attainment of equilibrium which in the later stages of expansion might involve both a lag in the transformation of a portion of the mass of matter into radiation as well as a lag in the escape of radiation from the matter. Such possibilities might be of interest for the actual universe.

Having found that a continued succession of irreversible expansions and contractions for our cosmological models would involve in the long run an increase in the proper energy of the elements of fluid therein when they return to the same volume, we must now examine the effects of such an increase on the character of later and later cycles. This can easily be done with the help of our equation for energy density (174.3), which gives

$$8\pi\rho_{00}\,e^{\frac{3}{2}g} = \frac{3}{R_0^2}e^{\frac{1}{2}g}+\tfrac{3}{4}e^{\frac{3}{2}g}\dot{g}^2, \tag{175.4}$$

as an expression which is proportional to the proper energy of any selected element of the fluid, having the proper volume

$$\delta v_0 = \frac{r^2\sin\theta\,e^{\frac{3}{2}g}}{[1+r^2/4R_0^2]^3}\,\delta r\delta\theta\delta\phi. \tag{175.5}$$

In accordance with these expressions we see that the volume of any element of the fluid will return to an earlier value when $g(t)$ so returns, and hence that the energy of the element can be greater at a later return only in case the square of the velocity, $\dot{g}^2$, has a greater value. This, however, is sufficient to indicate the general difference between the character of a given cycle and sufficiently later ones, as shown in Fig. 10, where the later cycle has larger values of $|\dot{g}|$ for a given value of $e^{\frac{1}{2}g}$, and hence also rises to a higher maximum.

Since the value of the energy density at the point of maximum expansion would be given by

$$8\pi\rho_{00} = \frac{3}{R_0^2}e^{-g}, \tag{175.6}$$

and the value of g at the maximum would ultimately increase without limit, we see that the energy density at this point would get smaller and smaller for later cycles. Hence, too, we may infer that the model might spend a greater and greater proportion of its period in a condition of lower density than that observed, for example at present in the actual universe, even though a return to higher densities would always occur. The above conclusion as to energy density at the point of maximum expansion, however, does not apply in general to the irreversible oscillations that could be obtained with other assumptions as to Λ and R_0^2.

In concluding this Part of Chapter X, it should be emphasized, at our present stage of very incomplete knowledge as to the actual behaviour of our surroundings over long periods of time, that the importance of the foregoing applications of thermodynamics to cosmological models lies primarily, not so much in providing immediate explanations for the phenomena of the real universe, as in indicating

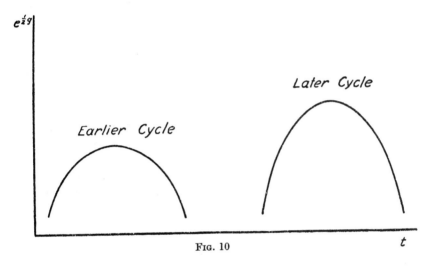

FIG. 10

the ultimate necessity for the use of relativistic rather than classical thermodynamics for a successful treatment of the problems of cosmology. Two considerations which might have a bearing on the problems of actual cosmology may, nevertheless, be mentioned.

In the first place, the foregoing discussion has suggested the possibility that the present state of the actual universe, or of that portion which lies within some 10^8 light years, may perhaps be the result of an irreversible expansion from an earlier state of exceedingly small volume, corresponding to the lower singular states that we have found in the case of some of our homogeneous models. In such a lower state of very small volume, the density, pressure, and temperature that we should have to assign to the fluid in the model would, however, be very high; and the conditions for thermodynamic equilibrium very much displaced, in the general direction for endothermic chemical reaction, as compared with those now prevailing on the average in our surroundings. In Chapter V, nevertheless, we have seen that the relative amounts of hydrogen and helium, of different isotopes, and perhaps also of matter and radiation, actually found in the universe, do show a great excess of endothermic substances when compared with present conditions for equilibrium. Hence it might be plausible to try to explain existing ratios as the result of slow irreversible changes which have taken place since an earlier state of high density and temperature.

In the second place, in connexion with the behaviour of the actual universe, some stress must be laid on the possibility found for a certain class of models to expand and contract irreversibly without ever reaching an unsurpassable state of maximum entropy. It would of course not be safe to conclude therefrom that the actual universe will never reach a state of maximum entropy, where further change would be impossible. Nevertheless, this finding in the case of certain kinds of model must be allowed to exert some liberalizing action on our general thermodynamic thinking. At the very least it would seem wisest, if we no longer dogmatically assert that the principles of thermodynamics necessarily require a universe which was created at a finite time in the past and which is fated for stagnation and death in the future.

X

APPLICATIONS TO COSMOLOGY (*contd.*)

176. Introduction

There are three different kinds of justification that can be given for our extensive consideration of the properties of non-static homogeneous cosmological models. In the first place, we have a natural interest and intellectual pleasure in trying to develop the consequences of any set of mathematical assumptions without reference to possible physical applications. Secondly, since we have based our treatment on acceptable physical theory, we have the right to expect that the theoretical behaviour of our models will at least inform and liberalize our thinking as to conceptual possibilities for the behaviour of the actual universe. In the third place, however, and this is perhaps most important of all, we have the right to hope that the models can be so constructed as to assist in the correlation and explanation of the observed phenomena of the actual universe, and indeed may even be sufficiently representative as to permit some cautious extrapolation forward and backward in time, which will give us not too fallacious ideas as to the past and future history of our surroundings.

It is this possibility of using non-static homogeneous models to correlate the phenomena of the actual universe, which has been only incidentally mentioned in the foregoing, to which we must now turn. In doing so, it is to be emphasized that we shall attempt as nearly as possible a phenomenological point of view. We shall regard the line element, which we have derived for our models, as an approximate expression which may cease to be even reasonably satisfactory when extrapolated to too great distances or over too long time intervals. Furthermore, we shall attempt to obtain information concerning the function $g(t)$, which occurs in the expression for the line element and hence determines the temporal behaviour of the models, not from hypothetical considerations as to the possible origin or fate of the universe, but by the more modest method of expanding this function as a power series in t around the present time, and then learning as much as we can about the coefficients in this series from actual observational data.

In the following rather long section, which is divided into several parts, we shall first consider the several kinds of observational information which are now available concerning the contents and structure of the real universe. This information has to do with the magnitudes, distances, spectra, diameters, masses, and distribution of the extra-galactic nebulae, since these are the only things outside our own galaxy concerning which we now have any certainty of knowledge. Our precise information as to the nature and behaviour of these objects is largely due to the work of Hubble and of Hubble and Humason. We shall not concern ourselves with the observational problems involved in obtaining the raw data, but shall present in some detail the methods used in the interpretation thereof. In later sections we shall then consider the correlation of the available information with the help of our non-static models.

177. The observational data

(a) The absolute magnitudes of the nearer nebulae. Of fundamental importance for our knowledge of the extra-galactic nebulae are the determinations of magnitude and distance made by Hubble and Humason.† The work divides itself into three parts. In the first place, we have the determination of the *mean absolute magnitude* for a considerable number of the *nearer nebulae* from direct observations on individual stars which they contain. In the second place, we have the observation of *apparent magnitudes* for more *distant nebulae* which are associated in groups or clusters. And in the third place, we then have the use of these results to calculate the *distances* to the clusters.

We may first consider the determination of absolute magnitudes for the nearer nebulae, and leave the apparent magnitudes and distances of the more distant nebulae to the next two parts of this section.

Types of stars which have been identified in the nearest extra-galactic nebulae include Cepheid variables, irregular variables, helium stars (B_0 and O), P Cygni stars, and novae. With the help of the observed magnitudes and periods of the Cepheid variables, actually found therein, Hubble and Humason have determined the distances to eight‡ of the nearest nebulae, using Shapley's 1930 zero point for

† Hubble and Humason, *Astrophys. Journ.* **74**, 43 (1931).

‡ This includes the two companions of M 31 which are assumed to be at the same distance as M 31 itself.

the period-luminosity relation for the Cepheids.† This together with Shapley's values for the two Magellanic clouds, then gives us at least fairly accurate values for the distances to ten extra-galactic nebulae. And these distances, moreover, are confirmed by the observed magnitudes for the other types of stars which can be recognized in these objects.

Combining this knowledge as to the distances of the ten nebulae, with their total observed visual magnitude as obtained from Hopmann's‡ correction of Holetschek's measures and from other sources, Hubble and Humason then find for the absolute visual magnitude of these objects—as they would appear at the standard distance of 10 parsecs—the mean value −14·7, with a total range of about 5·0 magnitudes, and an average residual of 1·5 around this mean.

In addition to these determinations, when the different types of star present can be distinguished, Hubble and Humason have also been able to use the luminosity of the brightest stars present to extend the determination of absolute magnitudes to any case where stars can be recognized at all. For eight of the ten objects considered above,§ the absolute magnitudes of the brightest stars therein were found to have the mean value −6·1, with a range of only about 1·8 magnitudes, and an average residual of 0·4 around the mean. Since the scatter in the magnitudes of the brightest stars is considerably less than the scatter in the magnitudes of the nebulae themselves, it seems rational to assume that the brightest stars in these objects have a reasonably constant absolute magnitude independent of the nebula in which they are located. Furthermore, the validity of this assumption is confirmed by the fact that all the data available show a scatter for the differences between the observed magnitudes for the nebulae as a whole and their brightest stars, which can apparently be accounted for by the scatter to be expected in the absolute magnitudes of the nebulae alone. Hence it seems justifiable to take −6·1 as a figure for the absolute magnitude of the brightest stars in any nebula where they can be seen at all, and then obtain the absolute magnitude of the nebula by adding the difference between the observed magnitudes of the nebula and of its brightest stars.

Proceeding along these lines, in the case of 40 nebulae where stars

---

† Shapley, *Star Clusters*, 1930, p. 189.
‡ Hopmann, *Astron. Nachr.* **214**, 425 (1921).
§ This excludes the two companions of *M* 31 as separate objects.

could be seen, Hubble and Humason have found for the mean difference between the observed magnitudes of the nebulae as a whole and their brightest stars the figure -8.88, with a range of 4·9 magnitudes, and an average residual around the mean of 0·77. The agreement between the scatter of 5·0 magnitudes for the ten nebulae first considered and the above scatter of 4.9 magnitudes will be noted.

Combining the figures, -6.1 for the absolute magnitudes of the brightest stars, and -8.88 for the mean difference in the magnitudes of the forty nebulae and such stars, the result -15.0 is found for the mean absolute magnitude of these objects, as compared with -14.7 for the original ten nebulae. The check is very satisfactory, and Hubble and Humason finally adopt

$$M_{\text{vis}} = -14.9, \tag{177.1}$$

for the mean absolute *visual* magnitude of extra-galactic nebulae.

For purposes of comparison with more distant nebulae, it is also desirable to have a figure for mean absolute *photographic* magnitude. The photographic magnitudes of nebulae were found to be best obtained by using extra-focal images larger than the focal dimensions of the nebulae. By comparing the photo-visual and photographic magnitudes corresponding to such images for sixty nebulae in the Virgo cluster, Hubble and Humason obtain the figure

$$CI = +1.10 \pm 0.02,$$

for the mean colour-index of not too distant nebulae, and this value was in reasonable agreement with other data available. Combining with (177.1), the figure

$$M_{pg} = -13.8, \tag{177.2}$$

is then obtained for the mean absolute photographic magnitude of extra-galactic nebulae, using extra-focal images.†

(b) **The corrected apparent magnitudes for more distant nebulae.** We now turn to the determinations by Hubble and Humason of the apparent photographic magnitudes of more distant nebulae where individual stars cannot be seen. The treatment involves several interesting corrections the nature of which may first be considered.

Since the light from the *more distant* nebulae is actually found to have suffered a shift in wave-lengths towards the red, these corrections

---

† Some revision in this value may result from the programme of work on photographic magnitudes now under way at Mount Wilson or from work elsewhere. Shapley gives the value $M_{pg} = -14.5$. *Proc. Nat. Acad.* **19**, 591 (1933).

must be applied to the immediately observed photographic magnitudes in order to make them comparable with the above absolute photographic magnitudes as determined for the *nearer* nebulae which show no appreciable red-shift. The general nature of the corrections will be seen from the following considerations: first, that the red-shift implies an actual decrease in the total rate at which energy is being delivered at the boundary of the earth's atmosphere; secondly, that the changed distribution of this energy in the spectrum implies a change in the fraction of it which will be absorbed in passing through the earth's atmosphere; thirdly, that the changed distribution of energy implies a change in the relation of its thermal to its visual effectiveness; and finally, that the changed distribution of energy also implies a change in the relation of its visual to its photographic effectiveness.

The detailed treatment of these corrections may be based on an equation which can be regarded as an empirical relation, connecting the photographic magnitude of a heavenly object as ordinarily measured, with its bolometric magnitude as would be determined from thermal measurements made *without* absorption by the earth's atmosphere. The equation may be written in the form

$$m_{pg} = m_b + \Delta m_r + HI + CI, \qquad (177.3)$$

where m_{pg} and m_b are the photographic and bolometric magnitudes, Δm_r is the empirical correction to be added to the bolometric magnitude to obtain the radiometric magnitude as measured thermally *after* absorption by the earth's atmosphere, HI is the empirical value of the so-called heat-index which must be added to the radiometric magnitude to obtain the visual magnitude, and CI is the empirical value of the so-called colour-index which must finally be added to the visual magnitude to obtain the photographic magnitude.

In accordance with this equation, we may then write

$$\Delta m_{pg} = \Delta m_b + \Delta(\Delta m_r) + \Delta(HI) + \Delta(CI) \qquad (177.4)$$

as an expression for the effect of the red-shift in increasing the photographic magnitude of the more distant nebulae, by producing changes in their bolometric magnitude and in the three empirically determined quantities that must be added thereto in order to obtain the photographic magnitude.

To calculate the direct effect of the red-shift in changing the bolometric magnitude, we must make use of what may be regarded as the equation of definition for magnitudes

$$m-M = 2 \cdot 5 \log L - 2 \cdot 5 \log l, \qquad (177.5)$$

where m and M are the observed and absolute magnitudes for a heavenly object, l is the observed luminosity of the object, and L the luminosity which it would have at the standard distance of 10 parsecs.

The effect on luminosity of superimposing a red-shift on the radiation from a distant object can be twofold. In the first place, it is evident that the frequency, and hence energy, associated with each individual photon coming to the observer will be decreased in the ratio $\lambda:(\lambda+\delta\lambda)$ of the original to the increased wave-length. In the second place, if the red-shift is actually due to a Doppler effect, the rate at which photons arrive at the observer will also be decreased in this same ratio. Since Hubble and Humason do not wish to assign any particular cause for the red-shift, they purposely allow for only the first of these effects, and hence write in accordance with (177.5).

$$\Delta m_b = 2 \cdot 5 \log \frac{\lambda+\delta\lambda}{\lambda}, \qquad (177.6)$$

as the change in bolometric magnitude due to the red-shift. If we allow for both effects, the change in bolometric magnitude would be twice as great. For the Leo cluster, nevertheless, which has the largest red-shift so far observed the additional correction would be within the limits of probable error.

To obtain the remaining quantities on the right-hand side of (177.4) we must consider the effect of red-shift in changing the apparent temperature of nebulae, since atmospheric absorption, heat-index, and colour-index are quantities which have been observationally related to spectral type and hence to apparent temperature. The spectral type of the nebulae may be taken as approximately dG_3 corresponding to a black-body temperature of the emitting source of 5,760° absolute. Hence if the light suffers a fractional shift in wave-length it is evident from Planck's law (65.6), that the new spectral distribution would correspond to an apparent temperature of emission given by

$$T = 5,760\left(\frac{\lambda}{\lambda+\delta\lambda}\right). \qquad (177.7)$$

Furthermore, as the empirical relations connecting temperature of emission with spectral type and hence also with atmospheric absorption, heat-index, and colour-index, we may take the tabular expression given by Hubble and Humason for the known data:—†

TABLE I

Temp. ° Abs.	Spectral Type	Effect of Atm.	Heat-Index HI	Colour-Index CI
6,500	F_5	0·44	0·30	0·62
6,000	dG_0	0·43	0·32	0·72
5,600	dG_5	0·41	0·39	0·83
5,100	dK_0	0·40	0·55	0·99
4,400	dK_5	0·48	1·10	1·26
3,400	dM	0·53	1·40	1·76

With the help of the foregoing equations, together with graphical interpolation of the data given by Table I, Hubble and Humason are then able to obtain the results shown in Table II for the effect of a given fractional shift in wave-length in producing—in accordance with (177.4)—a change in the photographic magnitudes of nebulae.

TABLE II

Effect of Red-Shift on Photographic Magnitude

Velocity km./sec.	Distance Parsecs	$\frac{\delta\lambda}{\lambda}$	Temp. °Abs.	Spectral Type	Δm_b	$\Delta(\Delta m_r + HI)$	$\Delta(CI)$	Δm_{pg}
..	..	0·0000	5,760	dG 3·0	0·000	0·000	0·000	0·000
1,000	$1·8 \times 10^6$	0·0033	5,740	3·3	0·003	0·004	0·008	0·015
4,000	7·2	0·0133	5,685	4·0	0·015	0·015	0·02	0·05
8,000	14·4	0·0267	5,615	4·7	0·03	0·03	0·04	0·10
12,000	21·6	0·0400	5,540	5·8	0·04	0·05	0·06	0·15
16,000	28·8	0·0533	5,470	6·5	0·06	0·06	0·08	0·20
20,000	36	0·0667	5,400	7·2	0·07	0·08	0·10	0·25
30,000	54	0·1000	5,235	8·8	0·11	0·13	0·16	0·40
40,000	72	0·133	5,080	dK 0·2	0·14	0·20	0·21	0·55
50,000	90	0·167	4,940	1·4	0·17	0·32	0·26	0·75
60,000	108	0·200	4,800	2·3	0·20	0·44	0·31	0·95

The *third* column in this table gives the actual fractional red-shift considered, while the first two columns give merely for convenience the velocity of recession which would correspond to this red-shift if we interpret it as due to an ordinary Doppler effect, together with the distance to the nebula which we shall later find observationally associated with that red-shift. The remaining columns give in order: the temperature corresponding to the red-shift as calculated from

† Relation of temperature to type from Russell, Dugan, and Stewart, *Astronomy*, 1927. Effect of atmosphere and heat index from Petit and Nicholson, *Astrophys. Journ.* **68**, 279 (1928). Colour-index from Seares, *Astrophys. Journ.* **55**, 165 (1922).

(177.7); the spectral type corresponding to this temperature as obtained from Table I; the change in bolometric magnitude caused by the red-shift as calculated from (177.6); the small change in Δm_r combined with that in heat-index as obtained from Table I; the change in colour-index as also obtained from that table; and finally the total change in photographic magnitude Δm_{pg} due to the red-shift as calculated in accordance with (177.4) by combining the figures in the three preceding columns. By subtracting the appropriate value for Δm_{pg}, it then becomes possible to correct the observed photographic magnitude for any nebula to that which would be expected if the light coming therefrom had suffered no red-shift.

The actual data of Hubble and Humason for the average photographic magnitudes and red-shifts found in eight clusters of nebulae, and for two groups composed of isolated nebulae having a moderate range in magnitude, are given in Table III, where the red-shifts are expressed in terms of the corresponding velocity of recession. The average figure given for the photographic magnitude, is the most frequent magnitude in the case of the clusters and the mean magnitude in the case of the two groups. The next to the last column in the table gives the correction $-\Delta m_{pg}$ which, in accordance with the preceding table, must be applied to the observed photographic magnitude to allow for the effects of the red-shift. It will be seen that this correction at the present time is actually negligible for all except the three most distant clusters. The last column in the table gives the distances to the clusters obtained by the method of calculation to be discussed below.

TABLE III

Cluster	Number Nebulae	Dia-meter Cluster	Number Red-shifts Measured	Mean Shift km/sec.	Average m_{pg}	Correction $-\Delta m_{pg}$	Distance Parsecs
Virgo	(500)	12°	7	890	12·5	..	$1 \cdot 8 \times 10^6$
Pegasus	100	1	5	3,810	15·5	..	7·25
Pisces	20	0·5	4	4,630	15.4	..	7
Cancer	150	1·5	2	4,820	16·0	..	9
Perseus	500	2·0	4	5,230	16·4	..	11
Coma	800	1·7	3	7,500	17·0	−0·10	13·8
Ursa Major	300	0·7	1	11,800	18·0	−0·15	22
Leo	400	0·6	1	19,600	19·0	−0·25	32
Group I	16	..	16	2,350	13·8	..	..
Group II	21	..	21	630	11·6	..	..

(c) **Nebular distances calculated from apparent magnitudes.** Taking the nebulae which appear in a cluster to be actually physically associated in a relatively restricted region, it now becomes possible to calculate the distance to the cluster by comparing the average observed photographic magnitude for the nebulae therein with the mean absolute photographic magnitude obtained from nearer nebulae. In a similar manner, it is also possible to calculate an *average* distance for the isolated nebulae which have been grouped together for purposes of treatment.

In order to make such calculations, Hubble and Humason make use of two interesting assumptions.

In the first place, it is assumed that the absolute magnitude to be expected on the average for any nebula is the same as that previously obtained for the nearer nebulae. Since the distance to the Leo cluster actually turns out to be a little more than 10^8 *light years*, this involves not only the assumption that nebulae in different parts of the universe tend to be alike at a given time, but also the assumption that the luminosity of a nebula would suffer little change in 10^8 *years*.

In the second place, it is assumed that the apparent luminosity of nebulae, making allowance for the effect of the red-shift, would be proportional to the inverse square of their distances, in the manner to be expected for stationary objects in ordinary Euclidean space.

To make use of these assumptions, we have the equation of definition for magnitude

$$m - M = 2 \cdot 5 \log L - 2 \cdot 5 \log l, \tag{177.8}$$

where m and M are the observed and absolute magnitudes for a heavenly object, l is the observed luminosity of the object, and L the luminosity which it would have at the standard distance of 10 parsecs. And we have the inverse square law for luminosities

$$\frac{l}{L} = \frac{D^2}{d^2}, \tag{177.9}$$

where d and D are the actual and standard distances. Combining the two equations and setting $D = 10$, we obtain

$$\log d = 0 \cdot 2(m - M) + 1, \tag{177.10}$$

for the distance d in parsecs in terms of apparent and absolute magnitude.

This result may be applied to the photographic magnitudes of Hubble and Humason in the form

$$\log d = 0 \cdot 2(m_{pg} - \Delta m_{pg} - M_{pg}) + 1,$$ (177.11)

where $-\Delta m_{pg}$ is the correction for the effect of red-shift already discussed. With the help of this equation and the figures for absolute and corrected apparent photographic magnitudes given by (177.2) and Table III, the distances to the various clusters may now be obtained, as already shown in the last column of Table III, where the estimated reliability of the result is roughly indicated by the number of significant figures presented.

A treatment of the relation between luminosity and distance, in which the assumption of flat space and stationary nebulae is not made will be given in § 179. It will be shown there that the calculated nebular distances d are related in a specially simple manner to the coordinate $\bar{r}$ which we have used in one of the later forms (149.5) in which we have expressed the original formula for the non-static line element (149.1).

(d) **Relation of observed red-shift to magnitude and distance.** In the case of extra-galactic nebulae, the nearly universal occurrence of spectral shifts towards the red was made evident at least as early as 1922 by the pioneer work of Slipher,† on the light from nearby nebulae; and, by employing the methods for determining nebular distances discussed above, an approximate linearity of red-shift with calculated distance out to 2×10^6 parsecs was established in 1929 by the work of Hubble.‡ With the present much more extended data of Hubble and Humason, it now becomes possible to obtain a very satisfactory treatment of the dependence of red-shift on observed magnitude and hence also on calculated distance.

Since the quantities actually observed are red-shift and apparent magnitude, we may first consider the values for these quantities as given in Table III for the case of the eight clusters and the two groups of isolated nebulae. The relation between the values given is shown in Fig. 11, taken from Hubble and Humason, where the logarithms of red-shift—expressed in terms of velocity v—have been plotted as ordinates, and the observed magnitudes are taken as abscissae. The relation is evidently closely linear, and was found to be satisfactorily

---

† See the table given by Eddington, *The Mathematical Theory of Relativity*, 1923, p. 162.   ‡ Hubble, *Proc. Nat. Acad.* **15**, 168 (1929).

expressed by the equation

$$\log v = 0.2m + 0.507, \qquad (177.12)$$

with an average deviation of 0·031 in $\log v$ and 0·15 in m over the range of interest.

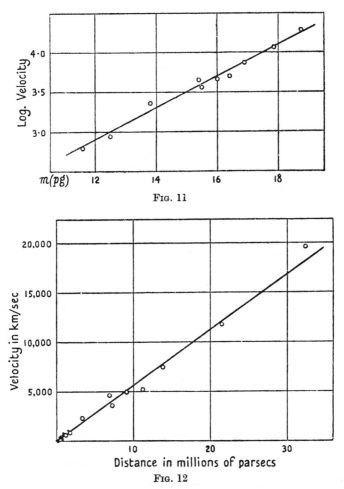

Fig. 11

Fig. 12

With the help of the equation connecting magnitude with distance given by (177.10), we can also examine the relation between red-shift and calculated distance. This is shown in Fig. 12, where the red-shift is again expressed in terms of velocity. The dots near the origin represent the data for nearby individual nebulae, and the circles represent the data given in Table III.

The direct relation between red-shift and calculated distance is seen from this plot to be closely linear, and by combining equations (177.11) and (177.12) and inserting the value for $M_{pg} = -13\cdot8$ can be expressed in the form

$$\frac{v}{d} = 5\cdot58 \times 10^{-4} \frac{\text{km./sec.}}{\text{parsec}}. \tag{177.13}$$

For our later purposes, it will be more useful to express the red-shift in terms of fractional change in the wave-length. We then obtain

$$\frac{\delta\lambda}{\lambda} = 1\cdot86 \times 10^{-9} d \tag{177.14}$$

with the distance d in *parsecs*, or

$$\frac{\delta\lambda}{\lambda} = 5\cdot71 \times 10^{-10} d \tag{177.15}$$

with d expressed in *light years*. It is believed that the uncertainty in the final result is definitely less than 20 per cent. and probably not more than 10 per cent.

In considering the significance of this remarkable discovery of a linear relation between red-shift and distance, it is pertinent to inquire into the constancy of the red-shift found for different nebulae in the same cluster. The Coma cluster may be taken as an example, since it shows a considerably wider range in red-shift than any of the other (nearer) clusters where more than one measurement has been made. At the time the data in Table III were assembled by Hubble and Humason, the red-shifts for four nebulae in this cluster had been measured. Three of these nebulae gave the values 6,700, 7,600, 7,900 km./sec., with a mean of 7,500 km./sec. when corrected for solar motion. The fourth nebula gave the value 5,000 km./sec. and was excluded from the treatment on the assumption that it was a superimposed object not belonging to the cluster. Since that time Humason has obtained the red-shifts, 6,600, 6,900, 6,900, 7,000, 8,500 km./sec. for five additional members of the cluster. The total range of 1,900 km./sec. may be somewhat exaggerated since each measurement depends on a single spectrogram with the small dispersion of 875 Å. per millimetre.

It is also of interest to inquire into the constancy of the fractional red-shift for different lines in the light from the same nebula. For this purpose Dr. Hubble has kindly placed at the writer's disposal data for ten lines in the spectrum of N.G. C. 1275 in the Perseus cluster, having unshifted values ranging from $\lambda = 3,727$ to $\lambda = 5,007$. The maximum and minimum values for $\delta\lambda/\lambda$ occur at $\lambda = 4,363$ and $\lambda = 5,007$ and differ by about 14 per cent. of their mean. Both of these lines, however, are labelled 'poor'. For the first and last lines in the spectrum labelled 'good', occurring at the much more widely separated positions $\lambda = 3,727$ and $\lambda = 4,861$, the values of $\delta\lambda/\lambda$ differ by only about 3 per cent. of their mean. Within the limits of accuracy of the present data the values of $\delta\lambda/\lambda$ may be regarded as independent of λ.

(e) **Relation of apparent diameter to magnitude and distance.** Assuming ordinary Euclidean space, populated with stationary nebulae all of which have the same actual dimensions, it is evident that the apparent diameters of these objects as measured by the subtended angle $\delta\theta$ could be taken inversely proportional to their distance d in accordance with the equation†

$$\delta\theta = \frac{\text{const.}}{d};\qquad (177.16)$$

and by combining this expression with the relation between distance and magnitude given by (177.10), we obtain

$$\log\delta\theta = -0\cdot2m + c,\qquad (177.17)$$

as a relation connecting the two immediately observable quantities apparent diameter and apparent magnitude, where c is a constant.

In applying this equation to actual observations, it was found by Hubble‡ that the value of c, although reasonably constant for nebulae of any given type, had different values as might be expected for different types of nebulae. Nevertheless, the values do not vary greatly, and in the case of regular nebulae show an interesting dependence on the sequence of types of elliptical, spiral, and barred spiral forms which can be distinguished. By reducing all the nebulae to a *standard type*, it then became possible to correlate all the available data as shown below in Fig. 13, where the logarithms of apparent diameter are plotted as abscissae and the total visual magnitudes of the nebulae as ordinates. The two highest points on the plot are for the Magellanic clouds. The equation for the representative line is

$$\log\delta\theta = -0\cdot2m + 2\cdot6,\qquad (177.18)$$

† For the angles and distances under consideration there is no need to distinguish between the subtended angle and the corresponding chord.

‡ Hubble, *Astrophys. Journ.* **64**, 321 (1926).

where $\delta\theta$ is the maximum apparent diameter in minutes of arc and m is the observed visual magnitude. The equation applies of course only after the reduction to a standard type.

The correlation shown by the above figure is sufficient to confirm our general idea as to the extra-galactic position of the objects considered. A treatment of the relation between luminosity and apparent

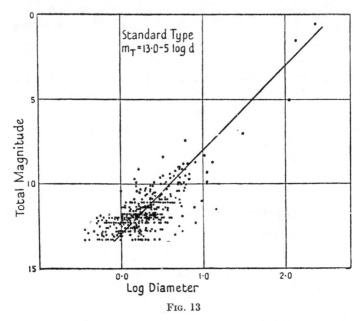

Fig. 13

diameter where the assumption of flat-space and stationary nebulae is not made will be given in § 180.

Equation (177.16) provides of course an alternative method for the determination of nebular distances, using apparent diameters instead of apparent magnitudes. In practice the method is complicated, however, not only by the presence of a wide variety in form, but also by the fact that determinations of apparent diameter are much more dependent on length of photographic exposure than those of apparent magnitude, owing to the high luminosity of the central regions of nebulae.

(f) **Actual diameters and masses of nebulae.** The actual diameters of nebulae can be calculated from their apparent diameters and distances. The figures of Hubble for the mean maximum diameters for different types of nebulae are given in Table IV.

TABLE IV

Type	Diameter in parsecs	Type	Diameter in parsecs
Elliptical Nebulae		Normal Spirals	
E_0	360	S_a	1,450
E_1	430	S_b	1,900
E_2	500	S_c	2,500
E_3	590	Barred Spirals	
E_4	700	SB_a	1,280
E_5	810	SB_b	1,320
E_6	960	SB_c	2,250
E_7	1,130	Irregular Nebulae	1,500

These figures would have to be diminished by about 15 per cent. in order to allow for the new zero point of the period-luminosity relation for the Cepheids. It is to be emphasized, however, that the values obtained are dependent on exposure time and must not be taken as definite. Present estimates[†] of the diameter of our own system as outlined by the globular clusters are of the order of 20,000 to 50,000 parsecs. Objects tentatively identified as globular clusters have recently been found in the Andromeda nebula by Hubble,[‡] and it seems probable that our own galaxy and the Andromeda nebula are of approximately the same size.

Estimates of the masses of nebulae may be made by combining figures for actual diameter with those for velocity of rotation determined with the spectroscope, by making the assumption of orbital rotation around the nucleus. They may also be obtained by Opik's method of assuming the same coefficient of emission for the material in the spirals as in our own galaxy. Using the somewhat meagre data available, Hubble[§] estimates

$$m = (6 \text{ to } 10) \times 10^8 \odot \qquad (177.19)$$

as a reasonable value for the mean mass of the nebulae, where the mass of the sun is
$$\odot = 1 \cdot 983 \times 10^{33} \text{ grammes.} \qquad (177.20)$$

(g) **Distribution of nebulae in space.** Assuming ordinary Euclidean space populated with a uniform distribution of stationary nebulae, it is evident that the number of nebulae N to be expected out to any distance d could be taken as proportional to the cube

† See Stebbins, *Proc. Nat. Acad.* **19**, 222 (1933).
‡ Hubble, *Astrophys. Journ.* **76**, 44 (1932).
§ Hubble, *Astrophys. Journ.* **79**, 8 (1934).

thereof, in accordance with the equation

$$N = \text{const.} \times d^3;$$ (177.21)

and by combining this expression with the relation between distance and magnitude given by (177.10), we obtain

$$\log N = 0 \cdot 6m + C,$$ (177.22)

where m is the limiting magnitude considered and C is a constant.

In applying this equation to actual observations, two interesting phenomena are encountered.

In the first place, there appears to be a practically complete lack of any extra-galactic nebulae at all in the plane of the Milky Way, the 'zone of avoidance' being somewhat irregular in shape but of the general order of 15° in width. The explanation of this phenomenon is doubtless to be found in the presence of a layer of obscuring material in our own galaxy. This explanation is strengthened by the presence of known clouds of material in the Milky Way which are even sufficient to obscure all but the nearer stars in our own system. The explanation is still further strengthened by data of Hubble[†] which show that nebular counts increase between the zone of avoidance and the galactic poles in the manner to be expected if the nebulae are actually seen through a layer of obscuring material.

The second phenomenon of interest is the irregularity in the density of nebular distribution, which is certainly found, unless sufficiently large ranges in depth and angular area are chosen for the individual counts. This is in any case partly due to the tendency for nebulae to be found in clusters. Thus, as emphasized by Shapley and Ames,[‡] the total number of nebulae observed out to magnitude 13 is twice as great in the northern hemisphere as in the southern hemisphere. This difference, however, can be entirely ascribed to the presence of the populous Virgo cluster in the northern hemisphere within that range of magnitudes. In addition to such effects of clustering, Shapley[§] finds out to magnitude 18·2, after correcting to uniform conditions, an excess for nebular counts in the northern as compared with the southern hemisphere. Hubble[†] finds, however, out to magnitude 20 no such difference between the two hemispheres.

Making due allowance for the obscuring effect in the Milky Way and

† Hubble, *Astrophys. Journ.* **79**, 8 (1934).
‡ Shapley and Ames, *Annals Harvard Observatory*, **88**, 43 (1932).
§ Shapley, *Proc. Nat. Acad.* **19**, 389 (1933).

the lack of uniformity corresponding to insufficient ranges in depth and area, Hubble[†] finds that the distribution of extra-galactic nebulae can be reasonably represented by an equation of the above form

$$\log N = 0\cdot6\,m - 9\cdot12, \qquad (177.23)$$

where N is the number per square degree, and m is the corrected apparent photographic magnitude.

A treatment of the density of nebular distribution which does not involve the assumption of flat space and stationary nebulae will be given in § 181.

(h) **Density of matter in space.** Making use of the best estimates now available for the mean mass of the nebulae and their density of distribution, Hubble[†] takes

$$\rho = (1\cdot3 \text{ to } 1\cdot6) \times 10^{-30} \text{ gm./cm.}^3 \qquad (177.24)$$

as an estimate for the averaged out density in space of the matter which composes the extra-galactic nebulae.

This is of course a lower limit for the actual density of matter in space, since we do not now know how much other matter may be present in the form of dust, gas, or moving particles associated with cosmic ray phenomena. Hubble estimates that the density of extra-galactic dust might be a thousand times the figure given above without having as yet been detected.

In addition to matter there is an unknown amount of radiation present in intergalactic space, including that which has come from the nebulae themselves and that which may be associated with cosmic ray phenomena. A uniform distribution of black-body radiation with a temperature of about $19°$ absolute would have a density of 10^{-30} gm./cm.^{3}

In view of the possibilities for other material besides the nebulae to be present in space, it is possible that the actual homogeneity of distribution may be much greater than would be concluded from the tendency for nebulae to occur in clusters.

For our later purposes, it will be advantageous to re-express (177.24) in the relativistic units of § 81. Doing so we can write

$$8\pi\rho = \frac{8\pi \times 10^{-30}}{1\cdot349 \times 10^{28}} \simeq 1\cdot86 \times 10^{-57} \text{ cm.}^{-2}, \qquad (177.25)$$

or changing to light years as the unit of distance

$$8\pi\rho \simeq 1\cdot7 \times 10^{-21} \text{ (yrs.)}^{-2}. \qquad (177.26)$$

Hubble, *Astrophys. Journ.* **79**, 8 (1934).

178. The relation between coordinate position and luminosity

We must now turn to the interpretation of the foregoing observational data with the help of our non-static models. In the next few sections it will first be necessary to derive a number of relations which will facilitate a comparison of the properties of such models with the data available, and a unified presentation of the correspondences between the behaviour of the model and the observed phenomena in the actual universe will have to be delayed until § 185.

In the present section we shall consider the relation between the coordinate positions of nebulae and the observed luminosities which would be expected on the basis of a non-static homogeneous model.

In carrying out the actual treatment, it will prove simplest not to use the line element in the first form that was obtained

$$ds^2 = -\frac{e^{g(t)}}{[1+r^2/4R_0^2]^2}(dr^2 + r^2\,d\theta^2 + r^2\sin^2\theta\,d\phi^2) + dt^2, \quad (178.1)$$

but in the later form

$$ds^2 = -e^{g(t)}\left(\frac{d\bar{r}^2}{1-\bar{r}^2/R_0^2} + \bar{r}^2\,d\theta^2 + \bar{r}^2\sin^2\theta\,d\phi^2\right) + dt^2, \quad (178.2)$$

which was obtained in § 149 by introducing the transformation

$$\bar{r} = \frac{r}{1+r^2/4R_0^2}, \quad (178.3)$$

Since this equation of transformation involves none of the coordinates except r, it is evident that all of our previous expressions, for example those for density and pressure, which do not depend on r, will be unchanged.

To obtain the desired relation between coordinate position and observed luminosity,† it will be simplest at the start to take the nebula as located at the origin of coordinates and the observer at the given coordinate distance of interest $\bar{r}$, both having no motion relative to the spatial coordinates in use, and hence in accordance with our previous considerations both permanently at rest with respect to the matter in their immediate neighbourhoods. As the definition of the observed luminosity l, we may take the rate which the observer finds for the energy received from the nebula in unit time and per unit area, using of course his own proper measures, and assuming no absorption between him and the nebula.

† Tolman, *Proc. Nat. Acad.* **16,** 511 (1930).

To undertake the calculation of this luminosity, let t_1 and t_2 be the respective times for the departure of light from the nebula at the origin and its arrival at the observer at the coordinate distance $\bar{r}$. In accordance with the expression for the velocity of light that corresponds to the line element (178.2), we can relate these two values of t to the distance travelled by the equation

$$\int_{t_1}^{t_2} e^{-\frac{1}{2}g(t)}\, dt = \int_{0}^{\bar{r}} \frac{d\bar{r}}{\sqrt{(1-\bar{r}^2/R_0^2)}}.$$

By differentiation—since the limits of integration of the right-hand side are constant—we then obtain

$$\frac{\delta t_2}{\delta t_1} = e^{\frac{1}{2}(g_2-g_1)} \tag{178.4}$$

as an expression which connects the time interval δt_1 between the departure of two electromagnetic disturbances from the source and the time interval δt_2 between their arrival at the observer with the values of $g(t)$ for the model, g_1 and g_2 at times t_1 and t_2.

Applying this result to the time interval between successive wave crests which leave the nebula, noting that our coordinate time t agrees with proper time both for the nebula and for the observer, we may evidently write

$$\frac{\lambda+\delta\lambda}{\lambda} = \frac{\nu}{\nu+\delta\nu} = e^{\frac{1}{2}(g_2-g_1)} \tag{178.5}$$

as an expression which relates the wave-length λ and frequency ν of light leaving the nebula with the shifted values $(\lambda+\delta\lambda)$ and $(\nu+\delta\nu)$ which it will exhibit to the observer. The expression agrees of course with our original treatment of the Doppler effect as shown by comparison with (155.7).

Furthermore, by applying the result given by (178.4) to the time interval between successive photons which we can regard as carrying energy away from the nebula, we obtain

$$\frac{z_2}{z_1} = e^{-\frac{1}{2}(g_2-g_1)} \tag{178.6}$$

as a connexion between the total rate z_1 at which photons leave the nebula, and their total observed rate of arrival z_2 at the surface around the origin defined by the coordinate distance $\bar{r}$.

Finally, moreover, it is evident, from the form of the line element

(178.2), that we may write

$$A_0 = 4\pi \bar{r}^2 e^{g_2}, \tag{178.7}$$

as an expression for the total proper area of this bounding surface defined by $\bar{r}$ through which the photons pass at time t_2.

Hence making use of these last three equations, and taking $h\nu_1$ as the mean energy of the photons that leave the nebula, it is evident that we can now write

$$l = \frac{z_1 h\nu_1 e^{g_1-g_2}}{4\pi \bar{r}^2 e^{g_2}} = \frac{z_1 h\nu_1}{4\pi \bar{r}^2 e^{g_2}} \left(\frac{\lambda}{\lambda+\delta\lambda}\right)^2 \tag{178.8}$$

as the desired expression for the observed luminosity of the nebula as defined above.

This result has been derived for simplicity taking the nebula as at the origin of coordinates and the observer at the coordinate distance $\bar{r}$. It has been shown, however, in § 149 (c) that the transformation to a new system of coordinates, having the same form of line element as given by (178.2) but having the observer at the origin, would place the nebula at the coordinate distance $\bar{r}$. Hence we may regard (178.8) as applying equally well when the observer is at the origin and the nebula at $\bar{r}$.

This now makes it easy to compare the observed luminosities for different nebulae which have different coordinate distances $\bar{r}$ but which are observed at the same time t_2 at the origin. Taking the intrinsic luminosities and hence z_1 and ν_1 as being the same for the different nebulae considered, we can immediately write from (178.8)

$$\frac{l}{l'} = \frac{\bar{r}'^2}{\bar{r}^2} \frac{(1+\delta\lambda'/\lambda')^2}{(1+\delta\lambda/\lambda)^2} \tag{178.9}$$

for the ratio of the observed luminosities of two identical nebulae located at the coordinate distances $\bar{r}$ and $\bar{r}'$, and exhibiting the fractional red-shifts $\delta\lambda/\lambda$ and $\delta\lambda'/\lambda'$ at the origin.

Introducing the transformation equation (178.3), this result can also be expressed in terms of our earlier coordinate r in the form

$$\frac{l}{l'} = \frac{r'^2}{r^2} \frac{(1+r^2/4R_0^2)^2}{(1+r'^2/4R_0^2)^2} \frac{(1+\delta\lambda'/\lambda')^2}{(1+\delta\lambda/\lambda)^2}. \tag{178.10}$$

It is the increased complexity of this form as compared with (178.9) which recommends the use of the coordinate system $(\bar{r}, \theta, \phi, t)$ for our present considerations instead of the original system (r, θ, ϕ, t) which was used in deriving the line element.

In obtaining these relations between luminosity and coordinate position, it will be noted that we assume, in addition to a homogeneous model, a constancy in the mean intrinsic luminosities of nebulae over the time intervals of the order of 10^8 years which will be involved in actual applications.

By solving (178.9) for the coordinate positions of nebulae $\bar{r}$ in terms of their luminosities l, we can evidently rewrite the result in the form

$$\bar{r} = \frac{\text{const.}}{\sqrt{l}}\left(\frac{\lambda}{\lambda+\delta\lambda}\right). \tag{178.11}$$

179. The relation between coordinate position and astronomically determined distance

With the help of the relation between luminosity and nebular position given by (178.9), we can now readily determine the relation between coordinate positions $\bar{r}$ and the computed distances d given by Hubble and Humason. To do this we must first replace luminosities by magnitudes, owing to the use of this latter quantity in the computations made by astronomers. Solving (178.9) for the ratio $\bar{r}$ to $\bar{r}'$ between the coordinate positions of two nebulae and taking logarithms we can write

$$\log\frac{\bar{r}}{\bar{r}'} = 0\cdot5\log\frac{l'}{l} + \log\frac{1+\delta\lambda'/\lambda'}{1+\delta\lambda/\lambda}, \tag{179.1}$$

and introducing the definition of magnitudes in the form

$$m-m' = 2\cdot5\log\frac{l'}{l},$$

the above can be written in the form

$$\log\frac{\bar{r}}{\bar{r}'} = 0\cdot2(m_b-m_b') + \log\frac{1+\delta\lambda'/\lambda'}{1+\delta\lambda/\lambda}, \tag{179.2}$$

where m_b and m_b' are the observed bolometric magnitudes, in agreement with the definition which we have given for luminosity in the preceding section.

To compare this expression with that used by Hubble and Humason, we must now change to photographic magnitudes by introducing the empirical relation between these two kinds of magnitude given by (177.3). Doing so, we obtain

$$\log\frac{\bar{r}}{\bar{r}'} = 0\cdot2\{m_{pg}-\Delta m_r-(HI)-(CI)-m_{pg}'+\Delta m_r'+$$
$$+(HI)'+(CI)'\}+\log\frac{1+\delta\lambda'/\lambda'}{1+\delta\lambda/\lambda}. \tag{179.3}$$

Taking $\bar{r}'$ as the coordinate for a nebula at the standard distance of 10 parsecs, and for convenience choosing the mesh system so that we can then put $\bar{r}' = 10$, it is evident—since the red-shift and its effects are negligible at that distance—that the above can be rewritten in the form

$$\log \bar{r} = 0 \cdot 2\{m_{pg} - \Delta(\Delta m_r) - \Delta(HI) - \Delta(CI) - M_{pg}\} - \log\left(\frac{\lambda + \delta\lambda}{\lambda}\right) + 1,$$

(179.4)

where m_{pg}' has been replaced by the absolute photographic magnitude M_{pg}. Noting, however, in accordance with (177.4) and (177.6), the expression for Δm_{pg} taken by Hubble and Humason as the effect of the red-shift on photographic magnitudes, this result can again be rewritten in the form

$$\log \bar{r} = 0 \cdot 2(m_{pg} - \Delta m_{pg} - M_{pg}) + 1 - 0 \cdot 5 \log\left(\frac{\lambda + \delta\lambda}{\lambda}\right),$$ (179.5)

which can be immediately compared with the expression (177.11)

$$\log d = 0 \cdot 2(m_{pg} - \Delta m_{pg} - M_{pg}) + 1$$ (179.6)

used in the calculation of distances by Hubble and Humason.

As a consequence we may now write

$$\bar{r} = d \Big/ \sqrt{\left(\frac{\lambda}{\lambda + \delta\lambda}\right)} \qquad d = \bar{r} \Big/ \sqrt{\left(\frac{\lambda + \delta\lambda}{\lambda}\right)}$$ (179.7)

as the desired relation between the distances d to the nebulae as computed by Hubble and Humason and the coordinate positions $\bar{r}$ which would be assigned to them on the basis of the line element (178.2), provided we choose the coordinate meshes for convenience so that $\bar{r} = 10$ at the standard distance of 10 parsecs at the time of interest.

The appearance of the factor $\sqrt{(1 + \delta\lambda/\lambda)}$ in these expressions is due to the fact, that the expanding model definitely assigns a Doppler effect as the cause of the red-shift and hence allows for a change in the frequency of arrival as well as in the intrinsic frequency associated with the photons which reach the observer, while the considerations of Hubble and Humason purposely allowed only for the latter of these two effects. The non-appearance of terms in R_0^2 is due to the properties of the coordinate system $(\bar{r}, \theta, \phi, t)$ which we have selected.

Since for the Leo cluster, the farthest yet examined, the difference between $\bar{r}$ and d would only be about 3 per cent, we can regard these quantities as the same within the observational error until further data are available.

180. The relation between coordinate position and apparent diameter

We may next consider the relation between the coordinate positions of nebulae and their apparent diameters,† again using the line element in the form

$$ds^2 = -e^{g(t)}\left(\frac{d\bar{r}^2}{1-\bar{r}^2/R_0^2} + \bar{r}^2\,d\theta^2 + \bar{r}^2\sin^2\theta\,d\phi^2\right) + dt^2. \quad (180.1)$$

For the purposes of the discussion we shall take the observer as permanently located at the origin of coordinates and the nebula at the coordinate distance $\bar{r}$. Furthermore, we shall take t_1 and t_2 as the times when the light which is observed leaves the nebula and arrives at the origin, this light travelling radially inward in accordance with § 154. Taking the diameter of interest as lying in the direction of $d\theta$, we can then write, in accordance with the form of the line element,

$$\delta l_0 = \bar{r}e^{\frac{1}{2}g_1}\,\delta\theta, \quad (180.2)$$

as an expression for the proper diameter δl_0 of the nebula at the time t_1 when the light is emitted, where g_1 is the value of $g(t)$ at that time, and $\delta\theta$ is the angular diameter for the nebula which will be observed at the origin.

Assuming δl_0 actually the same for the different nebulae which are being observed at the same time t_2 at the origin, we can then rewrite (180.2) in the form

$$\delta\theta = \frac{\text{const.}}{\bar{r}}\,e^{\frac{1}{2}(g_2 - g_1)}, \quad (180.3)$$

since the value g_2 for $g(t)$ at the time of observation will be the same for these different nebulae. And introducing the expression for the red-shift (178.5), this can be rewritten in the form

$$\delta\theta = \frac{\text{const.}}{\bar{r}}\left(\frac{\lambda + \delta\lambda}{\lambda}\right). \quad (180.4)$$

By combining (180.4) with the relation (178.11) between observed luminosities and coordinate position we can write

$$\frac{\delta\theta}{\sqrt{l}} = \text{const.}\left(\frac{\lambda + \delta\lambda}{\lambda}\right)^2, \quad (180.5)$$

as a relation between observed diameters, luminosities, and red-shift,

† Tolman, *Proc. Nat. Acad.* **16**, 511 (1930).

which could provide a direct empirical test of the hypothesis that
the red-shift is due to an actual expansion.

By combining (180.4) with the relation (179.7) between coordinate
position $\bar{r}$ and astronomically determined distance d, we obtain

$$\delta\theta = \frac{\text{const.}}{d}\left(\frac{\lambda+\delta\lambda}{\lambda}\right)^{\frac{3}{2}}, \tag{180.6}$$

as compared with the earlier expression (177.16) obtained by assuming stationary nebulae in ordinary Euclidean space.

181. The relation between coordinate position and counts of nebular distribution

We now turn to a treatment of the number of nebulae to be expected, from counts made out to a given coordinate distance $\bar{r}$, on
the assumption of a homogeneous expanding model. To obtain this
we may let n_0 be the number of nebulae per unit proper volume at
some selected initial time t_0, when $g(t)$ for the model has the value g_0.
In accordance with the expression for proper volume corresponding
to the form of the line element (180.1), it is then evident that we may
write for the number of nebulae between the coordinate positions
$\bar{r}$ and $\bar{r}+d\bar{r}$

$$dN = n_0\, dv_0 = \frac{4\pi n_0\, e^{\frac{1}{2}g_0}\bar{r}^2\, d\bar{r}}{\sqrt{(1-\bar{r}^2/R_0^2)}}. \tag{181.1}$$

We have shown, however, in § 153 that particles at rest with respect
to the spatial coordinates $(r,\ \theta,\ \phi)$ and hence also with respect to
$(\bar{r},\ \theta,\ \phi)$ would remain permanently so. Hence there will be no loss
or gain by a net passage of nebulae past the boundaries $\bar{r}$ and $\bar{r}+d\bar{r}$,
and (181.1) will give for all times the number of nebulae in the selected
coordinate range. Hence we may now write

$$dN = \text{const.}\frac{\bar{r}^2\, d\bar{r}}{\sqrt{(1-\bar{r}^2/R_0^2)}} \tag{181.2}$$

as a general expression for the change in nebular counts as we go to
greater and greater coordinate distances $\bar{r}$. Furthermore, from the
relation of coordinate position to luminosity and red-shift given by
(178.11), it is evident that this expression could provide means for a
direct empirical test of the actual homogeneity of nebular distribution.

Knowing the value of R_0 which can be real, infinite, or imaginary,
equation (181.2) can be integrated to give the total count out to any
given value of $\bar{r}$. From our later information as to the possible limits

for R_0^2, see (183.14), we shall find that $\bar{r}^2/R_0^2$ could hardly be greater than 2 per cent. even at the distance of the Leo cluster at some 10^8 light years. Hence it will be sufficient for many purposes to take

$$N = \text{const.}\,\bar{r}^3, \tag{181.3}$$

as an expression for the expected number of nebulae out to any given value of $\bar{r}$ so far considered. By substituting equations (178.11), (179.7), and (180.4), this result can also be written in the variety of forms

$$N = \text{const.}\,\bar{r}^3,$$

$$N = \text{const.}\,d^3\left(\frac{\lambda}{\lambda+\delta\lambda}\right)^{\frac{3}{2}},$$

$$N = \frac{\text{const.}}{l^{\frac{3}{2}}}\left(\frac{\lambda}{\lambda+\delta\lambda}\right)^3, \tag{181.4}$$

$$N = \frac{\text{const.}}{(\delta\theta)^3}\left(\frac{\lambda+\delta\lambda}{\lambda}\right)^3,$$

for the expected number of nebulae out to a given value of the coordinate $\bar{r}$, astronomically determined distance d, bolometric luminosity l, or apparent diameter $\delta\theta$, where $\delta\lambda/\lambda$ is the observed red-shift for nebulae at that limit.† The second of these expressions is to be compared with the earlier expression (177.21) obtained by assuming stationary nebulae in ordinary Euclidean space.

182. The relation between coordinate position and red-shift

We may next consider the relation between coordinate position and observed red-shift, still using the coordinates $(\bar{r},\ \theta,\ \phi,\ t)$ which have been found specially convenient for the correlation of astronomical data, and which correspond to the line element for the homogeneous model when written in the form

$$ds^2 = -e^{g(t)}\left(\frac{d\bar{r}^2}{1-\bar{r}^2/R_0^2}+\bar{r}^2\,d\theta^2+\bar{r}^2\sin^2\theta\,d\phi^2\right)+dt^2. \tag{182.1}$$

In accordance with the general treatment of the Doppler effect for homogeneous models given in § 155 (see equation 155.8), or the incidental treatment given in § 178 (see equation 178.5), we can write

$$\frac{\delta\lambda}{\lambda} = e^{\frac{1}{2}(g_2-g_1)}-1 \tag{182.2}$$

---

† The possibility of using such expressions for nebular counts to test the Einstein-de Sitter model with $R_0^2 = \infty$ and $\Lambda = 0$, was presented in a lecture given by Professor Einstein at the California Institute of Technology in the winter of 1932.

as an expression for the fractional red-shift in the wave-length of nebular light as observed at the origin, where g_1 is the value of $g(t)$ for the model when the light leaves the nebula at time t_1 and g_2 is its value when the light is observed at the origin at time t_2.

In applying this equation to observational data, it is evident that g_2 can be treated as a constant, since we actually observe different nebulae at our own location, which we take as the origin, all at the same time. The quantity g_1, however, will have to be regarded as a variable since by going to more and more distant nebulae we go to earlier and earlier times of emission and hence to changed values of g_1. It is hence evident that the relation between red-shift and nebular distance will depend on the form of the relation between g and t.

In order to have a definite expression for the form of $g(t)$, we could, of course, select a model having some one of the various possible types of time-behaviour discussed in Part II of this chapter, and then use the corresponding expression for $g(t)$. Such a selection, however, would have to be made at the present time mainly on the basis of metaphysical predilections. For our present purposes, it will be better to adopt a much more phenomenological point of view and endeavour to obtain what information we can as to an appropriate form for $g(t)$ by comparison with observational data.

To undertake this it will be most convenient to regard $g(t)$ as developed into a power series in t around the present time $t_2 = 0$, which we take for convenience as the starting-point for temporal measurements.† This form of development seems reasonable in view of the obvious rationality of taking $g(t)$ as a continuous function, and in view of the known approximate linearity of red-shift with distance. We may then write $g(t)$ as the series

$$g(t) = 2(kt+lt^2+mt^3+...),\qquad(182.3)$$

where $k, l, m,...$, are constant coefficients, the factor 2 has been introduced to avoid later fractions, and higher terms will for the present be neglected.

The omission from the series of a constant term in t^0 evidently involves no loss in generality, and by giving $g(t)$ the convenient value $g_2 = 0$ at the present time $t_2 = 0$, makes the line element (182.1) reduce to the special relativity form in the neighbourhood of the origin and at the present time, and makes it possible to rewrite the

† Tolman, *Proc. Nat. Acad.* **16**, 409 (1930).

expression (182.2) for the red-shift observed at the origin in the form

$$\frac{\delta\lambda}{\lambda} = e^{-\frac{1}{2}g_1} - 1 = e^{-(kt_1 + lt_1^2 + mt_1^3 + \dots)} - 1, \qquad (182.4)$$

where t_1 is the time in the past when the observed light left the nebula. To compare with astronomical data, however, it will be more convenient to have the red-shift expressed, not in terms of the time t_1 of light emission from the nebula, but as a power series in terms of the coordinate distance $\bar{r}$ to the nebula in question. To obtain such an expansion we shall need the values at $\bar{r} = 0$ for the successive derivatives of $\delta\lambda/\lambda$ with respect to $\bar{r}$.

For the first derivative we can write in accordance with (182.4)

$$\frac{d}{d\bar{r}}\left(\frac{\delta\lambda}{\lambda}\right) = -\tfrac{1}{2}e^{-\frac{1}{2}g_1}\frac{dg_1}{dt_1}\frac{dt_1}{d\bar{r}}, \qquad (182.5)$$

where $dt_1/d\bar{r}$ is the change in the time with change in the coordinate position of the nebula considered. And in accordance with the expression for the velocity of light which corresponds to the line element (182.1) we can evidently write therefor

$$\frac{dt_1}{d\bar{r}} = -\frac{e^{\frac{1}{2}g_1}}{\sqrt{(1 - \bar{r}^2/R_0^2)}}, \qquad (182.6)$$

which on substitution into (182.5) gives us

$$\frac{d}{d\bar{r}}\left(\frac{\delta\lambda}{\lambda}\right) = \frac{1}{2}\frac{1}{\sqrt{(1 - \bar{r}^2/R_0^2)}}\frac{dg_1}{dt_1} \qquad (182.7)$$

in agreement with our previous equation (156.6).

By similar treatments we may obtain the higher derivatives of $\delta\lambda/\lambda$. Doing so, introducing the expression for $g(t)$ given by (182.3), and taking the values for the derivatives at $r = 0$, we then finally obtain

$$\left[\frac{d}{d\bar{r}}\left(\frac{\delta\lambda}{\lambda}\right)\right]_{\bar{r}=0} = k$$

$$\left[\frac{d^2}{d\bar{r}^2}\left(\frac{\delta\lambda}{\lambda}\right)\right]_{\bar{r}=0} = -2l$$

$$\left[\frac{d^3}{d\bar{r}^3}\left(\frac{\delta\lambda}{\lambda}\right)\right]_{\bar{r}=0} = \frac{k}{R_0^2} + 2kl + 6m, \qquad (182.8)$$

$$\cdot \quad \cdot \quad \cdot \quad \cdot \quad \cdot \quad \cdot \quad \cdot \quad \cdot$$

$$\cdot \quad \cdot \quad \cdot \quad \cdot \quad \cdot \quad \cdot \quad \cdot \quad \cdot$$

where it is specially pleasurable to note that terms depending on the spatial curvature corresponding to R_0^2 do not appear until the third derivative.

With the help of these expressions we may now express the red-shift as a function of the coordinate position of the nebula in the form of the Maclaurin's series

$$\frac{\delta\lambda}{\lambda} = k\bar{r}-l\bar{r}^2+\left(\frac{k}{6R_0^2}+\tfrac{1}{3}kl+m\right)\bar{r}^3+\ldots. \qquad (182.9)$$

In applying this result to the observational data for the actual universe, taking the light year as the unit of distance, we may evidently put

$$k = 5\cdot71\times10^{-10} \text{ (yrs.)}^{-1} \qquad (182.10)$$

for the coefficient of the first term in the series, in accordance with the observations of Hubble and Humason as given by equation (177.15) where d may be replaced by $\bar{r}$ within the limits of observational error.

Furthermore, since the red-shift is actually found within the limits of error to increase approximately linearly with $\bar{r}$ out to the Leo cluster at about 10^8 light years, it is evident that we can place some restriction on the range of permissible values of the coefficients of the following terms. If we take the plot of observed red-shift against distance given by Fig. 12, as indicating that the deviations from a simple linear formula $\delta\lambda/\lambda = k\bar{r}$ should not exceed 1 per cent. at 10^7 light years, should not greatly exceed 3 per cent. at 3×10^7 light years, and should not exceed 18 per cent. at 10^8 light years, we are led to assign

$$|l| < 5\times10^{-19} \text{ (yrs.)}^{-2}, \qquad (182.11)$$

and

$$\left|\frac{k}{6R_0^2}+\tfrac{1}{3}kl+m\right| < 5\times10^{-27} \text{ (yrs.)}^{-3} \qquad (182.12)$$

as reasonable upper limits for the values of these coefficients without reference to sign.

These upper limits would produce the following percentage deviations from the simple formula $\delta\lambda/\lambda = k\bar{r}$ at the various distances given.

TABLE V

Distance in light years	1×10^7	2×10^7	3×10^7	6×10^7	10×10^7
Term in $\bar{r}^2$ / Term in $\bar{r}$	0·9%	1·8%	2·6%	5·3%	8·8%
Term in $\bar{r}^3$ / Term in $\bar{r}$	0·1%	0·4%	0·8%	3·2%	8·8%

183. The relation of density to spatial curvature and cosmological constant

We turn now to a comparison of the estimated density, as given by (177.26), for the matter in the universe in the form of nebulae

$$8\pi\rho = 1{\cdot}7 \times 10^{-21} \text{ (yrs.)}^{-2} \qquad (\rho = 10^{-30} \text{ gm./cm.}^3), \qquad (183.1)$$

with our expressions for pressure, total density, and density of matter in the model as given by (150.7), (150.8), and (150.10). In terms of our series expansion for $g(t)$, as given by (182.3), these expressions can be written as applying at the present time $t = 0$ in the form

$$8\pi p_0 = -\frac{1}{R_0^2} - 4l - 3k^2 + \Lambda, \qquad (183.2)$$

$$8\pi\rho_{00} = \frac{3}{R_0^2} + 3k^2 - \Lambda, \qquad (183.3)$$

and

$$8\pi\rho_m = \frac{6}{R_0^2} + 12l + 12k^2 - 4\Lambda, \qquad (183.4)$$

where the density of matter is taken as $\rho_m = \rho_{00} - 3p_0$ on the approximate basis discussed in § 150, which would be entirely valid if we could regard the pressure in the model as due solely to the radiation present.

These expressions are a little difficult to handle owing to our meagre observational information and to the simultaneous appearance of the two quantities R_0^2 and Λ concerning which we have as yet no information. Nevertheless, since the pressure in the model cannot be less than zero, and the density of matter cannot be less than that actually seen in the form of nebulae, we may write

$$0 < -\frac{1}{R_0^2} - 4l - 3k^2 + \Lambda, \qquad (183.5)$$

and

$$1{\cdot}7 \times 10^{-21} < \frac{6}{R_0^2} + 12l + 12k^2 - 4\Lambda, \qquad (183.6)$$

and by eliminating first Λ and then R_0^2 from these inequalities and combining with our previous knowledge as to the value of k and the limits imposed on l as given in the preceding section, we can readily obtain as pretty reliable lower limits

$$-1 \times 10^{-18} < \frac{1}{R_0^2} \qquad (183.7)$$

and

$$-2 \times 10^{-18} < \Lambda. \qquad (183.8)$$

The upper limits for these quantities are more uncertain. It would seem reasonable, however, to assume that the total density of matter and radiation present could hardly be greater than 1,000 times the value given for the density of matter in the nebulae which would give us

$$\frac{3}{R_0^2}+3k^2-\Lambda < 1{\cdot}7\times 10^{-18}.\tag{183.9}$$

And since radiation has the highest possible ratio of pressure to density we can also evidently write

$$-\frac{3}{R_0^2}-12l-9k^2+3\Lambda < \frac{3}{R_0^2}+3k^2-\Lambda,\tag{183.10}$$

and by using these inequalities together with our previous information as to k and l can set the upper limits

$$\frac{1}{R_0^2} < 2{\cdot}1\times 10^{-18},\tag{183.11}$$

and $$\Lambda < 5{\cdot}7\times 10^{-18}.\tag{183.12}$$

Furthermore, making use of the information we now have as to k, l, and $(1/R_0^2)$ in connexion with our previous expression (182.12), we now find that the limits for m itself would be given by

$$-5{\cdot}3\times 10^{-27} < m < 5{\cdot}2\times 10^{-27}.\tag{183.13}$$

For convenience of reference we may collect the information obtained in this and the preceding section as to permissible values in the form

$$k = 5{\cdot}71\times 10^{-10}\,(\text{yrs.})^{-1}$$
$$-5\times 10^{-19} < l < 5\times 10^{-19}\,(\text{yrs.})^{-2}$$
$$-5{\cdot}3\times 10^{-27} < m < 5{\cdot}2\times 10^{-27}\,(\text{yrs.})^{-3}\tag{183.14}$$
$$-1\times 10^{-18} < \frac{1}{R_0^2} < 2{\cdot}1\times 10^{-18}\,(\text{yrs.})^{-2}$$
$$-2\times 10^{-18} < \Lambda < 5{\cdot}7\times 10^{-18}\,(\text{yrs.})^{-2}.$$

It is interesting to note that the range of possible values given by (183.14) is such that we should not be justified in assuming that the original de Sitter line element was necessarily a good approximation for the behaviour of the actual universe. In accordance with (142.10), the de Sitter line element is a special case of the line element (182.1) now being used, which can be obtained by taking $1/R_0^2$, l, m, ... equal to zero. In the expressions for density and pressure, however, terms of the order Λ and $1/R_0^2$ occur additively and it is not evident from (183.14) that the latter could be neglected in comparison with the former.

184. The relation between red-shift and rate of disappearance of matter

In § 152 we have derived an expression for the fractional rate at which the mass of matter in a non-static model would be disappearing as the result of emission of radiation from the nebulae, or of processes of synthesis or annihilation of matter in internebular space which might be leading to the production of a radiational component of the cosmic rays. The expression obtained (152.7) was an approximate one to the extent that the density of matter ρ_m was taken as the total density ρ_{00} minus the density of radiation which was assigned the value $3p_0$, but was otherwise exact.

Using our present series expansion for $g(t)$

$$g(t) = 2(kt + lt^2 + mt^3 + ...)\qquad (184.1)$$

we can now write the previous expression (152.7), for the rate at which mass would be disappearing, in the form

$$-\frac{1}{M}\frac{dM}{dt} = 3\left[\frac{\rho_{00} + \tfrac{5}{8}p_0}{\rho_m} - \frac{1}{4\pi\rho_m}\left(4l + \frac{6m}{k}\right)\right]k,\qquad (184.2)$$

where higher coefficients in the series than k, l, m would in any case not occur. It is evident that this result might imply some restriction on the values of l and m in addition to those already found.

We may first consider the case of a perfectly linear expression for $g(t)$ with l and m equal to zero. In accordance with (182.9) and the limits which we have found for $(1/R_0^2)$ this would also imply a very closely linear expression for the red-shift as a function of distance. Under these circumstances with

$$l = m = 0,\qquad (184.3)$$

the rate of decrease in the mass of matter present would at least be as great as

$$-\frac{1}{M}\frac{dM}{dt} = 3k,\qquad (184.4)$$

since the numerator of the first term on the right-hand side of (184.2) can in any case not be less than the denominator.

Giving k its known observational value, however, this would imply a very rapid rate of decrease in the mass associated with matter.

This is illustrated by the following table[†] in which the value of $3k$ is compared with the known rates for the loss of mass by the emission of radiation from different types of stars.

TABLE VI

Generation of Energy by Typical Stars

Star	Ergs per Gramme per Second	$-\dfrac{1}{M}\dfrac{dM}{dt}\,(Yrs.)^{-1}$
H.D. 1337 A	15,000	$5\cdot3 \times 10^{-10}$
B.D. 6° 1309 A	(11,000)	$3\cdot9 \times 10^{-10}$
V Puppis A	1,100	$3\cdot9 \times 10^{-11}$
Betelgeux	(300)	$1\cdot1 \times 10^{-11}$
Capella A	48	$1\cdot7 \times 10^{-12}$
Sirius A	29	$1\cdot0 \times 10^{-12}$
Sun	$1\cdot90$	$6\cdot6 \times 10^{-14}$
α Centauri B	$0\cdot90$	$3\cdot2 \times 10^{-14}$
60 Kruger B	$0\cdot02$	$7\cdot0 \times 10^{-16}$
$3k$	50,000	$17\cdot1 \times 10^{-10}$

Hence unless we should be willing to allow the possibility of a higher average rate for the general transformation of the mass of matter into radiation, even than that observed in the star H.D. 1337A which at present has the highest known ratio of luminosity to mass, we should have to conclude that the dependence of $g(t)$ on t could not be strictly linear. It is conceivable, nevertheless, that a high rate of transformation of internebular matter into radiation might be connected with the production of cosmic rays.

It is interesting to note that the above conclusion, that $g(t)$ could not be an exactly linear function of t, also implies in accordance with (182.9) that the fractional red-shift $\delta\lambda/\lambda$ could not be expected to be an exactly linear function of the coordinate distance to the nebulae $\bar{r}$.

In addition, it is of interest to realize that our present considerations might also imply a further complication in regarding the original de Sitter line element, for an empty model, as providing an approximately satisfactory representation of the phenomena of the actual universe. To see this, we note from (142.10), as remarked in the preceding section, that the de Sitter line element is a special case of the line element (182.1) now being used, which can be obtained by taking $(1/R_0^2)$ as equal to zero and setting $g(t)$ exactly equal to $2kt$. With

† The first two columns are from Jeans, *Astronomy and Cosmogony*.

$g(t)$ linear, however, we should have the exceedingly high rate of annihilation given by the last line in Table VI for that matter which must be regarded as actually present, even though we use an empty model as our first approximation.

Having seen that the values of l and m cannot be taken as exactly zero unless we are willing to allow an extraordinarily high rate of transformation, we may now return to (184.2) and examine into the values of l and m which would be necessary to permit as low a rate of transformation as might be demanded by observational considerations. Since it is safe to assume that the density of radiation in the universe could hardly be more than of the same order as the density of matter, it is evident from (184.2) that we could reduce the rate of transformation down to the value zero

$$-\frac{1}{M}\frac{dM}{dt} = 0,$$

if we are allowed values of l and m large enough so that we could take

$$4l + \frac{6m}{k} \simeq 8\pi\rho_m. \tag{184.5}$$

Referring to (183.1), however, we should hardly wish to set $8\pi\rho_m$ greater than a thousand times the observed minimum corresponding to the mass of the nebulae, which would give us

$$4l + \frac{6m}{k} \simeq 1\cdot7 \times 10^{-18}, \tag{184.6}$$

and in accordance with (183.14) we can take

$$4l \leqslant 2 \times 10^{-18}, \tag{184.7}$$

and

$$\frac{6m}{k} \leqslant 5\cdot5 \times 10^{-17}, \tag{184.8}$$

as upper limits without disagreeing with the observed extent to which the red-shift has been found linear with distance.

Hence we can conclude that the approximate equality (184.6) could be satisfied, and as low a rate be assigned to the transformation of the matter in our model into radiation as may be found empirically necessary without controverting any observational data so far established. The desirability of more precise information as to the actual values of l and m is of course evident.

185. Summary of correspondences between model and actual universe

This completes the derivation of special relations needed for comparing the properties of non-static homogeneous models with the phenomena of the actual universe, and we may now undertake a unified presentation of the correspondences which can be established. In a general way it can be said that there are no essential conflicts between model and reality, and that the specific correspondences which can be presented are sufficient to make the model appear quite helpful in interpreting the behaviour of the actual universe at least out to some 10^8 light years.

To present these correspondences we may write the line element for the model in the form which we have found convenient

$$ds^2 = -e^{2(kt+lt^2+mt^3+\ldots)}\left(\frac{d\bar{r}^2}{1-\bar{r}^2/R_0^2}+\bar{r}^2\,d\theta^2+\bar{r}^2\sin^2\theta\,d\phi^2\right)+dt^2, \quad (185.1)$$

where $g(t)$ is expressed by a series expansion around the present time $t = 0$, and we take ourselves for convenience as located at the origin $\bar{r} = 0$. Furthermore, taking the light year and the year as the units of distance and time, we may assign in accordance with (183.14) as appropriate numerical values to consider in correlating the model with observational data,

$$k = 5\cdot71\times10^{-10}\,(\text{yrs.})^{-1}$$

$$-5\times10^{-19} < l < 5\times10^{-19}\,(\text{yrs.})^{-2}$$

$$-5\cdot3\times10^{-27} < m < 5\cdot2\times10^{-27}\,(\text{yrs.})^{-3} \qquad (185.2)$$

$$-1\times10^{-18} < \frac{1}{R_0^2} < 2\cdot1\times10^{-18}\,(\text{yrs.})^{-2}$$

$$-2\times10^{-18} < \Lambda < 5\cdot7\times10^{-18}\,(\text{yrs.})^{-2}.$$

As the first satisfactory feature of the model, we have the spatial isotropy and homogeneity which it exhibits. This is in agreement with our present observational findings, which on a large scale show no outstanding dependence on direction and indicate no preferred properties for our own location in the universe. With more extensive information, the change to a non-homogeneous model may become necessary as will be emphasized in the next section.

As a second feature of the model, we have its agreement with the findings of Hubble as to the relation between the computed distances to the nebulae and their apparent diameter and observed density of distribution. To show this, we have the relation between coordinate position $\bar{r}$ in the model and the computed distances d to the nebulae as obtained by Hubble and Humason

$$\bar{r} = d \sqrt{\{\lambda/(\lambda+\delta\lambda)\}}, \qquad (185.3)$$

where $\delta\lambda/\lambda$ is the observed red-shift in the light from the nebula under consideration; and we have—below at the left and right respectively —the theoretical expressions for the observed diameter $\delta\theta$ at a given position and the nebular count N out to a given position, together with the empirical expressions taken by Hubble as approximately fitting the observations.

$$\delta\theta = \frac{\text{const.}}{\bar{r}} \left(\frac{\lambda+\delta\lambda}{\lambda}\right) \qquad\qquad \delta\theta = \frac{\text{const.}}{d} \qquad (185.4)$$

$$N = \text{const.} \int_0^{\bar{r}} \frac{\bar{r}^2\, dr}{\sqrt{(1-\bar{r}^2/R_0^2)}} \qquad\qquad N = \text{const.}\, d^3. \qquad (185.5)$$

Owing to the small values of $\delta\lambda/\lambda$ and $\bar{r}^2/R_0^2$, even at 10^8 light years, and the approximate character of the observational data, we may regard the agreement between theory and observation as entirely satisfactory.

As a third very important feature of the model, we have its unstrained explanation of the observed red-shift in the light from the nebulae as due to a mutual recession of these objects. As the theoretical expression for this red-shift we have

$$\frac{\delta\lambda}{\lambda} = k\bar{r} - l\bar{r}^2 + \left(\frac{k}{6R_0^2} + \tfrac{1}{3}kl + m\right)\bar{r}^3 + \dots \qquad (185.6)$$

as compared with the empirical expression of Hubble and Humason

$$\frac{\delta\lambda}{\lambda} = kd. \qquad (185.7)$$

As shown in § 182, the range in possible numerical values given by (185.2) for the coefficients of the higher order terms is such, that the two expressions agree within a reasonable estimate as to the accuracy of the empirical formula.

As a fourth feature of the model, we have the conclusion to be drawn from (185.6) that the fractional red-shift in the light from any given nebula should be independent of the particular wave-length examined. This agrees with the data available as discussed at the end of § 177 (d).

As a fifth very satisfactory feature, we have the necessary presence of matter in the models. The numerical values given by (185.2) are such that the density of this matter could not be less than the 10^{-30} gm./cm.^{3} which may be estimated as the averaged-out density of matter actually seen in the form of nebulae, such that the total density of all matter and radiation could not be greater than 1,000 times this value, and such that the pressure could not be less than zero.

Finally, as a sixth feature of the model, it has been found that the numerical values allowed by (185.2) are such, that the rate at which the mass of matter in the model is decreasing in favour of free radiation could be assigned any value, from zero up to and beyond that for the star having the highest known ratio of luminosity to mass, as may be made necessary by further observational information. Thus the model permits the flow of radiation from the stars, and indeed must be non-static if this occurs, but prescribes no impossible figure for the amount of the flow.

In addition to these direct correspondences between the properties of the model and observed phenomena, it should not be overlooked that the basis upon which the model has been constructed is furnished by the relativistic theory of gravitation, which—over smaller distances than those now involved—has itself received excellent confirmation. Furthermore, it may be emphasized again that this theory has in any case indicated the impossibility of constructing a stable static model of the universe, so that some red-shift or violet-shift in the light from distant objects is at least to be expected.

It will be seen from the foregoing, that the degree of correspondence between the properties of the model and observed phenomena and the lack of any essential conflict are sufficient to give us considerable confidence in a cautious use of our theory in interpreting the behaviour of the actual universe.

A number of obvious suggestions present themselves as to further observational research.

It is of course very desirable to extend the observations as to large-scale homogeneity as far as possible. The establishment of a significant difference between near and far parts of the universe or between the northern and southern hemispheres, as to density of nebular distribution, or as to the relation between red-shift and distance would be very important and if found might provide the empirical basis for change to a non-homogeneous model.

A verification of the exact form of the predicted relation (180.5) between apparent diameters and luminosities

$$\frac{\delta\theta}{\sqrt{l}} = \text{const.}\left(\frac{\lambda+\delta\lambda}{\lambda}\right)^2 \qquad (185.8)$$

would lend strong support to the hypothesis of nebular recession, since this relation would not necessarily hold with other explanations of the red-shift. The test would be complicated by the difficulties of handling the data on diameters.

Similarly a verification of the relation between luminosity and nebular counts provided by the two equations (178.11) and (181.2)

$$\bar{r} = \frac{\text{const.}}{\sqrt{l}}\left(\frac{\lambda}{\lambda+\delta\lambda}\right) \qquad (185.9)$$

and

$$N = \text{const.}\int_0^{\bar{r}} \frac{\bar{r}^2\,d\bar{r}}{\sqrt{(1-\bar{r}^2/R_0^2)}} \qquad (185.10)$$

would test both the theory of recession and the hypothesis of homogeneous distribution. Both this and the foregoing test might be complicated when carried to the needed distances by the effect of intervening obscuration, or by the failure of the hypothesis that the nebulae have properties which can be regarded as constant over the time intervals involved. Indeed the main result of the tests might be to establish the probability of such effects.

Further investigation of the red-shift as a function of distance will be exceedingly important. At present we do not even know the sign of the second term in the series expression for $\delta\lambda/\lambda$ as a function of $\bar{r}$, and hence cannot say whether the rate of the mutual recession of the nebulae is increasing or decreasing with time. The answer to this question might be made possible by the use of the two-hundred inch reflector now under construction.

More information as to the contents of the universe in addition to the visible nebulae will also be important. As already indicated, the presence of intergalactic gas or dust may sometime be detected from the obscuration that they produce, and increased knowledge concerning the source and nature of the cosmic rays may soon be available. With more complete information as to the presence of internebular material, our limits for the possible values of $1/R_0^2$ and Λ might be considerably narrowed.

Concerning some of these questions and concerning others now less obvious, we can—in the near future—confidently expect increased observational knowledge. And it is observation rather than hypothesis that must dictate the final nature of our cosmological theory.

186. Some general remarks concerning cosmological models

In the present section we shall make some general remarks concerning the homogeneity, spatial curvature, and temporal behaviour to be ascribed to cosmological models. In the preceding section we have emphasized the specific correspondences that can be established between observational data and the properties of a model appropriately constructed in accordance with the principles of relativistic mechanics. In the present section, on the other hand, we shall be more impressed by the lack of sufficient observational data to permit a unique determination of all the characteristics for a reasonably successful cosmological picture that we might wish to know.

(a) **Homogeneity.** We may first consider the justification for ascribing spatial isotropy and hence also—as we have seen in § 148—spatial homogeneity to the models that we have investigated. A very practical justification for this procedure lies in the definiteness and mathematical tractability of the models that we thereby obtain. And a more real justification lies in the high degree of large-scale homogeneity actually observed.

On the other hand, from a smaller scale point of view, it is evident that there is a great tendency for the nebulae to occur in clusters. Hence the finer details of cosmic behaviour could not in any case be represented by a perfectly homogeneous model. Thus, for example, it should be clearly appreciated that the lower singular state of exactly zero radius, which might be thought of as occurring in the case of an oscillatory time behaviour, must be regarded as the attribute of a certain class of homogeneous models, and not as a state that would necessarily accompany an oscillating expansion and contraction of the whole or parts of the real universe.

Furthermore, even from a large-scale point of view, it is evident that we have no knowledge as to conditions in the actual universe beyond some 10^8 light years. Hence it is entirely possible, that other densities of distribution, or contraction instead of expansion, may be present in portions of the universe beyond the reach of our present telescopes. An investigation of the forces of control which distant parts of the universe could exert on each other would be very important. It is possible that these forces would not be sufficient to maintain uniform conditions throughout, a point which has also been emphasized in conversation by the writer's colleague, Professor Zwicky. The use of homogeneous models must hence be regarded as commendable on grounds of mathematical convenience for obtaining a suitable first approximation, as inappropriate, nevertheless, for the treatment of finer details, and as subject to possible important modification when data on more distant portions of the universe become available.†

(b) **Spatial curvature.** Adopting a homogeneous model as a satisfactory first approximation, the limits placed on the possible values of $1/R_0^2$ by known observational data are sufficiently wide as shown by (185.2), so that this quantity might actually be positive, zero, or negative. Hence we do not now have sufficient information to distinguish definitely between the three cases of a model which is closed, open and spatially uncurved, or open and spatially curved.

Even if we introduce the special but reasonable assumption that the cosmological constant Λ is to be assigned the value zero, we cannot definitely determine the sign of $1/R_0^2$. Making this assumption, the expression for density could be written in the form

$$8\pi\rho_{00} = \frac{3}{R_0^2} + 3k^2. \tag{186.1}$$

For $8\pi\rho_{00}$, however, we have felt it necessary to take the range of possible values from $1\cdot7\times10^{-21}$ to $1\cdot7\times10^{-18}$, while $3k^2$ has the approximate value 1×10^{-18}. It is interesting to see from this, nevertheless, that $1/R_0^2$ would have to be negative and the model open, unless the actual density were considerably larger than that which can be observed in the form of nebular material.

It is further evident that the known data do not conflict with the proposal of Einstein and de Sitter to take $1/R_0^2$ and Λ both equal to zero, as discussed in § 164. It is also interesting to note that the specific Einstein-de Sitter model, which could be obtained by taking the pressure equal to zero as in equation (164.5), can be shown to lead to values for $\ddot{g}$ and $\dddot{g}$ such that l and m would lie within the range given by (185.2) as compatible with the accuracy of the linear relation between red-shift and distance.

Although we do not have the necessary observational data to decide

† Compare Tolman, *Proc. Nat. Acad.* **20**, 169 (1934).

between open and closed models, two remarks of a somewhat metaphysical character may be made in connexion with the problem. On the one hand, it might be urged, as has been done at least in conversation by Professor Lemaître, that the hypothesis of a closed and hence finite model was an 'optimistic' one to make, since an infinite universe could not be regarded in its totality as an object susceptible to scientific treatment. On the other hand, it might be equally urged, nevertheless, that there has been nothing in the whole past history of scientific endeavour to indicate that the field of its investigations would ever be exhausted. Indeed, the goal of science has always appeared to present the character of a receding horizon. Hence on *a priori* grounds an open model might perhaps seem equally probable.

(c) **Temporal behaviour.** The observational data as summarized by (185.2) are also insufficient to make any decision as to the kind of temporal behaviour that should be ascribed to the model over long periods of time. We can, to be sure, assert with some confidence that the universe in our immediate neighbourhood is now undergoing an expansion. Until we have information as to the sign of the second derivative of the red-shift as a function of distance, however, we cannot say whether the rate of expansion is increasing with time as we might expect for a model which will ultimately arrive in the empty de Sitter state, or is decreasing with time as we might expect for a model which is undergoing oscillations.

Indeed, by making the specific hypothesis that the pressure in the model can be taken as zero, it can be shown that the extreme cases—of the Lemaître model (161.11) with $\Lambda = \Lambda_E$, $R_0^2 > 0$ which expands from an original static state, of the Einstein model (163.3) with $\Lambda = 0$, $R_0^2 > 0$ which oscillates between a lower singular state and a maximum, and of the Einstein-de Sitter model (164.6) with $\Lambda = 0$, $R_0^2 = \infty$ which expands from a singular state—can all three be adjusted to give a behaviour at the present time which would lie within the limits which we have assigned as possible for the density of matter, and for the accuracy of the linear relation of red-shift to distance. Hence we cannot now distinguish between the various possible types of time behaviour which were discussed at the end of Part II of this chapter, and must regard those discussions as presenting different conceptual possibilities rather than as immediately applicable.

We have shown in §§ 163 and 164, for the two cases of the Einstein model and the Einstein-de Sitter model, that the elapsed time of the expansion since the singular state would have to be short, being given by

$$\Delta t \leqslant \frac{4}{3\dot{g}},$$

or

$$\Delta t \leqslant \frac{2}{3k}, \qquad (186.2)$$

in terms of our series expansion. Hence for these models the elapsed time since the singular state could not be much greater than 10^9 years, which is of the order of the age of the earth. Furthermore, from the known value of the red-shift and its approximate linearity, it appears roughly in general that the major part of the past expansion has quite probably taken place in a past time of the order of 10^9 to 10^{10} years. In view of the much longer time scale, of the order of 10^{12} years, usually regarded as necessary for stellar evolution, some discussion of the short time thus probably involved in cosmic expansion is necessary.

In the first place in connexion with this apparent difficulty as to time scales,† it is to be emphasized that the highly idealized homogeneous models which we have employed can hardly be regarded as adequate for drawing any exact conclusions as to the precise state of the actual universe say 10^9 years ago. Thus, as already mentioned earlier in this section, it is evident that the unique singular state at the lower limit of volume from which the expansion would appear to start in the case of certain models must be regarded as a property of the homogeneous model rather than a character that could actually be found in the real universe. Furthermore, since we do not know the behaviour of the universe at distances beyond our own neighbourhood out to some 10^8 light years, it is evident that calculations of the exact time when the expansion for some given model started cannot be regarded as having a precise application to the real universe, and we can merely roughly conclude that the time of expansion for our own neighbourhood might well be of the general order of 10^9 to 10^{10} years.

† It should be appreciated that the disagreement as to time scales is not to be resolved by some trick of substituting a new time-like variable in place of our present coordinate t. In accordance with § 149 (d), the coordinate t itself would agree with the time measurements made on a natural clock at rest in our own galaxy and hence be equally suitable for recording stellar evolution or the approach and recession of other nebulae.

In the second place, it is to be emphasized, as has been done particularly by de Sitter,† that there is no necessity for regarding the beginning of the expansion as in any sense the beginning of the universe, and no reason for expecting an identity between the time scales for stellar evolution and nebular expansion. Indeed de Sitter would regard the unhomogeneous structure of the nebulae, their high velocities of rotation, and the apparent date of the birth of our own planetary system, as all being pieces of evidence in agreement with a close approach of pre-existing nebulae or galaxies some 10^9 to 10^{10} years ago.

The difference between the time scales for stellar evolution and nebular expansion suggests that no definiteness could now be attached to any idea as to *the* beginning of the physical universe. Indeed, it is difficult to escape the feeling that the time span for the phenomena of the universe might be most appropriately taken as extending from minus infinity in the past to plus infinity in the future. The classical thermodynamic arguments against such a view must certainly be somewhat modified in the light of the increased possibilities of behaviour provided by relativistic thermodynamics, and would be subject to even more serious modification if the principle of energy conservation should fail within the interior of stars as suggested possible by Bohr.

187. Our neighbourhood as a sample of the universe as a whole

It is evident from the foregoing that our present data are insufficient to provide a precise cosmological model which would necessarily correspond to the actual universe in all regions and over all time intervals. It is hence best to regard the line element which we have used for investigating the behaviour of the universe

$$ds^2 = -e^{2(kt+ll^2+ml^3+\dots)}\left(\frac{d\bar{r}^2}{1-\bar{r}^2/R_0^2}+\bar{r}^2\,d\theta^2+\bar{r}^2\sin^2\theta\,d\phi^2\right)+dt^2$$

$$(187.1)$$

as a first approximation, suitable for treating events not too far distant from our own location, at times not too remote from the present.

On the basis of the ideas that we thus gain, it seems reasonable to conclude that the expansion of the universe—for which we find evi-

† de Sitter, *Proc. Amsterdam Acad.* **35**, 596 (1932).

dence at the present time and in our own neighbourhood—is a phenomenon which has progressed during a past time at least of the order of 10^8 years and which will presumably continue for a comparable time in the future. Furthermore, we are reasonably safe in believing that the density of nebular distribution and the rate of expansion will be found to persist, with roughly unchanged values, perhaps to several times the distances of the order of 10^8 light years already investigated.

For the treatment of the whole universe in all its regions and during all of time we have, nevertheless, no adequate model, and to obtain ideas as to its complete nature can only rely on the roughest methods of scientific induction. To apply such methods we must proceed by regarding that portion of the universe which we have already studied as a fair sample, but not as an exact sample of the whole at all times and places.

Having discovered an expanding distribution of nebulae as far as our telescopes can penetrate, we may reasonably regard the presence of matter in relative motion as a typical feature of the universe. Nevertheless, to ascribe to this matter everywhere the same density and stage of evolutionary development which we now find in our own neighbourhood, and to exclude the possibility at all times and places of motions of contraction which are mechanically as simple as those of expansion, would be to regard our own present neighbourhood not only as a fair sample but quite unjustifiably as an exact replica of the whole.

It may seem somewhat ironic to conclude our elaborate treatment of the properties and temporal behaviour of specific cosmological models, with words which disparage their applicability to the actual universe. Their study, however, has certainly informed us as to conceptual possibilities, and has provided a provisional and approximate theoretical background which has already been successful in correlating a considerable number of the phenomena of the real universe.

As a final remark it is desirable to emphasize the special necessity in the field of cosmology of avoiding the evils of autistic or wish-fulfilling thinking. In the first place, the problems of cosmology are necessarily extensive and intricate and must be attacked in the light of very meagre information. Hence, we must be careful not to substitute the comfortable certainties of some simple mathematical model in place of the great complexities of the actual universe. In the second place, it is evident that the past history of the universe and the future fate of man are involved in the issue of our studies. Hence we must be specially careful to keep our judgements uninfected by the demands of theology and unswerved by human hopes and fears. The discovery of models, which start expansion from a singular state of zero volume, must not be confused with a proof that the actual universe was created at a finite time in the past. And the discovery of models, which could expand and contract irreversibly without ever coming to a final state of maximum entropy and rest, must not be confused with a proof that the actual universe will always provide a stage for the future role of man.

It is appropriate to approach the problems of cosmology with feelings of respect for their importance, of awe for their vastness, and of exultation for the temerity of the human mind in attempting to solve them. They must be treated, however, by the detailed, critical, and dispassionate methods of the scientist.

SYMBOLS FOR QUANTITIES

A subscript $_0$ or superscript ^{0} attached to a symbol usually designates a proper quantity as measured by a local observer. (Note exception in case of $R = R_0 e^{ig(t)}$.)

Scalar quantities (Italic type).

a	Stefan-Boltzmann constant.
A	Free energy. Number of molecules in a mol.
c	Velocity of light. Concentration.
d	Distance as determined astronomically.
e	Electric charge. Base of natural logarithms.
E	Energy.
F	Thermodynamic potential.
$g(t), g$	Function giving time dependence of line element for homogeneous cosmological models.
h	Planck's constant.
i	$\sqrt{-1}$.
k	Boltzmann's constant. Newton's constant of gravitation.
l	Luminosity of heavenly object.
m	Mass. Magnitude of heavenly object.
n	Number of mols.
N	Number of molecules.
p	Pressure.
Q	Heat.
r	Radial coordinate.
R	Gas constant.
$R_0 e^{ig(t)} = R$	Radius of cosmological model.
S	Entropy.
t	Time.
T	Temperature.
u	Velocity. Density of radiation.
U	Energy.
v	Volume. Velocity.
δv_0	Element of proper spatial volume.
V	Relative velocity of coordinate axes.
W	Work.
x, y, z	Spatial coordinates.

α Degree of dissociation.

ϵ Dielectric constant.

η Integrating factor.

θ, ϕ, χ Polar coordinates.

$\delta\theta$ Apparent diameter of a nebula.

κ Gravitational constant connecting energy-momentum tensor with contracted Riemann-Christoffel tensor.

λ Wave-length.

Λ Cosmological constant.

μ Magnetic permeability.

ν Frequency.

ρ Density.

ρ_{00} Proper macroscopic density of energy.

ρ_0 Proper density of electric charge.

σ Electrical conductivity.

τ Period.

ϕ Scalar potential. Entropy density.

ψ Newtonian gravitational potential.

Vector quantities (Clarendon type).

A   Vector potential.

B   Magnetic induction.

C   Density of conduction current.

D   Electric displacement.

E   Electric field strength.

F   Force.

f   Force acting on a unit cube.

g   Density of momentum.

G   Total momentum.

H   Magnetic field strength.

J   Current density.

M   Angular momentum. Magnetic polarization.

P   Electric polarization.

s   Density of energy flow.

u   Velocity.

Tensors (Italic type with indices).

Latin indices i, j, k, etc., assume values 1, 2, 3.

Greek indices $\alpha, \beta,..., \mu, \nu,...$, etc., assume values 1, 2, 3, 4.

ds Invariant interval.

$\delta_{\mu\nu}$ Galilean values of metrical tensor.

F^μ Minkowski force.

$F^{\mu\nu}$ Field tensor, electron theory.

$F^{\mu\nu}, H^{\mu\nu}$ Field tensors, macroscopic theory.

$g_{\mu\nu}$ Fundamental metrical tensor.

g Determinant $|g_{\mu\nu}|$.

$h_{\mu\nu}$ Deviations from Galilean values of $g_{\mu\nu}$.

J^μ Generalized current. Components of momentum and energy.

p_{ij} Components of (absolute) stress.

$R^\tau_{\mu\nu\sigma}$ Riemann-Christoffel tensor.

$R_{\mu\nu}$ Contracted Riemann-Christoffel tensor.

R Invariant obtained from Riemann-Christoffel tensor.

t_{ij} Components of (relative) stress.

$T^{\mu\nu}$ Energy-momentum tensor.

Tensor densities (German type).

$\mathfrak{F}^{\mu\nu}$ Electric field tensor density.

$\mathfrak{g}^{\mu\nu} = g^{\mu\nu}\sqrt{-g}$.

$\mathfrak{g}^{\mu\nu}_\alpha = \dfrac{\partial}{\partial x^\alpha}(g^{\mu\nu}\sqrt{-g})$.

$\mathfrak{J}^\mu$ Current vector density.

$\mathfrak{L}$ Lagrangian function (a pseudo scalar).

$\mathfrak{t}^\nu_\mu$ Pseudo tensor density of potential energy and momentum.

$\mathfrak{T}^\nu_\mu$ Tensor density of material energy and momentum.

APPENDIX II

SOME FORMULAE OF VECTOR ANALYSIS

Unit vectors parallel to axes   i, j, k.   (1)

Unit vector normal to a surface   n.   (2)

Resolution of vector into components:

$$\mathbf{F} = F_x\mathbf{i} + F_y\mathbf{j} + F_z\mathbf{k}. \tag{3}$$

Inner product of vectors:

$$(\mathbf{A \cdot B}) = A_xB_x + A_yB_y + A_zB_z = AB\cos(\mathbf{AB}). \tag{4}$$

Outer product of vectors:

$$[\mathbf{A \times B}] = (A_yB_z - A_zB_y)\mathbf{i} + (A_zB_x - A_xB_z)\mathbf{j} + (A_xB_y - A_yB_x)\mathbf{k}. \tag{5}$$

Normal component of vector:

$$A_n = (\mathbf{A \cdot n}) = A \cos(\mathbf{An}). \qquad (6)$$

The vector operator del:

$$\nabla = \left(\mathbf{i}\frac{\partial}{\partial x} + \mathbf{j}\frac{\partial}{\partial y} + \mathbf{k}\frac{\partial}{\partial z} \right). \qquad (7)$$

$$\operatorname{grad}\phi = \nabla\phi = \mathbf{i}\frac{\partial\phi}{\partial x} + \mathbf{j}\frac{\partial\phi}{\partial y} + \mathbf{k}\frac{\partial\phi}{\partial z}. \qquad (8)$$

$$\operatorname{div}\mathbf{A} = (\nabla\cdot\mathbf{A}) = \frac{\partial A_x}{\partial x} + \frac{\partial A_y}{\partial y} + \frac{\partial A_z}{\partial z}. \qquad (9)$$

$$\operatorname{curl}\mathbf{A} = [\nabla\times\mathbf{A}] = \left(\frac{\partial A_z}{\partial y} - \frac{\partial A_y}{\partial z}\right)\mathbf{i} + \left(\frac{\partial A_x}{\partial z} - \frac{\partial A_z}{\partial x}\right)\mathbf{j} + \left(\frac{\partial A_y}{\partial x} - \frac{\partial A_x}{\partial y}\right)\mathbf{k}. \qquad (10)$$

$$\operatorname{div}\operatorname{curl}\mathbf{A} = 0. \qquad (11)$$

The Laplacian operator:

$$\nabla^2 = \nabla\cdot\nabla = \left(\frac{\partial^2}{\partial x^2} + \frac{\partial^2}{\partial y^2} + \frac{\partial^2}{\partial z^2}\right). \qquad (12)$$

$$\operatorname{curl}\operatorname{curl}\mathbf{F} = \operatorname{grad}\operatorname{div}\mathbf{F} - \nabla^2\mathbf{F}. \qquad (13)$$

Gauss's theorem:

$$\int_{\text{vol}} (\nabla\cdot\mathbf{A})\,dv = \int_{\text{surf}} A_n\,d\sigma.$$

$$\int_{\text{vol}} \left(\frac{\partial A_x}{\partial x} + \frac{\partial A_y}{\partial y} + \frac{\partial A_z}{\partial z}\right) dv$$
$$= \int_{\text{surf}} \{A_x\cos(nx) + A_y\cos(ny) + A_z\cos(nz)\}\,d\sigma. \qquad (14)$$

Stokes's theorem:

$$\int_{\text{line}} \mathbf{A}\cdot d\mathbf{s} = \int_{\text{surf}} [\operatorname{curl}\mathbf{A}]_n\,d\sigma. \qquad (15)$$

Green's theorem:

$$\int_{\text{vol}} (\phi\nabla^2\psi - \psi\nabla^2\phi)\,dv = \int_{\text{surf}} (\phi\nabla\psi - \psi\nabla\phi)_n\,d\sigma. \qquad (16)$$

Another integral theorem:

$$\int_{\text{vol}} (\mathbf{A}\cdot\operatorname{curl}\mathbf{B} - \mathbf{B}\cdot\operatorname{curl}\mathbf{A})\,dv = -\int_{\text{surf}} [\mathbf{A}\times\mathbf{B}]_n\,d\sigma. \qquad (17)$$

The Dalembertian operator:

$$\left(\nabla^2 - \frac{1}{c^2}\frac{\partial^2}{\partial t^2}\right) = \left(\frac{\partial^2}{\partial x^2} + \frac{\partial^2}{\partial y^2} + \frac{\partial^2}{\partial z^2} - \frac{1}{c^2}\frac{\partial^2}{\partial t^2}\right). \qquad (18)$$

Solution of 'wave equation':

$$\left(\nabla^2 - \frac{1}{c^2}\frac{\partial^2}{\partial t^2}\right)\psi = \omega,$$

$$\psi(x, y, z, t) = -\frac{1}{4\pi}\int \frac{[\omega]}{r}\,dv, \tag{19}$$

where $[\omega]$ is the value of ω at location of volume element dv and at time $t - r/c$.

APPENDIX III

SOME FORMULAE OF TENSOR ANALYSIS

(a) GENERAL NOTATION.

Indices α, β,..., μ, ν,..., etc., assume values 1, 2, 3, 4. $\qquad$ (1)

Covariant indices as subscripts; contravariant indices as superscripts.

Coordinate systems. $\qquad\qquad\qquad\qquad\qquad\qquad\qquad\qquad\qquad\qquad$ (2)

$$x^\mu = x^1, x^2, x^3, x^4$$
$$x'^\mu = x'^1, x'^2, x'^3, x'^4$$
$$\text{etc.}$$

where $\qquad\qquad x'^\mu = x'^\mu(x^1, x^2, x^3, x^4).$

Summation convention for dummy indices. $\qquad\qquad\qquad\qquad\qquad$ (3)

$$A^\alpha B_\alpha = \sum_{\alpha=1}^{\alpha=4} A^\alpha B_\alpha = A^1 B_1 + A^2 B_2 + A^3 B_3 + A^4 B_4$$

$$A^{\alpha\beta} B_{\alpha\beta} = \sum_{\alpha=1}^{\alpha=4}\sum_{\beta=1}^{\beta=4} A^{\alpha\beta} B_{\alpha\beta} = A^{11} B_{11} + A^{12} B_{12} + ... + A^{44} B_{44},$$

etc. One of a pair of dummies always covariant and the other contravariant.

Definition of a tensor.

A collection of 4^r components (with the rank r equal to the total number of indices α, β,..., μ, ν,..., etc.) which are associated with a given point x^μ in the manifold, and are transformed to new values on a transformation of coordinates in accordance with the rule

$$T'^{\mu\nu...}_{\rho\sigma...} = \frac{\partial x'^\mu}{\partial x^\alpha}\frac{\partial x'^\nu}{\partial x^\beta}\frac{\partial x^\gamma}{\partial x'^\rho}\frac{\partial x^\delta}{\partial x'^\sigma}...T^{\alpha\beta...}_{\gamma\delta...}. \tag{4}$$

Examples.

Tensor of rank zero (scalar invariant):

$$S' = S. \tag{5}$$

Contravariant tensor of rank one (vector):

$$A'^{\mu} = \frac{\partial x'^{\mu}}{\partial x^{\alpha}} A^{\alpha}. \tag{6}$$

Covariant tensor of rank one:

$$A'_{\mu} = \frac{\partial x^{\alpha}}{\partial x'^{\mu}} A_{\alpha}. \tag{7}$$

Mixed tensor of rank two:

$$T'^{\nu}_{\mu} = \frac{\partial x'^{\nu}}{\partial x^{\alpha}} \frac{\partial x^{\beta}}{\partial x'^{\mu}} T^{\alpha}_{\beta}. \tag{8}$$

Symmetrical tensor: $\qquad T^{\mu\nu} = T^{\nu\mu}. \tag{9}$

Antisymmetrical tensor:

$$F^{\mu\nu} = -F^{\nu\mu}. \tag{10}$$

(b) The Fundamental Metrical Tensor and its Properties.

The metrical tensor: $\qquad g_{\mu\nu} = g_{\nu\mu}. \tag{11}$

The infinitesimal difference in coordinate position:

$$dx^{\mu} = dx^1, dx^2, dx^3, dx^4. \tag{12}$$

The scalar interval ds corresponding to dx^{μ}:

$$ds^2 = g_{\mu\nu} dx^{\mu} dx^{\nu}. \tag{13}$$

The determinant formed from the components $g_{\mu\nu}$:

$$g = |g_{\mu\nu}|. \tag{14}$$

The normalized minor:

$$g^{\mu\nu} = \frac{|g_{\mu\nu}|_{\text{minor}}}{g}. \tag{15}$$

The mixed tensor:

$$g^{\nu}_{\mu} = \delta^{\nu}_{\mu} = \begin{cases} 1 & \mu = \nu \\ 0 & \mu \neq \nu. \end{cases} \tag{16}$$

The Galilean values of the $g_{\mu\nu}$:

$$\delta_{\mu\nu} = \pm 1, 0. \tag{17}$$

The Christoffel three-index symbols:

$$[\mu\nu, \sigma] = \tfrac{1}{2}\left(\frac{\partial g_{\mu\sigma}}{\partial x^{\nu}} + \frac{\partial g_{\nu\sigma}}{\partial x^{\mu}} - \frac{\partial g_{\mu\nu}}{\partial x^{\sigma}}\right),$$
$$\{\mu\nu, \sigma\} = \tfrac{1}{2} g^{\sigma\lambda}\left(\frac{\partial g_{\mu\lambda}}{\partial x^{\nu}} + \frac{\partial g_{\nu\lambda}}{\partial x^{\mu}} - \frac{\partial g_{\mu\nu}}{\partial x^{\lambda}}\right). \tag{18}$$

The Riemann-Christoffel tensor:

$$R^{\tau}_{\mu\nu\sigma} = \{\mu\sigma, \alpha\}\{\alpha\nu, \tau\} - \{\mu\nu, \alpha\}\{\alpha\sigma, \tau\} + \frac{\partial}{\partial x^{\nu}}\{\mu\sigma, \tau\} - \frac{\partial}{\partial x^{\sigma}}\{\mu\nu, \tau\}. \tag{19}$$

The equation for a geodesic:

$$\delta \int ds = 0 \text{ is equivalent to } \frac{d^2 x^\sigma}{ds^2} + \{\mu\nu, \sigma\} \frac{dx^\mu}{ds} \frac{dx^\nu}{ds} = 0. \quad (20)$$

(c) TENSOR MANIPULATIONS.

The raising, lowering, and change of indices (examples):

$$A^\nu = g^{\nu\alpha} A_\alpha \quad (21)$$

$$A_\mu = g_{\mu\alpha} A^\alpha \quad (22)$$

$$A^\nu = g^\nu_\alpha A^\alpha. \quad (23)$$

Contraction (examples):

$$T = T^\nu_\nu = g_{\nu\alpha} T^{\nu\alpha} = T^1_1 + T^2_2 + T^3_3 + T^4_4 \quad (24)$$

$$R_{\mu\nu} = \{\mu\sigma, \alpha\}\{\alpha\nu, \sigma\} - \{\mu\nu, \alpha\}\{\alpha\sigma, \sigma\} + \frac{\partial}{\partial x^\nu}\{\mu\sigma, \sigma\} - \frac{\partial}{\partial x^\sigma}\{\mu\nu, \sigma\}. \quad (25)$$

Addition (example):

$$A_\mu = B_\mu + C_\mu = (B_1 + C_1), (B_2 + C_2), (B_3 + C_3), (B_4 + C_4). \quad (26)$$

Outer product (example):

$$\begin{aligned} A^\nu_\mu = B_\mu C^\nu = \quad & B_1 C^1 \quad B_1 C^2 \quad B_1 C^3 \quad B_1 C^4 \\ & B_2 C^1 \quad B_2 C^2 \quad B_2 C^3 \quad B_2 C^4 \\ & B_3 C^1 \quad B_3 C^2 \quad B_3 C^3 \quad B_3 C^4 \\ & B_4 C^1 \quad B_4 C^2 \quad B_4 C^3 \quad B_4 C^4. \end{aligned} \quad (27)$$

Inner product (example):

$$A = A^\nu_\nu = B_\nu C^\nu = B_1 C^1 + B_2 C^2 + B_3 C^3 + B_4 C^4. \quad (28)$$

Covariant differentiation (examples):

$$(A^\mu)_\nu = A^\mu_\nu = \frac{\partial A^\mu}{\partial x^\nu} + \{\alpha\nu, \mu\} A^\alpha \quad (29)$$

$$(A_\mu)_\nu = A_{\mu\nu} = \frac{\partial A_\mu}{\partial x^\nu} - \{\mu\nu, \alpha\} A_\alpha \quad (30)$$

$$(T^{\mu\nu})_\sigma = T^{\mu\nu}_\sigma = \frac{\partial T^{\mu\nu}}{\partial x^\sigma} + \{\alpha\sigma, \mu\} T^{\alpha\nu} + \{\alpha\sigma, \nu\} T^{\mu\alpha} \quad (31)$$

$$(T^\nu_\mu)_\sigma = T^\nu_{\mu\sigma} = \frac{\partial T^\nu_\mu}{\partial x^\sigma} + \{\alpha\sigma, \nu\} T^\alpha_\mu - \{\mu\sigma, \alpha\} T^\nu_\alpha \quad (32)$$

$$(T^{..\nu...}_{.\mu...})_\sigma = \frac{\partial T^{..\nu...}_{..\mu...}}{\partial x^\sigma} + \{\alpha\sigma, \nu\} T^{..\alpha...}_{..\mu...} \text{ for each contravariant index}$$
$$ - \{\mu\sigma, \alpha\} T^{..\nu...}_{..\alpha...} \text{ for each covariant index.} \quad (33)$$

Divergence: $\qquad\qquad (T^{..\nu...}_{......})_\nu. \quad (34)$

(d) MISCELLANEOUS FORMULAE.

$$(g_{\mu\nu})_\sigma = 0. \tag{35}$$

$$\{\mu\nu, \sigma\} = \{\nu\mu, \sigma\}. \tag{36}$$

$$\{\alpha\sigma, \sigma\} = \frac{\partial}{\partial x^\alpha} \log \sqrt{-g}. \tag{37}$$

$$T^{\alpha\beta} \, dg_{\alpha\beta} = -T_{\alpha\beta} \, dg^{\alpha\beta}. \tag{38}$$

$$\frac{dg}{g} = g^{\alpha\beta} \, dg_{\alpha\beta} = -g_{\alpha\beta} \, dg^{\alpha\beta}. \tag{39}$$

$$(\phi_\mu)_\nu - (\phi_\nu)_\mu = \frac{\partial \phi_\mu}{\partial x^\nu} - \frac{\partial \phi_\nu}{\partial x^\mu}. \tag{40}$$

$$(F_{\mu\nu})_\sigma + (F_{\nu\sigma})_\mu + (F_{\sigma\mu})_\nu = \frac{\partial F_{\mu\nu}}{\partial x^\sigma} + \frac{\partial F_{\nu\sigma}}{\partial x^\mu} + \frac{\partial F_{\sigma\mu}}{\partial x^\nu},$$

provided $F_{\mu\nu} = -F_{\nu\mu}$. $\tag{41}$

$$\phi_{\mu\nu\sigma} - \phi_{\mu\sigma\nu} = \phi_\epsilon R^\epsilon_{\mu\nu\sigma},$$

with $\phi_{\mu\nu\sigma} = ((\phi_\mu)_\nu)_\sigma$ and $\phi_{\mu\sigma\nu} = ((\phi_\mu)_\sigma)_\nu$. $\tag{42}$

$$(T_{..\mu...})_{\nu\sigma} - (T_{..\mu...})_{\sigma\nu} = \sum T_{..\epsilon...} \, R^\epsilon_{\mu\nu\sigma},$$

where the summation $\sum$ is for all the original indices μ. $\tag{43}$

(e) FORMULAE INVOLVING TENSOR DENSITIES.

$$\mathfrak{T} = T\sqrt{-g}. \tag{44}$$

$$\mathfrak{T}^{..\nu...}_{..\mu...} = T^{..\nu...}_{..\mu...} \sqrt{-g}. \tag{45}$$

$$(A^\mu)_\mu \sqrt{-g} = \mathfrak{A}^\mu_\mu = \frac{\partial}{\partial x^\mu}(A^\mu\sqrt{-g}) = \frac{\partial \mathfrak{A}^\mu}{\partial x^\mu}. \tag{46}$$

$$(T^\nu_\mu)_\nu \sqrt{-g} = \mathfrak{T}^\nu_{\mu\nu} = \frac{\partial \mathfrak{T}^\nu_\mu}{\partial x^\nu} - \tfrac{1}{2}\mathfrak{T}^{\alpha\beta}\frac{\partial g_{\alpha\beta}}{\partial x^\mu} = \frac{\partial \mathfrak{T}^\nu_\mu}{\partial x^\nu} + \tfrac{1}{2}\mathfrak{T}_{\alpha\beta}\frac{\partial g^{\alpha\beta}}{\partial x^\mu},$$

provided $T^{\mu\nu} = T^{\nu\mu}$. $\tag{47}$

$$F^{(\mu\nu)}_\nu \sqrt{-g} = \mathfrak{F}^{\mu\nu}_\nu = \frac{\partial \mathfrak{F}^{\mu\nu}}{\partial x^\nu},$$

provided $F^{\mu\nu} = -F^{\nu\mu}$. $\tag{48}$

(f) FOUR-DIMENSIONAL VOLUME. PROPER SPATIAL VOLUME.

When limits of integration correspond to a given four-dimensional region, we have the invariant

$$I = \iiiint \sqrt{-g'} \, dx'^1 dx'^2 dx'^3 dx'^4 = \iiiint \sqrt{-g} \, dx^1 dx^2 dx^3 dx^4. \tag{49}$$

When region is small enough to permit natural coordinates x, y, z, t or proper coordinates x_0, y_0, z_0, t_0, we have

$$\delta I = \iiiint dx\,dy\,dz\,dt = \iiiint dx_0\,dy_0\,dz_0\,dt_0$$
$$= \iiiint \sqrt{-g}\; dx^1 dx^2 dx^3 dx^4. \quad (50)$$

Hence we can take

$$\delta I = \delta v \delta t = \delta v_0\, \delta t_0 = \delta v_0\, \delta s = \sqrt{-g}\; \delta x^1 \delta x^2 \delta x^3 \delta x^4, \quad (51)$$

where δv and δv_0 are elements of spatial volume and δt and $\delta t_0 = \delta s$ elements of time in natural and in proper coordinates respectively.

APPENDIX IV

USEFUL CONSTANTS*

Stefan-Boltzmann constant $\quad a = 7{\cdot}623_7 \times 10^{-15}$ erg cm.$^{-3}$ deg.$^{-4}$

Avogadro's number $\quad\quad\quad\quad A = 6{\cdot}064_{36} \times 10^{23}$ mol.$^{-1}$

Velocity of light $\quad\quad\quad\quad\quad c = 2{\cdot}99796 \times 10^{10}$ cm. sec.$^{-1}$

Charge of electron $\quad\quad\quad\quad e = 4{\cdot}770 \times 10^{-10}$ abs. e.s. units.

Specific charge of electron $\quad e/m = 5{\cdot}279_{41} \times 10^{17}$ abs. e.s. units gm.$^{-1}$

Planck's constant $\quad\quad\quad\quad\quad h = 6{\cdot}547 \times 10^{-27}$ erg sec.

Boltzmann's constant $\quad\quad\quad k = 1{\cdot}3708_9 \times 10^{-16}$ erg deg.$^{-1}$

Newton's constant (gravita- $\quad k = 6{\cdot}664 \times 10^{-8}$ dyne cm.^{2} gm.$^{-2}$
tion)

Gas constant $\quad\quad\quad\quad\quad\quad\quad R = 8{\cdot}3136_0 \times 10^7$ erg deg.$^{-1}$ mol.$^{-1}$
$\quad\quad\quad\quad\quad\quad\quad\quad\quad\quad = 1{\cdot}9864_3$ cal. deg.$^{-1}$ mol.$^{-1}$

Transformation from relativistic to c.g.s. units:

l, t, m in relativistic units.

L, T, M in c.g.s. units.

$L = l$ cm.

$$T = \frac{1}{2{\cdot}998 \times 10^{10}} t = 3{\cdot}335 \times 10^{-11}\, t \text{ sec.}$$

$$M = \frac{(2{\cdot}998 \times 10^{10})^2}{6{\cdot}664 \times 10^{-8}} m = 1{\cdot}349 \times 10^{28}\, m \text{ gm.}$$

1 parsec $= 3{\cdot}258$ light years $= 3{\cdot}084 \times 10^{18}$ cm.

1 light year $= 9{\cdot}463 \times 10^{17}$ cm.

1 sidereal year $= 3{\cdot}1558 \times 10^7$ sec.

* In part from Birge, *Phys. Rev.*, Supplement, **1**, 1 (1929).

SUBJECT INDEX

Boltzmann-Stefan equation, 139.
Boundary conditions for electrical media, 110.

Carnot cycle involving velocity, 159.
Christoffel symbols, 173.
Clock, effect of potential on rate, 192.
Clock paradox, 194.
Complete static system, 80.
Conservation of angular momentum, 77.
Conservation of electric charge, 85, 109, 264.
Conservation of mass, energy, and momentum: for electromagnetic systems, 89–93, 113–16.
for mechanical systems, 59–62, 74–7.
for particles, 42, 49, 51.
in general relativity, 225–9, 261, 285.
Constants, Table of, 497.
Constitutive equations, 102, 104, 108, 263.
Coordinates: co-moving, 301, 364; Galilean, 37; proper, 33, 180; natural, 180.
Cosmic rays, 58, 151, 382, 385, 418, 476.
Cosmological constant, 189, 191, 341, 344, 402, 412, 473.
Cosmological models, general remarks, 5, 10, 332, 361, 482.
Cosmological models (static), 331.
Einstein model: behaviour of particles and light rays, 341; comparison with actual universe, 344; density and pressure, 339; derivation of line element, 335; Doppler effect, 343; geometry, 337; instability, 405; thermodynamic equilibrium, 423.
de Sitter model: behaviour of particles and light rays, 349; comparison with actual universe, 359, 474, 476; density and pressure zero, 348; derivation of line element, 335; Doppler effect, 354; geometry, 346.
Cosmological models (non-static, homogeneous), 361.
Application of mechanics, 361; application of thermodynamics, 420; assumption of spatial isotropy, 362; behaviour of light rays, 387; behaviour of particles, 383; change in energy with time, 379; change in matter with time, 381; correspondence to actual universe, 478; density and pressure, 376; density of matter, 379; density related to spatial curvature and cosmological constant, 473; derivation of line element, 364; different expressions for line element and interpretation thereof, 370, 375; Doppler effect, 389, 392; Einstein-de Sitter open model, 415; Friedmann model with conservation of energy, 408; geometry, 371; Lemaître model with conservation of mass, 408; models which expand continuously from a static state, 409; models which expand continuously from a non-static state, 412; models with transformation of matter into radiation, 434, 441, 475; oscillatory behaviour, 401, 402, 404, 412, 435, 439; periodic behaviour, 401, 429; reversible and irreversible behaviour 417, 424, 426, 427, 432, 439; time dependence for closed models, 394; time dependence for open models, 403; transfer of origin of coordinates, 372, 464; transformation of matter into radiation, 362, 417, 434, 441, 475.
Current vector, 95, 103, 107, 258, 262.

Dalembertian, 268.
Deflexion of light in gravitational field, 209, 285.
Density of matter in space, 461.
Dimensions of space-time, 29.
Doppler effect: general treatment, 288; in Einstein model, 343; in non-static model, 389, 392; in de Sitter model, 354.

Electrodynamics, 84, 258.
Electromagnetic field tensors, 96, 103, 259.
Emission theories of light, 16.
Energy content, 120.
Energy-momentum tensor, for matter, 71, 189, 215; for electricity 99, 115, 261; for perfect fluid, 216; for radiation, 217, 269, 273.
Entropy content, 121.

Entropy vector, 164, 294.
Equilibrium: between hydrogen and helium, 140; between matter and radiation, 146; chemical, 129, 311; in Einstein universe, 405, 423; thermal, 130, 312, 315; thermodynamic, 125, 127, 307, 308.
Euclidean space, 31.

Field: corresponding to flow of radiation, 273; electromagnetic, 84, 95, 102, 105, 258, 262; gravitational, 185–91; of charged particle, 265; spherical, 239, 241, 250; weak, 236.
Final state of a system, 134, 326.
First law of thermodynamics, 120, 152, 292.
First postulate of relativity, 12.
Free energy, 123.

Galilean transformation, 21.
General relativity, 165.
Geodesic, 172.
Geometry corresponding to space-time, 30.

Heat content, 123.
Heat, relativistic interpretation, 297.

Interval, 31, 169, 181.
Isotopes, 144.

Joule heat, 112.

Kennedy-Thorndike experiment, 14.
Kinetic energy, 47.

Lagrangian function, 222.
Light rays and particles, interaction of, 285.
Line element: Einstein, 335; non-static homogeneous universe, 364; Schwarzschild, 202, 349; de Sitter, 335.
Lorentz-Fitzgerald contraction, 13.
Lorentz rotation, 32.
Lorentz transformation, 18; for acceleration, 27; for contraction factor, 27; for electromagnetic densities and stresses, 92; for electromagnetic field, 87, 106; for energy, 154; for entropy, 157; for force, 46; for heat, 157; for mass, 45; for mechanical densities and stresses, 64, 68, 69; for pressure, 154; for temperature, 158; for velocity, 25; for volume, 153; for work, 156.

Macroscopic density, 68.
Macroscopic electrodynamics, 261.
Magnitudes, see Nebulae.
Mass, energy and momentum, relations between, 48.
Mass: longitudinal, 55; of electron, 53; of particle, 43; of radiation, 271; transverse, 55.
Maxwell-Lorentz field equations, 84, 258.
Maxwell's equations, 101.
Mechanics, 42, 214.
Metric and gravitation, 176.
Michelson-Morley experiment, 13.
Minkowski force, 52.

Nebulae: actual diameters and masses, 458; distances, 453; distribution in space, 459; magnitudes, 446, 448; relation of coordinate position, to apparent diameter 467, to distance 465, to luminosity 462, to nebular counts 468, to red-shift 469; relation of magnitude, to apparent diameter 457, to distance 453, to nebular counts 461, to red-shift 454.
Newton's theory as a first approximation, 198.

Pencil of light, 274.
Perfect fluid, behaviour of, 218.
Perfect gas, 136.
Perihelion, advance of, 208.
Planck law, 140.
Planetary motion, 205.
Poisson's equation, 185, 188, 199.
Potential: generalized electromagnetic, 96, 258; gravitational, 183; Newtonian, 199; scalar, 86; thermodynamic, 123; vector, 86.
Poynting vector, 90.
Principle of covariance, 166.
Principle of equivalence, 174.
Principle of Mach, 184.
Proper coordinates, 33, 180.
Proper quantities, use of, 7.
Proper volume, 496.
Pulse of light, 279.

Radiation: black body, 139; dynamics of black body, 161; energy-momentum tensor, 217, 269, 272; flow of, 272; mass, 271.
Red-shift, see Nebulae.
Relativity, of uniform motion, 12; of all kinds of motion, 176.
Reversibility and irreversibility, 121, 294, 296, 424.

Reversibility and rate, 132, 319.
Riemann-Christoffel tensor, 185; contracted 187.
Right-angled lever, 79.

Sakur-Tetrode equation, 138.
Sampling of universe, 486.
Schur's theorem, 368, 372.
Second law of thermodynamics, 121, 152, 162, 293, 296.
Second postulate of relativity, 15.
Signature of line element, 31.
Space and time, ideas as to, 17.
Space-time continuum, 28.
Spatial contraction, 22.
Spatial isotropy, 362, 364.
Special relativity, 12.
Stresses: electromagnetic 91, 115; mechanical, 60, 69.
Symbols for quantities, 489.

Tensor analysis, 34, 493.
Tensors: electromagnetic field tensors, 96, 103, 259; current vector, 95, 103, 107, 258, 262; energy-momentum tensor, for matter 71, 189, 215, for electricity 99, 115, 261, for perfect fluid 216, for radiation 217, 269, 273; entropy vector, 164, 294; metrical tensor, 36, 183, 494; pseudo-tensor of potential energy and momentum, 224; Riemann-Christoffel tensor,185.
Thermodynamic potential, 123.
Thermodynamics, 118, 291.
Third law of thermodynamics, 122.
Time dilation, 22.
Time scale, 412, 414, 416, 485.
Trajectories of particles and light rays, 171, 182.

Units used in general relativity, 201, 497.

Vector analysis, 491.

Waves: electromagnetic, 85, 267; gravitational, 239.
Wave-length, gravitational shift in, 211, 286.
Weight and mass, proportionality, 192.
Work, 47, 120, 156.

NAME INDEX

Adams, 212.
Ames, 460.
Anderson, 58.

Bainbridge, 58.
Birge, 138, 139, 142, 201, 497.
Birkhoff, 252.
Blackett, 58.
Bohr, 382, 486.
Boltzmann, 352, 353.
Born, 100.
Bradley, 144.

Campbell, 211.
Chazy, 209.
Comstock, 16.

Dallenbach, 100.
Dingle, 252, 253.
Dugan, 451.

Eddington, 182, 205, 222, 267, 362, 403, 411, 454.
Ehrenfest, 269, 285, 316, 319.
Eichenwald, 116.
Einstein, 1, 2, 3, 4, 7, 9, 10, 12, 15, 19, 110, 152, 165, 167, 168, 169, 185, 187, 188, 189, 198, 199, 213, 214, 225, 232, 236, 239, 291, 319, 333, 335, 337, 339, 341, 344, 412, 415, 438, 469.
Eötvös, 179, 192.

Fitzgerald, 23.
Friedmann, 337, 362, 408, 412.

Galileo, 3, 179, 192, 213.
Gerlach, 54.
Gibbs, 124.

Heckmann, 408.
Helmholtz, 124.
Holetschek, 447.
Hopmann, 447.
Hubble, 332, 344, 345, 356, 359, 363, 446, 447, 448, 450, 451, 452, 453, 454, 456, 457, 458, 459, 460, 461, 465, 466, 472, 479.

Humason, 345, 356, 359, 446, 447, 448, 450, 451, 452, 453, 454, 456, 465, 466, 472, 479.
Hupka, 57.

Illingworth, 13.

Jeans, 476.
Jüttner, 6, 119, 136.

Kaufmann, 53.
Kennedy, 9, 13, 14.
Kinsey, 58.
Kretschmann, 3, 168.

Lanczos, 348.
La Rosa, 17.
Laub, 14, 110.
Laue, 8, 59, 64, 108, 289.
Lemaître, 240, 251, 347, 362, 407, 408, 410, 411, 412.
Lense, 239.
Lenz, 424.
Levi-Civita, 168.
Lewis, 8, 44, 123, 124.
Lipschitz, 186.
Lorentz, 4, 18, 19, 23, 84, 85, 88, 94, 100, 258.

Mach, 3, 185.
Marjorana, 16.
Maxwell, 84, 91, 100, 101, 258.
McCrea, 411.
McVittie, 411.
Michelson, 9, 13.
Miller, 13, 17.
Milne, 364.
Minkowski, 4, 9, 28, 52, 101.
Morley, 9, 13.
Mosengeil, 162.

Neddermeyer, 58.
Nernst, 122, 123.
Neumann, 322, 344.
Newton, 3, 12, 15, 42, 46, 73.
Nicholson, 451.
Noble, 83.

Occhialini, 58.
Oliphant, 58.
Oppenheimer, 58, 150, 151.

Pauli, 115, 228.
Petit, 451.
Planck, 2, 122, 152, 291.
Plesset, 58, 150.
Podolsky, 251, 252, 285.

Ricci, 168.
Ritz, 16.
Robertson, 298, 337, 347, 348, 356, 362, 395.
Roentgen, 116.
Rowland, 108, 116.
Russell, 451.
Rutherford, 58.

Schwarzchild, 202, 245.
Sears, 451.
Seeliger, 322.
Shapley, 446, 447, 448, 460.
de Sitter, 9, 16, 17, 333, 336, 337, 346, 348, 349, 359, 408, 410, 412, 413, 415, 418, 419, 486.
Slipher, 356, 454.
Southerns, 179, 192.
Stebbins, 459.
Stern, 147.
Stewart, J. Q., 451.
Stewart, O. M., 16.
St. John, 212.

Thirring, 239.
Thomson, J. J., 16.
Thorndike, 14.
Tolman, 6, 8, 16, 17, 27, 44, 54, 57, 64, 88, 119, 136, 137, 140, 147, 229, 235, 247, 269, 285, 292, 294, 298, 314, 316, 337, 356, 362, 373, 381, 413, 424, 429, 430, 434, 435, 440, 462, 467, 470, 483.
Trouton, 83.
Trumpler, 211.

Urey, 144.

Ward, 435.
Weyl, 104, 348, 356.
Wilson, H. A., 116.
Wilson, M., 116.

Zwicky, 287, 483.

A CATALOG OF SELECTED
DOVER BOOKS
IN SCIENCE AND MATHEMATICS

A CATALOG OF SELECTED
DOVER BOOKS
IN SCIENCE AND MATHEMATICS

QUALITATIVE THEORY OF DIFFERENTIAL EQUATIONS, V.V. Nemytskii and V.V. Stepanov. Classic graduate-level text by two prominent Soviet mathematicians covers classical differential equations as well as topological dynamics and ergodic theory. Bibliographies. 523pp. 5⅜ × 8½. 65954-2 Pa. $10.95

MATRICES AND LINEAR ALGEBRA, Hans Schneider and George Phillip Barker. Basic textbook covers theory of matrices and its applications to systems of linear equations and related topics such as determinants, eigenvalues and differential equations. Numerous exercises. 432pp. 5⅜ × 8½. 66014-1 Pa. $10.95

QUANTUM THEORY, David Bohm. This advanced undergraduate-level text presents the quantum theory in terms of qualitative and imaginative concepts, followed by specific applications worked out in mathematical detail. Preface. Index. 655pp. 5⅜ × 8½. 65969-0 Pa. $13.95

ATOMIC PHYSICS (8th edition), Max Born. Nobel laureate's lucid treatment of kinetic theory of gases, elementary particles, nuclear atom, wave-corpuscles, atomic structure and spectral lines, much more. Over 40 appendices, bibliography. 495pp. 5⅜ × 8½. 65984-4 Pa. $12.95

ELECTRONIC STRUCTURE AND THE PROPERTIES OF SOLIDS: The Physics of the Chemical Bond, Walter A. Harrison. Innovative text offers basic understanding of the electronic structure of covalent and ionic solids, simple metals, transition metals and their compounds. Problems. 1980 edition. 582pp. 6⅛ × 9¼. 66021-4 Pa. $15.95

BOUNDARY VALUE PROBLEMS OF HEAT CONDUCTION, M. Necati Özisik. Systematic, comprehensive treatment of modern mathematical methods of solving problems in heat conduction and diffusion. Numerous examples and problems. Selected references. Appendices. 505pp. 5⅜ × 8½. 65990-9 Pa. $12.95

A SHORT HISTORY OF CHEMISTRY (3rd edition), J.R. Partington. Classic exposition explores origins of chemistry, alchemy, early medical chemistry, nature of atmosphere, theory of valency, laws and structure of atomic theory, much more. 428pp. 5⅜ × 8½. (Available in U.S. only) 65977-1 Pa. $10.95

A HISTORY OF ASTRONOMY, A. Pannekoek. Well-balanced, carefully reasoned study covers such topics as Ptolemaic theory, work of Copernicus, Kepler, Newton, Eddington's work on stars, much more. Illustrated. References. 521pp. 5⅜ × 8½. 65994-1 Pa. $12.95

PRINCIPLES OF METEOROLOGICAL ANALYSIS, Walter J. Saucier. Highly respected, abundantly illustrated classic reviews atmospheric variables, hydrostatics, static stability, various analyses (scalar, cross-section, isobaric, isentropic, more). For intermediate meteorology students. 454pp. 6⅛ × 9¼. 65979-8 Pa. $14.95

RELATIVITY, THERMODYNAMICS AND COSMOLOGY, Richard C. Tolman. Landmark study extends thermodynamics to special, general relativity; also applications of relativistic mechanics, thermodynamics to cosmological models. 501pp. 5⅜ × 8½. 65383-8 Pa. $12.95

APPLIED ANALYSIS, Cornelius Lanczos. Classic work on analysis and design of finite processes for approximating solution of analytical problems. Algebraic equations, matrices, harmonic analysis, quadrature methods, much more. 559pp. 5⅜ × 8½. 65656-X Pa. $13.95

SPECIAL RELATIVITY FOR PHYSICISTS, G. Stephenson and C.W. Kilmister. Concise elegant account for nonspecialists. Lorentz transformation, optical and dynamical applications, more. Bibliography. 108pp. 5⅜ × 8½. 65519-9 Pa. $4.95

INTRODUCTION TO ANALYSIS, Maxwell Rosenlicht. Unusually clear, accessible coverage of set theory, real number system, metric spaces, continuous functions, Riemann integration, multiple integrals, more. Wide range of problems. Undergraduate level. Bibliography. 254pp. 5⅜ × 8½. 65038-3 Pa. $7.95

INTRODUCTION TO QUANTUM MECHANICS With Applications to Chemistry, Linus Pauling & E. Bright Wilson, Jr. Classic undergraduate text by Nobel Prize winner applies quantum mechanics to chemical and physical problems. Numerous tables and figures enhance the text. Chapter bibliographies. Appendices. Index. 468pp. 5⅜ × 8½. 64871-0 Pa. $11.95

ASYMPTOTIC EXPANSIONS OF INTEGRALS, Norman Bleistein & Richard A. Handelsman. Best introduction to important field with applications in a variety of scientific disciplines. New preface. Problems. Diagrams. Tables. Bibliography. Index. 448pp. 5⅜ × 8½. 65082-0 Pa. $12.95

MATHEMATICS APPLIED TO CONTINUUM MECHANICS, Lee A. Segel. Analyzes models of fluid flow and solid deformation. For upper-level math, science and engineering students. 608pp. 5⅜ × 8½. 65369-2 Pa. $13.95

ELEMENTS OF REAL ANALYSIS, David A. Sprecher. Classic text covers fundamental concepts, real number system, point sets, functions of a real variable, Fourier series, much more. Over 500 exercises. 352pp. 5⅜ × 8½. 65385-4 Pa. $10.95

PHYSICAL PRINCIPLES OF THE QUANTUM THEORY, Werner Heisenberg. Nóbel Laureate discusses quantum theory, uncertainty, wave mechanics, work of Dirac, Schroedinger, Compton, Wilson, Einstein, etc. 184pp. 5⅜ × 8½. 60113-7 Pa. $5.95

INTRODUCTORY REAL ANALYSIS, A.N. Kolmogorov, S.V. Fomin. Translated by Richard A. Silverman. Self-contained, evenly paced introduction to real and functional analysis. Some 350 problems. 403pp. 5⅜ × 8½. 61226-0 Pa. $9.95

PROBLEMS AND SOLUTIONS IN QUANTUM CHEMISTRY AND PHYSICS, Charles S. Johnson, Jr. and Lee G. Pedersen. Unusually varied problems, detailed solutions in coverage of quantum mechanics, wave mechanics, angular momentum, molecular spectroscopy, scattering theory, more. 280 problems plus 139 supplementary exercises. 430pp. 6½ × 9¼. 65236-X Pa. $12.95

CATALOG OF DOVER BOOKS

ASYMPTOTIC METHODS IN ANALYSIS, N.G. de Bruijn. An inexpensive, comprehensive guide to asymptotic methods—the pioneering work that teaches by explaining worked examples in detail. Index. 224pp. 5⅜ × 8½. 64221-6 Pa. $6.95

OPTICAL RESONANCE AND TWO-LEVEL ATOMS, L. Allen and J.H. Eberly. Clear, comprehensive introduction to basic principles behind all quantum optical resonance phenomena. 53 illustrations. Preface. Index. 256pp. 5⅜ × 8½.
65533-4 Pa. $7.95

COMPLEX VARIABLES, Francis J. Flanigan. Unusual approach, delaying complex algebra till harmonic functions have been analyzed from real variable viewpoint. Includes problems with answers. 364pp. 5⅜ × 8½. 61388-7 Pa. $8.95

ATOMIC SPECTRA AND ATOMIC STRUCTURE, Gerhard Herzberg. One of best introductions; especially for specialist in other fields. Treatment is physical rather than mathematical. 80 illustrations. 257pp. 5⅜ × 8½. 60115-3 Pa. $6.95

APPLIED COMPLEX VARIABLES, John W. Dettman. Step-by-step coverage of fundamentals of analytic function theory—plus lucid exposition of five important applications: Potential Theory; Ordinary Differential Equations; Fourier Transforms; Laplace Transforms; Asymptotic Expansions. 66 figures. Exercises at chapter ends. 512pp. 5⅜ × 8½. 64670-X Pa. $11.95

ULTRASONIC ABSORPTION: An Introduction to the Theory of Sound Absorption and Dispersion in Gases, Liquids and Solids, A.B. Bhatia. Standard reference in the field provides a clear, systematically organized introductory review of fundamental concepts for advanced graduate students, research workers. Numerous diagrams. Bibliography. 440pp. 5⅜ × 8½. 64917-2 Pa. $11.95

UNBOUNDED LINEAR OPERATORS: Theory and Applications, Seymour Goldberg. Classic presents systematic treatment of the theory of unbounded linear operators in normed linear spaces with applications to differential equations. Bibliography. 199pp. 5⅜ × 8½. 64830-3 Pa. $7.95

LIGHT SCATTERING BY SMALL PARTICLES, H.C. van de Hulst. Comprehensive treatment including full range of useful approximation methods for researchers in chemistry, meteorology and astronomy. 44 illustrations. 470pp. 5⅜ × 8½. 64228-3 Pa. $11.95

CONFORMAL MAPPING ON RIEMANN SURFACES, Harvey Cohn. Lucid, insightful book presents ideal coverage of subject. 334 exercises make book perfect for self-study. 55 figures. 352pp. 5⅜ × 8¼. 64025-6 Pa. $9.95

OPTICKS, Sir Isaac Newton. Newton's own experiments with spectroscopy, colors, lenses, reflection, refraction, etc., in language the layman can follow. Foreword by Albert Einstein. 532pp. 5⅜ × 8½. 60205-2 Pa. $9.95

GENERALIZED INTEGRAL TRANSFORMATIONS, A.H. Zemanian. Graduate-level study of recent generalizations of the Laplace, Mellin, Hankel, K. Weierstrass, convolution and other simple transformations. Bibliography. 320pp. 5⅜ × 8½. 65375-7 Pa. $8.95

CATALOG OF DOVER BOOKS

THE ELECTROMAGNETIC FIELD, Albert Shadowitz. Comprehensive undergraduate text covers basics of electric and magnetic fields, builds up to electromagnetic theory. Also related topics, including relativity. Over 900 problems. 768pp. 5⅜ × 8¼. 65660-8 Pa. $18.95

FOURIER SERIES, Georgi P. Tolstov. Translated by Richard A. Silverman. A valuable addition to the literature on the subject, moving clearly from subject to subject and theorem to theorem. 107 problems, answers. 336pp. 5⅜ × 8½. 63317-9 Pa. $8.95

THEORY OF ELECTROMAGNETIC WAVE PROPAGATION, Charles Herach Papas. Graduate-level study discusses the Maxwell field equations, radiation from wire antennas, the Doppler effect and more. xiii + 244pp. 5⅜ × 8½. 65678-0 Pa. $6.95

DISTRIBUTION THEORY AND TRANSFORM ANALYSIS: An Introduction to Generalized Functions, with Applications, A.H. Zemaniah. Provides basics of distribution theory, describes generalized Fourier and Laplace transformations. Numerous problems. 384pp. 5⅜ × 8½. 65479-6 Pa. $9.95

THE PHYSICS OF WAVES, William C. Elmore and Mark A. Heald. Unique overview of classical wave theory. Acoustics, optics, electromagnetic radiation, more. Ideal as classroom text or for self-study. Problems. 477pp. 5⅜ × 8½. 64926-1 Pa. $12.95

CALCULUS OF VARIATIONS WITH APPLICATIONS, George M. Ewing. Applications-oriented introduction to variational theory develops insight and promotes understanding of specialized books, research papers. Suitable for advanced undergraduate/graduate students as primary, supplementary text. 352pp. 5⅜ × 8½. 64856-7 Pa. $8.95

A TREATISE ON ELECTRICITY AND MAGNETISM, James Clerk Maxwell. Important foundation work of modern physics. Brings to final form Maxwell's theory of electromagnetism and rigorously derives his general equations of field theory. 1,084pp. 5⅜ × 8½. 60636-8, 60637-6 Pa., Two-vol. set $21.90

AN INTRODUCTION TO THE CALCULUS OF VARIATIONS, Charles Fox. Graduate-level text covers variations of an integral, isoperimetrical problems, least action, special relativity, approximations, more. References. 279pp. 5⅜ × 8½. 65499-0 Pa. $7.95

HYDRODYNAMIC AND HYDROMAGNETIC STABILITY, S. Chandrasekhar. Lucid examination of the Rayleigh-Benard problem; clear coverage of the theory of instabilities causing convection. 704pp. 5⅜ × 8¼. 64071-X Pa. $14.95

CALCULUS OF VARIATIONS, Robert Weinstock. Basic introduction covering isoperimetric problems, theory of elasticity, quantum mechanics, electrostatics, etc. Exercises throughout. 326pp. 5⅜ × 8½. 63069-2 Pa. $8.95

DYNAMICS OF FLUIDS IN POROUS MEDIA, Jacob Bear. For advanced students of ground water hydrology, soil mechanics and physics, drainage and irrigation engineering and more. 335 illustrations. Exercises, with answers. 784pp. 6⅛ × 9¼. 65675-6 Pa. $19.95

CATALOG OF DOVER BOOKS

NUMERICAL METHODS FOR SCIENTISTS AND ENGINEERS, Richard Hamming. Classic text stresses frequency approach in coverage of algorithms, polynomial approximation, Fourier approximation, exponential approximation, other topics. Revised and enlarged 2nd edition. 721pp. 5⅜ × 8½.
65241-6 Pa. $14.95

THEORETICAL SOLID STATE PHYSICS, Vol. I: Perfect Lattices in Equilibrium; Vol. II: Non-Equilibrium and Disorder, William Jones and Norman H. March. Monumental reference work covers fundamental theory of equilibrium properties of perfect crystalline solids, non-equilibrium properties, defects and disordered systems. Appendices. Problems. Preface. Diagrams. Index. Bibliography. Total of 1,301pp. 5⅜ × 8½. Two volumes.
Vol. I 65015-4 Pa. $14.95
Vol. II 65016-2 Pa. $14.95

OPTIMIZATION THEORY WITH APPLICATIONS, Donald A. Pierre. Broad-spectrum approach to important topic. Classical theory of minima and maxima, calculus of variations, simplex technique and linear programming, more. Many problems, examples. 640pp. 5⅜ × 8½.
65205-X Pa. $14.95

THE CONTINUUM: A Critical Examination of the Foundation of Analysis, Hermann Weyl. Classic of 20th-century foundational research deals with the conceptual problem posed by the continuum. 156pp. 5⅜ × 8½.
67982-9 Pa. $5.95

ESSAYS ON THE THEORY OF NUMBERS, Richard Dedekind. Two classic essays by great German mathematician: on the theory of irrational numbers; and on transfinite numbers and properties of natural numbers. 115pp. 5⅜ × 8½.
21010-3 Pa. $4.95

THE FUNCTIONS OF MATHEMATICAL PHYSICS, Harry Hochstadt. Comprehensive treatment of orthogonal polynomials, hypergeometric functions, Hill's equation, much more. Bibliography. Index. ?22pp. 5⅜ × 8½.
65214-9 Pa. $9.95

NUMBER THEORY AND ITS HISTORY, Oystein Ore. Unusually clear, accessible introduction covers counting, properties of numbers, prime numbers, much more. Bibliography. 380pp. 5⅜ × 8½.
65620-9 Pa. $9.95

THE VARIATIONAL PRINCIPLES OF MECHANICS, Cornelius Lanczos. Graduate level coverage of calculus of variations, equations of motion, relativistic mechanics, more. First inexpensive paperbound edition of classic treatise. Index. Bibliography. 418pp. 5⅜ × 8½.
65067-7 Pa. $11.95

MATHEMATICAL TABLES AND FORMULAS, Robert D. Carmichael and Edwin R. Smith. Logarithms, sines, tangents, trig functions, powers, roots, reciprocals, exponential and hyperbolic functions, formulas and theorems. 269pp. 5⅜ × 8½.
60111-0 Pa. $6.95

THEORETICAL PHYSICS, Georg Joos, with Ira M. Freeman. Classic overview covers essential math, mechanics, electromagnetic theory, thermodynamics, quantum mechanics, nuclear physics, other topics. First paperback edition. xxiii + 885pp. 5⅜ × 8½.
65227-0 Pa. $19.95

CATALOG OF DOVER BOOKS

HANDBOOK OF MATHEMATICAL FUNCTIONS WITH FORMULAS, GRAPHS, AND MATHEMATICAL TABLES, edited by Milton Abramowitz and Irene A. Stegun. Vast compendium: 29 sets of tables, some to as high as 20 places. 1,046pp. 8 × 10½.   61272-4 Pa. $24.95

MATHEMATICAL METHODS IN PHYSICS AND ENGINEERING, John W. Dettman. Algebraically based approach to vectors, mapping, diffraction, other topics in applied math. Also generalized functions, analytic function theory, more. Exercises. 448pp. 5⅜ × 8¼.   65649-7 Pa. $9.95

A SURVEY OF NUMERICAL MATHEMATICS, David M. Young and Robert Todd Gregory. Broad self-contained coverage of computer-oriented numerical algorithms for solving various types of mathematical problems in linear algebra, ordinary and partial, differential equations, much more. Exercises. Total of 1,248pp. 5⅜ × 8½. Two volumes.   Vol. I 65691-8 Pa. $14.95
                                                        Vol. II 65692-6 Pa. $14.95

TENSOR ANALYSIS FOR PHYSICISTS, J.A. Schouten. Concise exposition of the mathematical basis of tensor analysis, integrated with well-chosen physical examples of the theory. Exercises. Index. Bibliography. 289pp. 5⅜ × 8½.
                                                        65582-2 Pa. $8.95

INTRODUCTION TO NUMERICAL ANALYSIS (2nd Edition), F.B. Hildebrand. Classic, fundamental treatment covers computation, approximation, interpolation, numerical differentiation and integration, other topics. 150 new problems. 669pp. 5⅜ × 8½.   65363-3 Pa. $15.95

INVESTIGATIONS ON THE THEORY OF THE BROWNIAN MOVEMENT, Albert Einstein. Five papers (1905-8) investigating dynamics of Brownian motion and evolving elementary theory. Notes by R. Fürth. 122pp. 5⅜ × 8½.
                                                        60304-0 Pa. $4.95

CATASTROPHE THEORY FOR SCIENTISTS AND ENGINEERS, Robert Gilmore. Advanced-level treatment describes mathematics of theory grounded in the work of Poincaré, R. Thom, other mathematicians. Also important applications to problems in mathematics, physics, chemistry and engineering. 1981 edition. References. 28 tables. 397 black-and-white illustrations. xvii + 666pp. 6⅛ × 9¼.
                                                        67539-4 Pa. $16.95

AN INTRODUCTION TO STATISTICAL THERMODYNAMICS, Terrell L. Hill. Excellent basic text offers wide-ranging coverage of quantum statistical mechanics, systems of interacting molecules, quantum statistics, more. 523pp. 5⅜ × 8½.   65242-4 Pa. $12.95

ELEMENTARY DIFFERENTIAL EQUATIONS, William Ted Martin and Eric Reissner. Exceptionally clear, comprehensive introduction at undergraduate level. Nature and origin of differential equations, differential equations of first, second and higher orders. Picard's Theorem, much more. Problems with solutions. 331pp. 5⅜ × 8½.   65024-3 Pa. $8.95

STATISTICAL PHYSICS, Gregory H. Wannier. Classic text combines thermodynamics, statistical mechanics and kinetic theory in one unified presentation of thermal physics. Problems with solutions. Bibliography. 532pp. 5⅜ × 8½.
                                                        65401-X Pa. $12.95

CATALOG OF DOVER BOOKS

ORDINARY DIFFERENTIAL EQUATIONS, Morris Tenenbaum and Harry Pollard. Exhaustive survey of ordinary differential equations for undergraduates in mathematics, engineering, science. Thorough analysis of theorems. Diagrams. Bibliography. Index. 818pp. 5⅜ × 8½. 64940-7 Pa. $16.95

STATISTICAL MECHANICS: Principles and Applications, Terrell L. Hill. Standard text covers fundamentals of statistical mechanics, applications to fluctuation theory, imperfect gases, distribution functions, more. 448pp. 5⅜ × 8½. 65390-0 Pa. $11.95

ORDINARY DIFFERENTIAL EQUATIONS AND STABILITY THEORY: An Introduction, David A. Sánchez. Brief, modern treatment. Linear equation, stability theory for autonomous and nonautonomous systems, etc. 164pp. 5⅜ × 8¼. 63828-6 Pa. $5.95

THIRTY YEARS THAT SHOOK PHYSICS: The Story of Quantum Theory, George Gamow. Lucid, accessible introduction to influential theory of energy and matter. Careful explanations of Dirac's anti-particles, Bohr's model of the atom, much more. 12 plates. Numerous drawings. 240pp. 5⅜ × 8½. 24895-X Pa. $6.95

THEORY OF MATRICES, Sam Perlis. Outstanding text covering rank, non-singularity and inverses in connection with the development of canonical matrices under the relation of equivalence, and without the intervention of determinants. Includes exercises. 237pp. 5⅜ × 8½. 66810-X Pa. $7.95

GREAT EXPERIMENTS IN PHYSICS: Firsthand Accounts from Galileo to Einstein, edited by Morris H. Shamos. 25 crucial discoveries: Newton's laws of motion, Chadwick's study of the neutron, Hertz on electromagnetic waves, more. Original accounts clearly annotated. 370pp. 5⅜ × 8½. 25346-5 Pa. $10.95

INTRODUCTION TO PARTIAL DIFFERENTIAL EQUATIONS WITH AP-PLICATIONS, E.C. Zachmanoglou and Dale W. Thoe. Essentials of partial differential equations applied to common problems in engineering and the physical sciences. Problems and answers. 416pp. 5⅜ × 8½. 65251-3 Pa. $10.95

BURNHAM'S CELESTIAL HANDBOOK, Robert Burnham, Jr. Thorough guide to the stars beyond our solar system. Exhaustive treatment. Alphabetical by constellation: Andromeda to Cetus in Vol. 1; Chamaeleon to Orion in Vol. 2; and Pavo to Vulpecula in Vol. 3. Hundreds of illustrations. Index in Vol. 3. 2,000pp. 6⅛ × 9¼. 23567-X, 23568-8, 23673-0 Pa., Three-vol. set $41.85

CHEMICAL MAGIC, Leonard A. Ford. Second Edition, Revised by E. Winston Grundmeier. Over 100 unusual stunts demonstrating cold fire, dust explosions, much more. Text explains scientific principles and stresses safety precautions. 128pp. 5⅜ × 8½. 67628-5 Pa. $5.95

AMATEUR ASTRONOMER'S HANDBOOK, J.B. Sidgwick. Timeless, compre-hensive coverage of telescopes, mirrors, lenses, mountings, telescope drives, micrometers, spectroscopes, more. 189 illustrations. 576pp. 5⅜ × 8¼. (Available in U.S. only) 24034-7 Pa. $9.95

CATALOG OF DOVER BOOKS

SPECIAL FUNCTIONS, N.N. Lebedev. Translated by Richard Silverman. Famous Russian work treating more important special functions, with applications to specific problems of physics and engineering. 38 figures. 308pp. 5⅜ × 8½.
60624-4 Pa. $8.95

OBSERVATIONAL ASTRONOMY FOR AMATEURS, J.B. Sidgwick. Mine of useful data for observation of sun, moon, planets, asteroids, aurorae, meteors, comets, variables, binaries, etc. 39 illustrations. 384pp. 5⅜ × 8¼. (Available in U.S. only)
24033-9 Pa. $8.95

INTEGRAL EQUATIONS, F.G. Tricomi. Authoritative, well-written treatment of extremely useful mathematical tool with wide applications. Volterra Equations, Fredholm Equations, much more. Advanced undergraduate to graduate level. Exercises. Bibliography. 238pp. 5⅜ × 8½.
64828-1 Pa. $7.95

POPULAR LECTURES ON MATHEMATICAL LOGIC, Hao Wang. Noted logician's lucid treatment of historical developments, set theory, model theory, recursion theory and constructivism, proof theory, more. 3 appendixes. Bibliography. 1981 edition. ix + 283pp. 5⅜ × 8½.
67632-3 Pa. $8.95

MODERN NONLINEAR EQUATIONS, Thomas L. Saaty. Emphasizes practical solution of problems; covers seven types of equations. ". . . a welcome contribution to the existing literature. . . ."—*Math Reviews.* 490pp. 5⅜ × 8½. 64232-1 Pa. $11.95

FUNDAMENTALS OF ASTRODYNAMICS, Roger Bate et al. Modern approach developed by U.S. Air Force Academy. Designed as a first course. Problems, exercises. Numerous illustrations. 455pp. 5⅜ × 8½. 60061-0 Pa. $9.95

INTRODUCTION TO LINEAR ALGEBRA AND DIFFERENTIAL EQUATIONS, John W. Dettman. Excellent text covers complex numbers, determinants, orthonormal bases, Laplace transforms, much more. Exercises with solutions. Undergraduate level. 416pp. 5⅜ × 8½. 65191-6 Pa. $10.95

INCOMPRESSIBLE AERODYNAMICS, edited by Bryan Thwaites. Covers theoretical and experimental treatment of the uniform flow of air and viscous fluids past two-dimensional aerofoils and three-dimensional wings; many other topics. 654pp. 5⅜ × 8½. 65465-6 Pa. $16.95

INTRODUCTION TO DIFFERENCE EQUATIONS, Samuel Goldberg. Exceptionally clear exposition of important discipline with applications to sociology, psychology, economics. Many illustrative examples; over 250 problems. 260pp. 5⅜ × 8½. 65084-7 Pa. $7.95

LAMINAR BOUNDARY LAYERS, edited by L. Rosenhead. Engineering classic covers steady boundary layers in two- and three-dimensional flow, unsteady boundary layers, stability, observational techniques, much more. 708pp. 5⅜ × 8½.
65646-2 Pa. $18.95

LECTURES ON CLASSICAL DIFFERENTIAL GEOMETRY, Second Edition, Dirk J. Struik. Excellent brief introduction covers curves, theory of surfaces, fundamental equations, geometry on a surface, conformal mapping, other topics. Problems. 240pp. 5⅜ × 8½. 65609-8 Pa. $8.95

ROTARY-WING AERODYNAMICS, W.Z. Stepniewski. Clear, concise text covers aerodynamic phenomena of the rotor and offers guidelines for helicopter performance evaluation. Originally prepared for NASA. 537 figures. 640pp. 6⅛ × 9¼.
64647-5 Pa. $15.95

DIFFERENTIAL GEOMETRY, Heinrich W. Guggenheimer. Local differential geometry as an application of advanced calculus and linear algebra. Curvature, transformation groups, surfaces, more. Exercises. 62 figures. 378pp. 5⅜ × 8½.
63433-7 Pa. $8.95

INTRODUCTION TO SPACE DYNAMICS, William Tyrrell Thomson. Comprehensive, classic introduction to space-flight engineering for advanced undergraduate and graduate students. Includes vector algebra, kinematics, transformation of coordinates. Bibliography. Index. 352pp. 5⅜ × 8½. 65113-4 Pa. $8.95

A SURVEY OF MINIMAL SURFACES, Robert Osserman. Up-to-date, in-depth discussion of the field for advanced students. Corrected and enlarged edition covers new developments. Includes numerous problems. 192pp. 5⅜ × 8½.
64998-9 Pa. $8.95

ANALYTICAL MECHANICS OF GEARS, Earle Buckingham. Indispensable reference for modern gear manufacture covers conjugate gear-tooth action, gear-tooth profiles of various gears, many other topics. 263 figures. 102 tables. 546pp. 5⅜ × 8½. 65712-4 Pa. $14.95

SET THEORY AND LOGIC, Robert R. Stoll. Lucid introduction to unified theory of mathematical concepts. Set theory and logic seen as tools for conceptual understanding of real number system. 496pp. 5⅜ × 8¼. 63829-4 Pa. $12.95

A HISTORY OF MECHANICS, René Dugas. Monumental study of mechanical principles from antiquity to quantum mechanics. Contributions of ancient Greeks, Galileo, Leonardo, Kepler, Lagrange, many others. 671pp. 5⅜ × 8½.
65632-2 Pa. $14.95

FAMOUS PROBLEMS OF GEOMETRY AND HOW TO SOLVE THEM, Benjamin Bold. Squaring the circle, trisecting the angle, duplicating the cube: learn their history, why they are impossible to solve, then solve them yourself. 128pp. 5⅜ × 8½. 24297-8 Pa. $4.95

MECHANICAL VIBRATIONS, J.P. Den Hartog. Classic textbook offers lucid explanations and illustrative models, applying theories of vibrations to a variety of practical industrial engineering problems. Numerous figures. 233 problems, solutions. Appendix. Index. Preface. 436pp. 5⅜ × 8½. 64785-4 Pa. $10.95

CURVATURE AND HOMOLOGY, Samuel I. Goldberg. Thorough treatment of specialized branch of differential geometry. Covers Riemannian manifolds, topology of differentiable manifolds, compact Lie groups, other topics. Exercises. 315pp. 5⅜ × 8½. 64314-X Pa. $9.95

HISTORY OF STRENGTH OF MATERIALS, Stephen P. Timoshenko. Excellent historical survey of the strength of materials with many references to the theories of elasticity and structure. 245 figures. 452pp. 5⅜ × 8½. 61187-6 Pa. $11.95

CATALOG OF DOVER BOOKS

GEOMETRY OF COMPLEX NUMBERS, Hans Schwerdtfeger. Illuminating, widely praised book on analytic geometry of circles, the Moebius transformation, and two-dimensional non-Euclidean geometries. 200pp. 5⅜ × 8¼.
63830-8 Pa. $8.95

MECHANICS, J.P. Den Hartog. A classic introductory text or refresher. Hundreds of applications and design problems illuminate fundamentals of trusses, loaded beams and cables, etc. 334 answered problems. 462pp. 5⅜ × 8½. 60754-2 Pa. $9.95

TOPOLOGY, John G. Hocking and Gail S. Young. Superb one-year course in classical topology. Topological spaces and functions, point-set topology, much more. Examples and problems. Bibliography. Index. 384pp. 5⅜ × 8¼.
65676-4 Pa. $9.95

STRENGTH OF MATERIALS, J.P. Den Hartog. Full, clear treatment of basic material (tension, torsion, bending, etc.) plus advanced material on engineering methods, applications. 350 answered problems. 323pp. 5⅜ × 8½. 60755-0 Pa. $8.95

ELEMENTARY CONCEPTS OF TOPOLOGY, Paul Alexandroff. Elegant, intuitive approach to topology from set-theoretic topology to Betti groups; how concepts of topology are useful in math and physics. 25 figures. 57pp. 5⅜ × 8½.
60747-X Pa. $3.50

ADVANCED STRENGTH OF MATERIALS, J.P. Den Hartog. Superbly written advanced text covers torsion, rotating disks, membrane stresses in shells, much more. Many problems and answers. 388pp. 5⅜ × 8½. 65407-9 Pa. $9.95

COMPUTABILITY AND UNSOLVABILITY, Martin Davis. Classic graduate-level introduction to theory of computability, usually referred to as theory of recurrent functions. New preface and appendix. 288pp. 5⅜ × 8½. 61471-9 Pa. $7.95

GENERAL CHEMISTRY, Linus Pauling. Revised 3rd edition of classic first-year text by Nobel laureate. Atomic and molecular structure, quantum mechanics, statistical mechanics, thermodynamics correlated with descriptive chemistry. Problems. 992pp. 5⅜ × 8½. 65622-5 Pa. $19.95

AN INTRODUCTION TO MATRICES, SETS AND GROUPS FOR SCIENCE STUDENTS, G. Stephenson. Concise, readable text introduces sets, groups, and most importantly, matrices to undergraduate students of physics, chemistry, and engineering. Problems. 164pp. 5⅜ × 8½. 65077-4 Pa. $6.95

THE HISTORICAL BACKGROUND OF CHEMISTRY, Henry M. Leicester. Evolution of ideas, not individual biography. Concentrates on formulation of a coherent set of chemical laws. 260pp. 5⅜ × 8½. 61053-5 Pa. $6.95

THE PHILOSOPHY OF MATHEMATICS: An Introductory Essay, Stephan Körner. Surveys the views of Plato, Aristotle, Leibniz & Kant concerning proposi- tions and theories of applied and pure mathematics. Introduction. Two appen- dices. Index. 198pp. 5⅜ × 8½. 25048-2 Pa. $7.95

THE DEVELOPMENT OF MODERN CHEMISTRY, Aaron J. Ihde. Authorita- tive history of chemistry from ancient Greek theory to 20th-century innovation. Covers major chemists and their discoveries. 209 illustrations. 14 tables. Bibliog- raphies. Indices. Appendices. 851pp. 5⅜ × 8½. 64235-6 Pa. $18.95

CATALOG OF DOVER BOOKS

THE FOUR-COLOR PROBLEM: Assaults and Conquest, Thomas L. Saaty and Paul G. Kainen. Engrossing, comprehensive account of the century-old combinatorial topological problem, its history and solution. Bibliographies. Index. 110 figures. 228pp. 5⅜ × 8½.   65092-8 Pa. $6.95

CATALYSIS IN CHEMISTRY AND ENZYMOLOGY, William P. Jencks. Exceptionally clear coverage of mechanisms for catalysis, forces in aqueous solution, carbonyl- and acyl-group reactions, practical kinetics, more. 864pp. 5⅜ × 8½.   65460-5 Pa. $19.95

PROBABILITY: An Introduction, Samuel Goldberg. Excellent basic text covers set theory, probability theory for finite sample spaces, binomial theorem, much more. 360 problems. Bibliographies. 322pp. 5⅜ × 8½.   65252-1 Pa. $8.95

LIGHTNING, Martin A. Uman. Revised, updated edition of classic work on the physics of lightning. Phenomena, terminology, measurement, photography, spectroscopy, thunder, more. Reviews recent research. Bibliography. Indices. 320pp. 5⅜ × 8¼.   64575-4 Pa. $8.95

PROBABILITY THEORY: A Concise Course, Y.A. Rozanov. Highly readable, self-contained introduction covers combination of events, dependent events, Bernoulli trials, etc. Translation by Richard Silverman. 148pp. 5⅜ × 8¼.
63544-9 Pa. $5.95

AN INTRODUCTION TO HAMILTONIAN OPTICS, H. A. Buchdahl. Detailed account of the Hamiltonian treatment of aberration theory in geometrical optics. Many classes of optical systems defined in terms of the symmetries they possess. Problems with detailed solutions. 1970 edition. xv + 360pp. 5⅜ × 8½.
67597-1 Pa. $10.95

STATISTICS MANUAL, Edwin L. Crow, et al. Comprehensive, practical collection of classical and modern methods prepared by U.S. Naval Ordnance Test Station. Stress on use. Basics of statistics assumed. 288pp. 5⅜ × 8½.
60599-X Pa. $6.95

DICTIONARY/OUTLINE OF BASIC STATISTICS, John E. Freund and Frank J. Williams. A clear concise dictionary of over 1,000 statistical terms and an outline of statistical formulas covering probability, nonparametric tests, much more. 208pp. 5⅜ × 8½.   66796-0 Pa. $6.95

STATISTICAL METHOD FROM THE VIEWPOINT OF QUALITY CONTROL, Walter A. Shewhart. Important text explains regulation of variables, uses of statistical control to achieve quality control in industry, agriculture, other areas. 192pp. 5⅜ × 8½.   65232-7 Pa. $7.95

THE INTERPRETATION OF GEOLOGICAL PHASE DIAGRAMS, Ernest G. Ehlers. Clear, concise text emphasizes diagrams of systems under fluid or containing pressure; also coverage of complex binary systems, hydrothermal melting, more. 288pp. 6½ × 9¼.   65389-7 Pa. $10.95

STATISTICAL ADJUSTMENT OF DATA, W. Edwards Deming. Introduction to basic concepts of statistics, curve fitting, least squares solution, conditions without parameter, conditions containing parameters. 26 exercises worked out. 271pp. 5⅜ × 8½.   64685-8 Pa. $8.95

CATALOG OF DOVER BOOKS

DE RE METALLICA, Georgius Agricola. The famous Hoover translation of greatest treatise on technological chemistry, engineering, geology, mining of early modern times (1556). All 289 original woodcuts. 638pp. 6¾ × 11.

60006-8 Pa. $18.95

SOME THEORY OF SAMPLING, William Edwards Deming. Analysis of the problems, theory and design of sampling techniques for social scientists, industrial managers and others who find statistics increasingly important in their work. 61 tables. 90 figures. xvii + 602pp. 5⅜ × 8½.

64684-X Pa. $15.95

THE VARIOUS AND INGENIOUS MACHINES OF AGOSTINO RAMELLI: A Classic Sixteenth-Century Illustrated Treatise on Technology, Agostino Ramelli. One of the most widely known and copied works on machinery in the 16th century. 194 detailed plates of water pumps, grain mills, cranes, more. 608pp. 9 × 12.

28180-9 Pa. $24.95

LINEAR PROGRAMMING AND ECONOMIC ANALYSIS, Robert Dorfman, Paul A. Samuelson and Robert M. Solow. First comprehensive treatment of linear programming in standard economic analysis. Game theory, modern welfare economics, Leontief input-output, more. 525pp. 5⅜ × 8½.

65491-5 Pa. $14.95

ELEMENTARY DECISION THEORY, Herman Chernoff and Lincoln E. Moses. Clear introduction to statistics and statistical theory covers data processing, probability and random variables, testing hypotheses, much more. Exercises. 364pp. 5⅜ × 8½.

65218-1 Pa. $9.95

THE COMPLEAT STRATEGYST: Being a Primer on the Theory of Games of Strategy, J.D. Williams. Highly entertaining classic describes, with many illustrated examples, how to select best strategies in conflict situations. Prefaces. Appendices. 268pp. 5⅜ × 8½.

25101-2 Pa. $7.95

MATHEMATICAL METHODS OF OPERATIONS RESEARCH, Thomas L. Saaty. Classic graduate-level text covers historical background, classical methods of forming models, optimization, game theory, probability, queueing theory, much more. Exercises. Bibliography. 448pp. 5⅜ × 8¼.

65703-5 Pa. $12.95

CONSTRUCTIONS AND COMBINATORIAL PROBLEMS IN DESIGN OF EXPERIMENTS, Damaraju Raghavarao. In-depth reference work examines orthogonal Latin squares, incomplete block designs, tactical configuration, partial geometry, much more. Abundant explanations, examples. 416pp. 5⅜ × 8¼.

65685-3 Pa. $10.95

THE ABSOLUTE DIFFERENTIAL CALCULUS (CALCULUS OF TENSORS), Tullio Levi-Civita. Great 20th-century mathematician's classic work on material necessary for mathematical grasp of theory of relativity. 452pp. 5⅜ × 8½.

63401-9 Pa. $9.95

VECTOR AND TENSOR ANALYSIS WITH APPLICATIONS, A.I. Borisenko and I.E. Tarapov. Concise introduction. Worked-out problems, solutions, exercises. 257pp. 5⅜ × 8¼.

63833-2 Pa. $7.95

CATALOG OF DOVER BOOKS

TENSOR CALCULUS, J.L. Synge and A. Schild. Widely used introductory text covers spaces and tensors, basic operations in Riemannian space, non-Riemannian spaces, etc. 324pp. 5⅜ × 8¼. 63612-7 Pa. $8.95

A CONCISE HISTORY OF MATHEMATICS, Dirk J. Struik. The best brief history of mathematics. Stresses origins and covers every major figure from ancient Near East to 19th century. 41 illustrations. 195pp. 5⅜ × 8½. 60255-9 Pa. $7.95

A SHORT ACCOUNT OF THE HISTORY OF MATHEMATICS, W.W. Rouse Ball. One of clearest, most authoritative surveys from the Egyptians and Phoenicians through 19th-century figures such as Grassman, Galois, Riemann. Fourth edition. 522pp. 5⅜ × 8½. 20630-0 Pa. $10.95

HISTORY OF MATHEMATICS, David E. Smith. Nontechnical survey from ancient Greece and Orient to late 19th century; evolution of arithmetic, geometry, trigonometry, calculating devices, algebra, the calculus. 362 illustrations. 1,355pp. 5⅜ × 8½. 20429-4, 20430-8 Pa., Two-vol. set $23.90

THE GEOMETRY OF RENÉ DESCARTES, René Descartes. The great work founded analytical geometry. Original French text, Descartes' own diagrams, together with definitive Smith-Latham translation. 244pp. 5⅜ × 8½. 60068-8 Pa. $7.95

THE ORIGINS OF THE INFINITESIMAL CALCULUS, Margaret E. Baron. Only fully detailed and documented account of crucial discipline: origins; development by Galileo, Kepler, Cavalieri; contributions of Newton, Leibniz, more. 304pp. 5⅜ × 8½. (Available in U.S. and Canada only) 65371-4 Pa. $9.95

THE HISTORY OF THE CALCULUS AND ITS CONCEPTUAL DEVELOPMENT, Carl B. Boyer. Origins in antiquity, medieval contributions, work of Newton, Leibniz, rigorous formulation. Treatment is verbal. 346pp. 5⅜ × 8½. 60509-4 Pa. $8.95

THE THIRTEEN BOOKS OF EUCLID'S ELEMENTS, translated with introduction and commentary by Sir Thomas L. Heath. Definitive edition. Textual and linguistic notes, mathematical analysis. 2,500 years of critical commentary. Not abridged. 1,414pp. 5⅜ × 8½. 60088-2, 60089-0, 60090-4 Pa., Three-vol. set $29.85

GAMES AND DECISIONS: Introduction and Critical Survey, R. Duncan Luce and Howard Raiffa. Superb nontechnical introduction to game theory, primarily applied to social sciences. Utility theory, zero-sum games, n-person games, decision-making, much more. Bibliography. 509pp. 5⅜ × 8½. 65943-7 Pa. $12.95

THE HISTORICAL ROOTS OF ELEMENTARY MATHEMATICS, Lucas N.H. Bunt, Phillip S. Jones, and Jack D. Bedient. Fundamental underpinnings of modern arithmetic, algebra, geometry and number systems derived from ancient civilizations. 320pp. 5⅜ × 8½. 25563-8 Pa. $8.95

CALCULUS REFRESHER FOR TECHNICAL PEOPLE, A. Albert Klaf. Covers important aspects of integral and differential calculus via 756 questions. 566 problems, most answered. 431pp. 5⅜ × 8½. 20370-0 Pa. $8.95

CHALLENGING MATHEMATICAL PROBLEMS WITH ELEMENTARY SOLUTIONS, A.M. Yaglom and I.M. Yaglom. Over 170 challenging problems on probability theory, combinatorial analysis, points and lines, topology, convex polygons, many other topics. Solutions. Total of 445pp. 5⅜ × 8½. Two-vol. set.
Vol. I 65536-9 Pa. $7.95
Vol. II 65537-7 Pa. $6.95

FIFTY CHALLENGING PROBLEMS IN PROBABILITY WITH SOLUTIONS, Frederick Mosteller. Remarkable puzzlers, graded in difficulty, illustrate elementary and advanced aspects of probability. Detailed solutions. 88pp. 5⅜ × 8½.
65355-2 Pa. $4.95

EXPERIMENTS IN TOPOLOGY, Stephen Barr. Classic, lively explanation of one of the byways of mathematics. Klein bottles, Moebius strips, projective planes, map coloring, problem of the Koenigsberg bridges, much more, described with clarity and wit. 43 figures. 210pp. 5⅜ × 8½.
25933-1 Pa. $5.95

RELATIVITY IN ILLUSTRATIONS, Jacob T. Schwartz. Clear nontechnical treatment makes relativity more accessible than ever before. Over 60 drawings illustrate concepts more clearly than text alone. Only high school geometry needed. Bibliography. 128pp. 6⅛ × 9¼.
25965-X Pa. $6.95

AN INTRODUCTION TO ORDINARY DIFFERENTIAL EQUATIONS, Earl A. Coddington. A thorough and systematic first course in elementary differential equations for undergraduates in mathematics and science, with many exercises and problems (with answers). Index. 304pp. 5⅜ × 8½.
65942-9 Pa. $8.95

FOURIER SERIES AND ORTHOGONAL FUNCTIONS, Harry F. Davis. An incisive text combining theory and practical example to introduce Fourier series, orthogonal functions and applications of the Fourier method to boundary-value problems. 570 exercises. Answers and notes. 416pp. 5⅜ × 8½.
65973-9 Pa. $9.95

THE THEORY OF BRANCHING PROCESSES, Theodore E. Harris. First systematic, comprehensive treatment of branching (i.e. multiplicative) processes and their applications. Galton-Watson model, Markov branching processes, electron-photon cascade, many other topics. Rigorous proofs. Bibliography. 240pp. 5⅜ × 8½.
65952-6 Pa. $6.95

AN INTRODUCTION TO ALGEBRAIC STRUCTURES, Joseph Landin. Superb self-contained text covers "abstract algebra": sets and numbers, theory of groups, theory of rings, much more. Numerous well-chosen examples, exercises. 247pp. 5⅜ × 8½.
65940-2 Pa. $7.95

Prices subject to change without notice.
Available at your book dealer or write for free Mathematics and Science Catalog to Dept. GI, Dover Publications, Inc., 31 East 2nd St., Mineola, N.Y. 11501. Dover publishes more than 175 books each year on science, elementary and advanced mathematics, biology, music, art, literature, history, social sciences and other areas.